Interfacial Rheology

Progress in Colloid and Interface Science

VOLUME 1

Interfacial Rheology

Edited by

R. Miller
L. Liggieri

CRC Press
Taylor & Francis Group
Boca Raton London New York

CRC Press is an imprint of the
Taylor & Francis Group, an **informa** business

First published 2009 by VSP

Published 2019 by CRC Press
Taylor & Francis Group
6000 Broken Sound Parkway NW, Suite 300
Boca Raton, FL 33487-2742

First issued in paperback 2019

No claim to original U.S. Government works

ISBN-13: 978-0-367-44605-5 (pbk)
ISBN-13: 978-90-04-17586-0 (hbk)
ISSN 1877-8569

Visit the Taylor & Francis Web site at
http://www.taylorandfrancis.com

and the CRC Press Web site at
http://www.crcpress.com

FOREWORD

PROGRESS IN COLLOID AND INTERFACE SCIENCE
(PCIS)

In this book series monographs will be published dedicated to all relevant topics in the field of colloid and interface science. The main aim is to present the most recent research developments and progress made on particular topics. The different volumes will cover fundamental and/or applied subjects.

All colloid and interface topics will be covered by this series, such as liquid/gas, liquid/liquid, solid/gas, solid/liquid interfaces, their theoretical description and experimental analysis, including adsorption processes, mechanical properties, catalysis reactions, emulsification, foam formation, nanoscience; the formation and characterization of all kinds of disperse systems, such as aerosols, foams, emulsions dispersions, and their stability; bulk properties of solutions, including micellar solutions and microemulsions; experimental methodologies, such as tensiometry, ellipsometry, contact angle, light scattering, electrochemical methods, bulk and interfacial rheology; self-assembling materials at interfaces, such as surfactants, polymers, proteins, (nano-)particles, and all their mixtures; applications in various fields, such as agriculture, biological systems, chemical engineering, coatings, cosmetics, detergency, flotation processes, food processing, household products, materials processing, environmental systems, oil recovery, pulp and paper technologies, pharmacy.

Authors interested in writing books that fit into the described fields of science are invited to contribute to this series.

Reinhard Miller and Libero Liggieri
Series Editors

FOREWORD

In his novel "Flatland" published in 1884 Edwin A. Abbott describes people living in a two-dimensional world [1]. As a masterpiece of literature this novella is respected for its satire of the social hierarchy in the past, but at the same time it is an introduction to the geometry of higher dimensions. Abbott was not the first to describe a two-dimensional universe inhabited by flat beings, but he was the first to explore what it would mean for such individuals to interact with phenomena of a dimension higher than their own.

Interfacial rheology is the science of the response of mobile two-dimensional phases to deformation. Requests to this rather young scientific field came first from theories for the conventional bulk rheology, where a so-called "free interface" has to be specified as boundary condition for the normal component of the bulk stress tensor. This specific boundary condition, which does not involve any mass, generally introduces two-dimensional tensile forces, as well as viscous and elastic forces, which have a localized action upon a two-dimensional, moving and deforming interface.

The history of interfacial rheology started in the 19th century. In 1838 Ascherson [2] submitted a paper to the Paris Academy of Sciences about his studies on a skin forming processes at the interface between a protein solution and an oil. He designated these thin skins as haptogen-membranes and this notation was universally used in the past. Hagen [3] was possibly the first who ascertained in 1845 that the viscosity of the interfacial region of a liquid differs from that of the adjoining bulk phase. Maybe the first recorded experiment in interfacial rheology was performed in 1869 by Plateau [4]. He compared the damping of an oscillating magnetic needle immersed in a liquid with one floating on the surface of that liquid. He thus confirmed that the surface of a liquid presents a higher resistance to deformation than that of the bulk liquid and demonstrated in this way the excess surface shear viscosity of the liquid surface. Plateau was not aware that liquid surfaces are almost always covered by an adsorption layer of surface active impurities. It was Marangoni [5, 6, 7] who recognized this fact and explained Plateaus' observations by the presence of such a layer. The movement of the needle causes surface compression in front of it and dilation on the other side. This creates a gradient in interfacial tension which opposes the movement of the needle. Another example is the calming of the sea by spreading oil on it. Early observations of water wave damping by oil films were already described since ancient times, dating back to Aristotle, Plutarch and Pliny the Elder [8]. The first interpretation of the deformational resistance of interfacial layers was proposed by Gibbs [9] and later improved by Lord Rayleigh [10].

In the early 20th century Boussinesq [11] introduced the terminology of modern interfacial rheology by proposing a two-dimensional analogue of the Newtonian fluid which was needed to explain the retardation of the terminal velocity of drops and bubbles and to characterize the dynamics of interfaces by using the concept of surface shear and dilational viscosities. Later Ericksen [12], Oldroyd [13], Scriven [14] and many other extended the theory developed by Boussinesq leading to the interfacial rheology of today.

Veniamin G. Levich can be possibly seen as the founder of the dilational interfacial rheology. He was the first to present a theory in which surface tension gradients were discussed in terms of surface dilational elasticity [15]. In 1951 Dorrestein [16] presented the first details on capillary waves the damping of which is influenced by the surface dilational elasticity. Starting from Levich's findings on capillary wave damping, the advent of the modern dilational rheology is surely the theory developed by Hansen [17] and the Dutch school [18, 19]. From his theory of capillary wave damping, Lucassen discovered a second type of surface waves, the called longitudinal waves, which turned out to be very suitable for studies of the surface dilational rheology. Shortly later, in 1970 Kretzschmar and Lunkenheimer [20] proposed a new methodology for measuring the dilational rheology via a harmonically oscillating bubble. Since this proposal, it took more than 25 years until a first commercial instrument was available for routine experiments of the dilational surface elasticity [21], based on oscillating drops and bubbles.

The shear and dilational rheology of self-assembled interfacial layers at liquid/air and liquid/liquid interfaces was understood as a relevant property in a wide range of technical applications such as mass transfer, monolayers, foaming, emulsification oil recovery, or high speed coating. The impact on the stability of foams and emulsions, for example, was recently emphasised in several reviews [22, 23, 24]. The role of film rheology in foam stabilisation was first emphasised by Gibbs [9] and later further studied for example in [25, 26]. Most recent studies also include Pickering systems [27], even under microgravity conditions at the International Space Station [28].

It is the purpose of this book to describe in detail the theoretical background of interfacial rheology, experimental tools and a well selected number of experimental examples. In the past, only reviews and few book chapters were dedicated to the rheology of interfacial layers, such as by Joly [29], Lucassen [30], Edwards and Wasan [31], Slattery

[32], Edwards et al. [33]. Over the years, also various special aspects of interfacial rheology were reviewed [34, 35, 36, 37, 38, 39, 40, 41, 42].

This is possibly the first book completely devoted to interfacial rheology. The first two chapters give an introduction into the field. Further, the main experimental methods of dilational and shear rheology are presented together with selected examples on interfacial layers formed by adsorption of surfactants, polymers, and mixed polyelectrolyte-surfactant and protein surfactant layers, and of spread monolayers. Additional chapters are dedicated to special topics, such as importance of 2D rheology in flotation, in medicine, in food technology. The book is completed by a chapter on micro-rheology of surface layers.

When writing this book, Valery Krotov as one of the authors passed away. We want to dedicate this book to him and include an obituary.

February 2009　　　　　Reinhard Miller and Libero Liggieri
　　　　　　　　　　　　　　(Editors)

References
1.　E.A. Abbott, Flatland, Basil Blackwell, Oxford, 1884.
2.　F.M. Ascherson, Archiv Anat. Physiol. u. wiss. Med., 1840, 44.
3.　G.H.L. Hagen, Abh. Königl. Akad. Wissenschaften zu Berlin (Phys. Math. Kl.), Berlin 1845, 41.
4.　J.A.F. Plateau, Phil. Mag. Ser. 4, 38 (1869) 445.
5.　C.G.M. Marangoni, Ann. Phys. (Poggendorf), 142 (1871) 337.
6.　C. Marangoni, Nuovo Cimento, 3 (1870) 50.
7.　C. Marangoni and P. Stefanelli, Nuovo Cimento, 9 (1873) 236.
8.　J.C. Scott, "The historical development of theories of wave-calming using oil", History of Technology, Vol.3, 163-186, 1978.
9.　J.W. Gibbs, The Collected Work of J.W. Gibbs, Vol. 1, Longmans Green, New York, 1931.
10.　Rayleigh, Proc. Roy. Soc. (London), 47 (1890) 281, 364.
11.　M. J. Boussinesq, Ann. Chim. Phys. Ser. 8, 29 (1913) 349, 357, 364.
12.　J.L. Erickson, J. Ration. Mech. Anal., 1 (1952) 521.
13.　J.G. Oldroyd, Proc. Roy. Soc. (London), A 232 (1955) 567.
14.　L.E. Scriven, Chem Eng. Sci., 12 (1960) 98.
15.　V.G. Levich, Acta Physicochim., 14 (1941) 307
16.　R. Dorrestein, Koninkl. Ned. Akad. Wet.., Proc., B, 54 (1951) 260
17.　R.S. Hansen, J. Applied Phys., 35 (1964) 1983
18.　M. van den Tempel and R.P. van de Riet, J. Chem. Phys., 42 (1965) 2769
19.　J. Lucassen, Trans. Faraday Soc., 64 (1968) 2221
20.　G. Kretzschmar and K. Lunkenheimer, Ber. Bunsenges. Phys. Chem. 74 (1970) 1064.

21. J. Benjamins, A. Cagna and E.H. Lucassen-Reynders, Colloids Surfaces A, 114 (1996) 245.
22. D. Langevin, Adv. Colloid Interface Sci. 88 (2000) 209.
23. J. Maldonado-Valderrama, A. Martin-Rodriguez, M.J. Galvez-Ruiz, R. Miller, D. Langevin and M.A. Cabrerizo-Vilchez, Colloids Surfaces A 323 (2008) 116.
24. B.S. Murray, E. Dickinson and Y. Wang, Food Hydrocolloids, 23 (2009) 1198
25. V.V. Krotov and A.I. Rusanov, Kolloidn. Zh., 34 (1972) 297.
26. Y.H. Kim, K. Koczo and D.T. Wasan, J. Colloid Interface Sci., 187 (1997) 29.
27. F. Ravera, M. Ferrari, L. Liggieri, G. Loglio, E. Santini and A. Zanobini, Colloids Surfaces A 323 (2008) 99.
28. J. Banhart, F. Garcia-Moreno, S. Hutzler, D. Langevin, L. Liggieri, R. Miller, A. Saint-Jalmes and D. Weaire, Europhysics News, 39 (2008) 26.
29. M. Joly, In: Recent Progress in Surface Science, J.F. Danielli, K.G.A. Pankhurst and A.C. Riddiford (eds.), Vol. 1, pp. 1-48, Academic Press, New York, 1964.
30. J. Lucassen, In: Surfactant Science Series, E.H. Lucassen-Reynders (ed.), Vol. 11, pp.217-265, Marcel Dekker, Basel, 1981.
31. D.A. Edwards and D.T. Wasan, J. Rheology, 32 (1988) 429.
32. J.C. Slattery, Interfacial Transport Phenomena, Springer-Verlag, New York, 1990.
33. D.A. Edwards, H. Brenner and D.T. Wasan, Interfacial Transport Processes and Rheology, Butterworth-Heinemann, Boston, 1991.
34. F.C. Goodrich, In: Progress in Surface and Membrane Science, J.F. Danielli, M.D. Rosenberg and D.A. Cadenhead (eds.), Vol. 7, pp. 151-181, Academic Press, Ney York, 1973.
35. L. Gupta and D.T. Wasan, Ind. Eng. Chem. Fundam., 13 (1974) 26.
36. F.C. Goodrich, In: Solution Chemistry of Surfactants, K.L. Mittal (ed.), Vol. 2, pp. 733-748, Plenum Press, New York, 1979.
37. R. Miller, R. Wüstneck, J. Krägel and G. Kretzschmar, Colloids Surf. A, 111 (1996) 75.
38. B. Warburton, In: Rheological Measurement, A.A. Collyer and D.W. Clegg (eds.), pp. 723-754, Chapman and Hall, London, 1998.
39. B.S. Murray and E. Dickinson, Food Sci. Technol. Int., 2 (1996) 131.
40. B.S. Murray, In: Studies in Interface Science Series, Proteins at Interfaces, in "Studies in Interface Science", Vol. 7, D. Möbius and R. Miller (eds.), pp. 179-220, Elsevier, Amsterdam, 1998.
41. M.A. Bos and T. van Vliet, Adv. Colloid Interface Sci., 91 (2001) 437.
42. J. Krägel, S.R. Derkatch and R. Miller, Adv. Colloid Interface Sci., 144 (2008) 38

OBITUARY

Professor Valery V. Krotov

* 23 September 1931 in Kirov
† 06 January 2007 in St.Petersburg

Valery Vladimorovich Krotov was born in 1931 in Kirov and spent a part of his childhood and youth in the Ural region where he was evacuated together with his mother during the Second World War. In 1949 he graduated from high school with a gold medal and entered the physical faculty of Leningrad (now St.Petersburg) State University. In 1951 he was transferred to Moscow Institute of Engineering Physics where he studied chemical physics. The corresponding member of the Academy of Sciences of USSR Veniamin Levich (later founder of the Levich Institute for Physico-Chemical Hydrodynamics at City University of New York), and the full member of the Academy of Sciences of USSR Bruno Pontecorvo were among his supervisors.

After graduation in 1956 V.V. Krotov started investigations on problems of nuclear synthesis in the Physico-Technical Institute of the Academy of Sciences of USSR. In 1965 Anatoly I. Rusanov invited him to become member of the solution theory department at the Chemical Faculty of Leningrad (St.Petersburg) State University. It was at the university where Valery Krotov realized his outstanding analytical and pedagogical abilities. He began to deliver lectures on theoretical mechanics for chemists and prepared new lecture courses on "Traditional theory of capillarity", "Physico-Chemical hydrodynamics" and "Physico-Chemical mechanics" for students specialized in colloid science. Since 1967 V.V. Krotov worked in the laboratory of surface phenomena of the Leningrad (St.Petersburg) State University and headed this laboratory after 1988. Since 1988 he was also nominated member of the editorial board of Russian Colloid Journal.

Valery Krotov was among the first scientists who made precise measurements of the Gibbs elasticity of foam films. These results were included later in his PhD thesis (1972). In his Dr. Sci. thesis (1984) he solved a few problems of the physico-chemical hydrodynamics of foams and proposed the first theory of foam syneresis. Among other results of the thesis one can mention the generic equation of state of insoluble monolayers, which generalizes a few earlier derived relationships. Later Valery Krotov continued his theoretical and experimental studies of the structure, hydrodynamics and optical properties of foams. At the same time he began to investigate the flow of super thin liquid jets, predicted and discovered experimentally the effect of capillary self-inhibition of a jet during its disintegration. Among other important results are the five laws of capillary disintegration of jets.

V.V. Krotov is the author of more than 130 papers published in Russian and international scientific journals. Together with A.I. Rusanov he also published a monograph "Physico-Chemical hydrodynamics of capillary systems" (1999, Imperial College Press).

Despite his brilliant scientific achievements, Valery Krotov was an extremely modest person and never forced his ideas upon other people. At the same time he always enthusiastically defended his opinions without compromise. His honesty and extreme delicacy in all his deeds gained respect of all the colleagues he was associated with.

V.V. Krotov was well-read and interesting to talk to. He had unique abilities to see bright unusual phenomena and events in the ambient world, which could appear insignificant and trivial for many people. He strived from his student years to discover the paradoxes in this world, to formulate the corresponding physical problems and to seek for their clear and well-founded solutions.

The bright memory of Valery Krotov – a kind person, outstanding scientist and tutor will remain with his students, colleagues and friends who were happy to know him.

February 2009   Boris A. Noskov   Reinhard Miller

CONTENT

BASICS OF INTERFACIAL RHEOLOGY

V.V. Krotov

St. Petersburg State University, Department of Chemistry, 198504 St. Petersburg, Petrodvorets, University Avenue 2, Russia

Contents

A. INTRODUCTION

Three-dimensional (bulk) rheology is one of the branches in continuum mechanics (hydrodynamics). However, while the main topics studied by the hydrodynamics is the determination of the spatial and time dependencies of the fields of velocity and contact forces (in the simplest case – the pressure) at given boundary conditions, the rheology studies a preparatory and more special problem: how to express the dependence of forces which arise in the infinitesimal element of a continuum on the deformations and the rate of deformation of this element. The forces relevant to this case are those which are transferred via a contact, and, therefore, are proportional to the contact area. Therefore the inertia of an element (similarly to other field-induced forces) is irrelevant: the inertia is proportional to a unit volume and therefore its contribution in the limiting case considered should be neglected. Thus, the approaches similar to that adopted in [1], where the inertia is involved into the rheological schemes, are incorrect in principle.

In the interfacial (surface) rheology developed recently, the continuum is considered to be two-dimensional, the contact forces are proportional to the contact line length, and the field forces are proportional to a unit area. Therefore, similarly to the three-dimensional case, the two-dimensional rheology disregards the field forces. While, in general case, these forces do exist (if a surfactant is present in the system, these forces in the gravitation field lead to the formation of certain, however small, equilibrium gradients in surface tension [2]), the two-dimensional hydrodynamics, formulated as a closed set of equations for the interfacial layer, could hardly be justified (this point of view was, in particular, expressed in [3]). This is because in such a layer large 'excesses' of forces, and of surfactants, can exist, as compared with bulk phases, while these layers cannot contain any appreciable excesses of mass, because at least one of the adjacent phases always exists as a condensed phase, i.e., its density is almost maximum (with respect to normal conditions). Therefore, the three-dimensional hydrodynamic equations at the interface are degenerated, and become just specific boundary conditions which do not involve any mass. Here both the tangential boundary conditions (for which the surface curvature is immaterial) and normal boundary conditions (for which this curvature is essential) are relevant. It is assumed in this case that the thickness of the surface layer is small as compared with the thickness of the regions separated by this layer.

On the contrary, the two-dimensional rheology of an interfacial (surface) layer is, in fact, more diverse than the three-dimensional rheology. This is primarily determined by the fact that monolayers of surfactants are characterised by an enormous (per unit of mass) contact area with adjacent phases. It is especially true for the 'uniform' (in this case – two-dimensional) compressibility-related deformations. It is well known (see e.g., [4, 5]) that the compressibility of condensed insoluble monolayers is immeasurably higher than the compressibility of three-dimensional condensed bodies, and soluble monolayers have no direct analogues in the three-dimensional case. In simple terms: because no fourth dimension exists in a real three-dimensional world which could be treated similarly to the third dimension supplementary to the two-dimensional case. The drastic simplification in two-dimensional systems is related to the fact that, as the area of contact with the adjacent condensed phases is enormous (per unit mass of interfacial layer), the two-dimensional rheology generally deals with isothermal processes.

Finally it should be noted that certain geometric complication of the two-dimensional rheology arises when the interfaces considered cannot be transformed into a planar surface. However, this fact is insignificant for the main problems of 2D rheology, because it deals only with the properties of an infinitesimally small element, for which the Euclidian geometry is relevant. From a practical point of view, this means that the 'planar' or two-dimensional or 2D rheology could be applied only if the dimensions of the surface elements considered are small as compared with the radius of curvature of the sections normal to the surface.

B. ADEQUATE DESCRIPTION OF SIMPLEST INTERFACIAL RHEOLOGY IN TENSOR TERMS

The interfacial rheology could mathematically be formulated considering the examples of a purely elastic and purely viscous two-dimensional continuum, which can be used as good approximations for monolayers of insoluble surfactant in the case when the viscosity of the two fluid phases is sufficiently small. The generalisation of a two-dimensional system starting from a three-dimensional one implies a formal replacement of 2 by 3 in the expressions for the dilational deformations, and does not involve any changes for shear deformations, because the form of the shear deformation remains the same. With respect to shear deformation, a fundamental difference exists only for one-dimensional systems (lines of contacts between three phases), as such type of deformation does not exist, in contrast to the one-dimensional dilational deformation.

1. Purely elastic two-dimensional continuum

Let us consider a solid monolayer of insoluble surfactants. This monolayer can be characterised by four values of tension, expressed in a matrix form:

$$\left\| \begin{matrix} \gamma_{xx}, \gamma_{xy} \\ \gamma_{yx}, \gamma_{yy} \end{matrix} \right\| \equiv \left\| \begin{matrix} \gamma_{11}, \gamma_{12} \\ \gamma_{21}, \gamma_{22} \end{matrix} \right\| \tag{1}$$

Here x and y are the Cartesian coordinates located in the monolayer surface (the second matrix is the alternative form with $x_1 \equiv x$, and $x_2 \equiv y$). The components γ_{ik} determine the forces in the x_i directions (which necessarily acts from the side of larger x_i values) per unit length of the contact line perpendicular to the x_k direction (while the force $-\gamma_{ik}$ per unit length of the same line acts from the side of smaller x_i values, according to third Newton's law).

The matrix (1) actually contains only three (not four) independent values: the main diagonal elements γ_{xx} and γ_{yy} can differ from one another (depending on the boundary conditions involved), while the off-diagonal elements γ_{xy} and γ_{yx} are equal to each other; such a matrix is called symmetric. To show that the matrix (1) is symmetric we consider Fig. 1, where the rectangle at the interface is shown with sides equal to a and b. The forces $b\gamma_{yx}$ and $-b\gamma_{yx}$ acting together create the moment of forces with the absolute value $ab\gamma_{yx}$, which rotates the considered rectangular element of surface in Fig. 1 counter clockwise, while the forces $a\gamma_{xy}$ and $-a\gamma_{xy}$ rotate the same element clockwise due to the moment of forces $ab\gamma_{xy}$ created by these forces. In equilibrium state these moments should compensate each other (e. $\gamma_{xy} = \gamma_{yx}$), because this compensation does not happen due to any external boundary conditions. This matrix in Eq. (1) is also symmetric under dynamic conditions, however, its proof is more complicated. The Newton equation for the rotation of the rectangular element considered can be derived by accounting for inertial forces of the surfactant monolayer itself and for the viscosity of the bulk phases. Proceeding to the limit $ab \rightarrow 0$ and assuming $\gamma_{xy} \neq \gamma_{yx}$ one obtains the infinite value of the angular acceleration of the element, which is in fact impossible.

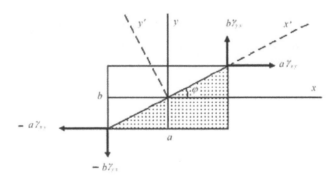

Fig. 1.   The definition of main properties of two-dimensional stress tensor

From similar considerations one can show that the sum of all the forces which act on the sides of the right-angled triangle shown in Fig. 1 should be zero. This enables us to derive the transformation rules of the components of matrix (1) for the rotation of the reference coordinate system with angle φ:

$$\gamma'_{ik} = \sum_{j,m=1,2} v_{ji} v_{mk} \gamma_{jm}, \qquad (i,k = 1,2) \tag{2}$$

where $v_{11} = v_{22} = \cos\varphi$ and $v_{21} = -v_{12} = -\sin\varphi$. The matrix γ_{ik} which transforms according to Eq. (2) is called tensor. Therefore, the values γ_{ik} are the components of the "surface tension tensor" $\hat{\gamma}$.

In the following, we will consider only small perturbations, because only for this case simplest linear rules can be derived, and the superposition principle applies. Therefore, the tensor $\hat{\gamma}$ is represented as the sum of the non-perturbed constituent:

$$\left\| \begin{matrix} \gamma_0, 0 \\ 0, \gamma_0 \end{matrix} \right\| = \gamma_0 \cdot \left\| \begin{matrix} 1, 0 \\ 0, 1 \end{matrix} \right\| \equiv \gamma_0 \cdot \hat{1} \tag{3}$$

($\hat{1}$ is the two-dimensional unit tensor), and the perturbation constituent:

$$\left\| \begin{matrix} \gamma_{11} - \gamma_0, \gamma_{12} \\ \gamma_{21}, \gamma_{22} - \gamma_0 \end{matrix} \right\| = \Delta\hat{\gamma} \tag{4}$$

(the sum rule for tensors as the summation of monadic components is being considered as the consequence of the tensor operations definition).

The tensor (e., the value which obeys the rule (2)) can be non-symmetric. However, such tensor can always be decomposed into a symmetric and an anti-symmetric constituent (the tensor $\hat{b}$ is called anti-symmetric if $b_{ik} = -b_{ki}$):

$$\left\| \begin{matrix} a_{11}, a_{12} \\ a_{21}, a_{22} \end{matrix} \right\| = \left\| \begin{matrix} a_{11}, (a_{12} + a_{21})/2 \\ (a_{12} + a_{21})/2, a_{22} \end{matrix} \right\| + \left\| \begin{matrix} 0, (a_{12} - a_{21})/2 \\ (a_{21} - a_{12})/2, 0 \end{matrix} \right\| \tag{5}$$

(the main diagonal elements of an anti-symmetric tensor are zero by the definition). It is easily seen that, according to (2), three conservation laws are held true: (i) $\sum_i a_{ii} = const$ (tensor spur conservation);

(ii) $a_{ik} - a_{ki} = const$ for $i \neq k$ (non-diagonal elements difference conservation); (iii) conservation of all elements of the anti-symmetric constituent $\hat{a}$ of a tensor during coordinate system rotation.

Let us consider now the definition of the deformation tensor. The displacements of the points at a plane surface are determined by the displacement vector $\vec{u}(x, y)$. To eliminate the transfer of the surface element as a whole (for which physical deformations are irrelevant) one has to differentiate the displacement vector components u_x and u_y:

$$du_x = \frac{\partial u_x}{\partial x}dx + \frac{\partial u_x}{\partial y}dy$$

$$du_y = \frac{\partial u_y}{\partial x}dx + \frac{\partial u_y}{\partial y}dy \qquad (6)$$

and compose the displacement tensor:

$$\left\| \begin{matrix} \dfrac{\partial u_x}{\partial x}, \dfrac{\partial u_x}{\partial y} \\[2mm] \dfrac{\partial u_y}{\partial x}, \dfrac{\partial u_y}{\partial y} \end{matrix} \right\| \equiv \hat{u} \qquad (7)$$

(one can use the definition (2) to show that (7) is a tensor).

Applying the Eq. (6) to the operation of a rotation of the two-dimensional continuum element as a whole in the plane of this element, one obtains for the tensor:

$$\left\| \begin{matrix} 0, \dfrac{1}{2}\left(\dfrac{\partial u_x}{\partial y} - \dfrac{\partial u_y}{\partial x}\right) \\[3mm] -\dfrac{1}{2}\left(\dfrac{\partial u_x}{\partial y} - \dfrac{\partial u_y}{\partial x}\right), 0 \end{matrix} \right\| \equiv \hat{u}_a \qquad (8)$$

This operation does not alter the physical state of the two-dimensional continuum; therefore, the anti-symmetric tensor $\hat{u}_a$ should be eliminated from the displacement tensor $\hat{u}$. The subtraction of monadic components of the tensor (8) from (7) yields:

$$\hat{e} \equiv \left\| \begin{matrix} \dfrac{\partial u_x}{\partial x}, \left(\dfrac{\partial u_x}{\partial y} + \dfrac{\partial u_y}{\partial x}\right)\Big/2 \\[3mm] \left(\dfrac{\partial u_x}{\partial y} + \dfrac{\partial u_y}{\partial x}\right)\Big/2, \dfrac{\partial u_y}{\partial y} \end{matrix} \right\| \qquad (9)$$

This tensor is the so-called deformation tensor. In the three-dimensional elasticity theory the components of this tensor are usually represented as $\left(\dfrac{\partial u_i}{\partial x_k} + \dfrac{\partial u_k}{\partial x_i}\right)\Big/2$ [6].

In general case, no direct proportionality between the tensors (9) and (4) can exist, because such direct proportionality would imply that the elastic medium is characterised by one parameter only, while it is known from experiments that two parameters exist, which correspond to the deformations (and, consequently, to the tensors) of different kind. Note that only in the one-dimensional case, for a line separating two-dimensional phases, the elasticity could be described by one parameter, the Young modulus. It becomes also clear from experimental data, into which qualitatively different constituents each tensor should be decomposed in order to establish a simple proportionality between the corresponding constituents.

The perturbed constituent $\Delta\hat{\gamma}$ (4) of the tension tensor can be decomposed as:

$$\Delta\hat{\gamma} = \Delta\hat{\gamma}_i + \Delta\hat{\gamma}_d \tag{10}$$

where $\Delta\hat{\gamma}_i$ is 'responsible' for the isotropic part of the tensor $\Delta\hat{\gamma}$, while $\Delta\hat{\gamma}_d$ is the deviator tensor, corresponding to the non-isotropic residual of the tensor $\Delta\hat{\gamma}$:

$$\Delta\hat{\gamma}_i \equiv \left\| \begin{matrix} \left(\Delta\gamma_{xx} + \Delta\gamma_{yy}\right)/2, 0 \\ 0, \left(\Delta\gamma_{xx} + \Delta\gamma_{yy}\right)/2 \end{matrix} \right\| \tag{11}$$

$$\Delta\hat{\gamma}_d \equiv \left\| \begin{matrix} \left(\Delta\gamma_{xx} - \Delta\gamma_{yy}\right)/2, \gamma_{xy} \\ \gamma_{yx}, \left(\Delta\gamma_{yy} - \Delta\gamma_{xx}\right)/2 \end{matrix} \right\| \tag{12}$$

As the rotation of the coordinate system within the surface is determined by a single parameter, and $\gamma_{xy} = \gamma_{yx}$, thus one can apply the rotation to let the two non-diagonal components of the tensor in (12) vanish, and, therefore, the corresponding components of surface tension tensor $\hat{\gamma}$. This operation is called the diagonalisation of the tensor.

In a similar way for the deformation tensor one has:

$$\hat{e} = \hat{e}_i + \hat{e}_d \tag{13}$$

where $\hat{e}_i$ and $\hat{e}_d$ are the isotropic and deviator constituents of the deformation tensor, respectively, which are defined by the expressions:

$$\hat{e}_i \equiv \left\| \begin{matrix} \left(\dfrac{\partial u_x}{\partial x} + \dfrac{\partial u_y}{\partial y}\right)/2, 0 \\ 0, \left(\dfrac{\partial u_x}{\partial x} + \dfrac{\partial u_y}{\partial y}\right)/2 \end{matrix} \right\| \tag{14}$$

$$\hat{e}_d \equiv \left\| \begin{array}{cc} \left(\dfrac{\partial u_x}{\partial x} - \dfrac{\partial u_y}{\partial y} \right) \Big/ 2, & \left(\dfrac{\partial u_x}{\partial y} + \dfrac{\partial u_y}{\partial x} \right) \Big/ 2 \\[3mm] \left(\dfrac{\partial u_x}{\partial y} + \dfrac{\partial u_y}{\partial x} \right) \Big/ 2, & \left(\dfrac{\partial u_y}{\partial y} - \dfrac{\partial u_x}{\partial x} \right) \Big/ 2 \end{array} \right\|$$

(15)

For purely elastic surfaces the expressions:

$$\Delta \hat{\gamma}_i = k_i \hat{e}_i$$

(16)

$$\Delta \hat{\gamma}_d = k_d \hat{e}_d$$

(17)

are valid, where k_i and k_d are two independent constants which characterise the elastic properties of a two-dimensional continuum subjected to small perturbations. For $k_i = \infty$, $k_d = \infty$ one obtains the ideal solid body (the so called Euclidian body).

The constants k_i and k_d can be related to the known parameters traditionally used to characterise the elastic properties of two-dimensional continuum. Let us consider a small square of the solid monolayer of a surfactant (at the interface between two phases) and assume the side of the square equal to L and the isotropic deformation value is ΔL. Then the square area $A = L^2$ is varied by $\Delta A = 2L\Delta L$, and, therefore, $\Delta L / L = \Delta A / 2A$. On the other hand, for this deformation one has $\dfrac{\partial u_x}{\partial x} = \dfrac{\partial u_y}{\partial y} = \dfrac{\Delta L}{L}$. Then, it follows from (13) and (14) that

$$\hat{e} = \hat{e}_i = \left\| \begin{array}{cc} \dfrac{\Delta A}{2A}, & 0 \\[3mm] 0, & \dfrac{\Delta A}{2A} \end{array} \right\| = \dfrac{\Delta A}{2A} \cdot \hat{1}$$

(18)

Similarly, one obtains from (10), (11):

$$\Delta \hat{\gamma} = \Delta \hat{\gamma}_i = \left\| \begin{array}{cc} \Delta \gamma, & 0 \\ 0, & \Delta \gamma \end{array} \right\| = \Delta \gamma \cdot \hat{1}$$

(19)

It follows from (16), (18), (19) that:

$$\Delta \hat{\gamma} = k_i \dfrac{\Delta A}{2A}$$

(20)

Comparing Eq. (20) with the known definition of the surface elasticity modulus $E \equiv \dfrac{\Delta \gamma}{\Delta A / A}$ [4, 5] one obtains

$$k_i = 2 E \tag{21}$$

Let us consider now a surface monolayer of surfactant which is subject to pure shear along the x axis: $\dfrac{\partial u_x}{\partial y} \neq 0$ with $\dfrac{\partial u_y}{\partial x} = 0$; $\dfrac{\partial u_x}{\partial x} - 0$; $\dfrac{\partial u_y}{\partial y} = 0$. According to (13), (15) one obtains:

$$\hat{e} = \hat{e}_d = \begin{Vmatrix} 0, & \dfrac{\partial u_x}{2 \partial y} \\ \dfrac{\partial u_x}{2 \partial y}, & 0 \end{Vmatrix} = \frac{1}{2} \frac{\partial u_x}{\partial y} \begin{Vmatrix} 0,1 \\ 1,0 \end{Vmatrix} \tag{22}$$

By the definition of this deformation, and using (16), one obtains from (11):

$$\Delta \hat{\gamma}_i = \begin{Vmatrix} (\Delta \gamma_{xx} + \Delta \gamma_{yy})/2, 0 \\ 0, (\Delta \gamma_{xx} + \Delta \gamma_{yy})/2 \end{Vmatrix} = \frac{\Delta \gamma_{xx} + \Delta \gamma_{yy}}{2} \hat{1} = 0 \tag{23}$$

which leads to $\Delta \gamma_{xx} = 0$, $\Delta \gamma_{yy} = 0$. Therefore one obtains from (10) and (12):

$$\Delta \hat{\gamma} = \Delta \hat{\gamma}_d = \begin{Vmatrix} 0, \Delta \gamma_{xy} \\ \Delta \gamma_{yx}, 0 \end{Vmatrix} = \Delta \gamma_{xy} \begin{Vmatrix} 0,1 \\ 1,0 \end{Vmatrix} \tag{24}$$

due to the tension tensor symmetry ($\gamma_{xy} = \gamma_{yx}$). Now from (17), (22), (24) one obtains

$$\Delta \gamma_{xy} = \frac{k_d}{2} \frac{\partial u_x}{\partial y} \tag{25}$$

Using the definition of the shear stress modulus G [4, 5]:

$$\Delta \gamma_{xy} = G \frac{\partial u_x}{\partial y} \tag{26}$$

one obtains from (25) and (26)

$$k_d = 2G \tag{27}$$

By using Eqs. (16), (17), (21), (27), one can show that the rheological tensor equations for the two-dimensional Hooke body are:

$$\Delta\hat{\gamma}_i = 2E\hat{e}_i \tag{28}$$

$$\Delta\hat{\gamma}_d = 2G\hat{e}_d \tag{29}$$

The shear deformation is determined by two dimensions only, therefore, in the three-dimensional case Eq. (29) remains valid, and the factor 2 in Eq. (28) should be replaced by 3, because of this extra dimension.

The constants k_i and k_d can alternatively be expressed via another pair of parameters of the two-dimensional continuum: Young's surface modulus [7]:

$$I \equiv \left(\frac{\Delta\gamma_{xx}}{\Delta L/L}\right)_{\Delta\gamma_{yy}=0} \tag{30}$$

and the Poisson coefficient:

$$v \equiv -\left(\frac{\Delta l/l}{\Delta L/L}\right)_{\Delta\gamma_{yy}=0} \tag{31}$$

where Δl is the change in cross-dimension of the two-dimensional solid strip caused by an elongation of ΔL. This problem can also be easily solved in the framework of the phenomenological rheology.

In fact, from (11), (14), (16) for $\Delta\gamma_{yy} = 0$ one obtains:

$$\left\|\begin{matrix}\Delta\gamma_{xx}/2,0\\0,\Delta\gamma_{xx}/2\end{matrix}\right\| = k_i\left\|\begin{matrix}\left(\dfrac{\partial u_x}{\partial x}+\dfrac{\partial u_y}{\partial y}\right)\Big/2,0\\0,\left(\dfrac{\partial u_x}{\partial x}+\dfrac{\partial u_y}{\partial y}\right)\Big/2\end{matrix}\right\| \tag{32}$$

It follows from (32) that:

$$\Delta\gamma_{xx} = k_i\left(\frac{\partial u_x}{\partial x}+\frac{\partial u_y}{\partial y}\right) = k_i\left(\frac{\Delta L}{L}+\frac{\Delta l}{l}\right) \tag{33}$$

Now from Eq. (33) and the definitions (30) and (31) one obtains:

$$k_i = \frac{I}{1-\nu} \tag{34}$$

It follows from Eq. (12) for $\Delta\gamma_{yy} = 0$:

$$\Delta\hat{\gamma}_d = \left\| \begin{matrix} \Delta\gamma_{xx}/2, \Delta\gamma_{xy} \\ \Delta\gamma_{yx}, -\Delta\gamma_{xx}/2 \end{matrix} \right\| \tag{35}$$

Then from Eq. (15) one can express the product of the deformation tensor deviation by the coefficient k_d as:

$$k_d\hat{e}_d = \left\| \begin{matrix} \dfrac{k_d}{2}\left(\dfrac{\partial u_x}{\partial x} - \dfrac{\partial u_y}{\partial y}\right), \dfrac{k_d}{2}\left(\dfrac{\partial u_x}{\partial y} - \dfrac{\partial u_y}{\partial x}\right) \\ \dfrac{k_d}{2}\left(\dfrac{\partial u_x}{\partial y} - \dfrac{\partial u_y}{\partial x}\right), -\dfrac{k_d}{2}\left(\dfrac{\partial u_x}{\partial x} - \dfrac{\partial u_y}{\partial y}\right) \end{matrix} \right\| \tag{36}$$

According to Eq. (17), the diagonal components of expressions (35) and (36) should be equal to each other; therefore:

$$\Delta\gamma_{xx} = k_d\left(\frac{\partial u_x}{\partial x} - \frac{\partial u_y}{\partial y}\right) \equiv k_d\left(\frac{\Delta L}{L} - \frac{\Delta l}{l}\right) \tag{37}$$

From Eq. (37) and definitions (30), (31) one obtains:

$$k_d = \frac{I}{1+\nu} \tag{38}$$

Combining Eqs. (34) and (38) yields:

$$I = 2k_i k_d/(k_i + k_d) \tag{39}$$

$$\nu = (k_i - k_d)/(k_i + k_d) \tag{40}$$

For $k_i = \infty$ one obtains: $k_d = \dfrac{E}{2}$, $\nu = 1$ (which is the largest ν value in the theoretical rheology) for a two-dimensional incompressible solid monolayer (the two-dimensional 'rubber'). For the two-dimensional 'cork' $\nu = 0$, $I = k_i = k_d$. It is interesting to note that the theoretical rheology does not impose any restriction on negative values of the Poisson coefficient, and even at the limiting value of this coefficient $\nu = -1$.

2. Purely viscous two-dimensional continuum

We now turn to the tensor formulation of rheological equations for two-dimensional viscous liquid. The simplest case, the purely viscous liquid, is defined as the continuum in which the arising forces depend only on the velocities of perturbation. In the two-dimensional case the velocities are defined by the components v_x and v_y, which replace the displacement vector components u_x and u_y in Eqs. (6) and (7):

$$dv_x = \frac{\partial v_x}{\partial x} dx + \frac{\partial v_x}{\partial y} dy$$

$$dv_y = \frac{\partial v_y}{\partial x} dx + \frac{\partial v_y}{\partial y} dy \tag{41}$$

$$\left\| \begin{matrix} \dfrac{\partial v_x}{\partial x}, & \dfrac{\partial v_x}{\partial y} \\[2mm] \dfrac{\partial v_y}{\partial x}, & \dfrac{\partial v_y}{\partial y} \end{matrix} \right\| \equiv \hat{v} \tag{42}$$

Applying Eqs. (41) to the rotation operation of two-dimensional element as a whole within the plane of this element, one obtains:

$$\hat{v} = \left\| \begin{matrix} 0, & \dfrac{1}{2}\left(\dfrac{\partial v_x}{\partial y} - \dfrac{\partial v_y}{\partial x} \right) \\[3mm] -\dfrac{1}{2}\left(\dfrac{\partial v_x}{\partial y} - \dfrac{\partial v_y}{\partial x} \right), & 0 \end{matrix} \right\| \tag{43}$$

This operation does not alter the physical state of the element of two-dimensional continuum (we assumed earlier that the centrifugal forces, which are the inertial forces, are disregarded), and therefore the anti-symmetric tensor (43) should be eliminated from the tensor (42) similarly to what was made for the deformation tensor. Subtraction of monadic components of tensor (43) from (42) yields the expression which would also be obtained by differentiation of Eq. (9) with respect to time:

$$\hat{e} \equiv \left\| \begin{array}{cc} \dfrac{\partial v_x}{\partial x}, & \left(\dfrac{\partial v_x}{\partial y} + \dfrac{\partial v_y}{\partial x} \right)\Big/ 2 \\[4mm] \left(\dfrac{\partial v_x}{\partial y} + \dfrac{\partial v_y}{\partial x} \right)\Big/ 2, & \dfrac{\partial v_y}{\partial y} \end{array} \right\|$$

(44)

This tensor is called the deformation velocity tensor.

Similarly to the discussion presented in the preceding section, the tensor (44) should be decomposed into two tensors, which are qualitatively different with regard to the physical influence on the continuum, cf. Eqs. (13) - (15):

$$\hat{v} = \hat{v}_i + \hat{v}_d \equiv \hat{e}_i + \hat{e}_d$$

(45)

where

$$\hat{v}_i \equiv \left\| \begin{array}{cc} \left(\dfrac{\partial v_x}{\partial x} + \dfrac{\partial v_y}{\partial y} \right)\Big/ 2, & 0 \\[4mm] 0, & \left(\dfrac{\partial v_x}{\partial x} + \dfrac{\partial v_y}{\partial y} \right)\Big/ 2 \end{array} \right\|$$

(46)

$$\hat{v}_d \equiv \left\| \begin{array}{cc} \left(\dfrac{\partial v_x}{\partial x} - \dfrac{\partial v_y}{\partial y} \right)\Big/ 2, & \left(\dfrac{\partial v_x}{\partial y} + \dfrac{\partial v_y}{\partial x} \right)\Big/ 2 \\[4mm] \left(\dfrac{\partial v_x}{\partial y} + \dfrac{\partial v_y}{\partial x} \right)\Big/ 2, & -\left(\dfrac{\partial v_x}{\partial x} - \dfrac{\partial v_y}{\partial y} \right)\Big/ 2 \end{array} \right\|$$

(47)

Only now we can formulate the expressions which are the analogues of Eqs. (16), (17) applicable to the simplest purely viscous continuum. Combining Eqs. (11), (12) and (46), (47) yields:

$$\Delta \hat{\gamma}_i = m_i \hat{v}_i$$

(48)

$$\Delta \hat{\gamma}_d = m_d \hat{v}_d$$

(49)

where m_i and m_d are constants. These constants of theoretical rheology could be related to the parameters known from practice: the surface shear viscosity η and the surface dilational viscosity η_d (η is sometimes called the first viscosity, and η_d the second viscosity). The first viscosity is defined by the expression:

$$\eta \equiv \Delta \gamma_{xy} \Big/ \dfrac{\partial v_x}{\partial y}$$

(50)

(with v_y set zero). The surface dilational viscosity was defined by Boussinesq [8] as:

$$\eta_d = \Delta\gamma \left/ \frac{\dot{A}}{A} \right.$$
(51)

where A is the surface area.

Let us first consider the deformation of pure dilation, when

$$\frac{\partial v_x}{\partial y} = \frac{\partial v_y}{\partial x} = 0, \qquad \frac{\partial v_x}{\partial x} = \frac{\partial v_y}{\partial y} = \frac{\dot{L}}{L} = \frac{\dot{A}}{2A}$$
(52)

where L is the side of a square element at the surface, subject to the uniform two-dimensional expansion (compression). It follows from Eqs. (45), (46) that:

$$\hat{v} = \hat{v}_i = \left\|\begin{array}{c} \dfrac{\partial v_x}{\partial x}, 0 \\ 0, \dfrac{\partial v_y}{\partial y} \end{array}\right\| = \left\|\begin{array}{c} \dfrac{\dot{A}}{2a}, 0 \\ 0, \dfrac{\dot{A}}{2A} \end{array}\right\| = \frac{\dot{A}}{2A}\hat{1}$$
(53)

Combining Eqs. (48), (51), (19) and (53) one obtains:

$$m_i = 2\eta_d$$
(54)

For a pure shear deformation the perturbation tensor is represented by (24). According to Eq. (50) (the flow along the x axis) one obtains from (45), (47):

$$\hat{v} = \hat{v}_d = \left\|\begin{array}{c} 0, \dfrac{1}{2}\dfrac{\partial v_x}{\partial y} \\ \dfrac{1}{2}\dfrac{\partial v_x}{\partial y}, 0 \end{array}\right\| = \frac{1}{2}\frac{\partial v_x}{\partial y}\left\|\begin{array}{c} 0,1 \\ 1,0 \end{array}\right\|$$
(55)

Combining (49), (50), (24) and (55) one obtains:

$$m_d = 2\eta$$
(56)

Now expressions (48) and (49) for the purely viscous two-dimensional continuum can be represented as:

$$\Delta\hat{\gamma}_i = 2\eta_d\hat{v}_i \left(\equiv 2\eta_d\hat{\dot{e}}_i\right)$$
(57)

$$\Delta\hat{\gamma}_d = 2\eta\hat{v}_d \left(\equiv 2\eta\hat{\dot{e}}_d\right)$$
(58)

As only two dimensions are necessary to define the shear deformation both in the three-dimensional, and in the two-dimensional case, Eq. (58) remains valid also in the three-dimensional case. At the same time, for the three-dimensional case the factor 2 in Eq. (57) should be replaced by 3 [6].

The liquid which obeys the rheological equations (57) or (58) is called Newtonian liquid. For the three-dimensional case, no physical continuum is known which obeys Eqs. (57) and (58) simultaneously. At the same time, there exist purely elastic three-dimensional solids which simultaneously obey Eqs. (28) and (29). The behaviour of real three-dimensional liquids with respect to the dilational deformations is different from their response to shear deformations. With respect to dilational deformation, a real liquid exhibits a behaviour similar to that of a solid, and could be treated theoretically as an incompressible fluid, which obeys the Euclidian solid rheology. On the other hand, with respect to shear deformation a real fluid in the simplest case behaves like a Newtonian viscous liquid. This is also true for fluids characterised by extremely low viscosity, known from the classic experiments by Kapitza. Even a gas, due to its low shear viscosity, could be treated in many cases like an Euclidian solid with respect to dilation and, at the same time, like a viscous liquid with respect to shear deformation. The above conclusions refer to the two-dimensional case, while it is known from experiments that rheology in this case is more diverse, which is primarily attributed to the fact that the exchange of a surfactant is possible between the interface and the two (or one) adjacent bulk phase(s).

C. RHEOLOGICAL EQUATIONS AND MECHANICAL MODELS OF INTERFACIAL CONTINUA

It is known from experiments that with respect to dilational and shear deformation, three-dimensional and two-dimensional continua exhibit both elastic (reversible processes) and viscous (irreversible processes) properties. This fact is obvious from the point of view of general thermodynamics, because each process which takes place with finite velocity is accompanied by a dissipation of energy. The perturbation of the tension of the continuum can be caused by two factors acting simultaneously: the existence of deformations of different kinds (28) and (29), and the existence of deformation rates of different kinds (57) and (58). As the perturbations are assumed to be small, and, therefore, the superposition principle for the deformation should be valid, the resulting stress is described by the tensors:

$$\Delta \hat{\gamma}_i = 2E\hat{e}_i + 2\eta_d \hat{\dot{e}}_i \qquad\qquad (59)$$

$$\Delta \hat{\gamma}_d = 2G\hat{e}_d + 2\eta \dot{\hat{e}}_d \tag{60}$$

All components of the tensors $\Delta \hat{\gamma}_i(t)$ and $\Delta \hat{\gamma}_d(t)$ are determined by Eqs. (59) and (60) if the functions $\hat{e}_i(t)$ and $\hat{e}_d(t)$ are defined by the experimental conditions. And, *vice versa*, if the $\Delta \hat{\gamma}_i(t)$ and $\Delta \hat{\gamma}_d(t)$ are given, then Eqs. (59) and (60) should be considered as the differential equations which determine $\hat{e}_i(t)$ and $\hat{e}_d(t)$, respectively.

Starting from the same initial expressions for the isotropic tensors and deviators, which characterise the elastic and viscous properties of the continuum, one can develop an alternative version for possible rheological properties. In fact, the integration of (57) and (58) yields:

$$\hat{e}_i = \frac{1}{2\eta_d} \int_0^t \Delta \hat{\gamma}_i dt \tag{61}$$

$$\hat{e}_d = \frac{1}{2\eta} \int_0^t \Delta \hat{\gamma}_d dt \tag{62}$$

Then, adding Eq. (28) divided by 2E to (61), and adding Eq. (29) divided by $2G$ to (62), one obtains the expressions:

$$\hat{e}_i = \frac{\Delta \hat{\gamma}_i}{2E} + \frac{1}{2\eta_d} \int_0^t \Delta \hat{\gamma}_i dt \tag{63}$$

$$\hat{e}_d = \frac{\Delta \hat{\gamma}_d}{2G} + \frac{1}{2\eta} \int_0^t \Delta \hat{\gamma}_d dt \tag{64}$$

which determine the tensors $\hat{e}_i(t)$ and $\hat{e}_d(t)$ via the given tensors $\Delta \hat{\gamma}_i(t)$ and $\Delta \hat{\gamma}_d(t)$. Also, the expressions (63), (64) can be represented as differential equations:

$$\Delta \dot{\hat{\gamma}}_i + (E/\eta_d)\Delta \hat{\gamma}_i = 2E\dot{\hat{e}}_i \tag{65}$$

$$\Delta \dot{\hat{\gamma}}_d + (G/\eta)\Delta \hat{\gamma}_d = 2G\dot{\hat{e}}_d \tag{66}$$

The two-dimensional continuum which can be described by Eqs. (59) and (60) is called the Kelvin-Voigt solid body, and the two-dimensional continuum which can be described by Eqs. (65) and (66) is called Maxwellian liquid. It should be stressed at this point once more that a continuum could behave itself as a solid body with respect to the isotropic tensors and as a fluid with respect to the deviators, and *vice versa*.

The rheological equations derived above can be adequately interpreted using mechanical models. We consider first the simplest rheological equations (28) and (29). The proportionality between tension deformation and displacement in Eqs. (28) and (29) can be represented by the behaviour of a spring (see Fig. 2) described by the equation:

$$\Delta \gamma = \varepsilon \varphi \quad \text{or} \quad \Delta \gamma = G \varphi \tag{67}$$

where $\Delta \gamma$ stands for the acting force, and φ for the corresponding elongation of the spring. Both these values in Eq. (67) at any time could be replaced by the tensors from Eqs. (28) and (29) with ε and G as the proportionality coefficients, respectively. It is implied by Eqs. (67) that the force arises instantly: no relaxation process takes place.

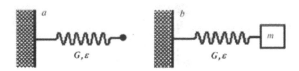

Fig. 2.   Hooke's solid body models: a - static (correct) and b - 'dynamic' (erroneous)

The model of a spring for a purely elastic continuum illustrated by Fig. 2a, is called the static Hooke model, which should be distinguished from the dynamic Hooke model where the element with finite mass is attached to the moving end of the spring, see Fig. 2b [1]. The latter model, however, can have only a qualitative but not a quantitative sense, in contrast to the model illustrated by Fig. 2a. It was shown in the Introduction section that, if pure rheology is considered, the influence of mass becomes negligibly small with the infinite decrease of the dimensions of the surface element. But, even if one does not consider the rheological limit, it should be kept in mind that the mass is distributed along the surface of the element (and, similarly, the viscous forces acting from the adjacent phases), in contrast to the rheological forces which act at the edges of the element. In Fig. 2 this edge is symbolically represented by the point at the moving end of the spring; in this case the force depends not on the spring length, but on its physical state, while the mass is proportional to the length of the spring, and its influence vanishes with the unlimited decrease of the length, similarly to the situation which takes place with the surface element. The mass of the spring should be accounted for in another way – when the equation of motion of the spring is derived; it should be stressed at this point that this equation is the equation in partial derivatives for the waves which propagate along the spring. A similar situation exists for the two-dimensional continuum, and does not depend on its rheological complexity, as shown below.

Similarly to equations (67), the tensor rheological equations (57) and (58) can be symbolically presented by:

$$\Delta\gamma = \zeta\dot{\varphi}, \qquad\qquad \Delta\gamma = \eta\dot{\varphi} \qquad\qquad (68)$$

These equations, valid for a 'Newtonian fluid', can be adequately described by the ideal damper (cf. Fig. 3) considered as the element of the Newtonian non-relaxing viscosity, when corresponding forces vanish immediately after the motion is ceased $(\dot{\varphi} = 0)$.

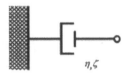

Fig. 3. Single-element model of a Newtonian fluid

Unlike the elasticity element, for which the analogy becomes obvious from Fig. 2, for the viscous element the analogy illustrated by Fig. 3 is less obvious. This viscous element is usually treated as a piston combined with an incompressible fluid, which inflows or outflows through the gap due to the motion of the plate (right) inside the immobile 'piston' (left). However, for this kind of motion the main resistance to the displacement of the right element in Fig. 3 is essentially the head resistance, which is by no means proportional to the velocity, and is related to the ejection of the jet from the closed gap (this resistance is obviously nonzero even if the viscosity of the liquid which is ejected or sucked in is negligible). The resistance force, which is proportional to the velocity, arises when a motion of one cylindrical surface coaxially to another cylindrical surface takes place (this motion is not accompanied by the ejection of jets). In this case, strictly speaking, in addition to the absence of ram pressure, two more conditions should be satisfied: the length of the gap between two the cylindrical surfaces should be much larger than the thickness of the fluid layer in the gap h; and the characteristic time of the acceleration of motion of internal cylindrical surface should be much larger than the time of viscous relaxation of the fluid h^2/v, here v is the kinematic viscosity of the fluid.

The relation between the model definitions of dilational rheological parameters by Eqs. (67), (68), and the physical definitions of these parameters as given in the previous section, is expressed as $\varepsilon = E/A$, $\zeta = \eta_d/A$, while the shear-related physical parameters are the same as the model ones, and therefore it is not necessary to introduce new notations here.

It is easy to understand that the continuum described by the rheological equations (59), (60) [with the forces corresponding to Eqs. (28), (29), (57) and (58) being summed, as these forces are created by the deformations of different kinds, according to the superposition principle which is valid for small perturbations] could be symbolically characterised by a block with a spring and damper in parallel, Fig. 4. According to (67) and (68), the displacements, and also the velocities of the displacement, and the forces which arise due to displacements, can be described in this model as:

$$\Delta\gamma = \varepsilon\varphi + \zeta\dot{\varphi} \quad \text{or} \quad \Delta\gamma = G\varphi + \eta\dot{\varphi} \tag{69}$$

which is called the Kelvin-Voigt model. The in-parallel connection of several springs and dampers does not alter the scheme if the sum of the elasticities is ε (or G), and the sum of the viscosities is ζ (or η). In the following, for sake of brevity only a pair of elements ε and ζ will be used.

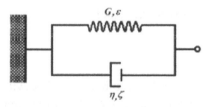

Fig. 4. Kelvin-Voigt solid body model: upper and lower branches correspond to the retarding elasticity and relaxing viscosity, respectively

It follows from Eq. (69) that the function $\Delta\gamma(t)$ is determined explicitly if the $\varphi(t)$ is defined. However, to calculate $\varphi(t)$ for a given $\Delta\gamma(t)$ one should solve a differential equation. Dividing the first of the equations (69) by ζ and introducing the integrating factor $\exp(\varepsilon t/\zeta)$ one obtains:

$$d\left[\varphi(t)\exp\left(\frac{\varepsilon t}{\zeta}\right)\right] = \frac{\Delta\gamma(t)}{\zeta}\exp\frac{\varepsilon t}{\zeta}dt \,, \tag{70}$$

for which the general solution reads

$$\varphi(t) = \exp\left(-\frac{\varepsilon t}{\zeta}\right)\left[\frac{1}{\zeta}\int_0^t \Delta\gamma(t)\exp\left(\frac{\varepsilon t}{\zeta}\right)dt + \varphi\Big|_{t=0}\right] \tag{71}$$

Let us consider the elementary process: for the position $\varphi\big|_{t=0} = 0$ let the constant perturbation be $\Delta\gamma(t) = \Delta\gamma_0$. Integration of (71) yields:

$$\varphi(t) = \frac{\Delta\gamma_0}{\varepsilon} - \frac{\Delta\gamma_0}{\varepsilon}\exp\left(-\frac{\varepsilon t}{\zeta}\right) \tag{72}$$

For $t \to \infty$, with $\Delta\gamma_0$ as defined above, the system acquires the equilibrium state $\varphi_0 = \frac{\Delta\gamma_0}{\varepsilon}$ (this state can also be thought of as the state of static perturbation). The transition to this state is described by the exponential dependence:

$$\Delta\varphi(t) = \varphi_0 \exp\left(-\frac{t}{\theta}\right) \tag{73}$$

where $\Delta\varphi(t) \equiv \varphi_0 - \varphi$ is the deviation of the displacement from the equilibrium state for given $\Delta\gamma_0$, and

$$\theta \equiv \frac{\zeta}{\varepsilon} \tag{74}$$

is the time during which the amplitude of this deviation decays e times. In rheology this time is called the retardation time. The higher is the viscosity (ζ or η), the higher is the retardation time; but the higher is the elasticity (ε or G), the lower is the retardation time.

Let us assume now that, when the equilibrium (for the given $\Delta\gamma_0$) displacement φ_0 is acquired, the load $\Delta\gamma_0$ becomes zero. Considering this initial time moment ($t = 0$), one obtains from (71) for $\varphi\big|_{t=0} = \varphi_0$:

$$\varphi(t) = \varphi_0 \exp\left(-\frac{t}{\theta}\right) \tag{75}$$

i.e., the system approaches the equilibrium state in a similar way in both cases considered above: for the statically disturbed equilibrium state, Eq. (73), and for the statically non-disturbed state, given by Eq. (75).

The behaviour of the Kelvin-Voigt solid body is similar to that of the Euclidean body with respect to fast deformations taking place during the time much shorter than θ; but for slow deformations taking place during the time much longer than θ this body behaves purely elastic. Therefore, the elasticity applied in parallel to the viscosity is called the retarded elasticity, while the viscosity in this case is called the relaxing viscosity.

In the framework of the Kelvin-Voigt model it becomes possible, in a first approximation, to account for a weak viscosity characteristic for elastic bodies: the acoustic adsorption in solids and liquids (in the case of liquids the dilational deformation should be relevant, which can also be accompanied by a more or less significant shear deformation), the decay of vibrations of rods and plates [6]. This model was also successfully used to study the effect of dissipative processes in solid bodies during its free rotation in vacuum unaffected by external fields [9]. In this case, the process is governed by the angular momentum conservation law, and for the body other than the spherical top, a certain part of its kinetic energy (the proportion depends on the initial conditions) can be dissipated within the body, which finally assumes the state of rotation around the axis corresponding to the largest moment of inertia. For example, a thin disk ('coin'), which rotates initially around the axis lying in its plane, at $t \to \infty$ will lose half of its kinetic energy and assumes the state of rotation around its symmetry axis. This is a good example of an unexpected result predicted by rheology.

As was shown above, the parallel connection of several springs and dampers could be treated by the Kelvin-Voigt model and, therefore, by the same type of rheological equation. In contrast, if the elastic element is connected in series with the Kelvin-Voigt model system (as is shown in Fig. 5) the rheological equation should be reconsidered as follows.

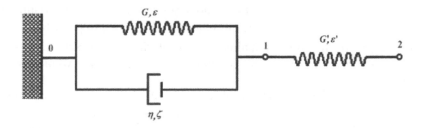

Fig. 5.   A-model of standard solid body: serial connection of Kelvin-Voigt model (0,1) and Hooke's model (1,2)

The serial connection of the Kelvin-Voigt model (0,1) and Hooke's model (1,2) (in what follows we refer to this system as to the A-Model) implies the additivity of the displacement $\varphi_{0,1}$ described by Eq. (71), and the displacement $\varphi_{1,2} = \dfrac{\Delta\gamma}{\varepsilon'}$, given by Eq. (67). Therefore, we have

$$\varphi(t) = \frac{\Delta\gamma(t)}{\varepsilon'} + \exp\left(\frac{\varepsilon t}{\varsigma}\right)\left[\frac{1}{\varsigma}\int_0^t \Delta\gamma(t)\exp\left(\frac{\varepsilon t}{\varsigma}\right)dt + \varphi_0\right] \quad (76)$$

It should be noted that, as the elements connected in series [Kelvin-Voigt (0,1) element and Hooke's element (1,2)] do not possess any inertial properties, the $\Delta\gamma(t)$ function which enters the first term is the same as that entering the integrand in Eq. (76). As the differential equation is easier solved (and, in particular, classified analytically) than the integral equation, it would be convenient to derive the differential equation which describes the model shown in Fig. 5. To eliminate the constant φ_0, one should multiply Eq. (76) by $\exp\left(-\dfrac{\varepsilon t}{\varsigma}\right)$, and differentiate the result by t. Then the result should be divided by the same factor, and one finally obtains the differential expression for the model:

$$\Delta\dot{\gamma} + \Delta\gamma(\varepsilon + \varepsilon')/\varsigma = \varepsilon'(\dot{\varphi} + \varphi\varepsilon/\varsigma) \quad (77)$$

or, in tensor form:

$$\Delta\hat{\dot{\gamma}} + \Delta\hat{\gamma}(E + E')/Z = E'(2\hat{\dot{\varphi}} + 2\hat{\varphi}E/Z) \quad (78)$$

Let us analyse now the rheological equation (77) to determine the characteristic processes. The unrestricted return to the statically non-perturbed state (e. for $\Delta\gamma = 0$) is the same process that follows from Eq. (69) for the Kelvin-Voigt model illustrated by Fig. 4. This fact is obvious from Fig. 5: the spring ε' does not contribute to the variation of φ, because when the force is eliminated at the statically perturbed state, the length of the spring remains constant during the relaxation towards the statically non-perturbed (equilibrium) state of the whole three-element unit.

The state of the Kelvin-Voigt model at $\varphi = 0$ is by definition a non-perturbed state, and, in particular, the statically non-perturbed. In contrast, the model defined in Fig. 5 at $\varphi = 0$ could exist in a perturbed state, because the force can be non-zero. It is seen from Fig. 5 that the point 1 can be displaced from the initial equilibrium state of the left unit (e., $\Delta\gamma \neq 0$) even if the point 2 remains in the initial equilibrium state. This can be easily made, for example, by two successive operations: first, one should displace the point 2 and then, after the whole three-element system assumes the statically perturbed state, the point 2 should be instantly returned to its initial position. In this case it would require some time until the left unit returns to its initially non-perturbed state and, therefore, the length of the right spring assumes its non-perturbed value, i.e. at $\varphi = 0$ initially we will have $\Delta\gamma \neq 0$.

For $\varphi(t) - 0$ Eq. (77) becomes:

$$\Delta\dot\gamma + \Delta\gamma(\varepsilon + \varepsilon')/\zeta = 0 \qquad (79)$$

which is similar to Eq. (69) for $\dot\varphi$, but corresponds to the relaxation of the force $\Delta\gamma$ (not the displacement) with the time required for an e-times decay of the initial $\Delta\gamma$ value:

$$\tau = \zeta/(\varepsilon + \varepsilon') \qquad (80)$$

In rheology, the value τ is called the relaxation time [7]. It follows from (74) and (80) that:

$$\tau \leq \theta \qquad (81)$$

It is seen from (80) that the equality in (81) corresponds to the limit $\varepsilon' \ll \varepsilon$.

Both for very fast and very slow processes the viscosity ζ can be eliminated from the rheological equation (77). In fact, for times much shorter than θ and τ, in Eq. (77) one should account for the terms which involve the time derivatives only. This leads to the relation:

$$\Delta\dot\gamma = \varepsilon'\dot\varphi \qquad (82)$$

Integration of this equation from the initial (non-perturbed) state yields:

$$\Delta\gamma = \varepsilon'\varphi \qquad (83)$$

The spring ε' in the model shown in Fig. 5 represents the instant elasticity. In contrast, if the characteristic time of the deformation is much higher than θ and τ, one can neglect the terms containing the derivatives in (77). This leads to the expression which determines the quasi-equilibrium elasticity:

$$\Delta \gamma = \varphi \frac{\varepsilon \varepsilon'}{\varepsilon + \varepsilon'} \tag{84}$$

As the viscosity ζ does not enter into both Eqs. (83) and (84), it can be concluded that the dissipative processes are essentially relevant for the rheology only if the characteristic deformation time is comparable with θ or τ.

The model which is described by the rheological equation (77) is called the standard solid body model. In fact, the model reproduces satisfactorily all the basic (both dilational and shear) visco-elastic properties of a real solid body. For a liquid this model provides a good description of the dilational properties only, which for the liquid are basically less important than shear properties, because usually liquids are treated as incompressible media. However, in the two-dimensional case, for interfaces between two immiscible fluid phases covered by a surfactant, the standard solid body model is of large practical significance (see section 3).

However, we will continue now with liquids. The simplest single-element model which describes a liquid was presented in Fig. 3 above. Similarly to the two-element solid, the two-element liquid is modelled by a single combination of visco-elastic elements presented in Fig. 6, with a serial connection of a viscous and elastic element, a damper and a spring, respectively. This model could be derived from the model shown in Fig. 5, if one eliminates the elastic element (ε or G) connected in parallel to the damper. Assuming $\varepsilon = 0$ in Eq. (77) one obtains the rheological equation:

$$\Delta \dot{\gamma} + \frac{\varepsilon'}{\zeta} \Delta \gamma = \varepsilon' \dot{\varphi} \tag{85}$$

The corresponding tensor equation (for the Maxwellian fluid) was derived above, Eq. (65). It is seen from Eq. (85) that for the model described by Fig. 6 the relaxation time is:

$$\tau = \frac{\zeta}{\varepsilon'} \tag{86}$$

which is naturally different from the relaxation time of the standard solid body, Eq. (80).

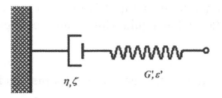

Fig. 6.   Two element model of a Maxwellian fluid: serial connection of
damper (relaxing viscosity) and spring (instant elasticity)

For deformations with a relaxation time much lower than that defined
by Eq. (86), the model described by Fig. 6 behaves itself similarly to a
Hooke elastic body. However, in contrast to the standard solid body, the
model represented by Fig. 6 does not exhibit any equilibrium elasticity.
The elasticity ε' shown in Fig. 6 is the instant elasticity, while the
viscous element in this figure describes the relaxing viscosity. The
continuum which corresponds to this model is called the Maxwellian
liquid. An example of a two-dimensional Maxwellian liquid is the
surface layer of soluble surfactants with respect to dilational deformation
(see section 3). In the three-dimensional case, the continua which obey
the Maxwellian liquid model with respect to shear deformation are pitch,
beeswax, polymers and their concentrated solutions, and also ice cream.
The shear relaxation time characteristic for water is ca. 10^{-13} s, for the
polymethyl siloxane (the 'jumping putty') this time is several seconds,
while for the Earth this time is ca. 300 years.

The two-dimensional models of visco-elastic bodies involve only the
connection of these elements in parallel (Fig. 4) and in series (Fig. 6).
However, for the standard solid body there are two options to connect the
three elements. To explain this, we turn to Eq. (77) and divide $\Delta\gamma$ into
two constituents:

$$\Delta\gamma = \Delta\gamma_{el} + \Delta\gamma_{rel} \tag{87}$$

where $\Delta\gamma_{el}$ is the elastic portion of $\Delta\gamma$, which corresponds to the
equilibrium relation (84):

$$\Delta\gamma_{el} = \varphi\varepsilon\varepsilon'/(\varepsilon+\varepsilon') \equiv \varepsilon_{el}\varphi, \ \Delta\dot{\gamma}_{el} = \varepsilon_{el}\dot{\varphi} \tag{88}$$

and $\Delta\gamma_{rel}$ is the relaxing force (similar to the relaxing pressure in
liquids [10]). Introducing Eq. (87) into Eq. (77) and using Eq. (88) one
obtains the rheological equation for $\Delta\gamma_{rel}$:

$$\Delta\dot{\gamma}_{rel} + (\varepsilon+\varepsilon')\Delta\gamma_{rel}/\zeta = \varepsilon'^{2}\dot{\varphi}/(\varepsilon+\varepsilon') \tag{89}$$

Comparing Eq. (89) with Eq. (85) one can see that Eq. (89) is the equation for a Maxwellian liquid with instantaneous elasticity:

$$\varepsilon_{rel} = \varepsilon'^2 / (\varepsilon + \varepsilon') \tag{90}$$

The relaxation time is the same as that for the standard solid body:

$$\tau_{rel} = \frac{\varsigma}{(\varepsilon + \varepsilon')} \tag{91}$$

In compliance with Eq. (85), the relaxing viscosity which corresponds to Eq. (89) is the instantaneous elasticity multiplied by the relaxation time:

$$\varsigma_{rel} = \varepsilon_{rel}\tau_{rel} = \varsigma\left[\varepsilon'/(\varepsilon + \varepsilon')\right]^2 \tag{92}$$

Thus, all elements in Fig. 7 for this B-model which is alternative to the standard solid body A-model (Fig. 5) are defined. It should be stressed that both models are described by the same rheological equation (77). It is easy to see that the instantaneous elasticity of this system, which is just the sum of ε_{rel} and ε_{el} according to the scheme presented in Fig. 7, is equal to the instantaneous elasticity ε' of the model described by Fig. 5.

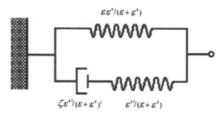

Fig. 7.   B-model of standard solid body alternative to the A-model shown in Fig. 5: parallel connection of the Maxwellian model and the Hooke model (the dilational characteristics of B-model are expressed via the dilational characteristics of A-model)

Let us finally present two more complicated models of visco-elastic fluid. The three-element model shown in Fig. 8 consists of a non-relaxing viscous element connected in parallel with an elastic element, and a relaxing viscosity element which is connected in series to the two elements above.

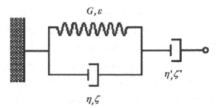

Fig. 8.   Three-element model of visco-elastic fluid: the serial connection in
the Kelvin-Voigt and Newton models

This model possesses, similarly to the Kelvin-Voigt model, the
retarding elasticity and the non-relaxing viscosity, however, these
models should be connected in series with the viscosity element of the
stationary flow, resulting in a model for a liquid, according to the
definition. Unfortunately, however, the important and general property of
the continuum, the instantaneous elasticity, is disregarded in this model:
the liquid described by the model in Fig. 8 is the Euclidean solid body
with respect to instantaneous perturbations.

To eliminate this deficiency, one has to include the elasticity element
in series to the system shown in Fig. 8; the resulting scheme presented in
Fig. 9 assumes the instant elasticity of the continuum, while this
continuum still remains to be a liquid. This four-element model involves
a serial connection of a standard solid body model and a Maxwellian
liquid model. The medium defined in this way is called Burger's visco-
elastic liquid. This model allows to describe the rheology of liquids at the
same accuracy level which is characteristic for the standard solid body
model with respect to the description of real solid bodies.

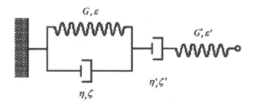

Fig. 9.   Four-element model of a Burger fluid: the serial connection of the
three-element visco-elastic liquid and Hooke's solid body

Now the elasto-plastic and visco-plastic models will be mentioned. In
rheological models, the plasticity is introduced via the special 'dry
friction element' which is also called 'slider', using the Saint-Venant

model. This element acts according to the idealised dry friction law: its behaviour corresponds to the Euclidean solid body below the plasticity threshold, while above this threshold a constant force F_0 independent of the velocity is created. The slider is represented schematically in Fig. 10.

Two versions of elasto-plastic bodies exist, with a serial or parallel connection of the elastic and plastic elements. In the first case (see Fig. 11,a), the model provides a pure elasticity with respect to shear or dilation deformation below the plasticity threshold, i.e. it behaves itself like an elastic solid body. Above this threshold such continuum exhibits specific liquid properties: its fluidity becomes infinitely large. In the second case (see Fig. 11,b) the model possesses the properties of an Euclidean solid body below the plasticity threshold, and of a specific elastic solid body above the threshold.

Fig. 10. Slider model (Saint-Venant plasticity model)

Fig. 11. Elasto-plasticity models with serial (a) and parallel (b) connection of plasticity and elasticity elements, respectively

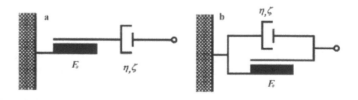

Fig. 12. Visco-plasticity models with serial (a) and parallel (b) connection of plasticity and viscosity elements, respectively

Similarly, only two types of visco-plastic bodies are possible: the model which involves a serial connection of viscous and plastic elements (Fig. 12,a) which demonstrates a purely viscous behaviour below the plasticity threshold and a visco-plastic behaviour above this threshold, and a model with parallel connection of these elements (Fig. 12,b) which behaves itself as an Euclidean solid body below the plasticity threshold and as a visco-plastic liquid above this threshold.

In the end, let us mention the three-element visco-elasto-plastic model, the so called Bingham model, Fig. 13, which behaves like a purely elastic solid body below the plasticity threshold, and like a liquid above this threshold. However, this liquid cannot be reduced to those considered above, because the existence of the plasticity threshold implies a non-linear behaviour of the model. Bingham's model could be formally reduced to the Maxwellian liquid model only for $F \gg F_0$, where F is the total acting force.

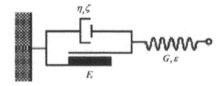

Fig. 13. Visco-elasto-plastic Bingham model

In many cases the shear behaviour of continua could be described by more complicated rheological models. One example is the five-element Schofield-Scott-Blair model (see [11]) which involves a serial connection of the Bingham model unit with the Kelvin-Voigt model unit, or, equivalently, the two-unit model with a parallel connection of the visco-elastic model and the standard solid body model.

All models considered above imply both small deformations and, in a strict sense, yet low enough deformation rates: it is only for this case the elasticity and viscosity of real continuum could be assumed constant. If the deformation rate is too high, the linear theory presented above becomes inapplicable.

D. RHEOLOGICAL EQUATIONS FOR IDEAL INTERFACIAL SURFACTANT LAYERS

If the variations of the interfacial area A are not too fast, the variations of the interfacial tension $\Delta\gamma$ under isothermal conditions are determined only by the amount of surfactants adsorbed at the interface. Therefore, the derivation of rheological equations is necessarily based on kinetic equations which describe the adsorption of a surfactant under dynamic conditions.

For insoluble surfactants, the differentiation of the two-dimensional conservation law:

$$\Gamma'A = const \tag{93}$$

(where Γ' is the adsorption of the insoluble component) yields:

$$\frac{d\Gamma'}{dt} = -\Gamma'\frac{dA}{Adt} \tag{94}$$

Under dynamic conditions, the adsorption Γ of the soluble surfactant depends on three factors. For the area variation factor, obviously, the differential contribution is the same as that for the insoluble surfactants:

$$\left(\frac{d\Gamma}{dt}\right)^{*} = -\Gamma\frac{dA}{Adt} \tag{95}$$

In the framework of the simplest (barrier-governed) adsorption kinetics, for a localised (Langmuir) adsorption regime, with the number of local adsorption sites per unit interfacial area Γ_∞ it was shown in [12] that

$$\left(\frac{d\Gamma}{dt}\right)^{**} = \left(\beta_1 c_1 + \beta_2 c_2\right)\left(1 - \frac{\Gamma' + \Gamma}{\Gamma_\infty}\right) \tag{96}$$

where c_1 and c_2 are the bulk concentrations of the surfactants in the two phases (in the framework of barrier-governed adsorption kinetics these values are constant), and β_1 and β_2 are the corresponding adsorption coefficients. The term in the second parenthesis in Eq. (96) represents the fraction of the interface free from adsorbed surfactants. Finally, assuming ideal surfactant monolayers, for which only the interaction of surfactant molecules with the solvent is relevant [13], the contribution of the desorption factor is:

$$\left(\frac{d\Gamma}{dt}\right)^{***} = -\frac{\Gamma}{\tau_0} \tag{97}$$

where $\dfrac{1}{\tau_0}$ is the desorption coefficient, and τ_0 is the time during which the number of 'labelled' molecules of the surfactants at the interface is decreased due to the desorption to the two phases by a factor of e (here τ_0 is the average lifetime of the soluble surfactant molecule at the interface, or simply the 'adsorption time'). Therefore, the kinetic equation for the adsorption of soluble surfactants for dilational deformation of the interface is (see [14]):

$$\frac{d\Gamma}{dt} = \left(\beta_1 c_1 + \beta_2 c_2\right)\left(1 - \frac{\Gamma' + \Gamma}{\Gamma_\infty}\right) - \left(\frac{1}{\tau_0} + \frac{dA}{Adt}\right)\Gamma \tag{98}$$

To derive the kinetic equation for small deviations from adsorption equilibrium (this equation will be used to formulate the rheological equation), one should present Eq. (98) in a form which corresponds to the equilibrium state:

$$0 = \left(\beta_1 c_1 + \beta_2 c_2\right)\left(1 - \frac{\Gamma' + \Gamma}{\Gamma_\infty}\right) - \frac{\Gamma_0}{\tau_0} \tag{99}$$

Subtraction of Eq. (99) from Eq. (98) yields:

$$\frac{d\Gamma}{dt} + \left(\beta_1 c_1 + \beta_2 c_2\right)\frac{\Gamma' - \Gamma_0' + \Gamma - \Gamma_0}{\Gamma_\infty} + \frac{\left(\Gamma - \Gamma_0\right)}{\tau_0} + \Gamma\frac{dA}{Adt} = 0 \tag{100}$$

Introducing small (with respect to Γ, Γ', A) finite perturbations of adsorptions at the surface:

$$\Delta\Gamma \equiv \Gamma - \Gamma_0, \quad \Delta\Gamma' \equiv \Gamma' - \Gamma_0', \quad \Delta A \equiv A - A_0, \tag{101}$$

one can represent Eq. (100) in the form:

$$\frac{d\Delta\Gamma}{dt} + \frac{\Delta\Gamma}{\tau} + \frac{\beta_1 c_1 + \beta_2 c_2}{\Gamma_\infty}\Delta\Gamma' + \frac{\Gamma_0}{A_0}\frac{d\Delta A}{dt} = 0, \tag{102}$$

$$\frac{1}{\tau} \equiv \frac{1}{\tau_0} + \frac{\beta_1 c_1 + \beta_2 c_2}{\Gamma_\infty} \tag{103}$$

The ratio $\dfrac{\Gamma}{A}$ in the last term of Eq. (100) was substituted by the ratio

$\dfrac{\Gamma_0}{A_0}$, noting that, by definition, the values $\Delta\Gamma$, ΔA and $\dfrac{d\Delta A}{dt}$ are small but finite, and therefore the account for the difference between these ratios would involve the account for the values of higher order in Eq. (102). If the two-dimensional system, which initially exists in the state with arbitrary perturbation, is rapidly returned into the state of $\Delta A = 0$ (and, therefore, $\Delta\Gamma' = 0$), then the relaxation of the adsorption of dissolved surfactant will go on according to the equation:

$$\frac{d\Delta\Gamma}{dt} + \frac{\Delta\Gamma}{\tau} = 0 \tag{104}$$

which shows that the τ, defined by Eq. (103), is the characteristic time for the decrease of the perturbation of adsorption of soluble surfactant by a factor of e (in this case the non-labelled molecules are considered, see the definition of τ_0 above). As the relaxation process takes place at constant (non-perturbed) area, the value Γ' also remains constant (non-perturbed), and therefore it can be concluded that τ also represents the surface tension relaxation time. It follows from the argumentation presented above that, based on Eq. (94), one can write two expressions which involve small perturbations $\Delta\Gamma'$ and ΔA:

$$\frac{d\Delta\Gamma}{dt} + \frac{\Gamma_0'}{A_0}\frac{d\Delta A}{dt} = 0, \qquad \Delta\Gamma' + \frac{\Gamma_0'}{A_0}\Delta A = 0 \tag{105}$$

Using Eq. (105) it is straightforward to obtain the kinetic equation for the sum of adsorption-related perturbations of insoluble and soluble surfactant from Eq. (102):

$$\frac{d(\Delta\Gamma' + \Delta\Gamma)}{dt} + \frac{\Delta\Gamma' + \Delta\Gamma}{\tau} + \frac{\Gamma_0'}{A_0\tau_0}\Delta A + \frac{\Gamma_0' + \Gamma}{A_0}\frac{d\Delta A}{dt} = 0 \tag{106}$$

From the perturbation of $\Delta\Gamma' + \Delta\Gamma$ one can easily obtain the perturbation of interfacial surface tension $\Delta\gamma$. For ideal interfacial layers one can use the generalised van Laar relation for two surfactants:

$$\Lambda = \Lambda_* + RT\Gamma_\infty \ln\left(1 - \frac{\Gamma' + \Gamma}{\Gamma_\infty}\right) \tag{107}$$

where Λ_* is the interfacial tension in absence of a surfactant, R is the gas constant, T is the absolute temperature. Therefore, we obtain:

$$d\Lambda = -RT(d\Gamma' + d\Gamma)\Big/\left(1 - \frac{\Gamma' + \Gamma}{\Gamma_\infty}\right) \tag{108}$$

i.e., for small perturbations $\Delta\gamma$ of the interfacial tension one obtains, according to Eq. (101), the equation which relates $\Delta\Gamma' + \Delta\Gamma$ to the perturbation $\Delta\gamma$:

$$\Delta\gamma = -RT(\Delta\Gamma' + \Delta\Gamma)\Big/\left(1 - \frac{\Gamma_0' + \Gamma_0}{\Gamma_\infty}\right) \tag{109}$$

In what follows we omit the subscript '0' at Γ_0 and Γ_0'' (and also at A_0), because we do not need to differentiate these functions. From Eq. (109) the kinetic equation for the sum of adsorption perturbations can be derived in form of a rheological equation for a standard solid body [cf. Eq. (77)] valid for dilational deformations of interfacial layers comprised of a mixture of a soluble and an insoluble surfactant:

$$\frac{d\Delta\gamma}{dt} + \frac{\Delta\gamma}{\tau} = \frac{E_M}{A}\left(\frac{d\Delta A}{dt} + \frac{\Delta A}{\theta}\right) \tag{110}$$

Here the interfacial tension relaxation time τ is determined by Eq. (103), the non-equilibrium 'instantaneous' dilational elasticity (the Marangoni elasticity) is determined by the relation:

$$E_M = RT(\Gamma' + \Gamma)\Big/\left(1 - \frac{\Gamma' + \Gamma}{\Gamma_\infty}\right), \tag{111}$$

and the retardation time is determined by the relation:

$$\theta = \tau_0 \frac{\Gamma' + \Gamma}{\Gamma'} \tag{112}$$

According to Eqs. (111) and (112), for $\Gamma' \to 0$ one obtains:

$$E_M \to RT\Gamma\Big/\left(1 - \frac{\Gamma}{\Gamma_\infty}\right), \tag{113}$$

$$\theta \to \infty. \tag{114}$$

Therefore, the rheological equation for the standard solid body (110) turns into the equation of a Maxwellian liquid with the relaxation time τ equal to that of a standard solid body [cf. Eq. (103)]:

$$\frac{d\Delta\gamma}{dt} + \frac{\Delta\gamma}{\tau} = \frac{E_M}{A}\frac{d\Delta A}{dt}$$ (115)

Note, from Eq. (103) it follows that the relaxation time τ is independent of the presence of an insoluble surfactant. This independence does not contradict with the fact that the right hand sides of Eqs. (80) and (86) appear to be different: these expressions involve similar notations of parameters, but the values of the parameters depend on the model used.

The expression (103) for τ as applied to the Maxwellian rheological equation (115) for a single soluble surfactant can be simplified, if one introduces into Eq. (103) the expression for $\beta_1 c_1 + \beta_2 c_2$ obtained from the equilibrium relation (99) at $\Gamma' = 0$. This leads to the expression for the adsorption relaxation time τ [see Eq. (102) with $\Delta\Gamma' = 0$] and the time τ of mechanical relaxation [see Eq. (115)]:

$$\tau = \tau_0\left(1 - \frac{\Gamma}{\Gamma_\infty}\right)$$ (116)

i.e., the relaxation time is equal to the adsorption time multiplied by the fraction of the interfacial area not covered by adsorbed surfactant molecules. Comparing Eq. (115) with the definition (85) one can see that the dilational viscosity of the interfacial monolayer can be expressed as:

$$\eta_d = \tau E_M = \tau_0 RT\Gamma.$$ (117)

To obtain the second equation in (117) the expressions (113) and (116) were used. For ideal interfacial layer the simple formula (117) remains valid for any Γ. However, for the case $\Gamma \ll \Gamma_\infty$, i.e., for an ideal two-dimensional gas (which is not a model, but a model-independent approximation) the expression can be further simplified on a physical basis, noting that in this approximation $RT\Gamma = \Pi$ is just the two-dimensional pressure created by the surfactant molecules:

$$\eta_d = \tau_0\Pi$$ (118)

i.e., the viscosity in this case is equal to the adsorption time multiplied by the two-dimensional pressure of the interfacial layer comprised of the adsorbed surfactant molecules.

Relations (117) and (118) express the second (dilational) viscosity as a physical parameter. It was mentioned above that for monolayers which involves soluble surfactants this value was defined by Boussinesq [8] by Eq. (51). However, it was concluded by Levich who formulated the physico-chemical hydrodynamics that for soluble surfactant the dilational viscosity (as physical parameter) does not exist [15]. Here it is shown that for soluble surfactant the dilational viscosity (as physical parameter) does exist, if the two-element Maxwellian liquid is considered rather than a single-element Newtonian viscous fluid which was treated by Boussinesq and Levich.

As a next step we continue with a more complicated case: mixtures of soluble and insoluble surfactant, for which the rheological analysis can now be completed. Similarly to what was done before, cf. Eqs. (111) and (112), the relation (103) can be expressed in terms of adsorptions using Eq. (99):

$$\tau = \tau_0 \left(1 - \frac{\Gamma}{\Gamma'_\infty} \right) \equiv \tau_0 \left(1 - \frac{\Gamma}{\Gamma_\infty - \Gamma'} \right) \tag{119}$$

This expression for τ has, however, one deficiency: it is not obvious from this relation that τ is independent of Γ', while it follows from Eq. (103) that the variation of Γ' leads to the variation of the equilibrium value Γ for which the ratio $\Gamma/(\Gamma_\infty - \Gamma')$ remains constant if $c_1 = const$, $c_2 = const$. In contrast to τ, the retardation time θ depends on the amount of insoluble surfactant present in the system. At $\Gamma \to 0$, according to Eq. (112), the retardation time (which is always higher than the relaxation time) tends to its lower limit τ_0 if $\Gamma' = const$. In this case, according to Eq. (113) τ tends to its upper limit τ_0.

From the relations (110)-(112), (119) one can uniquely express the equilibrium (Gibbs') dilational elasticity of the interfacial layer which depends on the three coefficients of Eq. (110). This value is defined as $\Delta\gamma = E_G \dfrac{\Delta A}{A}$; according to Eq. (110) we get:

$$E_G = E_M \frac{\tau}{\theta} = RT\Gamma' \bigg/ \left(1 - \frac{\Gamma'}{\Gamma_\infty} \right) \tag{120}$$

(the equilibrium elasticity does not depend on the presence of the soluble surfactant).

However, in contrast to the values of E_G, E_M, τ, θ, the dilational viscosity of the interfacial layer for mixtures of a soluble and an insoluble surfactant is of alternative nature, because the rheological equation (110) for the standard solid body could be adequately described by two rheological models shown in Fig. 5 (A-Model) and Fig. 7 (B-Model).

According to Eq. (92), the viscosity in the Maxwell branch of the standard solid body model illustrated by Fig. 7 is determined via the parameters of the model shown in Fig. 5 by the relation:

$$\eta_B = \eta_A E'^2 / (E + E')^2 \tag{121}$$

where the relations between the uppercase and lowercase symbols are defined as $\eta_d = \zeta A$, $E = \varepsilon A$, $E' = \varepsilon' A$, see section 2. It is natural now to express the parameters of the model B, i.e., the values E and E' involved in Eq. (121) via the directly measurable (and model-independent) physical quantities E_M and E_G. It follows from obvious properties of the model B (Fig. 7) that:

$$E_M = \frac{EE'}{E + E'} + \frac{E'^2}{E + E'} = E', \qquad E_G = \frac{EE'}{E + E'} \tag{122}$$

From Eq. (122) one can express the pure model parameter E via E_M and E_G:

$$E = \frac{E_M E_G}{E_M - E_G}, \tag{123}$$

Therefore, from Eqs. (121)-(123) and (120) one can obtain the simple formula:

$$\eta_B = \eta_A \left(1 - \frac{E_G}{E_M}\right)^2 = \eta_A \left(1 - \frac{\tau}{\theta}\right)^2 \tag{124}$$

Finally, the remarkably simple expression for the dilational viscosity in the Model B via the model-independent parameters should be mentioned:

$$\eta_B = \tau (E_M - E_G) \tag{125}$$

This relation follows from Eqs. (122)-(124) and (91) (which for this case should be expressed as $\eta_A = \tau(E + E')$). Figure 14 illustrates the Model B of the standard solid body valid for interfacial layers comprised of a mixture of a soluble and insoluble surfactant. The visco-elastic characteristics of this model are expressed via the values which can be directly measured in an experiment.

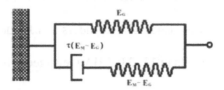

Fig. 14. B-model of standard solid body for the interfacial monolayer comprised of the mixture of soluble and insoluble surfactant

E. REFERENCES

1. N.N. Kruglitsky, Foundations of Physicochemical Mechanics, Vyssha Shkola, Kiev, 1975, P.267, (in Russian).
2. J.W. Gibbs, The Scientific Papers. Longmans and Green, New York, 1906.
3. B. Stuke, Fortschr. Kolloide Polymere, 55 (1971) 106.
4. M. Joly, Recent Progress in Surface Science, 1 (1964) 1-50.
5. M. Joly, Surface Colloid Science, 5 (1972) 1-193.
6. L.D. Landau and E.M. Lifshitz, Fluid Mechanics, Addison-Wesley, Reading, Mass., 1959.
7. V.V. Krotov and A.I. Rusanov, Physico-Chemical Hydrodynamics of Capillary Systems, Imperial College Press, London, 1999, pp.475.
8. M.J. Boussinesq, Ann. Chem. Phys., Ser.8, 29 (1913) 349-357.
9. V.V. Krotov, Zh. Tekh. Fiz., 58 (1988) 1581-1583, (in Russian).
10. F.M. Kuni, Teor. Mat. Fiz., 21 (1974) 233-246, (in Russian).
11. R. Wüstneck, V.V. Krotov and M. Ziller, Colloid Polymer Sci., 262 (1984) 67-76.
12. T. Shimbashi and T. Shiba, Bull. Chem. Soc. Japan, 38 (1965) 581-588.
13. J. Frenkel, Zh. Phys., 26 (1924) 117-138, (in Russian).
14. V.V. Krotov, in "Problems of Thermodynamics of Heterogeneous Systems and Theory of Surface Phenomena", ed. A.V. Storonkin and V.T. Zharov, Leningrad. Univ. Leningrad, 1979, Vol. 5, pp.146-203, (in Russian).
15. V.G. Levich, Physico-Chemical Hydrodynamics, Prentice-Hall, Englewood Gliffs, 1962.

SURFACE DILATIONAL RHEOLOGY: PAST AND PRESENT

E.H. Lucassen-Reynders and J. Lucassen

Mathenesselaan 11, 2343 HA Oegstgeest, the Netherlands

Contents

A. INTRODUCTION - EARLY HISTORICAL BACKGROUND

This Chapter aims at reviewing the interplay between surface rheology and liquid motion in systems containing surfaces or interfaces. For such systems, from foams and emulsions to lung alveoli, the presence of surface-active agents is a crucial factor in their behaviour and functionality. The focus will be on surface rheology in dilational deformation rather than in shear, as in many cases the former is found to have the larger effect on liquid motion. Both hydrodynamic and surface-chemical aspects will be approached from a phenomenological rather than a molecular point of view.

The earliest indirect references to the effect of surface properties on liquid motion are about the damping of waves and can be traced back to antiquity, from Aristotle in the fourth century BC to Pliny the elder in the year 77 AD [1]. Apparently, in those days it was common knowledge among sailors and fishermen that some kinds of oil poured on the surface of a rough sea had a calming effect on the waves. More than 17 centuries elapsed before Benjamin Franklin published the first attempt to put this phenomenon on a scientific basis [2]. Intrigued by the practice of Bermudian fishermen to "put oil on the water to smooth it when they would strike fish, which they could not see if the water was ruffled by the wind", he started his famous experiments on the pond of London's Clapham Common. It transpired that a teaspoonful of oil (probably olive oil [2]) was enough to still the waves on an area of half an acre (2000 m^2) of water. His conclusion – still generally accepted – was that the oil mainly affected the damping of waves with small wavelength, now called ripples, rather than that of long waves. He suggested that "there are continually rising on the back of every great wave, a number of small ones, which roughen its surface, and give the wind hold, as it were, to push it with greater force". Quantitative interpretations in terms of surface properties were not yet in sight as, at the time, the concept of surface tension had not really taken hold, although it had been introduced in 1751 by von Segner [3].

Two factors of major importance for wave damping could not be dealt with by the early authors. First is the liquid viscosity, which provides the only way to dissipate wave energy and so "to still the waves". Second, a very important factor to be accounted for is the effect of surface-active materials on surface wave properties. The olive oil used by Franklin is an example, not only because its triacylglycerol components spread out over an air/water surface, but also because such oil usually contains very surface-active contaminants, e.g., fatty acids and monoacylglycerols. Many other classes of surface actives are used in essential technological and biological processes, such as detergency, foaming, emulsification, distillation, extraction and respiration, to name but a few. Section B briefly reviews the hydrodynamic theory of viscous liquids bounded by a rippled surface. This theory was developed in the 19th century for small-amplitude surface waves on water, and was initially restricted to 'uncontaminated' (= surfactant-free) systems. For such 'clean' surfaces, the 19th century also saw quantitative relationships developed for the wave damping by Stokes [4] and for the wave length by Kelvin [5] (see Section B).

The effect of surface actives was first addressed by Reynolds [6]. In a large step forward, he proposed that on a contaminated surface wave damping increases because local surface compressions and extensions produce gradients in surface tension. These result in an alternating tangential drag on the water, which promotes the dissipation of energy, thereby increasing wave damping through liquid viscosity. Reynolds also derived an expression for the damping in the limiting case where surface actives render the surface completely incompressible (see Section B.3.). Independently, Marangoni demonstrated that elastic forces in a surface layer can oppose motion of a needle lying in the surface [7]. The first more complete theory on the effect of surface-tension gradients was presented in 1941 by Levich [8] in terms of surface dilational elasticity. The basic assumptions of later treatments are the same as in Levich's theory. However, the complexity of the equations was such that a quantitative analysis for the entire elasticity range was not yet feasible. Nevertheless, an intriguing feature of wave damping was discovered in 1951 by Dorrestein [9]. His theory showed that damping does not increase continuously with surface elasticity up to the Reynolds value. Instead, as Dorrestein proved, damping must pass through a maximum that is approximately twice as high as the Reynolds value, at an intermediate elasticity value.

In the 1960s, further theories corrected and extended Levich's treatment with various types of surface dilational viscosity [10-12] and with a second type of surface wave [13,14]. Unlike the transverse ripples of the sea wave problem, the newly discovered wave proved to be longitudinal, and particularly suited to the measurement of surface dilational (visco)elasticity. Aspects of these theories are considered in Sections B and C, in terms of a phenomenological treatment which provides relationships between wave properties (wavelength and damping) and measurable surface parameters. Prominent among the latter are surface tension and surface dilational elasticity (see Section D). Combination of the hydrodynamic treatment with the boundary conditions imposed by the surface has resulted in quantitative links between wave characteristics and surface parameters, under experimentally relevant conditions. Early reviews were published on waves generated by mechanical means [15,16]. Section E briefly surveys experimental methods and some major results of experimental developments.

B. QUANTITATIVE THEORY OF WAVE CHARACTERISTICS IN RELATION TO SURFACE PROPERTIES

1.   Hydrodynamics of wave propagation

Waves on water are easily observed and widely known, but they are by no means a simple example of waves. Their treatment is more complex than that of many other waves, e.g., sound and light, as was noted and explained by Feynman [17]. The reason for their complexity lies in the essential incompressibility of water. Water cannot simply move up and down as it travels between wave crests and troughs: being incompressible, it must also move sideways. Therefore, liquid motion during wave passage is a mixture of transverse and longitudinal displacements. Even in the simple case of negligible viscosity and a clean surface, water particles do not move in straight lines but in circles, as demonstrated by the hydrodynamic theory given below.

In combination with the boundary conditions, hydrodynamic theory yields quantitative relations for the wave characteristics as a function of surface properties (including surface elasticity) and liquid viscosity. The importance of the boundary conditions imposed by the surface cannot be overemphasised since, in some cases, the wave characteristics can directly be linked to the surface properties without going through the hydrodynamic equations. In most cases, however, we do need hydrodynamic theory to arrive at this link. The hydrodynamic theory is a phenomenological treatment in which liquid density and viscosity are considered constant right up to a mathematical plane separating two fluid bulk phases. This plane is the 'surface of tension', introduced by Gibbs in his thermodynamic treatment of heterogeneous systems [18]. Combined with the boundary conditions imposed by the surface, the hydrodynamic theory completely describes the state of motion of an incompressible liquid if the three components of the velocity vector $\mathbf{v}$ and the hydrostatic pressure p as a function of time and spatial coordinates are known. Two relations are required to find these functions. First, the continuity equation states that the volume of an incompressible liquid element is constant:

$$\partial v_x/\partial x + \partial v_y/\partial y + \partial v_z/\partial z = 0. \tag{1}$$

(See List of Symbols). Second, the Navier-Stokes equation expresses Newton's second law, according to which the product of mass and acceleration for a small element of fluid is equal to the sum of the forces acting upon the element (see textbooks, e.g., [19-23]). After conversion of mass into density:

$$\rho\, \partial v / \partial t + (v\ grad)\, v = -grad\ p + \eta \nabla^2 v + \rho g \tag{2}$$

The left-hand side reflects Newton's second law in a convective coordinate system, where the convective term, (v grad) v, renders the hydrodynamic equations non-linear. This is the only term quadratic in the velocity, so it can be neglected for sufficiently low liquid velocities. As all velocities decrease with decreasing wave amplitude, Eq (2) can be made linear by choosing sufficiently small amplitude/wavelength ratios. On the right hand side of Eq (2), the first contribution reflects the variations in hydrostatic pressure with distance, the second is the viscous pressure due to momentum being transferred from faster to slower layers and in the third term g is the gravitational acceleration. The specific case considered is that of plane waves in the x direction, such as can be generated by a rod or bar placed at the surface in the y direction and subjected to vibrations. In that case all motion in the y direction vanishes, so that the linearised Navier-Stokes equation and the continuity equation have components only in the x and z directions:

$$\rho\, \partial v_x/\partial t = -\partial p / \partial x + \eta\, [\partial^2 v_x/\partial x^2 + \partial^2 v_x/\partial z^2] \tag{3}$$

$$\rho\, \partial v_z/\partial t = -\partial p / \partial z + \eta\, [\partial^2 v_z/\partial x^2 + \partial^2 v_z/\partial z^2] - \rho g \tag{4}$$

$$\partial v_x/\partial x + \partial v_z/\partial z = 0 \tag{5}$$

Solving Eqs (3) to (5) is made much easier by considering that any liquid velocity field in space can be expressed as the sum of a rotation-free field and a divergence-free field: $v = v_1 + v_2$, with curl. $v_1=0$ and div $v_2=0$ (see textbooks, e.g., [6, 8, 10]). The rotation-free field is characterised by the potential function φ, a scalar, and the divergence-free field by the vorticity or stream function, ψ, a vector. The velocity components and the pressure p are then expressed by:

$$v_x = -\partial\varphi/\partial x - \partial\psi/\partial z , \tag{6}$$

$$v_z = -\partial\varphi/\partial z + \partial\psi/\partial x . \tag{7}$$

$$p = p_0 - \rho g z + i\omega\rho\, \varphi \tag{8}$$

The last equation reflects the boundary condition at rest (z=0), where all velocities vanish and p is equal to the reference pressure p_0.

Wave damping caused by liquid viscosity can occur only through the vorticity function (ψ), while the potential function (φ) does not depend on viscosity. Therefore, these often employed functions are very useful in the study of damped surface waves. An example of their application is the information on particle trajectories to be given below.

All wave functions vary not only with x and ω, but also in the vertical direction, z. For sufficiently deep liquid layers all motion vanishes at the bottom of the layer, i.e., for $-kz \gg 1$, where the wave number k equals the inverse penetration depth of the potential function φ. (In fact, the viscosity of an aqueous layer is so low that the layer depth does not have to be much greater than 1/k for the potential flow to be nearly zero.) The inverse penetration depth of ψ is measured by a different number, m. Introducing the periodicity of simple harmonic oscillations in complex-number notation, two choices of experimental conditions are (1) stationary waves imposed by a constant frequency, where the frequency is a real number and damping is expressed by the imaginary part of the complex wave number and (2) decaying waves, where the wavelength is a real number and the damping with time resides in the imaginary part of the frequency. The two damping coefficients are simply correlated through the propagation speed of the wave [9]. Unless stated otherwise, this Chapter will focus on stationary waves with a constant, real, frequency and a complex wave number:

$$k = \kappa - i\beta \tag{9}$$

Equation (9) defines the damping coefficient β as a positive number. A special class of solutions for the potential and vorticity functions is obtained by integration of Eqs (6) and (7) over a wave period:

$$\varphi = \Phi e^{kz} e^{i(kx+\omega t)} \tag{10}$$

$$\psi = \Psi e^{mz} e^{i(kx+\omega t)} \tag{11}$$

subject to two conditions. The first is Eq (8), and the second relates m to k:

$$m^2 = k^2 + i\omega\rho/\eta \cong i\omega\rho/\eta . \tag{12}$$

The second equality in Eq (12) is valid to first order because of the relatively low viscosity of water: $k^2 \ll \omega\rho/\eta$ (at not too high wave frequencies). The relationship between k and m expressed in Eq (12) then implies that $m^2 \gg k^2$. Thus, the penetration depth of the vorticity function is much smaller than that of the potential function. This causes the liquid velocities, v_1 and. v_2, to differ considerably in character. We note that the Navier-Stokes equation is not only satisfied by v_1 and v_2 but also by all linear combinations of them.

For plane waves, the potential and vorticity components of the liquid velocity are, respectively:

$$v_{1x} = -\partial\varphi/\partial x = -ik\varphi \quad ; \quad v_{1z} = -\partial\varphi/\partial z = -k\varphi \tag{13}$$

$$v_{2x} = \partial\psi/\partial z = m\psi \quad\quad ; \quad v_{2z} = -\partial\psi/\partial x = -ik\psi \tag{14}$$

Interesting information on the orbits of surface particles during wave motion follows from Eqs (13) and (14). In first approximation, the velocity components at the surface are equal to the rate at which points at the surface are displaced from their position at rest in the z and x directions:

$$(v_z)_s = \partial\varsigma/\partial t + (v_x)_s\,\partial\varsigma/\partial x \cong \partial\varsigma/\partial t \quad ; \quad (v_x)_s \cong \partial\xi/\partial t \tag{15}$$

Figure 1 illustrates the large difference between the two types of motion by comparing the particle orbits after integration of Eq (13) for potential flow and Eq (14) for vorticity flow.

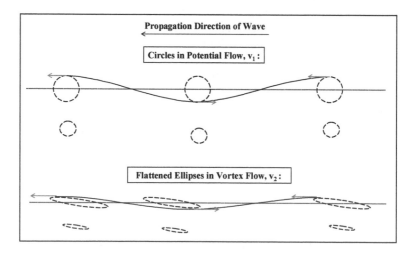

Figure 1. Orbits of liquid elements in potential and vorticity flow; schematically; arrows: direction of flow. Dotted lines: surface at rest.

In potential flow, on the one hand, this produces very simple expressions for the vertical and horizontal displacements, ξ_1 and ζ_1:

$$\xi_1 = -(k/\omega)\varphi ; \quad \zeta_1 = -(k/i\omega)\,\varphi . \tag{16}$$

It is seen that the displacements ξ_1 and ζ_1 are of equal amplitude, but 90° out of phase. In other words, the particle orbits are circles, as shown in Figure 1. (It is worth noting that the diminishing velocity in deeper layers is not caused by viscous damping but by transport of momentum, imposed by the requirement that all velocities must decrease with increasing distance from the disturbed surface). In vorticity flow, as described by Eq (11), on the other hand, the orbits are very different owing to the small penetration depth (1/m) resulting from Eq (12):

$$\xi_2 = (im/\omega)\psi ; \qquad \zeta_2 = (k/\omega)\psi . \qquad (17)$$

The particle orbits now are no longer circles but elongated ellipses lying nearly flat in the surface. The exact value of the phase difference between the two displacements depends on wave frequency and liquid viscosity. An illustrative discussion and a numerical example were given by Hansen and Ahmad [16]. Figure 1 also implies that the surface is being compressed in the first half wave after each crest, while it is being expanded in the second half wave. In the presence of surface-active material, the tension in the compressed surface will be lower and in the expanded part it will be higher. Figure 2 illustrates this point more clearly, and it also shows that the liquid above the undisturbed surface moves in the direction of the wave, while liquid below this level moves in the opposite direction. Thus, looking at each single wave, there is no net flow of liquid, which is why an object fallen into a pond and floating on the surface cannot be driven back to the shore by stones cast in the water behind it.

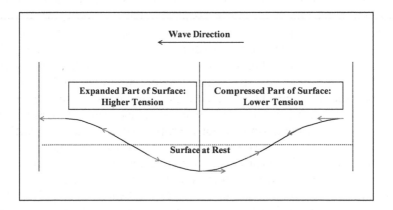

Figure 2. Elastic stress in surface with surface-active agent. Red arrows indicate motion of surface elements.

All equations derived so far hold not only for water in contact with a vapour – as in the sea-wave problem – but also for a water phase in contact with an immiscible liquid. Equations for the second liquid are similar to those above. For liquids less dense than water, the corresponding equations must be introduced with a negative sign in the exp (kz) and exp (mz) terms as the upper phase extends to +z » 1/k instead of –z » 1/k. Also, non-zero values of viscosity and density of the upper phase must be accommodated. The extension to clean liquid/water interfaces was introduced by Koussakov, who also examined the effects of finite depths of both liquids [24]. The analysis for surfactant-covered surfaces followed later [15].

At this point we need to consider the boundary conditions at the liquid surface to obtain quantitative information in terms of surface tension and surface rheological parameters. The boundary conditions will turn out to be numerically different for the air/water and liquid/water surfaces because air cannot support stress, but a liquid can and will.

2.   Surface stress boundary conditions and the wave dispersion equation

The normal and tangential boundary conditions for the surface stress follow from the same principle that produced the equations of motion in the bulk liquid: in both cases Newton's second law must be obeyed. The main difference is a fundamental one: the forces acting upon a surface element result not only from hydrostatic-pressure variations and viscous stresses in the liquid, but also from stresses existing in the deformed surface. The basic cause of the difference is that a surface element, unlike a liquid element, cannot be assumed to be incompressible. In other words, for the surface there is no equation of continuity such as Eq (1) for the liquid. Application of Eq (2) to a small element containing the surface implies that at the surface of tension, to which no mass and density are assigned, the stresses due to pressure and liquid viscosity must be exactly balanced by surface stresses arising from surface tension and surface elasticity. In the case of plane surface waves, the compression and expansion is not isotropic but uniaxial. In principle, therefore, we should account not only for dilational elasticity and viscosity, but also for shear elasticity and viscosity, as first described by Boussinesq [25]. (Subsequent analyses of the effect of surface shear on waves have been presented by Mann [27], Edwards et al [23] and in Chapter 7 of this volume). Accounting for surface rheology in both dilation and shear requires the use of tensor algebra, a tensor describing the relation between two vector fields. Its complete description requires six parameters in three dimensions, and four parameters in two dimensions. In practice, particularly in the case of small surfactant molecules, measured shear parameters generally are much smaller than their dilational counterparts. For such surfactants, a true reflection of reality is the early statement by Rayleigh that *what a water surface resists is not shearing but the creation of a gradient in surface tension* [26]. For macromolecules, however, surface shear properties are more pronounced than for small molecules, and the above simple picture must be re examined; see Chapter 7. For the present purpose, we shall avoid the complications of surface shear and consider only the effects of dilational elasticity and viscosity.

For plane small-amplitude waves on a surface with negligible shear properties, the surface stress, on the one hand, is isotropic and equal to the scalar surface tension. The pressure in a flowing viscous liquid, on the other hand, is not isotropic but must be expressed as a tensor, the components of which are:

$$p_{xx} = -p + 2\eta \, \partial v_x / \partial x \, ; \, p_{zz} = -p + 2\eta \, \partial v_z / \partial z \, ; \, p_{xz} = p_{zx} = \eta \, (\partial v_x / \partial z + \partial v_z / \partial x) \, (18)$$

In the present approximation, the normal stress boundary condition requires the stress arising from surface tension to be balanced by the normal viscous stress, p_{zz}, exerted by the flowing liquid at the surface:

$$\gamma \, \partial^2 \zeta / \partial x^2 - 2\eta \, (\partial v_z / \partial z)_s + (p - p_0) = 0 \qquad (19)$$

This is equivalent to the law of Laplace [28] for the pressure difference across a curved gas/water surface, after accounting for viscous flow in the liquid. The small amplitude/wavelength ratio implies a curvature small enough to neglect the term $p-p_0$ (given by Eq (8)).

The tangential stress boundary condition reflects the balance between the surface tension gradient and the tangential viscous stress, p_{xz}, from the liquid at the surface. The gradient in tension is related to the area change, dA, by the surface elasticity (also called surface dilational modulus) operationally defined by Gibbs [18]:

$$E = d\gamma / d \ln A \qquad (20)$$

For a plane wave in the x direction we have $d \ln A = \partial \xi / \partial x$. The gradient in tension is then expressed by $d\gamma / dx = E \, \partial^2 \xi / \partial x^2$, in the absence of surface shear effects. The resulting tangential stress boundary condition then is given by:

$$E \, \partial^2 \xi / \partial x^2 - \eta [(\partial v_x / \partial z)_s + (\partial v_z / \partial x)_s] = 0 \qquad (21)$$

The similarity of the two boundary conditions is evident from Eqs (19) and (21): the modulus E and the horizontal displacement ξ in Eq (21) correspond with the tension γ and the vertical displacement ζ in Eq (19). The difference is that, unlike the surface tension, the modulus can consist of not only an elastic part (E') but also a viscous part (E''), in case surface relaxation processes take place in the time scale of the wave $(1/\omega)$. The imaginary part (E'') defines the surface dilational viscosity, η_d:

$$E = E' + i E'' \equiv E' + i\omega\eta_d \tag{22}$$

The two stress boundary conditions can now be expressed in terms of the wave parameters (k and ω), the parameters of the water phase (η, ρ, Φ, Ψ and their counterparts in the upper liquid) and the surface characteristics γ and E, by eliminating the velocity components by means of Eq (15). Equations for the more general case of any fluid interface have been formulated first for very low concentrations where the adsorption is proportional to the concentration [29], and later also for higher concentrations [15]. Not surprisingly, the complete boundary conditions for any fluid interface are fairly cumbersome and do not readily permit examining the physical background of the solutions found. Therefore, we shall first consider the simpler case of the gas/water surface.

The normal stress boundary condition for this case, in first approximation, is given by:

$$[-i\omega\rho - 2\eta k^2 + ik(\gamma k^2 + \rho g)/\omega]\Phi + [2i\eta km + k(\gamma k^2 + \rho g)/\omega]\Psi = 0 \tag{23}$$

and the tangential stress boundary condition by:

$$[-2\,i\eta k^2 - E k^3/\omega]\Phi + [-\eta(k^2 + m^2) + i E k^2 m/\omega]\Psi = 0 \tag{24}$$

The wave dispersion equation follows after elimination of Φ and Ψ from Eqs (23) and (24):

$$\eta^2 (k - m)^2 + f(E).f(S) = 0 \tag{25}$$

Here the functions f (E) and f (S) for elasticity and surface tension, respectively, are defined for convenience [15]:

$$f(E) \equiv E k^2/\omega + i\eta(k + m) \tag{26}$$

$$f(S) \equiv \gamma k^2/\omega + \rho(g/\omega - \omega/k) + i\eta(k + m) \tag{27}$$

These two functions facilitate analysing the physical significance of the solutions to be obtained. Therefore, the dispersion equation as formulated in Eq (25) differs in form but not in content from the original literature [8-13].

At a liquid/liquid interface, as at air/water, the liquid velocity components must be continuous, i.e., no cracks or other discontinuities are allowed to exist. The difference with the air/water case is that now the velocities in the upper phase are not negligible. Their values depend on four parameters characterising the second liquid, viz., the known viscosity and density (η' and ρ') and the unknown amplitudes Φ' and Ψ'. This yields two additional relations between the two unknown parameters Φ' and Ψ':

$$ik(\Phi - \Phi') + m\Psi + m'\Psi' = 0 \quad and \quad \Phi + \Phi' + i(-\Psi + \Psi') = 0 \tag{28}$$

The resulting wave dispersion equation is more complicated than Eq (25-27) [15]:

$$[\eta(k-m) - \eta'(k-m')]^2 + f(E).f(S) = 0 \tag{29}$$

$$f(E) \equiv Ek^2/\omega + i[\eta(k+m) + \eta'(k+m')] \tag{30}$$

$$f(S) \equiv \gamma k^2/\omega + i[\eta(k+m) + \eta'(k+m')] + g(\rho - \rho')/\omega - (\rho + \rho')\omega/k) \tag{31}$$

These equations represent the dispersion relation for small-amplitude plane waves on any liquid interface in the absence of surface-shear effects. It contains the wave propagation characteristics wavelength (λ) and damping coefficient (β) as a function of the surface properties γ and E, the liquid viscosities (η and η'), the liquid densities (ρ and ρ') and the gravitational constant (g). In its general form it is, unfortunately, quite complicated. Numerical analysis requires splitting up Eqs (29-31) into real and imaginary parts to discover that the component equations are of too high a degree to permit exact solution without fairly advanced computer techniques. Therefore it may be illustrative first to look into the general nature of the results implicit in Eqs (29-31) and into the type of liquid flow caused by surface waves.

3. Limiting values for wave properties at the air/water surface

As early as 1845, Stokes deduced a limiting equation for wave damping, at a clean air/water surface in the absence of gravity terms (i.e., at sufficiently small wavelengths) [4]. He achieved this without recourse to the full dispersion equation, Eq (25). Instead, he used the hydrodynamic equations for the velocity components v_x and v_z (Eqs (13,

14)) in a low-viscosity liquid, such as water. In this way, Stokes found for the approximate damping coefficient at the clean air/water surface:

$$\beta_0 = 4\eta\kappa^3 / 3\rho\omega \tag{32}$$

The next step forward was the derivation of Kelvin's law [5]. For small enough wave length, he demonstrated that surface tension is the main force driving a rippled surface back to equilibrium, while waves with large wavelength are governed by gravity. The first approximate relation between wavelength (λ) and wave frequency (ω) for small-amplitude waves in terms of these forces was formulated by Kelvin for a pure liquid with zero viscosity, in contact with a gaseous phase. Under these simple conditions, the stress boundary condition normal to the surface prescribes that the pressure difference across the slightly tilted surface is balanced by the sum of Laplace pressure [28] and gravity pressure according to:

$$\gamma\kappa^3 + \rho g\kappa = \omega^2 \rho \tag{33}$$

The viscosity of water is low enough for Eq (33) to be a good approximation for the dispersion equation of capillary waves at not extremely high frequency. Therefore, the surface tension of clean water can be determined quite accurately from Eq (33). The approximation is still reasonable up to fairly high surfactant concentrations [15].

Kelvin's law is the first example of a surface wave dispersion equation, for a frictionless liquid. It does not tell us anything about wave damping, because damping can take place only through liquid viscosity. It also ignores that the presence of surface active material must cause gradients in tension during wave passage. Thus, Eq (33) contains no information on the role of surface rheology. However, the equation can serve to reformulate Stokes' formula by elimination of the wave number. If the effect of gravity is negligible (i.e., if $\kappa^2 \gg \rho g/\gamma$) the wave number is given by $\kappa^3 = \omega^2 \rho/\gamma_0$, and Eq (32) can be rewritten as

$$\beta_0 = 4\eta\omega / 3\gamma_0 \tag{34}$$

In this form, Stokes' equation illustrates that the damping coefficient for a clean surface is proportional to the liquid viscosity and the wave frequency, and inversely proportional to the surface tension.

Two other limiting cases of wave damping are worth noting. The first was studied by Reynolds [6], for a surface covered by a rigid, incompressible film. Like Stokes, Reynolds did not employ a boundary condition for the tangential <u>stress</u>, but for the horizontal <u>velocity</u> at the surface. For his completely rigid film, this velocity was taken to be zero, which resulted in the following expression:

$$\beta_\infty = \mbox{\small{1/6}} \, \kappa^2 \, (2\eta / \omega \rho)^{1/2} \tag{35}$$

Comparison of the rigid surface of Eq (35) with the clean surface of Eq (32) shows that damping in the former case greatly exceeds that in the latter. However, damping is not at its maximum for a completely rigid film. As first shown by Dorrestein [9], damping passes through a maximum that is roughly twice as high as the Reynolds damping. (This maximum in damping was missed by Levich [8,22] because of an error in sign of the elasticity term). Dorrestein's expression for time-damped waves is easily converted to distance-damped waves. For purely elastic surfaces the result is:

$$\beta_{max} \approx \mbox{\small{1/3}} \, \kappa^2 \, (2\eta / \omega \rho)^{1/2} \tag{36}$$

The maximum can occur at a rather low value of the elasticity, where the film is still far from rigid in many cases

$$E_{\beta = \beta_{max}} \approx (2\omega^3 \rho \eta)^{1/2} / \kappa^2 \approx 2\eta\omega / 3\beta_{max} \tag{37}$$

Eqs (36) and (37) provided a first, limited, opportunity to derive approximate elasticity values from measured ripple damping. More exact values of computed ripple damping using Eq (25) were first obtained for both purely elastic and viscoelastic surfaces by Hansen and Mann [10]. Examples are given in Figure 3, illustrating that damping constitutes a very small part of the complex wave number, in line with the minor part of vortex flow in ripple propagation.

This small part can serve to produce approximate values of pure elasticities in a restricted range around the maximum in damping. In view of the limited accuracy of damping measurements, the useful range in Figure 3 is restricted to $0.06 \leq E'/\gamma \leq 1$.

Thus, values above 50 mN/m (and values below 3 mN/m) cannot reliably be obtained from ripple damping in this example. For visco-elastic surfaces, the maximum damping is considerably lower, as a result of the relaxation process producing the visco-elasticity. Relaxation by diffusion interchange of surfactant between surface and bulk solution, in particular, effectively short-circuits gradients in surface tension at sufficiently low frequencies (see Section D). For visco-elastic surfaces, therefore, prospects of deriving elasticity values from ripple damping appear to be even more limited than for purely elastic surfaces.

Experimentally, the existence of a maximum in damping was discussed by Davies and Rideal in 1961 [31], and confirmed in many later studies. Figure 4 illustrates the orbits of surface particles for the three cases described by Eqs (34-36).

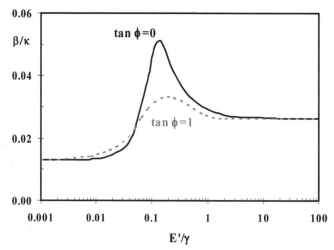

Figure 3. Examples of computed ripple damping for η=0.9 mPa s, γ=50 mN/m, at frequency ω/2π=545 Hz. Top curve: purely elastic surface (E″=0). Bottom curve: visco-elastic surface with E″= E′, according to [10].

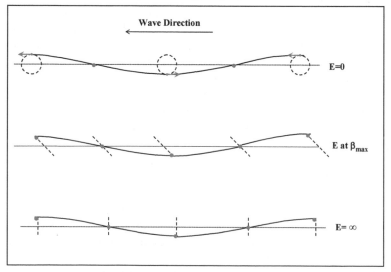

Figure 4. Effect of surface elasticity, E, on orbits of surface elements when surface wave passes by.

Strictly speaking, the circles in Figure 4 are valid only for a non-viscous liquid, but the viscosity of water is low enough for the circles to be only marginally distorted. Analysis of the complete dispersion relation, Eq (25) or (29) with the aid of fairly sophisticated computer techniques revealed that the maximum in damping is caused by the presence of a longitudinal wave unknown at the time [13,15]. The physical background of this wave is analysed in Section C.

C. ANALYSIS OF THE DISPERSION RELATION; THE LONGITUDINAL SURFACE WAVE

A full analysis of the wave-dispersion equation followed the advance of computer technology in the early sixties [10-13]. The presence of surface active material (E>0) results in a non-zero tangential surface stress that enforces a dissipative flow in the adjacent liquid layer. At relatively low elasticities, the effect on liquid flow is mainly related to the phase difference between the surface displacements in the horizontal and vertical directions, ξ and ζ. While the amplitudes of ξ and ζ remain essentially equal, their phase difference increases from 90° (at E=0) to 180°, at the value of E for which β is at its maximum. This means that in the middle picture of Figure 4 the surface particles no longer move in circular orbits but along straight lines at an angle of 45° with the wave direction. At higher surface elasticities the situation is different: here it is the amplitude of ξ that is affected rather than its phase [15]. Clearly, at the end of the elasticity range, in a completely incompressible surface (E=∞), all horizontal surface motion has stopped, and the orbits of surface particles are straight vertical lines, as in the bottom picture of Figure 4. The effect of surface elasticity on horizontal surface motion for ripples was demonstrated experimentally by Thiessen and Sheludko, who measured the amplitudes of small particles floating on the surface [32]. Qualitatively, it is evident that the change of the motion from straight-line orbits to the circular orbits prevailing in deeper liquid layers involves drastic variations in the liquid velocity vector. Such large velocity gradients result in higher dissipation of energy, i.e., higher damping of the ripple for the elasticity values at the maximum and at E=∞. Quantitatively, it was found that at the maximum in damping, the wavelength of a second wave of longitudinal character was equal to that of the ripple at the same frequency [13]. This strongly suggests a resonance-like additional damping as the root cause of the maximum in ripple damping.

Thus, the physical reason for this behaviour is connected with the ability of an elastic surface to conduct two types of surface waves, not only the capillary ripple described so far. Analysis of the complete dispersion relation, Eq (29), has revealed the following characteristics of the second type of wave [13,15], for an elastic air/water surface:

(1) the nature of the wave is mainly longitudinal, i.e., characterised by horizontal surface motion far in excess of vertical motion

(2) vorticity flow (ψ) plays a much greater part in the longitudinal wave than in the transverse ripple. As a result, the wave is far more heavily damped than the ripple: its damping coefficient is of the same order as its wave number;

(3) the wave is governed almost exclusively by the tangential surface stress boundary condition (Eq (21)) and, therefore, it depends on surface elasticity rather than on surface tension;

(4) in contrast to the transverse ripple, the longitudinal wave cannot exist on surfaces without elasticity.

Computed results are closely approximated if liquid motion is assumed to be dominated by vorticity near the surface, that is, for $|m|z \ll 1$. The dispersion equation of the longitudinal wave, for purely elastic surfaces (E″=0; E =E′), is approximated by [13]

$$ i E' \, k_L^2 \cong \eta \, m \, \omega = (i\eta\rho\omega^3)^{1/2} \tag{38} $$

This is equivalent to the tangential stress boundary condition (Eq (26)) for f(E)=0, with $|m| \gg |k|$. The measurable properties, i.e., wave number (κ_L) and damping coefficient (β_L), are expressed by

$$ \kappa_L \cong (E')^{-1/2} (\eta \rho \omega^3)^{1/4} \cos(\pi/8) \quad \text{(for E″= 0)} \tag{39} $$

$$ \beta_L \cong \tan(\pi/8)\kappa_L = (\sqrt{2}-1)\kappa_L \quad \text{(for E″= 0)} \tag{40} $$

For visco-elastic surfaces, the dispersion equation was derived in first-order approximation by Lucassen and Van den Tempel in 1972 [33]. In terms of the viscoelastic modulus |E| and the viscous phase angle ϕ, they derived for κ_L and β_L:

$$ \kappa_L \cong |E|^{-1/2} (\eta \rho \omega^3)^{1/4} \cos(\pi/8 + \phi/2) \tag{41} $$

$$ \beta_L \cong |E|^{-1/2} (\eta \rho \omega^3)^{1/4} \sin(\pi/8 + \phi/2) \tag{42} $$

Both κ_L and β_L are inversely proportional to $\sqrt{|E|}$, which is a much stronger dependence on surface elasticity than found for κ_T and β_T in the capillary-ripple case.

Figures 5 and 6 illustrate the basic differences between the two surface waves. The character of both waves, obeying the same dispersion equation, Eq (29), is generally quite complex, since the ripple is not purely transverse, as argued earlier [17], and the longitudinal wave is not purely longitudinal. However, the dependence of the ripple on surface elasticity is only weak, and that of the longitudinal wave on surface tension is even weaker. As is evident from Eqs (38) and (33), the information on surface elasticity contained in the longitudinal wave is far greater and more direct than in the ripple. Early elasticity values were derived from the measured wave properties (β_L and κ_L), at both high and low frequencies [14,33].

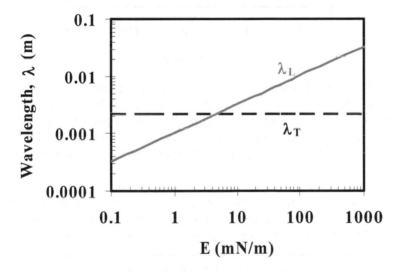

Figure 5. Wavelengths (λ_T and λ_L) of transverse and longitudinal waves for a purely elastic surface (E=E'; E''=0). Parameter values: $\omega/2\pi$=200 Hz; η=1 mPa s; ρ=1 g/cm^3; γ=70 mN/m; from Lucassen [13].

The longitudinal wave and its dispersion relation have long escaped detection, both in theory and experiment. One reason is the highly dissipative character of the wave motion: the wave is usually damped out over a much smaller distance than the ripple. Another reason is that the longitudinal wave cannot exist on a clean surface. For clean surfaces (E'=0), Eq (39) implies a total loss of the wave character as the wavelength is reduced to zero: the vorticity flow needed for the longitudinal wave requires the elasticity to exceed a critical value of $E'_{crit} = (\eta^3\omega/\rho)^{1/2}$ [15].

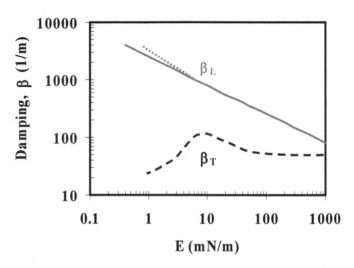

Figure 6. Damping coefficients (β_T and β_L) of transverse and longitudinal waves for a purely elastic surface ($E=E'$; $E''=0$). Parameter values as in Figure 5. Dotted line: full dispersion equation, Eq (29); drawn line: approximation of Eq (38). Redrawn from Lucassen [13]. Maximum ripple damping is observed at the E value for which $\lambda_L = \lambda_T$.

The existence of two types of wave on an elastic surface is strongly reminiscent of the behaviour of an elastic membrane under tension. The analogy between a surface and a membrane was noted in 1751 by Von Segner [3], but only much later analysed for surface waves by Lucassen [13] and for membrane waves by Kramer [34].

Finally, an interesting feature of liquid/liquid interfaces is revealed by the dispersion relation when the two liquids have equal values for both viscosity and density [15]. In that case, the first term of Eq (29) vanishes and the dispersion relation reduces to the simple form $f(E) \times f(S) = 0$. This yields two uncoupled solutions, one of which depends on the interfacial tension (but not on the elasticity) and the other on the elasticity (but not on the tension). The former is the capillary ripple (Eq (33)) and the latter is the longitudinal wave (Eq (38)). As the two solutions of Eq (29) are completely uncoupled, there is no horizontal surface motion in the ripple and no vertical surface motion in the longitudinal wave. In such a case, both waves have lost their dual character of being a mixture of transverse and longitudinal surface motion. Thus, there is no longer any reason for resonance-like extra damping. Theoretically, the expected disappearance of the maximum in ripple damping has been confirmed by computer calculation [15]. Figure 7 shows experimental support for this conclusion: at the heptane/water interface, where both viscosity and density are approximately equal for the two liquids, the maximum in ripple damping in fact has virtually disappeared.

In the absence of surfactant, and at constant wave number, Stokes' relation (Eq (32)) predicts damping at an interface between two equally viscous liquids to be twice as high as at air/water, where only one liquid is viscous. This factor of 2 is indeed found in the systems of Figure 5 (after accounting for the higher values of the wave number at the oil/water interface). In the presence of surfactant, the damping at oil/water remains higher over most of the frequency range. It is only in the neighbourhood of the maximum at air/water that the damping rises above that at oil/water, because of resonance-type additional damping at air/water but not at oil/water.

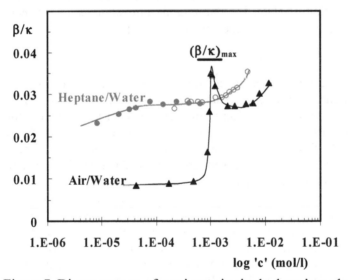

Figure 7. Disappearance of maximum in ripple damping when viscosity and density of the upper liquid are close to those of water. Example for elastic monolayers of dodecylammonium chloride at ripple frequency of 200 Hz. 'c'= mean ionic product of surfactant. At air/water surface: triangles [30]. At heptane/water interface: closed circles [29]; open circles: [15].

The main conclusion of this Section is that the study of ripples provides limited opportunity to obtain information on surface dilational properties, while the longitudinal wave does offer such information directly through the dispersion relation, Eq (38).

D. THE SURFACE DILATIONAL MODULUS

1. Purely elastic surfaces

In general, the surface dilational modulus defined by Eqs (20) is a complex number, with a real part equal to the surface elasticity and an imaginary part measuring the surface viscosity as defined in Eq (22). In the simplest experimentally relevant case the surface viscosity is negligible. This case of pure elasticity imposes two conditions on the time scale of the area change in a wave experiment:

(1) exchange of surfactant with the adjoining bulk solution must be negligible (i.e., $\Gamma \times A$ is constant);

(2) the surface tension must equilibrate with the changed value of Γ within the experimental time scale. In such a case, the modulus E is a pure elasticity with a limiting value, E_0, determined by the surface equation of state, i.e., the equilibrium relationship between surface pressure, Π, and surfactant adsorption, Γ:

$$E_0 = (d\Pi / d \ln \Gamma)_{eq} \tag{43}$$

Several models for the equation of state have been employed in order to compare theoretical results with measured curves of E_0 as a function of surfactant concentration at the surface or in the bulk solution. These versions range from very simple, such as the Szyszkowski-Langmuir model, to fairly complicated models of macromolecules. The resulting equations are broadly based on one of three models describing the relationship between Π and Γ:

(1) a phenomenological treatment based on thermodynamic equations without *a priori* assumptions about molecular geometry and interactions at the surface, combined with first-order approximations for non-ideal mixing at the surface [35-38]. Specifically, macromolecules are not assumed to be adsorbed as either flat disks or flexible polymer chains. In its simplest form, the equilibrium relation between Π and Γ depends on three constant parameters, with numerical values to be determined from the experimental results. An extension of this model accounted for multiple conformations with different molecular areas of adsorbed macromolecules, by adding a fourth phenomenological parameter [38].

(2) statistical-thermodynamic theories based on specific molecular models [39-42]. Most detailed is the self-consistent field theory developed originally for flexible-chain polymers [41]. So far, this type of model has not been applied to surface dilational elasticity.

(3) models based on scaling laws [43-45]. This approach assumes macromolecules to be adsorbed either in isolated flat 'pancakes' or in two or three 'semi-dilute' regimes, one characterised by overlap of the 'pancakes' and the other two by overlapping three-dimensional structures. These regimes are governed by different correlation lengths and by solvent quality of the interface. So far, no single analytical formula covering the various regimes has been reported.

The thermodynamic approach to the surface equation of state is based on the same phenomenological principles as the hydrodynamic theory described in the previous two sections, *viz.*, a Gibbs dividing surface and constant values of intensive parameters right up to that surface. In surface-wave studies, the thermodynamic treatment has been applied both to small surfactant molecules [15,46-48] and to macromolecules [38,49-51].

2. Visco-elastic surfaces

Deviations from the limiting case of Eq (39) occur when relaxation processes in or near the surface affect the tension γ within the time of order $1/\omega$ available in a wave experiment. Many different processes have been proposed as possible contributors to dilational surface viscosity. For soluble surfactants, diffusion exchange between surface and solution is inevitable at high enough concentrations. In the absence of other relaxation processes, the effect on surface dilational rheology has been evaluated quantitatively [46]. Other relaxation processes are, e.g., exchange of surfactant between micelles and the intermicellar solution [52], slow adsorption/desorption because of ad-/desorption barriers at the surface [53], and in-surface reactions. This last category includes processes of reorientation [54], reconformation by changes in molecular area [55] and aggregation [56]. The next three sub-sections will focus on cases for which both theoretical and experimental evidence is available.

2.1. Relaxation by diffusion in non-micellar solutions

In such cases, the measured modulus |E| is viscoelastic, with an elastic part accounting for the recoverable energy stored in the surface and a viscous contribution reflecting energy dissipation through relaxation processes. Much attention has been paid to diffusion relaxation in sub-micellar solutions, with surfactant concentrations high enough for exchange of molecules between surface and solution to occur [8,10,12]. In such cases, the surface tension gradients during surface contraction and expansion are to a certain extent short-circuited, and the value of |E| will drop below the limiting value, E_0. The reduced value of |E| can be evaluated as a function of frequency if any other relaxation processes are finished within the time scale of the area oscillations. This means that in-surface relaxations and adsorption 'barriers' are not considered here. In other words, equilibrium is assumed between the local values of γ, Γ and the sub-surface concentration immediately below the surface, $c_{z=0}$. For such pure diffusion relaxation, the only time-dependent phenomenon is transport of surfactant between the subsurface and deeper layers. The mass balance for the surface then reflects that the total amount of surfactant ($\Gamma \times A$) in a deformed surface element is no longer constant:

$$\partial(\Gamma A)/\partial t + AD(\partial c/\partial z)_{z=0} = 0 \tag{44}$$

The concentration gradients in the solution are governed by Fick's diffusion laws. In low-amplitude plane wave experiments, where convection is negligible and there are no gradients in the y direction, Fick's law is expressed by

$$\partial c/\partial t = D(\partial^2 c/\partial x^2 + \partial^2 c/\partial z^2) \tag{45}$$

In analogy with the treatment of vorticity flow, of which the penetration depth $1/m$ was introduced in Eq (11), the penetration depth of diffusion ($1/n$) is found to be given by

$$n^2 = k^2 + i\omega/D \cong i\omega/D \tag{46}$$

The analogy arises since vorticity flow and diffusion both are dissipative processes, with the diffusion coefficient D playing the part of the kinematic viscosity η/ρ in Eq (12). The result is the following expression for the complex modulus E:

$$E/E_0 = \left[1 + (1-i)(dc/d\Gamma)\sqrt{D/2\omega}\right]^{-1} \tag{47}$$

Here we consider relaxation by diffusion alone, so the parameter $dc/d\Gamma$ must reflect equilibrium between the surface and the sub-surface (where $c = c_{z=0}$). The elastic part E' and the viscous part E" are readily obtained from Eq (47), as are the equivalent expressions for the absolute value |E| and the ratio E"/ E'. The latter combination is often used in experiments since both quantities follow directly from a single experiment. For example, the ratio E"/ E' shows up as a time lag between the imposed area change and the surface-tension response. This is expressed in the viscous phase angle, ϕ:

$$E''/E' \equiv \tan\phi \tag{48}$$

The frequency spectrum of the two quantities is given by Eqs (50) and (51). To clarify the physical background of the equations, we first introduce a characteristic time scale of diffusion by equating Einstein's diffusion length − which is of order $(Dt)^{1/2}$ − to the depth of solution needed to re-establish adsorption equilibrium − which is of order $d\Gamma/dc$. A suitable definition of the diffusion time scale then is:

$$\tau_D \equiv D^{-1}(d\Gamma/dc)^2 \tag{49}$$

The final results for $|E|$ and E''/E' are fairly simple characteristic functions of the ratio of experimental and diffusional time scales, $1/\omega\tau_D$:

$$|E|/E_0 = \left[1 + \sqrt{2/\omega\tau_D} + 1/\omega\tau_D\right]^{-1/2} \tag{50}$$

$$E''/E' = \left[1 + \sqrt{2\omega\tau_D}\right]^{-1} \tag{51}$$

An equivalent formulation expresses Eqs (50) and (51) in terms of a characteristic frequency for diffusion, ω_D, related to τ_D by $2\omega_D \times \tau_D = 1$. Figure 8A demonstrates the frequency dependence of the viscoelastic modulus. In bi-logarithmic dimensionless form, the curve serves as a master curve to verify whether diffusion is the only active relaxation mechanism, because the curve has the same shape for any single surfactant, independent of its molecular characteristics. A notable feature of Figure 8B is that the viscous angle ϕ cannot exceed 45°: the elastic part of the modulus always remains at least equal to the viscous part. Remarkable is the absence of any frequency range with a constant value of the viscosity η_d. Thus, the viscosity described by E'' in Eq (51) is inherently non-Newtonian. Nevertheless, it is a true viscosity since it is directly proportional to the amount of energy dissipated by diffusion [57].

Finally, it is worth remembering that Gibbs applied his definition of surface elasticity, Eq (20), not to a wave experiment on a deep liquid, but to a thin liquid film [18]. Both experiments apply small changes in surface area, but the subsequent diffusion is limited in a thin film by the film thickness, and in a wave experiment by the frequency. Not surprisingly, the analogy between the two situations turns out to be very strong when values are compared at the same values of the adsorption distance, $d\Gamma/dc$ [58]. At both low and high frequencies the analogy was shown to be virtually complete, while in the intermediate range, around $\omega \approx \omega_D$, values of the modulus $|E|$ are less than 8% higher than the thin-film elasticity [48]. It is also stressed that only at high frequencies both elasticities can be equated to the static limiting modulus E_0 defined in Eq (43). This illustrates that E_0 represents only a limiting case of the more general elasticity defined by Gibbs. Therefore, it would be preferable to apply the term "Gibbs elasticity" to the dynamic modulus $|E|$ rather than to the static modulus E_0.

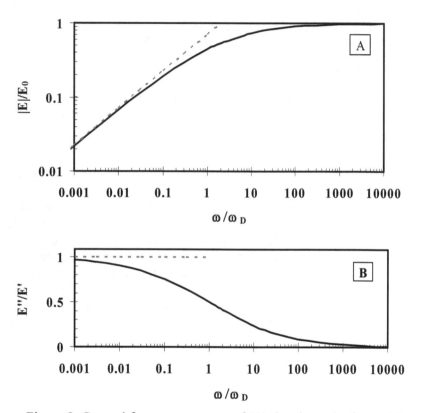

Figure 8. General frequency spectra of (A) the visco-elastic modulus $|E|/E_0$ and (B) the ratio E''/E' for pure diffusion relaxation according to Eqs (50,51). Dotted lines: limiting values for very soluble monolayers with $\omega \ll \omega_D$

2.2.   *Relaxation in micellar solutions*

Above the critical concentration for micelle formation, any change in the monomer concentration at the sub-surface resulting from a surface area change is bound to disturb the equilibrium between monomers and micelles. This will set off a chain of events producing micelle breakdown in surface expansion, and micelle formation in compression. Qualitative evidence for this was first noted when viscous phase angles exceeding 45° were measured by Lucassen and Giles [59]. A quantitative theory followed [52], based on a model with three assumptions, apart from the usual assumption of a sufficiently low amplitude-to-wavelength ratio:

(1) micelles are not surface active and therefore do not adsorb;
(2) the intrinsic surface properties (γ, Γ and E_0) remain constant above the critical concentration for micelle formation;
(3) relaxation of micelles follows the mechanism proposed by Kresheck et al. [60], in which the relaxation rate is governed by one slow decomposition step with a rate constant, k_2.

The resulting equations for the frequency spectra of $|E|$ and E''/E' are fairly complicated, depending on the number of monomers per micelle and the diffusion coefficients of monomers and micelles. A striking aspect is that the equation predicts viscous phase angles up to nearly 90°, i.e., far in excess of the phase angle of 45° for non-micellar solutions. Application of the equations to non-ionic surfactants is discussed in Section E.

2.3. Relaxation by exchange of monomers with surface aggregates

An early analysis of in-surface relaxation was proposed by Veer and Van den Tempel for the first stages of collapsing monolayers [61]. This is another case of mixed relaxation, with molecular exchange between multilayer particles and non-collapsed material at the surface combined with exchange between solution and surface. Before collapse, sparingly soluble monolayers, e.g., of C_{12} to C_{16} n-alkanols, have quite high elasticity and practically zero viscosity. After collapse had started, gel-like or mesomorphous particles resembling liquid crystals were observed in the surface. The elasticity of surfaces containing such multilayers was very high and, moreover, the viscous part of the modulus was high even at very low frequencies. The relaxation mechanism proposed was the formation/destruction of the three-dimensional particles, which exchange surfactant with the monolayer at a very low rate. A first-order rate process for the exchange leads to the following material balance for the monolayer part of the surface [61]:

$$-\partial \Gamma / \partial t = r(\Gamma - \Gamma_{z=0}) + D(\partial c / \partial z)_{z=0} + \Gamma d \ln A / dt \qquad (52)$$

Here r is a rate constant comparable to that for growth and evaporation of crystals from a vapour phase at low values of super- and sub-saturation. The flux resulting from this process was added to the diffusion flux. In other words, the two processes were treated as simultaneous, in contrast to the case of micelles where they were consecutive. This produced the following frequency spectra for the mixed relaxation:

$$|E|/E_0 = \left[(r/\omega + \sqrt{\omega_D/\omega})^2 + (1 + \sqrt{\omega_D/\omega})^2\right]^{-1/2} \qquad (53)$$

$$E''/E' = (r/\omega + \sqrt{\omega_D/\omega})/(1 + \sqrt{\omega_D/\omega}) \qquad (54)$$

Viscous phase angles greater than 45° are predicted by Eq (54), as in the case of micellar relaxation, but the limiting slope of the log |E| vs log ω plot now is not +0.5 but +1.

An interesting result further distinguishing this in-surface relaxation from micellar relaxation is that monomer exchange with gel-like surface particles was found to be much slower than with micelles in solution [61]. In combination with very high values of the modulus |E|, such slow relaxation implies that non-equilibrium ultra-low tensions will be much more persistent than in the micellar case [62].

E. SURVEY OF EXPERIMENTAL METHODS, PROBLEMS AND RESULTS

Small-amplitude surface waves have been the main vehicle used to measure the surface dilational modulus in most studies. Advantages of such waves are that (i) hydrodynamic coupling with the adjoining bulk liquid(s) can be accounted for, (ii) disturbing effects of convection in the flowing liquid can be avoided by choosing sufficiently small wave amplitudes and (iii) such waves can yield precise information on dilational properties at a given wave frequency, i.e., in a given time scale. Experimentally, wave methods differ in the nature of the surface area changes studied and in the techniques used to measure surface tension and surface dilational viscoelasticity. The area changes are either (i) generated externally by a vibrating probe touching the surface or (ii) spontaneously present on a liquid surface because of thermal roughness. In both cases, measuring the average surface tension by mechanical or optical means is fairly straightforward, but reliable values of the surface dilational modulus (E) are less easily obtained.

1. Externally excited surface waves

These waves have been studied over a large range of frequencies, roughly from 10^{-3} up to 10^4 Hz. The great majority of results were obtained at the air/water surface, but in recent years various oil/water interfaces have received increasing attention [48,63-66]. Up to 1968, the only type of wave considered was the capillary wave or ripple, generated by a vertically vibrating probe. The resulting surface wave is not very sensitive to surface elasticity, since it is mainly governed by the average value of the surface tension according to Eq (33). Quantitative information on the surface dilational elasticity can be derived from ripples only in a restricted range around the maximum damping coefficient, as illustrated in Figure 3. A wider window was opened with the discovery of the longitudinal surface wave, which depends directly on the surface dilational elasticity [13,14]. In principle, both waves will be generated by any generator vibrating in whichever direction (see Section B.1.) Nevertheless, oscillation in horizontal rather than vertical direction does increase the role played by longitudinal waves, which depend directly on the surface dilational modulus through the tangential surface stress.

A wave frequency range spanning seven decades requires methods with different response times to measure the surface tension variations caused by surface area variations. At low frequencies, – roughly from 10^{-3} to 1 Hz, depending on the value of $|E|$ – the method does not need to have a very short response time. For not too high liquid viscosity, a Wilhelmy plate is often quite suitable in longitudinal wave studies at these frequencies. The overwhelming majority of low-frequency results reported so far have been obtained at the air/water surface with a horizontally oscillating barrier and, usually, with a Wilhelmy plate to monitor the tension variations. This simple technique is most convenient if deformation of the area under consideration is uniform, i.e., if the tension variations measured by the plate do not depend on its distance from the barrier. Large errors can result if non-uniformity of the surface deformation is ignored. The tension variations are propagated from the oscillating barrier to the plate by means of the longitudinal surface wave generated by the barrier [33]. Uniform surface deformation requires wave damping to be low enough to result in multiple reflections of the wave between barrier and vessel wall [67]. As surface wave damping occurs primarily through viscous friction in the adjoining bulk phases, it is qualitatively evident that the condition of low wave damping is easier satisfied for a gaseous non-aqueous phase than for very viscous oils. Quantitatively, the effects of wave propagation have been evaluated from the dispersion equation of the longitudinal wave by Lucassen and Barnes [67]. The multiple reflections necessary for uniform surface deformation were found to impose the following condition on the effective length, L, over which the wave must propagate:

$$L \ll \beta^{-1} = \frac{\sqrt{|E|}}{(\sqrt{\eta\rho} + \sqrt{\eta'\rho'})^{1/2} \omega^{3/4} \sin(\pi/8 + \phi/2)} \qquad (55)$$

It follows that high liquid viscosity, high frequency and a low modulus all are factors that can require a much smaller length L than is experimentally feasible with a conventional trough and a Wilhelmy plate. Numerical examples of the effect of viscosity and the required length L have been evaluated [68] and measured [63]. A solution for high-viscosity oils was found by modifying a conventional technique that uses axisymmetric drop shape analysis to study the kinetics of adsorption at constant area. In the modified set-up, a small drop or bubble can be subjected to sinusoidal oscillations of its volume, and hence of its surface area [64]. As the diameter of the drop or bubble is only a few mm, the effective length, L, over which the wave must propagate in order to ensure uniform surface deformation is smaller than 10 mm and, therefore, much more likely to satisfy the condition of Eq (55) up to quite high liquid viscosity. This is especially relevant for food oils, such as triacylglycerols, which are far more viscous than single-chain alkanes and water.

2.   Spontaneous surface waves

This type of wave results from thermal molecular motion, and has been investigated by quasi-elastic surface light scattering [68-72]. Advantages of the technique are that (i) it can probe surface properties from roughly 10 kHz, a frequency range not accessible by other methods and (ii) it is non-invasive and non-perturbative. However, the latter advantage implies the disadvantage that the balance between dissipative and non-dissipative flow cannot be altered. Light scattering is primarily a response to vertical displacements of surface elements, as is the externally excited ripple. Thus, the method suffers from the same drawback as the excited ripple: it is not very sensitive to the surface dilational modulus. Moreover, a peculiar feature of the analysis often applied to the scattered light is that it has produced negative values of the surface dilational viscosity [69]. Such values are physically not possible since they imply that energy is not dissipated but generated in the experiments. The root cause of these puzzling results appears to be one assumption used in the processing of the experimental data: it is assumed that the surface tension can have a viscous component, expressed in the vertical shear viscosity first proposed by Goodrich [73]. This idea is controversial since surface tension is directly related to the *free* energy of the system and, therefore, it cannot contain a dissipative contribution. The assumption of such a 'third' surface viscosity fundamentally conflicts with the current wave theory outlined in Sections B and C, where the surface tension is a real quantity. (See also Boussinesq [25] and Scriven [75]). Moreover, calculations on the constitutive surface model by Buzza have shown that the vertical shear viscosity violates the principle of 'frame invariance' [74]. His re-analysis of experimental data for which negative viscosity values were reported revealed that the anomaly disappeared when surface tension was taken as real rather than complex. The conclusion must be that the vertical shear viscosity is unphysical, as are the negative dilational viscosities reported.

3. Examples of experimental results from longitudinal waves

Many experimental techniques are available for monitoring the surface response in different frequency ranges. For details we refer to several instructive reviews and regular articles [14,16,27,32,33,46,48,64,67,76-92]. This list is by no means complete; more information is to be found in other Chapters of this Volume. In the present Chapter, we choose two examples of experimental results, each illustrating more than one significant aspect. In many studies, the dilational modulus E has been measured only at or near adsorption equilibrium. However, useful additional information can be gained from the changing values of E and Π during the adsorption process by measuring tensioelastograms, as exemplified in Figure 9 [59].

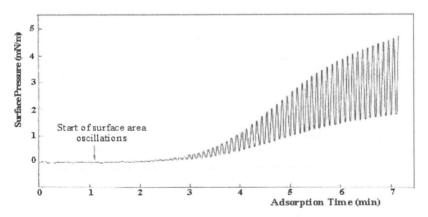

Figure 9. Example of surface tensioelastogram, for bovine plasma albumin at c=1 mg/l, with sinusoidal area oscillations at 1/6 Hz (ω=1.05 rad/s) and relative amplitude $\Delta \ln A$=0.058. From Lucassen and Giles [47]. During an initial "induction" time of 150 s, the surface tension is constant at γ_0 and $|E| \approx 0$. Only after this time does the dilational modulus gradually increase up to E≈50 mN/m.

The first example concerns small-area molecules in a trough with a Wilhelmy plate halfway between two moving barriers in order to avoid hydrodynamic drag on the plate [59]. Figure 10 shows values of $|E|(c)$ for two relaxation mechanisms, *viz.*, pure diffusion relaxation in sub-micellar solutions and mixed relaxation in micellar solutions by diffusion and micellar breakdown [52]. For the latter mechanism, a relaxation time of about 10^{-2} s was found to agree very well with the experimental data. As expected, measured viscous phase angles (ϕ, not shown here) in the micellar region were far above 45°. This and the sharp drop in the modulus values were also as expected: disintegration of micelles into monomers increases the reservoir from which diffusion can suppress the surface tension gradient.

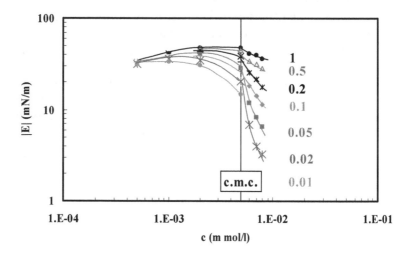

Figure 10. Surface dilational modulus ($|E|$) *vs* concentration (c) for solutions of tetradecyl hexaethylene glycol, below and above the critical micellar concentration (c.m.c.). Numbers indicate frequencies (Hz). Temperature: 25 °C. From Lucassen and Giles [52,59]. Drawn lines: theory for given set of parameter values [52]

The second example, in Figure 11, deals with large-area molecules adsorbed at three liquid interfaces and presented as E(Π) curves [51].

At the given wave frequency and within experimental error, all values of E at different concentrations are seen to coincide on a single E(Π) curve, up to $\Pi \approx 20$ mN/m. In this range, viscous phase angles were small (tan $\phi \leq 0.1$) indicating that, within the time scale of the surface wave (≈ 10 s), any relaxation processes either had not yet started or were essentially finished. At high Π, however, the curves did split up into branches (not shown here), proving the occurrence of slow relaxation; for the viscous phase angles see Chapter 7. A further aspect to note in Figure 11 is that the 'induction time' shown in Figure 9 is not caused by an adsorption barrier, but by a strongly non-ideal equation of state [51,62]. Such results have been interpreted with a four-parameter equation of state accounting for compression of macromolecules into smaller areas at higher pressures Π [38,51].

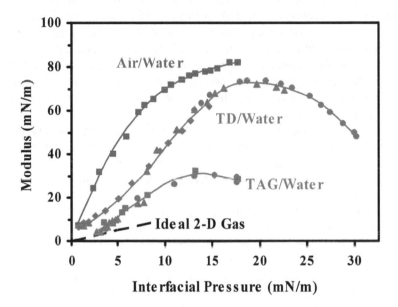

Figure 11. Interfacial dilational modulus as a function of surface pressure for ovalbumin at three interfaces. Apolar phases: Air, TD (*n*-tetradecane) and TAG (triacylglycerol). Diamonds: 3 mg/l; triangles: 10 mg/l; squares: 0.1 g/l; circles: 1 g/l. Frequency: 0.13 Hz. T = 25 °C.

A final point worthy of note in Figure 11 is that the $E(\Pi)$ curves strongly depend on the polarity of the interfaces as expressed in the values of their clean-interface tensions, γ_0. Strong non-ideality is most evident at the least polar interfaces, i.e., the air/water and TD/water interfaces, with the highest values of γ_0 (71 and 51 mN/m, respectively).

These two interfaces exhibit no discernible trace of ideal surface behaviour, while the most polar interface (TAG/water, with $\gamma_0 \approx 31$ mN/m) is seen to be much closer to ideality. It was noted as early as 1941 that the adsorption of small-area surfactants at oil/water interfaces generally increased with γ_0 [93]. The trends in Figure 11 indicate that this is also true for proteins. Thus, a low clean-interface tension directly results in small values of the adsorption and, hence, also in small surface dilational elasticities [51].

F. SUMMARY AND CONCLUSIONS

Small-amplitude surface waves are the most reliable vehicle used to measure the surface dilational modulus. Advantages of such waves are: (i) hydrodynamic coupling of the surface with the adjoining bulk liquid(s) can be accounted for, (ii) disturbing effects of convection in the flowing liquid can be avoided and (iii) such waves can yield precise information on dilational properties at a given wave frequency, i.e., in a given time scale. Ironically, more information on dilational surface rheology has been gained from longitudinal surface waves than from transverse waves, although the latter have been studied over a much longer time span (Sections A and B)

Both types of surface wave have limitations to their applicability. The transverse wave or ripple is only weakly dependent on dilational visco-elasticity. The longitudinal wave is quite sensitive to dilational properties, but its application is hindered by high damping of the wave (Section C).

Various relaxation mechanisms, e.g., molecular diffusion, micellar breakdown and in-surface exchange of molecules with surface aggregates, have been theoretically described in simple models to interpret measured viscous phase angles (Section D).

Experimentally, reliable results can be obtained mainly in two frequency ranges: low frequencies (up to ~1 Hz) and a smaller range of high frequencies for quite high dilational moduli. In the low- frequency range, results with small-area molecules were interpreted in terms of molecular diffusion and micellar breakdown. Macromolecules, such as proteins, were not affected by diffusion up to quite high surface pressures, owing to their low diffusion coefficients and low solution concentrations. In this range of purely elastic moduli, maximum information on static and dilational properties was obtained by including the changing values of modulus and pressure during the course of the adsorption process. Results with several proteins at different liquid interfaces revealed a strong influence of interface polarity, which is reflected in the clean-interface tension, γ_0 (Section E).

G. LIST OF SYMBOLS

A	area of surface under deformation
D	diffusion coefficient of surfactant
E	surface visco-elastic modulus defined in Eqs (20) and (22)
E', E''	elastic and viscous contributions to E, respectively
$\|E\|$	absolute value of E
E_0	limiting value of surface elasticity defined in Eq (41)
f (E)	elasticity function defined in Eq (30)
f (S)	surface tension function defined in Eq (31)
g	gravitational constant
k	complex wave number ($\equiv \kappa - i\beta$)
L	effective length required for uniform surface deformation in Eq (55)
m	inverse penetration depth of vorticity
n	inverse penetration depth of diffusion
p	hydrostatic pressure
r	rate constant of in-surface relaxation in Eqs (52-54)
v	liquid velocity vector
x, y, z	rectangular coordinates
β	distance damping coefficient ($\equiv -k_{Im}$)
γ	surface or interfacial tension
Γ	adsorption of surface active agent
η, η'	viscosity of lower and upper fluid, respectively
ζ	vertical displacement of surface element
ξ	horizontal displacement of surface element
κ	wave number ($= 2\pi/\lambda$)
λ	wavelength
Π	surface pressure ($\equiv \gamma_0 - \gamma$)
ρ, ρ'	density of lower and upper fluid, respectively
ϕ	viscous phase angle ($\equiv$ arc tan (E''/E'))
φ	potential function of liquid flow
ψ	vorticity or stream function of liquid flow
ω	angular frequency of wave
Λ^2	Laplace operator ($\equiv \partial^2/\partial x^2 + \partial^2/\partial y^2 + \partial^2/\partial z^2$) in Eq (2)
Φ, Φ'	amplitudes of potential function in water and upper fluid, respectively
Ψ, Ψ'	amplitudes of vorticity function in water and upper fluid, respectively

Subscripts

0	for clean interface; for interface at rest in Eq (8)
1	for contribution of potential flow
2	for contribution of vorticity flow
eq	for values at (local) equilibrium
L	for longitudinal surface wave
s	for values at surface
T	for transverse surface wave

H. REFERENCES AND NOTES

1. Gaius Plinius Secundus, in Historia Naturalis, Liber II, Caput 103*, AD 77
    *Different caput numbers are used in other versions of Pliny's manuscript.
2. B. Franklin, Phil. Trans.., 64 (1774) 445. Franklin did not specify the type of oil he used, but in his time the word 'oil' was employed for edible oils, generally olive oil and occasionally whale oil or fish oil.
3. J.A. von Segner, De Figuris Superficierum Fluidarum (Nature of Liquid Surfaces), Commentarii Societatis Regiae Scientiarum Gottengensis, Vol. I, (1751) 301-373. Von Segner was a Hungarian-German mathematician and physicist, who likened the liquid surface to a stretched membrane and attempted to give a mathematical description of capillary action in terms of surface tension. This attempt was unsuccessful, which may have contributed to the new concept initially being ignored. However, his view that minute attractive forces in the surface region are the root cause of surface tension has stood the test of time. It constituted the basis for the subsequent theory of surface tension.
4. G.G Stokes, Cambridge Trans., 8 (1845) 287
5. Lord Kelvin (W. Thomson), Phil. Mag., 42 (1871) 368
6. O. Reynolds, report presented to the British Association for the Advancement of Science, (1880) 489
7. C. Marangoni, Nuovo Cimento, Series 2, Vol. 5/6 (1872) 239
8. V.G. Levich, Acta Physicochim., 14 (1941) 307, 321
9. R. Dorrestein, Koninkl. Ned. Akad. Wetenschap., Proc., B, 54 (1951) 260, 350
10. R.S. Hansen and J.A. Mann, J Appl. Phys., 35 (1964) 152
11. R.S. Hansen, J. Applied Phys., 35 (1964) 1983
12. M. van den Tempel and R.P. van de Riet, J. Chem. Phys., 42 (1965) 2769
13. J. Lucassen, Trans. Faraday Soc., 64 (1968) 2221

14. J. Lucassen, Trans. Faraday Soc., 64 (1968), 2230
15. E.H. Lucassen-Reynders and J. Lucassen, Adv. Colloid Interface Sci, 2 (1969) 347
16. R.S. Hansen and J. Ahmad, Progress Surface Membrane Sci., 4 (1971) 1
17. R.P. Feynman, R.B. Leighton and M. Sands, The Feynman Lectures on Physics, Vol. 1, Chapter 51, Addison-Wesley, Reading, Massachusetts, 1965
18. J.W. Gibbs, "On the Equilibrium of Heterogeneous Substances", Trans. Connecticut Acad., III (1876) pp 108-248 and (1878) pp 343-524, as reprinted in The Scientific Papers of J. Willard Gibbs, Vol. I. Thermodynamics. Dover Publications, New York, 1961, p.302
19. H. Lamb, Hydrodynamics, Dover Publications, New York, NY, 1945, First Edition: Cambridge University Press, Cambridge (U.K.) 1879
20. G. Joos, Lehrbuch der theoretischen Physik, Akademische Verlagsgesellschaft, Leipzig, 1934. English translation: Theoretical Physics, Blackie & Son Ltd, London, Glasgow, 1958
21. L.D. Landau and E.M. Lifshitz, Fluid Mechanics, Pergamon Press, London, 1959
22. V.G. Levich, Physicochemical Hydrodynamics, Prentice-Hall, Englewood Cliffs, New Jersey, 1962
23. D.A. Edwards, H. Brenner and D.T. Wasan, Interfacial Transport Processes and Rheology, Butterworth-Heinemann, Boston (USA.), 1991
24. M. Koussakov, Acta Physicochimica U.R.S.S., 19 (1944) 286
25. J. Boussinesq, Comptes Rendus, 156 (1913) 983, 1035, 1124;
26. Lord Rayleigh (J.W. Strutt), Proceedings of the Royal Institution, 13 (1890) 85; Proc. Roy. Soc., Ser. A, 48 (1890) 127; Phil. Mag., 30 (1890) 386
27. J.A. Mann, Techniques of Surface and Colloid Chemistry and Physics, R.J. Good, R.R. Stromberg and R.L. Patrick (eds), Marcel Dekker, New York, 1972, p 77
28. Oeuvres Complètes de Laplace, Tome quatrième, Supplément au Livre X du Traité de Mécanique Céleste sur l'Action Capillaire. Gauthiers-Villars, Paris, 1880. According to an editor's note, the Supplement had already been published in 1806. Laplace's derivation of his law is very thorough, but not very easy to follow. His way of reasoning and choice of variables look unusual to a modern reader, and he does not mention the term surface tension anywhere. However, his parameter H/2, described as a constant characterising the peculiarities of the attractive forces near a fluid/fluid interface, does correspond to the surface tension.

29. R.S. Hansen, J. Lucassen, R.L. Bendure and G.P. Bierwagen, J. Colloid Interface Sci., 26 (1968) 198
30. J. Lucassen and R.S. Hansen, J. Colloid Interface Sci., 22 (1966) 32
31. J.T. Davies and E.K. Rideal, Interfacial Phenomena, New York, 1961
32. D. Thiessen and A.D. Sheludko, Kolloid-Z., 218 (1967) 139
33. J. Lucassen and M. van den Tempel, J. Colloid Interface Sci., 41 (1972) 491
34. L. Kramer, J. Chem. Phys., 55 (1971) 2097
35. E.H. Lucassen-Reynders and M. van den Tempel, International Congress on Surface Active Substances, Brussels, 1964
36. E.H. Lucassen-Reynders, Colloids Surfaces, 91 (1994) 79
37. V.B. Fainerman and R. Miller, Langmuir, 15 (1999) 1812
38. V.B. Fainerman, E.H. Lucassen-Reynders and R. Miller, Adv. Colloid Interface Sci., 106 (2003) 237-259
39. L. Ter Minassian-Saraga and I. Prigogine, Mém. Serv. Chimie Etat, 38 (1953) 109
40. H.L. Frisch and R. Simha, J. Chem. Phys., 24 (1956) 652
41. J.M.H.M. Scheutjens and G.J. Fleer, J. Phys. Chem., 83 (1979) 1619
42. G.J. Fleer, J.M.H.M. Scheutjens and M.A. Cohen Stuart, Polymers at Interfaces, Chapman & Hall, London, 1994
43. P. Cicuta and I. Hopkinson, J. Chem Phys., 114 (2001) 8659
44. A. Hambardzumyan, V. Aguié-Béghin, I. Panaïotov and R. Douillard, Langmuir, 19 (2003) 72
45. R. Douillard, M. Daoud and V. Aguié-Béghin, Current Opinion Colloid Interface Sci., 8 (2003) 380
46. J. Lucassen and M. van den Tempel, Chem. Eng. Sci., 27 (1972) 1283
47. E.H. Lucassen-Reynders, J. Lucassen, P.R. Garrett, D. Giles and F. Hollway, Advances in Chemistry Series, (E D Goddard, ed.) 144 (1975) 272
48. E.H. Lucassen-Reynders, A Cagna and J. Lucassen, Colloids Surfaces A, 186 (2001) 63
49. E.H. Lucassen-Reynders and J. Benjamins, Food Emulsions and Foams, E. Dickinson and J.M. Rodriguez Patino (eds), Special Publication No. 227, p 195, Royal Society of Chemistry, Cambridge, UK, 1999
50. E.H. Lucassen-Reynders, V.B. Fainerman and R. Miller, J. Phys. Chem., B, 108 (2004) 9173
51. J. Benjamins, J. Lyklema and E.H. Lucassen-Reynders, Langmuir, 22 (2006) 6181
52. J. Lucassen, Faraday Discussions Chemical Society, 59 (1975) 76

53. V.B. Fainerman, A.V. Makievski and R. Miller, Colloids Surfaces, A, 87 (1994) 6153
54. V.B. Fainerman, R. Miller, R. Wüstneck and A.V. Makievski, J. Phys. Chem., 100 (1996) 7669
55. G. Serrien, G. Geraerds, L. Ghosh and P. Joos, Colloids Surfaces A, 68 (1993) 210
56. V.B. Fainerman and R. Miller, Langmuir, 12 (1996) 6011
57. E.H. Lucassen-Reynders and J. Lucassen, Colloids Surfaces A, 85 (1994) 211
58. J. Lucassen, Anionic Surfactants: Physical Chemistry of Surfactant Action, E.H. Lucassen-Reynders (ed.), Marcel Dekker, New York, 1981, p 217
59. J. Lucassen and D. Giles, J. Chem. Soc., Faraday Trans. I, 71 (1975) 217
60. G.C. Kresheck, E. Hamori, G. Davenport and H.A. Scheraga, J. Amer. Chem. Soc., 88 (1966) 246
61. F.A. Veer and M. van den Tempel, J. Colloid Interface Sci., 42 (1973) 418
62. M. van den Tempel and E.H. Lucassen-Reynders, Adv. Colloid Interface Sci., 18 (1983) 281
63. J. Lucassen, unpublished measurements (1969)
64. J. Benjamins, A. Cagna and E.H. Lucassen-Reynders, Colloids Surfaces A, 114 (1996) 245
65. R. Wüstneck, B. Moser and G. Muschiolik, Colloids Surfaces B, 15 (1999) 263
66. E.M. Freer, K.S. Yim, G.G. Fuller and C.J. Radke, J. Phys. Chem B, 108 (2004) 3835
67. J. Lucassen and G.T. Barnes, J. Chem. Soc., Faraday Trans. I, 68 (1972) 2129
68. D. Byrne and J.C. Earnshaw, J. Phys. D, 12 (1979) 1145
69. J.C. Earnshaw, E. McCoo, C. Nugent and D.J. Sharpe, Lecture Notes Phys., 467 (1996) 166
70. D. Langevin, J. Colloid Interface Sci., 80 (1981) 412
71. D. Langevin, Light Scattering by Liquid Surfaces and Complementary Techniques, D. Langevin (ed.), Marcel Dekker, New York, 1992, p 161
72. P. Cicuta and I. Hopkinson, Colloids Surfaces A, 233 (2004) 97
73. F.C. Goodrich, J. Phys. Chem., 66 (1962) 1858
74. D.M.A. Buzza, Langmuir, 18 (2002) 8418
75. L.E. Scriven, Chem. Eng. Sci., 12 (1960) 98
76. D. Platikanov, I. Panajotov and A Scheludko, Abhandlungen der Deutschen Akademie der Wissenschaften zu Berlin, Nr. 6b, (1967) 773

77. J. Benjamins, J.A. de Feijter, M.T.A. Evans, D.E. Graham and M.C. Phillips, Faraday Trans. Chem. Soc., 59 (1975) 218
78. K. Lunkenheimer and G. Kretzschmar, Z. Physik. Chem. (Leipzig) 256 (1975) 593
79. C.H. Sohl, K. Miyano and J.B. Ketterson, Rev. Sci. Instrum., 49 (1978) 1469
80. H.C. Maru, D.T. Wasan, Chem. Eng. Sci., 34 (1979) 1295
81. C. Stenvot and D. Langevin, Langmuir, 4 (1988) 1179
82. G. Loglio, U. Tesei and R. Cini, Rev. Sci. Instrum., 59 (1988) 2045
83. J.J. Kokelaar, A. Prins and M. de Gee, J. Colloid Interface Sci., 146 (1991) 507
84. J. Benjamins and F. van Voorst Vader, Colloids Surfaces, 65 (1992) 161
85. Q. Jiang, Y.C. Chiew and J.E. Valentini, J. Colloid Interface Sci., 155 (1993) 8
86. A. Dussaud and M. Vignes-Adler, J. Colloid Interface Sci., 167 (1994) 256
87. B.S. Murray and P.V. Nelson, Langmuir, 12 (1996) 5973
88. K.D. Wantke, H. Fruhner, J. Fang and K. Lunkenheimer, J. Colloid Interface Sci., 208 (1998) 34
89. B.A. Noskov, D.A. Alexandrov and R. Miller, J. Colloid Interface Sci., 219 (1999)250
90. J. Benjamins, Static and dynamic properties of proteins adsorbed at liquid interfaces, PhD Thesis, Wageningen University, the Netherlands, 2000, p 68
91. J. Benjamins and E.H. Lucassen-Reynders, Food Colloids, Biopolymers and Materials, E. Dickinson (ed.), Royal Society of Chemistry, Cambridge, 2003, p 216
92. E.M. Freer, K.S. Yim, G.G. Fuller and C.J. Radke, Langmuir, 20 (2004) 10159
93. F.E. Bartell and J.K. Davis, J. Phys. Chem., 45 (1941) 1321

DETERMINATION OF THE DILATIONAL ELASTICITY AND VISCOSITY FROM THE SURFACE TENSION RESPONSE TO HARMONIC AREA PERTURBATIONS

S.A. Zholob[1], V.I. Kovalchuk[2], A.V. Makievski[3], J. Krägel[4],
V.B. Fainerman[1] and R.Miller[4]

[1] Physicochemical Centre, Donetsk Medical University, 16 Ilych Avenue, Donetsk 83003, Ukraine

[2] Institute of Biocolloid Chemistry, 42 Vernadsky Avenue, 03142 Kyiv (Kiev), Ukraine

[3] SINTERFACE Technologies, Volmerstr. 5-7, 12489 Berlin, Germany

[3] MPI for Colloids and Interfaces, 14424 Potsdam/Golm, Germany

Contents

A. INTRODUCTION

The majority of natural systems are dynamic, i.e. their parameters vary with time. The most common way to investigate such systems is to apply some disturbance to the system under study and monitor the response to this disturbance. For sufficiently small disturbances, the relation between disturbance and response is described by a transfer function [1].

Regarding the behavior of adsorption layers under dynamic conditions, a study of the interfacial tension response is often used to provide information about mechanisms of possible interfacial relaxation processes. A large variety of experimental methods were developed to study relaxation processes in adsorption layers many of which are based on harmonic perturbations. As an example, the damping of capillary waves belongs to the classical methods [2]. Another technique using harmonic disturbances is the oscillating bubble method [3], which belongs now to most frequently applied dilational rheology methods [4].

The information about relaxation properties of adsorption layer is given in the form of the complex quantity $E(i\omega)$ the so-called complex surface visco-elastic modulus (see Chapter 2),

$$E(i\omega) = E_s(\omega) + i\omega\eta_s(\omega) \tag{1}$$

where $E_s(\omega)$ and $\eta_s(\omega)$ are the dilatation elasticity and viscosity, respectively, which are functions of the angular frequency ω.

Knowledge about the dilational rheological properties is very useful for the selection of adequate theoretical models to describe quantitatively the behavior of adsorption layers. Therefore, the aim of many experimental studies is to determine the complex elasticity, however, this task is complicated by some methodological difficulties. This chapter deals with the different aspects of obtaining the complex elasticity from data of harmonic relaxation experiments. Also various ways of data processing are discussed.

For a systematic analysis some assumptions are needed. At first we assume that the interfacial layer is a linear system and a perturbation is generated by an area change

$$\Delta A = A(t) - A_0 \tag{2}$$

where A_0 is the initial surface area. The response is the surface tension change

$$\Delta\gamma = \gamma(t) - \gamma_{ini} \tag{3}$$

where γ_{ini} is the initial surface or interfacial tension.

There is a functional interrelation between perturbation and response of the system $\Delta\gamma = f(\Delta A)$. Linearity of the system assumes that two conditions are fulfilled:
1. Additivity – two perturbations are combined independently

$$f(\Delta A_1 + \Delta A_2) = f(\Delta A_1) + f(\Delta A_2) \tag{4}$$

2. Uniformity – condition of scale transformation

$$f(k\Delta A) = kf(\Delta A) \tag{5}$$

All perturbations are assumed to be small enough to meet the conditions of linearity

$$\Delta A \ll A_0; \Delta\gamma \ll \gamma_0 \tag{6}$$

These assumptions are made to simplify the following analysis.
When the interfacial area is subject to harmonic changes (oscillations)

$$A(t) = A_0 + \Delta A(t) = A_0 + \Delta A \sin(\omega t + \phi_A) \tag{7}$$

where A_0, ΔA, ϕ_A are the mean area, the amplitude and phase angle of area changes, respectively, ω is the angular frequency of oscillation. For linear system follows that changes of interfacial tension are also harmonic with a certain amplitude and possible phase shift [1]

$$\gamma(t) = \gamma_{ini} + \Delta\gamma \sin(\omega t + \phi_\gamma) \tag{8}$$

$$\gamma(t) = \gamma_0 + \Delta\gamma \sin(\omega t + \phi_\gamma)$$

where γ_0, $\Delta\gamma$, ϕ_γ are the mean surface tension value, the amplitude and phase shift of surface tension changes, respectively.

The complex visco-elasticity carries all information about change in amplitude as well as in phase shift of the oscillations. The most general definition is given by the Fourier transform of the response relative to the perturbation [1]

$$E(i\omega) = \frac{F[\Delta\gamma]}{F[\ln(\Delta A)]} \tag{9}$$

Accounting for condition (6) usually Eq. (9) is used in a form

$$E(i\omega) = A_0 \frac{F[\Delta\gamma]}{F[\Delta A]} \tag{10}$$

Any complex function can be represented in vector or exponential form, which yields: $E(i\omega) = E_r + iE_i$ or $E(i\omega) = |E| \exp(i\phi)$, where $|E| \equiv \dfrac{\Delta\gamma}{\Delta A / A_0}$ is the surface dilational modulus, and E and ϕ can be considered as visco-elasticity parameters.

Either of the Eqs. (7) or (8) can be written in a general form as

$$x(t) = x_a + \Delta x \sin(\omega t + \phi) \tag{11}$$

so that Eq. (10) leads to the definition of the visco-elasticity parameters through the oscillation parameters of surface tension and interfacial area

$$E = A_0 \frac{\Delta\gamma}{\Delta A}, \quad \phi = \phi_\gamma - \phi_A \tag{12}$$

This can be done in one of the two ways:
1. Each of the two functions is processed separately, the needed parameters of sinusoidal function are estimated, and then the visco-elasticity parameters are calculated from Eq. (12). A spectral analysis based on a discrete Fourier transform [5] is the most popular method for solving this problem.

2. Both data sets are processed simultaneously. An idea of this kind is mentioned in [4], where a plot of surface tension changes over the corresponding area change was proposed. The data points are fitted by the equation of an ellipse. Then, its tilt angle is a measure of dilatational elasticity, while the thickness is proportional to the dilatational viscosity (phase angle).

In each of these formulations the problem is closely related to a digital signal processing (DSP). The above mentioned discrete Fourier transform (DFT) is a fitting of the given data by a number of sinusoidal functions of certain frequencies. In experiments with harmonic oscillations the externally applied frequency is known and in general equal to the fundamental frequency of the system response. Nevertheless, there are reasons for which the fundamental frequency can slightly differ from the applied one. The most important reason is the imperfection of the experimental device which can lead to a non-constancy of time intervals between points or distortion of the oscillation shape from being sinusoidal. Therefore, besides DFT there were some alternative methods developed to estimate the most significant frequencies present in a set of data, as discussed in literature [6 - 8]. Nevertheless, the DFT based classic methods are most popular for several reasons, first of all, due to the solid theoretical and technical background. In this chapter the DFT was therefore used as well. In order to solve the problem of unevenly sampled data, polynomial smoothing is proposed. Details of numerical data processing with some examples are also discussed below.

B. NUMERICAL TREATMENT OF EXPERIMENTAL DATA

The problem of numerical processing of experimental data can be formulated as follows: given a set of points consisting of N pairs of values (A_k, γ_k), corresponding to the time moments t_k, perhaps unevenly spaced, one has to calculate the average value of the surface area A_0 and surface tension γ_0, amplitudes of changes of these values (ΔA and $\Delta \gamma$) or their ratio, and the phase difference between the two oscillations φ.
This stage is subdivided into three subsequent tasks:

1. Removal of a surface tension trend
2. Calculation of Fourier transforms for the sets of area and interfacial tension values, respectively.
3. Estimation of measurement accuracy.

1. Removal of a surface tension trend

The definition of the visco-elasticity parameters assumes that the data are achieved at a small deviation from the equilibrium. For substances with a high surface activity such a state can take several hours to be reached [4]. This inconvenience sometimes forces researchers to begin area oscillations without waiting until a real equilibrium state is reached. Therefore a slight surface tension decrease due to the on-going adsorption process or some other reasons may superimpose the surface tension oscillations. This situation is illustrated in Fig.1.
In this case, the Fourier transform $F[\Delta \gamma]$ also contains frequency components reflecting presence of a trend that can considerably reduce the accuracy of the definition of E by Eq. (10). It was proposed in [6] to remove a trend by using of specially designed digital filter. This way is rather complicated, and in our case it appears to be sufficient to use a simpler technique based on approximation by a polynomial of a low degree

$$f(t) = \sum_{k=0}^{n} a_k t^k \tag{13}$$

where ak are polynomial coefficients, t – time, n – polynomial degree between 1 and 3.

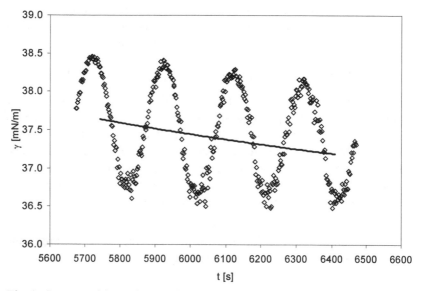

Fig.1. Superposition of a trend to a harmonic surface tension oscillation:
◊ - experiment, solid line - trend curve.

Usually, the points in between maximum and minimum are taken for this approximation as the effect of oscillation is smallest here. Although this algorithm is simple, it is not rigorous because the positions of the extrema are not exactly known. Hence, the application of some smoothing procedure would be useful before removing the trend. The procedure of a *moving polynomial approximation* described in Section D.1. is suitable for this purpose. It was validated against a number of artificial and real-time data that it does not add low-frequency components to the spectrum. After the definition of $f(t)$, the trend can be removed

$$\Delta\gamma(t) = \gamma(t) - f(t) \tag{14}$$

For $f(t) = \gamma_0$ we have the ideal situation where area oscillations start when system is very close to the equilibrium.

2. Calculation of the visco-elasticity modulus

Calculation of the visco-elasticity modulus by Eq. (10) uses relation of the discrete Fourier transform (DFT) of a series of experimental area values to that of interfacial tension. The discrete Fourier transform for a set of N experimental values (x_n) is defined as a sum of complex exponents weighted with x_n

$$F[k] = \sum_{n=0}^{N-1} x_n \exp\left(-i\frac{2\pi nk}{N}\right) \tag{15}$$

where F[k] is the kth complex component in the spectrum. Each component in the spectrum corresponds to specific frequency $f_k = \dfrac{k}{T_{exp}}$ (T_{exp} is the whole time of measurement) and according to the Nyquist theorem upper frequency is reached at k=N/2. Usually in experiments with harmonic area variation, for calculation parameters of oscillation in the form

$$x(t) = x_a + \Delta x \sin(\omega t + \phi) \tag{16}$$

the component with maximum amplitude is used, which should correspond to the applied frequency of oscillations. All other components are supposed to be noise. Calculation algorithm contains the following steps:

1. If a surface tension data series is considered then possible trend is removed. Also the use of some smoothing technique is possible if data contain too much noise or for some other reason (see below).
2. The DFT spectrum is calculated according to Eq. (15)
3. The number of the spectrum component k_m with a maximum amplitude value is defined. The maximum amplitude is looked from k=1 up to k=N/2 (usually N is even) because k=0 corresponds to average value $F[0] = \sum_{n=0}^{N-1} x_n = Nx_a$ as follows from Eq. (15). The amplitude of the given component $F[k] = X[k] + iY[k]$ is defined as

$$Z[k] = \sqrt{X[k]^2 + Y[k]^2} \tag{17}$$

4. The sampling interval Δ is defined as the coefficient of linear regression $t_n = n\Delta$ in coordinates $t_n = f(n)$.
5. Oscillation parameters in Eq. (16) are defined as follows:

$$\omega = \frac{2\pi k_m}{N\Delta}, \quad x_a = \frac{Z[0]}{N}, \quad \Delta x = \frac{2Z[k_m]}{N}, \quad \phi = \arctan\frac{Y[k_m]}{X[k_m]} \tag{18}$$

6. If analyzed time range does not span exactly the multiple of the oscillation period $T = 2\pi/\omega$ and/or sampling interval Δ is not exactly constant then calculation error (see below) will not be minimal. So oscillation parameters should be adjusted to reach minimum error according to Eq. (16).

7. Complex visco-elasticity modulus is defined by Fourier component
    with maximum amplitude for interfacial tension data set $F[k_m]_\gamma$ and
    that of area $F[k_m]_A$ according to Eq. (10):

$$E = A_0 \frac{F[k_m]_\gamma}{F[k_m]_A} \qquad\qquad (19)$$

leading to Eq. (12). These steps provide an improved calculation
algorithm as compared to the classic DFT algorithm.

3. Estimation of measurement accuracy and optimal amplitude of area change

The quality of an experiment is typically estimated by a comparison
with a theoretical dependence and the target function of this comparison
is the mean square deviation

$$\xi_x = \sqrt{\frac{1}{N}\sum_{k=1}^{N}(x_{theor} - x_{exp})^2} \qquad\qquad (20)$$

In the present case x_{theor} corresponds to one of the theoretical
functions, calculated either by Eq. (7) or by Eq. (8), and their parameters
estimated as discussed in Section C.2. The x_{exp} are the respective
experimental dependencies on time, a series of data of either area or
interfacial tension. As a natural criterion of experimental quality a
dimensionless ratio $\lambda_x = \xi_x/\Delta x$ could be proposed. However, the
oscillation amplitude Δx must be small as it should satisfy the linearity
condition, which is a disadvantage for this purpose.

An interesting way to estimate the degree of system linearity was
proposed in [9]. Note, the Fourier Transform yields typically the basic
frequency and a number of higher harmonics, often multiples of this
basic frequency. The suggested criterion is called Total Harmonic
Distortion (THD). It is determined as the ratio of the higher harmonics
amplitude to the amplitude at the measured basic frequency

$$THD = \frac{\sqrt{\sum_{k=2}^{n} a_k^{\,2}}}{a_1} \qquad\qquad (21)$$

where a_1 is the amplitude value of the basic frequency, a_2, a_3, ...,a_n are
amplitude values of the higher harmonics. Here all frequencies are sorted
in order of decreasing amplitude, so that a_1 is the maximum amplitude.
Usually n is at least five.

The value of THD is considered in [9] as a criterion of linearity, i.e. the lower THD the better is the system's linearity. However, the amplitude of area changes must not be too high due to Eq. (6), as this can be a reason for non-linear effects, resulting in higher order harmonics, and consequently, in high values of THD. The amplitude of area changes must also not be too low as the signal to noise ratio should be kept sufficiently high for the calculation of the amplitude with an acceptable accuracy. It follows from Eq. (21), that a decrease in amplitude of the basic frequency leads to an increase in THD if the numerator in Eq. (21) is assumed to be constant. Therefore, there is some optimum amplitude of area changes, or equivalently a ratio $\Delta A/A$. In our analysis it was found to be of the order of 7-8%, close to the value 10% as proposed in [9].

4. Factors influencing the measurement accuracy

Experimental errors may significantly influence the results of calculated oscillation parameters, especially of ϕ. The calculation algorithm is based upon the discrete Fourier transform which has two main requirements:
1. The sampling interval, i.e. time interval between measurements, should be constant. It is advantageous when the sampling frequency is a multiple of the basic frequency $f=\omega/2\pi$.
2. The data to be transformed should contain multiples of the fundamental frequency ω.

An extensive discussion of these two questions is presented in [4] where the components of DFT are computed from Eq. (22) with a slightly modified multiplier

$$G[k] = \frac{2\Delta}{T_{exp}} \sum_{n=0}^{N-1} x_n \exp\left(-i\frac{2\pi nk}{N}\right) \tag{22}$$

which is equivalent to

$$G[k] = \frac{2}{N} \sum_{n=0}^{N-1} x_n \exp\left(-i\frac{2\pi nk}{N}\right) \tag{23}$$

because $T_{exp} = N\Delta$. Note, the multiplier $2/N$ in Eq. (23) is the same as in Eqs. (18) for the estimation of the amplitude.

DFT requires that the sampling interval Δ would be constant during an experiment. But this requirement is not easily met under real experimental conditions. In oscillating drop experiments based on drop shape unequal intervals were observed at frequencies of 0.1 Hz or higher.

At such high frequencies the interval between two points should be set at a minimum, normally about 0.2 s, which is technically at the limits.

Due to this reason, there is a scattering of time intervals in some experiments performed at high frequencies. In such cases there are at least two methods for data handling. The first, proposed in [4], is to replace the multiplier Δ in Eq. (22) by the correct time interval as multiplier Δ_n in each term, leading to the following expression

$$G[k] = \frac{2}{T_{\exp}} \sum_{n=0}^{N-1} \Delta_n x_n \exp\left(-i\frac{2\pi nk}{N}\right) \tag{24}$$

where this correct time interval is defined by $\Delta_n = t_{n+1} - t_n$. Also the condition of periodicity leads to $\Delta_{N-1} = t_N - t_{N-1} = T_{\exp} - t_{N-1}$.

As a second way to solve this problem we propose to take values x_n at exactly the same intervals applying a procedure of moving polynomial approximation. This procedure takes a small number of experimental points N_{app} so that the interested time is nearly the centre of the approximation interval (except some first and last points), builds the best fitting polynomial in a form similar to Eq. (11) $x(t) = \sum_{k=0}^{n} a_k t^k$ and then estimates value of x (that is either A or γ) by the obtained polynomial at the desired time value. Then the approximation interval is shifted step by step down to the end of the whole measurement interval, repeating the smoothing procedure. The proposed smoothing procedure differs from a simple polynomial interpolation by a more intensive numerical calculation which is however of minor importance with modern fast computer technique. What is more important is that this procedure effectively removes noise out of data giving the possibility of more accurate calculations of oscillation parameters as it will be demonstrated below. Optimal smoothing parameters are: n=3...4, $N_{app} = (4...5) \times n$, while the best values depend slightly on $\omega\Delta$.

The procedure described above is a solution of the first question raised. There is a question on what happens with the spectrum of original data after the proposed smoothing procedure. Processing of various experimental data has shown that neither the fundamental frequency nor its amplitude change significantly. A combination of polynomial smoothing (*Smooth*) and DFT (*Fourier*) will be discussed later in Section C.1.

The second question, i.e. $T_{\exp}/T$ is not an integer number, means that the first and last part of measured data cannot be linked together without visual discontinuity. This is a violation of the condition of periodic continuation needed for the DFT. One of the possible solutions of the problem is to truncate the measured data such that T_{exp}/T is an integer, which however, decreases the frequency resolution due to the definition of the uncertain product $\Delta\omega T_{exp} = 2\pi$. The shorter the time we are allowed to inspect oscillations, the larger is our uncertainty regarding its true frequency. However, accurate knowledge of ω is important for the calculation of the dilatation viscosity in Eq. (1), and the characteristic frequency and limiting elasticity, if the applied frequency is unknown or known with high error.

To adjust oscillation parameters an optimisation is proposed by finding the best match between experimental data and theoretical oscillation curves as defined by Eq. (16) via minimization of the target function (20), for example using Marquardt method [10]. This optimization requires good initial parameters, so that the Fourier based algorithm yields almost optimal parameters. Only the phase shift can be far from the optimum value due to non-constancy of sample intervals. A combined technique consisting of DFT followed by non-linear adjustment process (*Adjust*) will be discussed later in Section C.1.

C. CALCULATION OF CHARACTERISTIC FREQUENCY AND LIMITING ELASTICITY

The dependence of the visco-elasticity on oscillation frequency $E(i\omega)$ provides important information on the equilibrium and dynamic properties of a surface layer. Assuming only one surfactant in solution and diffusion controlled adsorption kinetics, this dependence was derived first by Lucassen and van den Tempel [11, 12] as (see also Chapter 2)

$$E(i\omega) = \frac{E_0}{1 + \sqrt{\omega_0/i\omega}},$$
(25)

with the limiting elasticity and characteristic frequency, respectively

$$E_0 = d\Pi/d\ln\Gamma \text{ and } \omega_0 = D\left(\frac{dc}{d\Gamma}\right)^2.$$
(26)

D is here the diffusion coefficient and c is the surfactant bulk concentration. There are different ways of extracting E_0 and ω_0 from $E(i\omega)$.

1. Calculation of E_0 and ω_0 at a selected frequency

Separating the real and imaginary parts of $E(i\omega)$ in Eq. (25)

$$E(i\omega) = E_r + iE_i \tag{27}$$

gives the components of E in an explicit form [11]

$$E_r = E_0 \frac{1+\zeta}{1+2\zeta+2\zeta^2} \tag{28}$$

$$E_i = E_0 \frac{\zeta}{1+2\zeta+2\zeta^2} \tag{29}$$

where

$$\zeta = \sqrt{\frac{\omega_0}{2\omega}} \tag{30}$$

Kovalchuk et al. [13] suggested a transformation of Eqs. (28) and (29) such that for each value of complex elasticity the corresponding values of E_0 and ω_0 can be obtained:

$$\tan(\phi) = \frac{E_i}{E_r} = \frac{\zeta}{1+\zeta} \tag{31}$$

From given values of E_r and E_i, it is possible to define E_0

$$E_0 = \frac{E_r^2 + E_i^2}{E_r - E_i} = \frac{|E|}{\cos(\phi) - \sin(\phi)} \tag{32}$$

and the expression for ω_0 follows

$$\omega_0 = 2\omega \left[\frac{1}{\cot(\phi)-1} \right]^2 . \tag{33}$$

In this way the values of E_0 and ω_0 can be obtained separately at each frequency. Note, however, due to their physical meaning these values have to be independent of frequency [13].

2. Logarithmic procedure of Lucassen & Giles

The original procedure of finding E_0 and ω_0 was proposed by Lucassen and Giles [14]. It follows from Eqs. (28), (29) and (31) that

$$\frac{|E(i\omega)|}{E_0} = \frac{1}{\sqrt{1+2\zeta+2\zeta^2}} \tag{34}$$

$$\phi = \arctan\left[\frac{\zeta}{1+\zeta}\right] \tag{35}$$

Eqs. (34) and (35) allow to plot the dependences of $|E|/E_0$ and ϕ on ω/ω_0. Lucassen and Giles plotted these dependences in a logarithmic scale:

$$\phi = f_1(\log(\omega/\omega_0)) \tag{36}$$

(see Fig. 2) and

$$\log(|E(i\omega)|/E_0) = f_2(\log(\omega/\omega_0)) \tag{37}$$

(see Fig. 3), and suggested to determine E_0 and ω_0 as a parameters of the best fit (found by the least squares method) to corresponding experimental dependences.

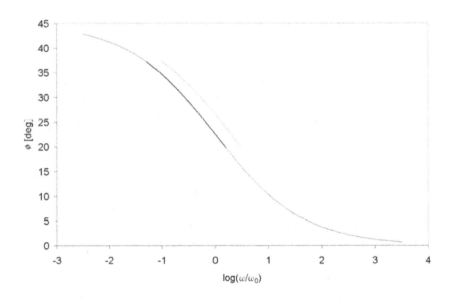

Fig. 2. Fitting of the theoretical dependence to "experimental" data $\phi = f_1(\log(\omega/\omega_0))$: 1 – "experiment" at ω_0=1 rad/s; 2 – theory ω_0=2 rad/s; 3 – "experiment" at ω_0=2 rad/s.

For example, in the Fig. 2 the "experimental" dependencies $\phi = f_{01}(\log(\omega))$ at $\omega_0=1$ rad/s and $\omega_0=2$ rad/s obtained by numerical simulations from Eq. (36), are shown. Evidently, $\omega_0= 2$ rad/s provides better fit than $\omega_0=1$ rad/s. After determination of the optimal ω_0, the dependence of $\log(|E|/E_0)$ on $\log(\omega/\omega_0)$ at some reasonable value of E_0 can be plotted. This finally gives an optimal value of E_0 (Fig. 3).

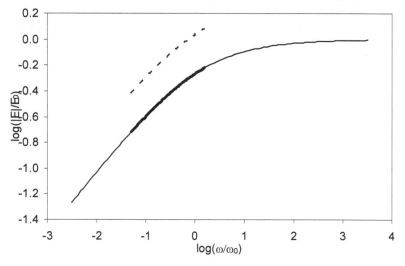

. Fig. 3. Fitting of theoretical to an "experimental" dependence of $\log(|E|/E_0)= f_2(\log(\omega/\omega_0))$: 1–"experiment" at $\omega_0 =2$ rad/s and $E_0=15$ mN/m; 2 - "experiment" at $\omega_0=2$ rad/s and $E_0=30$ mN/m; 3 – theory at $\omega_0=2$ rad/s and $E_0=30$ mN/m

3. Calculation of E_0 and ω_0 by regression in the real or complex domain

Eq. (25) can be presented in a form suitable for processing simultaneously a series of experiments with constant surfactant concentration i.e. when values E_0 and ω_0 should be constant, according to their physical meaning. To do this Eq. (32) is transformed as follows:

$$E_r^2 + E_i^2 = E_0(E_r - E_i).\tag{38}$$

The parameter E_0 can be seen as a factor k of a linear regression $y=kx$ in the coordinates $y = E_r^2 + E_i^2$ and $x = E_r - E_i$.

Eq. (33) can be solved in respect to ω:

$$\omega = \frac{1}{2}\omega_0 [\cot(\phi) - 1]^2 .$$

(39)

Similarly to E_0, the parameter ω_0 acts as a factor k of a linear regression $y=kx$ in the coordinates $y = \omega$, $x = \frac{1}{2}[\cot(\phi) - 1]^2$. But determination of E_0 and ω_0 is also possible from a regression equation in the complex domain. Let us introduce the variables $x \equiv \sqrt{i\omega}$, $y \equiv E(i\omega)$, and parameters $a \equiv E_0$ and $b \equiv \sqrt{\omega_0}$. Then their substitution into Eq. (25) results

$$y = \frac{a}{1 + b/x} \quad \text{or} \quad \frac{1}{y} = \frac{1 + b/x}{a} = \frac{b}{a}\frac{1}{x} + \frac{1}{a},$$

(40)

The values of E_0 and ω_0 are determined from parameters of a linear regression $y'=ax'+b$ in the coordinates $x' = 1/\sqrt{i\omega}$ and $y' = 1/E(i\omega)$.

The linear regression in the complex domain was proposed for example in [6]. As it is known parameters of a linear equation $y(x) = c x + d$ are estimated according to the least squares method using N points (x_i; y_i) as follows

$$c = \frac{S_x S_y - N S_{xy}}{S_x^2 - N S_{xx}}; \quad d = \frac{1}{N}(S_y - cS_x)$$

(41)

where

$$S_x = \sum_{k=1}^{N} x_k, \quad S_{xx} = \sum_{k=1}^{N} x_k^2, \quad S_y = \sum_{k=1}^{N} y_k, \quad S_{xy} = \sum_{k=1}^{N} x_k y_k$$

(42)

Usually, the values x_i, y_i, are assumed to be real, but they can also be complex [6]. However, as applied to the present case, only real values should be used, i.e. we take the real part of complex number for this purpose. According to Eqs. (40) and (41) we obtain:

$E_0 = \text{Re}(a) = \text{Re}(1/d)$

(43)

$\omega_0 = \text{Re}(b^2) = \text{Re}((c \times a)^2)$

(44)

It should be mentioned that the use of a linear regression for the determination of visco-elasticity parameters significantly improves the accuracy of calculation, but it makes sense only if three or more points per concentration are available.

D. RESULTS AND DISCUSSION

This paragraph is dedicated to the discussion of some examples demonstrating the applicability of the procedures proposed in Sections C.2 and C.3. The results on the visco-elasticity presented here are essentially made with the highly surface active surfactant $C_{14}EO_8$ using the advanced drop/bubble shape analysis tensiometer PAT-1 (SINTERFACE Technologies, Berlin) [15, 16].

1. Effect of polynomial smoothing

In Fig. 4 experimental and theoretical surface tension oscillation at a frequency of 0.1 Hz are shown. As mentioned above, this frequency requires a maximum data acquisition rate resulting in data points at non-equal distant times.

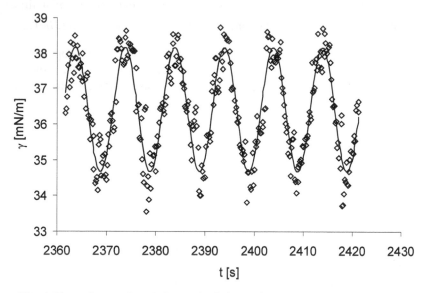

Fig.4. Experimental and theoretical dependences of surface tension oscillation at a frequency of 0.1 Hz: ◊ - experiment, solid line – theory.

In Fig. 5 the corresponding time intervals between two points each are shown for the data given in Fig. 4. As one can see there is a large scattering and the minimum and maximum intervals differ by more than a factor of two. Such a data set has to be subjected to a smoothing procedure in order to get an optimum in accuracy.

The experimental data shown in Fig. 4 have been selected on purpose to demonstrate the possibilities of different data processing techniques, presented in Table 1.

The smoothed curve obtained after the application of the moving polynomial approximation to the experimental dependence is shown in Fig. 6. As can be seen from Table 1 the application of the "Smooth" procedure only slightly changes Phase Shift and THD level. Moreover, application of the "Adjust" procedure does not change the oscillation parameters if the Smooth procedure was used before "Fourier".

For experiments having unevenly sampled intervals the results may differ more strongly. For example, the results shown in Table 2 are for such an experiment (see Fig. 7). Here, the minimum and maximum intervals differ by almost a factor of seven. Moreover, the distribution of intervals over the time range is not uniform. As a consequence, we observe a considerable change of the phase shift ϕ after application of smoothing.

It is seen from Table 2 that the "Smooth" procedure reduces the THD level by more than two times, while the "Adjust" procedure is necessary after "Fourier". These two examples allow the following conclusions:

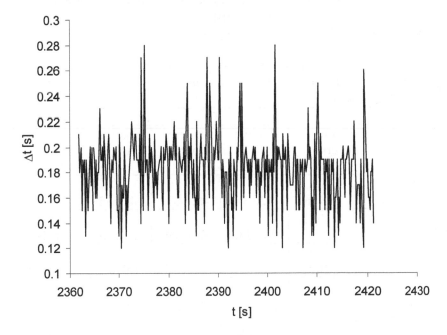

Fig.5. Dependence of interval between points on time for the experiment shown in Fig.4

Table 1. Comparison of different types of data processing for the experiment at external frequency 0.1 Hz (ω=0.62832 rad/s)

Oscillation parameters		Types of data processing			
		Fourier	Smooth+ Fourier	Fourier+ Adjust	Smooth+ Fourier+ Adjust
Average Value	γ_a, mN/m	36.417	36.424	36.418	36.424
Phase Shift	ϕ_γ, rad	0.3007	0.2668	0.2742	0.2668
Amplitude	$\Delta\gamma$, mN/m	1.74805	1.74212	1.74757	1.74212
Angular Frequency	ω, rad/s	0.62587	0.62587	0.62587	0.62587
Non-linearity	THD, %	15.950	13.570	Not defined	Not defined
Root mean square	ξ_γ, mN/m	0.50373	0.50277	0.50264	0.50277

Fourier – classic Fourier procedure as described above; **Smooth** – polynomial smoothing of data; **Adjust** – adjusting of oscillation parameters after Fourier procedure by non-linear optimization.

Table 2. Comparison of different types of data processing for an experiment with uneven time intervals at a frequency of 0.1Hz (ω=0.62832 rad/s)

Oscillation parameters		Types of data processing			
		Fourier	Smooth+ Fourier	Fourier+ Adjust	Smooth+ Fourier+ Adjust
Average Value	γ_a, mN/m	42.688	42.682	42.685	42.682
Phase Shift	ϕ_γ, rad	0.66838	-0.1006	0.10996	0.10065
Amplitude	$\Delta\gamma$, mN/m	1.81337	2.01255	1.99075	2.01255
Angular Frequency	ω, rad/s	0.621536	0.621536	0.621536	0.621536
Non-linearity	THD, %	48.95	21.46	Not defined	Not defined
Root mean square	ξ_γ, mN/m	1.02513	0.74917	0.69050	0.69072

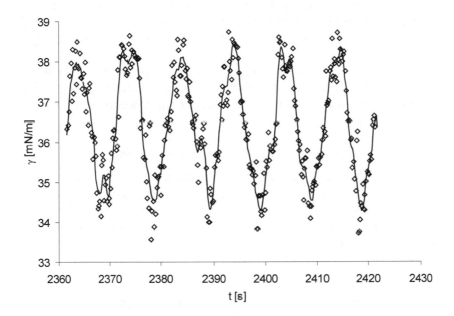

Fig. 6. Results of the moving polynomial approximation to experimental data given in Fig.4.: ◊ – experiment, solid line – smoothed curve.

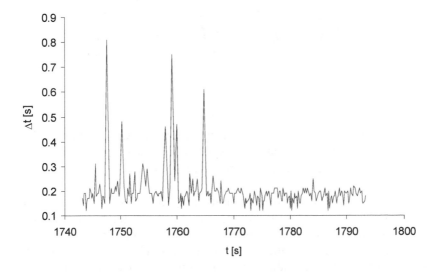

Fig.7. Distribution of time intervals between two points for the experiment with data acquisition at unevenly distributed time intervals

Application of the polynomial smoothing leads to oscillation parameters very close to the optimal values. The application of a final adjustment is desirable especially in the case of non-constant sample intervals.

Though the final adjustment gives best fit parameters, it can be strongly affected by some points far from theoretical curve. The "Smooth" procedure is designed to make the influence of such points as small as possible.

2. Influence of non-integer number of periods

The next example demonstrates the influence of a fractional number of periods of the initial data set. If the applied frequency is known no problem arises. For the example of drop area oscillations given in Fig. 8, which contains approximately 5.1 periods, it was tried to determine the fundamental frequency.

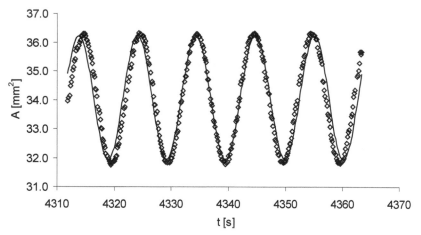

Fig. 8. Oscillation curve simulating an experiment with a number of points slightly more than an integer periods: ◊ – experiment, solid line – approximation curve.

The phase shift ϕ_A and root mean square ξ_A are most sensitive to non-integer number of periods: ϕ_A=0.403 rad, ξ_A=0.435 mm^2. Even a final adjustment did not allow to determine the optimal parameters. 10 trials were made when the length of the period was changed by 1/10. The length of period corresponding to the minimum error is then considered optimal. The results of calculation are shown in Fig. 9. Now we obtain ϕ_A= -0.119 rad, ξ_A=0.084 mm^2. As compared to the previous value, the root mean square decreased by more than 80% as confirmed by Fig. 9. Hence, truncation of data up to a maximum number of complete periods is strongly recommended. Note, all data were processed with the classic DFT algorithm plus improvements as described above.

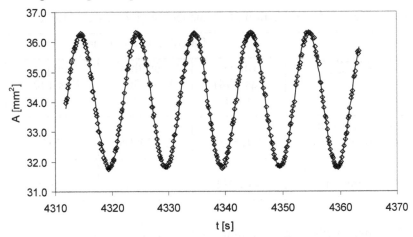

Fig. 9. Same as in Fig. 8 but simulated for an integer number of periods: ◇ – experiment, solid line – approximation curve.

3. Estimation of characteristic frequency and limiting elasticity

This example aims at showing the calculation of characteristic frequency and limiting elasticity by different methods. In Table 3 the results of oscillation experiments at six different frequencies for a 4 μmol/l $C_{14}EO_8$ solution are summarised.

Table 3. Results of oscillation experiments at six different frequencies

Number	ω, Hz	E_r, mN/m	E_i, mN/m	$|E|$, mN/m	ϕ, deg
1	0.200	37.3838	5.1918	37.7426	7.9066
2	0.100	35.3367	6.9424	36.0122	11.1151
3	0.050	31.0667	7.9199	32.0604	14.3019
4	0.020	26.4488	9.1376	27.9827	19.0591
5	0.010	21.7400	9.6940	23.8034	24.0323
6	0.005	17.0819	8.7525	19.1937	27.1299

The experiments were performed with a buoyant bubble at a temperature of 25 °C in a measuring cell with a volume of V = 20 ml. Using the data of Table 3 the characteristic frequency and limiting elasticity are calculated via three different algorithms (Table 4).

Table 4. Results of calculation of E_0 and ω_0 using the data of Table 3

Parameters of Eq. (25)	Calculation algorithm		
	Linear Eqs. (38)-(39)	Logarithmic Eqs. (36)-(37)	Complex Eqs. (40)-(44)
Characteristic frequency ω_0, rad/s	0.076533	0.070095	0.071762
Limiting elasticity E_0, mN/m	44.2911	47.0323	44.9651

The three methods give similar results. A more consistent comparison of the different methods for the calculation of E_0 and ω_0 can be made using experimental adsorption values, which will be presented in the next section.

4. Comparison of theory and experiments

From Eq. (26) it is possible to obtain the derivative $d\Gamma/dc$ from ω_0

$$\frac{d\Gamma}{dc} = \sqrt{D/\omega_0} \qquad (45)$$

Having a dependence of ω_0 on c it can be transformed into a dependence of $d\Gamma/dc$ on c. From an independent experimental or theoretical dependence $\Gamma(c)$ we can determine $d\Gamma/dc$ as a function of concentration as well. Thus, results of experiments, in which rheological and adsorption properties of surfactant are studied, can be compared with each other and with theoretical models. For this purpose, we compare the surface tension isotherms obtained by a drop shape and bubble shape method, to which a surfactant mass balance was applied to estimate the adsorbed amount of $C_{14}EO_8$ [17]. In Figs. 10 and 11 the dependences of the visco-elasticity modulus and phase angle, respectively, on the harmonic oscillations frequency at different $C_{14}EO_8$ concentration are shown. The comparison of theoretical and experimental values of $d\Gamma/dc$

for $C_{14}EO_8$ is shown in Fig. 12. The diffusion coefficient used in the calculations by Eq. (45) was 4×10^{-10} m^2/s.

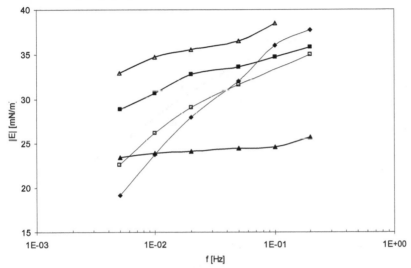

Fig. 10. Dependence of the surface dilational modulus on oscillations frequency of the bubble area for $C_{14}EO_8$ solutions at various concentrations: (▲) 0.1 μmol/l; (△) 0.4 μmol/l; (■) 1 μmol/l; (□) 2 μmol/l and (◆) 4 μmol/l; lines are guides for eyes.

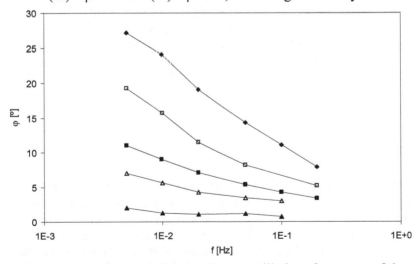

Fig. 11. Dependence of phase angle on oscillations frequency of the bubble area for the $C_{14}EO_8$ solutions at various concentrations; legend is the same as in Fig. 10.

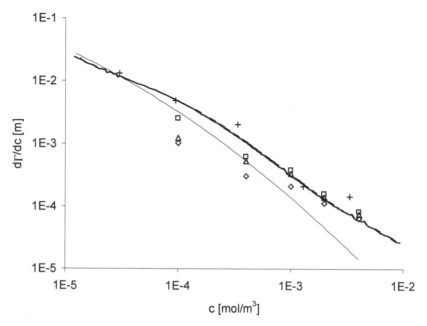

Fig.12. Comparison of theoretical and experimental values of $d\Gamma/dc$ for $C_{14}EO_8$ solutions. Data obtained from drop and bubble shape experiments by calculation ω_0 using different algorithms: $\square$ - linear, $\diamond$-logarithmic, $\triangle$-complex; (+) data obtained by numerical differentiation of experimental dependence $\Gamma(c)$ [17]; theoretical dependence $\Gamma(c)$ calculated via Frumkin model (solid line), or via a combined model assuming reorientation and intrinsic compressibility (thin line).

Numerical derivatives $d\Gamma/dc$ were obtained from data given in [17] by finite differences

$$\left[\frac{d\Gamma}{dc}\right]_i \approx \frac{\Gamma_i - \Gamma_{i-1}}{c_i - c_{i-1}} \tag{46}$$

These values were attributed to the concentrations in the centre of the interval $(c_i + c_{i-1})/2$. The theoretical dependencies $\Gamma(c)$ were calculated using the Frumkin model [18] and a combined reorientation / interfacial compressibility model [19, 20]. This last model assumes the ability of the oxyethylene group to be completely adsorbed at the solution/air interface at low surface coverage, and be partially desorbed with increasing surface pressure, i.e. reorientation of the EO groups. It also accounts for the internal compressibility of the monomolecular interfacial layer in the state with minimum molar area, i.e., variation in the orientation of the hydrocarbon chains. From Fig. 12 we can conclude that at high $C_{14}EO_8$ concentrations (above 1 µmol/l) all proposed calculation algorithms yield the same characteristic frequency ω_0 and $d\Gamma/dc$. At high concentrations the values of $d\Gamma/dc$ agree rather well with the proposed model, but at low concentrations differences between various algorithms and experimental values become distinguishable. The values of $d\Gamma/dc$ obtained at low concentrations using the rheological experiments lay a little bit below the values obtained by the direct method [17] (see Fig. 12). For a $C_{14}EO_8$ concentration of 0.1 µmol/l, very small phase angles, less than 5°, are observed. For such small phase angles the experimental error (close to 1°) can strongly influence accuracy of calculation of ω_0 and therefore the accuracy of $d\Gamma/dc$ by Eq. (45). This may be one of the reasons for the lower values of $d\Gamma/dc$ at concentration below 1 µmol/l as observed in Fig. 12. In any case, the linear algorithm shows the best agreement with the given experiments in all ranges of $C_{14}EO_8$ concentrations.

E. CONCLUSIONS

The discrete Fourier transform was used for the determination of the complex dilational elasticity modulus of adsorption layers. As applied to harmonic relaxation experiments it enables to calculate all parameters. It is known that the DFT provides good results if certain conditions are met. Preliminary smoothing of data with the moving polynomial approximation technique was proposed to solve the problem of unevenly sampled data. If the data set does not cover a multiple of the period of the fundamental frequency the appropriate truncation and non-linear optimization allows to adjust the oscillation parameters correctly. The experimental dependences of the visco-elasticity modulus and phase angle on the harmonic oscillations frequency for $C_{14}EO_8$ solutions are used to demonstrate different algorithms for estimation of the characteristic frequency and limiting elasticity.

ACKNOWLEDGEMENTS

The work was financially supported by projects of the German Space Agency (DLR) and the DFG SPP 1273 (Mi418/16).

F. REFERENCES

1. Loglio, G., Tesei, U., and Cini, R., J. Colloid Interface Sci. 71(1979)316.
2. B.A. Noskov, D.A. Alexandrov, R. Miller, J. Colloid Interface Sci. 219(1999) 250.
3. G. Kretzschmar and K. Lunkenheimer, Ber. Bunsenges. Phys. Chem. 74 (1970) 1064-1071.
4. D. Möbius and R. Miller (Eds.), Novel Methods to Study Interfacial Layers, in "Studies in Interface Science", Vol. 11, Elsevier Science, Amsterdam, 2001
5. R.M. Bracewell, The Fourier Transform and Its Applications, McGrow-Hill, New-York, 1965.
6. S.L. Marple, Digital spectral analysis with applications. Englewood Cliffs; Prentice-Hall. 1987.
7. S.L. Marple. Spectral Line Analysis by Pisarenko and Prony methods, Proceedings of the 1979 International Conference on Acoustics, Speech and Signal Processing, Washington D.C., pp.159-161
8. W.H. Press, G.B. Rybicki, Astrophysical Journal, 338(1989)277
9. G. Loglio, P. Pandolfini, R. Miller, A. Makievski, J. Krägel, and F. Ravera, Phys. Chem. Chem. Phys. 6(2004)1375.
10. D.W. Marquardt, J. Soc. Ind. & Appl. Math. 11(1963)431–441.
11. J. Lucassen, M. van den Tempel, J. Colloid Interface Sci. 41 (1972) 491.
12. J. Lucassen, M. van den Tempel, Chem. Eng. Sci. 27 (1972) 1283.
13. V.I. Kovalchuk, G. Loglio, V.B. Fainerman and R. Miller, J. Colloid Interface Sci., 270 (2004) 475.
14. J. Lucassen, D. Giles, J. Chem. Soc., Faraday Trans., 71(1975)217
15. S.A. Zholob, A.V. Makievski, R. Miller, V.B. Fainerman, Adv. Colloid Interface Sci., 134(2007)322.
16. R. Miller, C. Olak, A.V. Makievski, SÖFW-Journal, 130 (2004) 2.
17. V.B. Fainerman , S.A. Zholob , J.T. Petkov, R. Miller, Colloids Surf A, (2008), doi:10.1016/j.colsurfa.2007.09.019
18. A. Frumkin, Z. Phys. Chem. Leipzig, 116 (1925) 466-485
19. V.B. Fainerman, V.I. Kovalchuk, E.V. Aksenenko, M. Michel, M.E. Leser and R. Miller, J. Phys. Chem. B, 108 (2004) 13700-13705.
20. V.I. Kovalchuk, R. Miller, V.B. Fainerman and G. Loglio, Adv. Colloid Interface Sci., 114-115 (2005) 303-313.

CAPILLARY WAVES IN INTERFACIAL RHEOLOGY

Boris A. Noskov

St. Petersburg State University, Chemical Faculty, Universitetsky pr. 26,
198504 St. Petersburg, Russia

Contents

A. INTRODUCTION

The idea to apply capillary waves to studying surface rheological properties seems to be simple and natural. In the case of bulk liquids the determination of the dynamic compressibility and viscosity of compression (expansion) is mainly based on the application of another kind of mechanical waves – ultrasound waves [1-3]. The corresponding experimental methods are well-established in the physical chemistry [2, 3]. One can assume that surface waves, first of all capillary waves, have to play an analogous role in surface science of liquid systems. The influence of surfactants on the surface wave properties is indeed a well-known phenomenon, which reveals the possibility to extract information on the surface rheological properties from the characteristics of surface waves.

The realisation of this idea, however, proved to be difficult. It looks surprising but the first experimental studies, where the authors tried to determine the surface rheological properties from measurements of the damping and length of capillary waves, appeared only in the sixties of the past century [4-10]. Nowadays capillary waves are widely used in surface rheological studies. At the same time the number of systems, where the application of capillary waves leads to a detailed description of surface dilational rheology, is limited. Moreover, some observed effects still require interpretation.

The surfactant influence on the surface wave properties is probably one of the oldest scientific problems which have been ever discussed in the literature. The first descriptions of capillary waves arising when drops of oil fall onto the water surface were written in clay with a wedged shape stylus and date back to the 18th century B.C. [11]. Several ancient authors, Aristotle [12], Plutarch [13], Plinius the Elder [14] among them, described the attenuation of water waves by thin oil films ("slicks"). In closer time the influence of surfactant on water waves was a subject of a famous article by Benjamin Franklin, who did some experiments and proposed the first explanation of the observed effect [15]. Franklin thought that the oil film acted as a lubricant between water and moving air, and did not allow the wind to excite ripples (capillary waves). Many authors repeated later the experiments of Franklin [16-18] but the theory of these phenomena was not developed for more than 150 years. The difficulties were not connected with only fortuitous causes.

Unlike the case of acoustic waves, one can describe the propagation of surface waves in liquids by means of a simple wave equation only in few exceptional limit cases. In the general case one has to solve the boundary problem for the equations of liquid dynamics taking into account the bulk liquid viscosity. The effect of surfactants manifests itself in the mathematical description of the liquid motion mainly through a non-trivial change of the boundary conditions.

Capillary waves were frequently used to measure the surface tension of pure liquids [19-21] but the attempt to apply this method to surfactant solutions led to unexpected difficulties probably because of an inadequate theory [22, 23]. The foundations of the modern theory of surface wave damping by adsorbed and spread surfactant films were formulated only in the middle of the twentieth century. The surface elasticity was taken into account for the first time by Lamb who considered only the limit of inextensible surface films [24]. Levich proposed a more general theory and showed that one had to take into account tangential forces at the liquid surface in the boundary conditions [25]. In other words, the attenuation of capillary waves by surfactants is caused by the Plateau – Gibbs – Marangoni effect and connected with the dynamic surface elasticity. All subsequent theories of these phenomena are based on the same main ideas [26-32].

The results of Levich were independently obtained by Dorrestein who also predicted a local maximum of the damping coefficient as a function of the surface elasticity [26]. Other authors developed a more elaborated theory of capillary wave damping under the influence of surfactants and elucidated interrelations between various surface rheological properties and the capillary wave characteristics [27-32]. An important result was the prediction of longitudinal capillary waves [33], which followed almost immediately by their experimental discovery [34].

The power of the capillary wave method in experimental studies of surface rheology became clear already at the end of the sixties. The development of the corresponding experimental technique relates mainly to the subsequent decades. The advent of lasers allowed the determination of the spectral broadening of light scattered by thermal surface fluctuations (spontaneous capillary waves, surface ripplons) [35]. The spectral shape is connected directly with the damping coefficient of capillary waves and consequently with the surface rheological properties [36]. The main advantages of the surface quasielastic light scattering method (SQELS) consist in the non-invasive probing of the liquid surface and in the high upper limit of possible frequencies (up to about 2 MHz [37]). Various versions of the capillary wave method have been applied to surface films of different chemical nature and these results are described in several reviews [30, 36, 38, 39], the last of which published about 15 years ago. For the last years surface rheological studies by the capillary wave method really did not develop as fast as one could have been expected thirty or twenty years ago. The arising difficulties are connected with two main reasons. Firstly, the sensitivity of the damping coefficient of capillary waves to the surface rheological properties depends strongly on the absolute value of the complex dynamic surface elasticity E and the surface tension γ. The numerical results show that the sensitivity and, consequently, the precision of the surface elasticity determination by means of transverse capillary waves is low if $|E| > 0.16\gamma$ [40], which restricts seriously the applicability of this experimental method. The longitudinal capillary waves are sufficiently sensitive to the surface rheological properties almost in the whole range of their values with a minor exception ($|E|<1$). However, the high damping of longitudinal waves limits from above the frequency range of their applicability by a few hertz. Secondly, some results of SQELS led to negative dilational surface viscosities [40-52] in contradiction with the second law of thermodynamics [53]. The dynamic surface elasticity at low frequencies sometimes exceeds its value at high frequency which also contradicts the basic principles of rheology [54]. The observed discrepancies with expected values probably indicate that the dispersion equation of capillary waves does not take into account all possible effects [50, 53, 55]. This problem has been discussed in literature and some improvements have been proposed [54, 55, 56] but they do not allow us to exclude all the discrepancies [48-52, 54]. Moreover, the negative surface viscosity was also discovered for some systems by means of externally excited capillary waves at low frequencies [57].

Note that SQELS leads to a negative surface viscosity only for a limited number of surface films at the liquid-gas interface in a limited range of parameters. There are a lot of systems where the application of the capillary wave method leads to a reasonable agreement with the theory and the data of other experimental methods. This chapter contains a review of some recent results. The theory of capillary waves in a linear approximation and the existing problems are also briefly discussed. The damping and propagation of non-linear capillary waves [58, 59] is beyond the scope of this work.

B. THEORY OF CAPILLARY WAVES

1. Solution of hydrodynamic equations

The liquid motion in the case of surface waves propagating along the liquid-gas interface is described by the Navier – Stokes and continuity equations. If the liquid is isotropic, incompressible and Newtonian, these equations have the following form [1]:

$$\rho\left[\frac{\partial \vec{v}}{\partial t} + (\vec{v} \cdot \vec{\nabla})\vec{v}\right] = -\vec{\nabla}p + \mu\Delta\vec{v} + \rho\vec{g} \tag{1}$$

$$\vec{\nabla} \cdot \vec{v} = 0 \tag{2}$$

where $\vec{v}$ is the vector of liquid velocity, t is the time, p is the pressure, $\vec{g}$ is the gravitational acceleration, μ is the liquid viscosity, ρ is the liquid density.

In the general case the surfactants influence not only the bulk liquid properties ρ and μ but also the boundary conditions for Eqs. (1) and (2). On the other hand, the surfactant distribution in the solution bulk and at the surface depends on the liquid velocity. Therefore, one must solve the complex general boundary problem for the system of equations (Eqs. (1), (2) and the equation of convective diffusion), which cannot be decoupled into a separate hydrodynamic and diffusion problem. However, this does not lead to serious complications in the case of linear waves, when the wavelength λ exceeds significantly the wave amplitude A (A/λ « 1) and it is possible to neglect the second non-linear term in the square brackets in Eq. (1). In this case the boundary conditions influence only the parameters of the well-known wave solution of the remaining linear equations. Another important simplification arises if the viscosity is low. Then one can neglect the second term on the right-hand side of Eq. (1) and introduce the potential of velocity φ:

$$\vec{v} = \vec{grad}\varphi \tag{3}$$

This allows reducing the hydrodynamic equations to the Laplace equation [1]:

$$\Delta\varphi = 0 \tag{4}$$

This equation has a special solution corresponding to surface waves. In Cartesian coordinates x,y,z, where the x,y coordinate plane coincides with the interface, the harmonical surface wave travelling in the positive direction of x-axis takes the form of

$$\varphi = Ae^{kz}e^{i(kx-\omega t)} \tag{5}$$

where $k = 2\pi/\lambda$ is the wave number and ω is the angular frequency. The first exponential factor ensures the attenuation of the liquid motion with increasing distance to the interface ($z \rightarrow -\infty$).

Then it follows from relations (1) and (5):

$$p = -i\rho\omega\varphi - \rho gz \tag{6}$$

The substitution of (5) and (6) into the linearised boundary condition at the liquid-gas interface leads to the dispersion equation for surface waves in ideal liquid (the Kelvin equation):

$$\rho\omega^2 = \rho gk + k^3\gamma \tag{7}$$

where γ is the surface tension.

The two limit cases $k \rightarrow 0$ or $k \rightarrow \infty$, when one can neglect the second or the first term in the right-hand side, correspond to gravitational and capillary waves, respectively.

If ω is given, it follows from the Kelvin equation that k is a real number and the waves propagate without damping. This is natural because Eq. (7) assumes an ideal liquid. However, one can use the Kelvin equation as rough estimation of the wave number even for real liquids with surfactants [30, 38, 39, 60]. In order to describe the wave damping all terms must be kept in the right-hand side of Eq. (1). The contribution of low viscosity ($\mu \ll \omega\rho/k^2$) distorts only slightly the liquid motion in the surface wave [60]. Therefore one can seek the solution in the form of $\vec{v} = \vec{v}_1 + \vec{v}_2$, where $\vec{v}_1$ is described by relation (5) and $\vec{v}_2$ is a small correction. Then $\vec{v}_2$ satisfies the following equation:

$$\rho\frac{\partial\vec{v}_2}{\partial t} = \mu\Delta\vec{v}_2 \tag{8}$$

It follows from Eq. (2) that one can introduce the stream function ψ:

$$v_{2x} = -\frac{\partial\psi}{\partial z}; v_{2z} = \frac{\partial\psi}{\partial x} \tag{9}$$

Then Eq. (8) reduces to

$$\rho\frac{\partial\psi}{\partial t} = \mu\Delta\psi \tag{10}$$

which has a wave solution:

$$\psi = Be^{mz}e^{i(kx-wt)} \tag{11}$$

with $m^2 = k^2 - i\omega\rho/\mu$ and Real(m) > 0, B is a constant.

Relations (5) and (11) describe linear surface waves in the general case. All parameters in these equations depend on both bulk and surface properties of the liquid. One has to apply the boundary conditions at the liquid-gas interface to determine these dependencies and to find the general dispersion equation.

2. Boundary conditions

The assumptions of the preceding subsection are rather general and the applicability of solution (5), (11) have never been questioned in the case of linear waves, however, the boundary conditions to equations (1), (2) are widely discussed in literature [25-32, 38, 39, 54-56, 60-63]. It is just due to the boundary conditions that surface rheology comes into play. The solution (5), (11) contains two constant amplitudes and, consequently, it is necessary to use two conditions at the liquid-gas interface. These relations follow from the balance of normal and tangential forces at the moving curved interface.

The first condition is a generalisation of the well-known Laplace equation for the equilibrium capillary pressure where the equilibrium liquid pressure at the interface must be replaced by the corresponding component of the stress tensor in the flowing liquid σ_l:

$$\sigma_g - \sigma_l = \gamma\left(\frac{1}{R_1} + \frac{1}{R_2}\right) \tag{12}$$

where R_1 and R_2 are the principal radii of curvature and the normal component of the stress tensor in the gas phase σ_g is equal to the gas pressure with the opposite sign $-p_g$.

Eq. (12) can be significantly simplified if the surface waves are linear and planar (the front of waves is parallel to the x-axis). Then σ_l equals approximately the component σ_{zz} of the stress tensor in the coordinate system of the preceding subsection. Besides, $R_2 = 0$ and $1/R_1 \approx -\partial^2\zeta/\partial x^2$, where $\zeta(x,y)$ is the displacement of the liquid surface from the unperturbed level z = 0 [38]. Therefore on can write the following boundary condition at the liquid surface for the linearised hydrodynamic equations:

$$-\left(-p_l + 2\mu\frac{\partial \vec{v}_z}{\partial z}\right)_\zeta - p_g + \gamma\frac{\partial^2 \zeta}{\partial x^2} = 0 \tag{13}$$

where the subscript ζ indicates that the expression in brackets must be estimated at $z = \zeta$ and p_l is the liquid pressure.

It is just the condition (13) that has been intensely discussed in literature [32, 38, 39, 54, 56, 61-63]. The main problem is the meaning of γ in this equation. It is well known that in non-equilibrium systems the surface tension is not a scalar quantity but a tensor [64]. For plane surface waves one should use γ_{xx} instead of γ in the relations (12) and (13). However, in the approximation of linear waves the difference between these two quantities leads only to an infinitesimal contribution of higher order in the boundary condition. Consequently, in this approximation γ in relation (13) is the static surface tension.

Another problem arises as a consequence of the appearing surface rheological properties in the boundary conditions. The propagation of capillary waves leads to the distortion of the initially flat liquid surface. Consequently, in the general case it is necessary to take into account not only the usual two-dimensional surface rheological properties but also the rheology of surface bending. Goodrich and Kramer assumed that surface bending in the linear approximation could be taken into account by a single rheological parameter – the transverse surface viscosity μ'_N [32, 62, 63]. In this case the surface tension proves to be a complex quantity in relations (12) and (13):

$$\gamma = \gamma' + i\omega\mu'_N \tag{14}$$

Many authors tried to determine μ'_N experimentally for various systems [40, 44-47, 65 -68]. Typical values were of the order of 10^{-4} mN·s·m^{-1} or less. Therefore this quantity is negligible at least in the range of low-frequency capillary waves (< 1 kHz). Recently Buzza has shown that the rigorous boundary condition for the normal forces at the liquid-gas interface does not contain the transverse surface viscosity if the adsorption layer is almost two-dimensional (the surface layer thickness is much less than λ) and the surface tension does not take ultralow values [54]. One should take into account the bending modulus κ and its viscous counterpart (the bending viscosity) only if these conditions are violated at $(\kappa/\gamma)^{1/2} \geq \lambda$. At the air-liquid interface γ is finite and the bending rheological properties can be important only in rare extreme cases of macroscopically thick adsorption films. Therefore, in real cases it is necessary to use the relation (12), where γ is the static surface tension, as a boundary condition for hydrodynamic equations.

The second boundary condition is the balance of tangential forces at the interface. If the system does not contain surfactants, the tangential forces acting on a surface element from the liquid and gas phase respectively must compensate each other. The gas viscosity is negligible and this leads to the following condition in the linear approximation

$$\sigma_{xz} = -\mu\left(\frac{\partial v_x}{\partial z} + \frac{\partial v_z}{\partial x}\right)_\varsigma = 0 \qquad (15)$$

The application of Eq. (15) together with Eq. (13) allows the determination of the complex wave number $k = 2\pi/\lambda + i\alpha$, where α is the damping coefficient. Calculations of α reasonably agreed with experimental data for pure liquids but not with the results for surfactant solutions. This puzzle had no answer until the work of Levich [25], who assumed the existence of a special tangential surface force for surfactant solutions with a non-uniform distribution of the liquid velocity at the interface (Marangoni effect). This tangential surface force equals the surface tension gradient. Levich assumed that the surface tension depends only on the adsorbed amount Γ. Then this force per unit area has the following form for planar surface waves [60]:

$$\frac{\partial \gamma}{\partial x} = \frac{\partial \gamma(\Gamma)}{\partial \Gamma}\frac{\partial \Gamma}{\partial x} \qquad (16)$$

However, in the general case the surface tension can depend not only on the number of adsorbed molecules but also on their orientation or conformation. Besides, one can use γ_{xx} instead of γ in relation (16) for planar surface waves. In the linear approximation the tensor of surface tension must depend only on the relative surface deformation $d\xi/dx$ and its rate $d^2\xi/dxdt$ where ξ is the tangential surface displacement. These expressions assume an unidirectional surface displacement and a flat interface.

The application of Kelvin's rheological model (cf. Chapter 1 of this book) leads to the following expression for γ_{xx}:

$$\gamma_{xx} = \gamma + (E+G)\frac{d\xi}{dx} + (\eta_d + \eta)\frac{d^2\xi}{dxdt} \qquad (17)$$

where E is the dilational surface elasticity, G is the shear surface elasticity, η_d is the dilational surface viscosity, and η is the shear surface viscosity.

Note that Eq. (17) contains both dilational and shear surface rheological properties because of the unidirectional deformation. The use of complex numbers allows rewriting (17) in a simpler form in the case of harmonical perturbations with the angular frequency ω:

$$\gamma_{xx} = \gamma + \varepsilon(\omega)\frac{d\xi}{dt} \qquad (18)$$

where $\varepsilon = E + G - i\omega(\eta_d + \eta)$ is the complex dynamic longitudinal surface elasticity.

It is noteworthy that expression (18) is not necessarily connected with Kelvin's rheological model. It is more general than expression (17) and valid in the case of an arbitrary surface rheological model. The use of (18) leads to the following boundary condition for the hydrodynamic equations (1) and (2) [30, 38]:

$$-\mu\left(\frac{\partial v_x}{\partial z} + \frac{\partial v_z}{\partial x}\right)_\varsigma + \varepsilon\frac{d^2\xi}{dx^2} = 0$$

$$(19)$$

The two conditions (13) and (19) are sufficient for the solution of the hydrodynamic boundary problem.

3. Dispersion equation

The substitution of solution (5), (6), (11) into relations (13), (19) leads to two algebraic homogeneous equations relative to the two amplitudes A and B. The system of homogeneous algebraic equations has a non-trivial solution only if the determinant of the system equals zero. This condition gives the dispersion relation of surface waves [38]:

$$\left(\rho\omega^2 - \gamma k^3 - \rho g k\right)\left(\rho\omega^2 - mk^2\varepsilon\right) - \varepsilon k^3\left(\gamma k^3 + \rho g k\right)$$
$$+ 4i\rho\mu\omega^3 k^2 + 4\mu^2\omega^2 k^3(m - k) = 0 \qquad (20)$$

This complex relation connects the wave characteristics (k, ω) with the properties of the bulk phase (ρ, μ) and the surface properties (γ, ε). If ω is given, one can consider (20) as an algebraic equation relative to k. At least two roots of this equation have physical meaning. One can obtain a rough approximation of the first root by equating to zero the expression in the first brackets in (20). This leads to the Kelvin relation (7). Consequently, the first root corresponds to the usual transverse surface waves and Eq. (7) gives the first approximation for the real part of the wave number (the wavelength) in the general case. In order to find more exact values of the real part and also the imaginary part of the wave number (the damping coefficient), it is necessary to use the complete dispersion relation (20) and to apply numerical methods.

One can obtain an approximate value of the second root by equating to zero the expression in the second brackets of (20):

$$\rho\omega^2 - mk^2\varepsilon = 0 \tag{21}$$

In this case, unlike the transverse waves, liquid particles move along strongly stretched elliptical trajectories. The shape of the liquid surface deviates only slightly from a plane and therefore Eq. (21) is an approximate dispersion equation for longitudinal capillary waves. This wave mode was discovered by Lucassen [33, 34] and also applied to the determination of surface rheological properties together with transverse capillary waves [69-79].

Note that the capillary wave methods allow determination of only the sum of shear and dilational surface rheological properties (cf. Eq. (17)). However, in the case of solutions of conventional surfactants the dilational contributions to the longitudinal dynamic surface elasticity are several orders of magnitude larger than the shear ones [70]. This is also true for most of the polymer systems (Chapter 6) and, consequently, the experimental methods based on capillary waves give information mainly on the dilational surface rheology.

The application of transverse waves is more traditional in surface rheological studies [4-10, 27-32, 36-52, 57, 59, 65-67]. The damping coefficient of transverse capillary waves is a non-monotonic function of the surface elasticity modulus. It increases strongly with the increase of the modulus and at $|\varepsilon| \approx 0.16\gamma$ it reaches a maximum value. Any further increase of the surface elasticity leads to a decrease of damping, which tends asymptotically to the value [60]:

$$\alpha \approx \frac{1}{2\sqrt{2}} \frac{\mu k^{5/4}}{\rho^{3/4}\gamma^{1/4}} \tag{22}$$

The real part of the wave number depends only slightly on the surface elasticity and corrections to the results of calculations by Eq. (7) are of the order of $\leq 10\%$. Therefore the sensitivity of the damping coefficient of transverse capillary waves to the surface elasticity decreases strongly with the increase of the surface elasticity at $|\varepsilon| > 0.16\gamma$ and the accuracy of the corresponding experimental method in this range is low.

After the discovery of longitudinal surface waves, it was found numerically that a pit in the damping of these waves occurred concurrently with the peak in the damping coefficient of transverse capillary waves [80, 81]. The peak and pit correspond to the conditions when the phase velocities of the two wave modes coincide. Theoretical studies showed that these phenomena were a consequence of the resonant interactions between longitudinal and transverse capillary waves [82, 83]. Relation (21) shows that both the wavelength and the damping coefficient of longitudinal waves depend strongly on the dynamic surface elasticity. It is the interaction between these two wave modes that leads to the influence of the surface rheological properties on the characteristics of transverse capillary waves. Another phenomenon closely connected with the resonance between longitudinal and transverse waves is the mode mixing [82, 84]. For a certain combination of surface rheological properties the two wave modes can be either completely transverse or longitudinal. The continual change of the rheological properties leads to the two modes adopting the opposite wave mode behaviour. The range of the parameters where this change occurs coincides with the resonance conditions.

The theory briefly discussed above allows the calculation of the length of capillary waves with high precision (better than 1 %) [38]. Calculations of the damping coefficient are more difficult. Stone and Rice discovered deviations of about 1% from experimental data even for pure water [85].

The lack of information on some surface rheological properties makes a direct comparison of theoretical and experimental results difficult for adsorbed and spread of surfactant films. However, measurements of the surface wave characteristics are one of the main sources of information on the surface dilational viscosity and the only one at high frequencies ($\geq$ 10 kHz). The discovery of the negative surface dilational viscosity in surfactant solutions was connected at first with the destabilisation of surface longitudinal waves at the expense of the adsorption barrier [40-42]. The adsorption barrier can really lead to a negative viscosity but only at high deviations from equilibrium [53, 86-89]. It follows from the second law of thermodynamics that the surface viscosity must be positive at small deviations from equilibrium [53]. Note that although the surface viscosity is an excessive quantity [63] it does not influence this conclusion because the viscosity of a surfactant monolayer usually exceeds by orders of magnitude the viscosity of a water monolayer. The negative surface viscosity of insoluble monolayers was discovered a little later [44, 46]. These results probably indicate that the negative surface viscosity are an artefact connected with the inadequacy of the applied dispersion relation [53].

In order to improve the theory Hennenberg et al. proposed to take into account the convective term in the diffusion equation [55]. If there is a steady-state surfactant transfer through the interface, this term influences the dynamic surface elasticity and, consequently, the dispersion relation [53, 88, 90]. However, if the deviations from equilibrium are small, the convective term is a small quantity of the second order and can be neglected [53, 88]. The question arises what is the reason of the steady-state mass transfer in the studies [40-43, 50, 51, 57]. Buzza showed that the negative surface viscosity could appear as a result of erroneous application of the dispersion relation (cf. subsection B.2) [54]. If one substitutes the equilibrium surface tension in Eq. (20) and uses the condition $\mu_N' = 0$, the surface dilational elasticity becomes positive and the real part of the dynamic surface elasticity exceeds the static value for most of the systems [54]. Nevertheless, the application of SQUEL to surface films of block copolymers led to negative surface viscosities even when Buzza's approach had been used [51, 52]. The application of low-frequency waves also leads sometimes to an anomalous damping of transverse capillary waves [57, 91]. These results indicate that Eq. (20) probably does not take into account all possible effects at the air-water interface and makes necessary further studies of the capillary wave hydrodynamics in complex liquids.

Unlike the case of transverse capillary waves, the characteristics of longitudinal waves are more sensitive to the dynamic surface elasticity (relation (21)). Unfortunately, the longitudinal waves almost do not disturb the shape of the interface and, consequently, measurements of their properties are connected with more serious technical problems [78, 79]. Besides, the damping coefficient increases fast with frequency and measurements are possible only in the low frequency range (< 10 Hz).

4. Scattering of capillary waves

All results of the preceding sections were obtained under the assumption of a homogeneous surface film. However, many surface films usually contain two-dimensional aggregates, which influence the characteristics of surface waves [91-98]. Homogeneous films are, in fact, an exception rather than a general rule. Two-dimensional heterogeneity can be observed, for example, in the regions of coexistence of surface phases, where two-dimensional colloid systems arise [99, 100], as well as in solid-like monolayers, which usually contain two-dimensional crystallites [101]. In the latter case, heterogeneities of a larger scale can also occur, if separate regions of the monolayer have different surface tension values [102]. Finally, the monolayer can coexist with regions of a bilayer or with "islands" of a three-dimensional phase [103]. It has been discovered that the macroscopic structure of surface films influences the dependency of the damping coefficient of capillary waves on the area per molecule [91-94, 104]. While describing the propagation of capillary waves, not only the local values of surface tension but also the local dynamic surface elasticity must be taken into account, so that the standard dispersion equation for capillary waves may be inapplicable. If a surface film is heterogeneous, one has to take capillary wave scattering into account.

The first study of surface wave scattering by two-dimensional visco-elastic particles was published about fifteen years ago [95]. A somewhat different approach to the same problem based on the application of Green functions was developed later by Chou and Nelson [96]. The influence of surface tension fluctuations and bending rigidity on the capillary wave propagation was studied numerically by Chou et al. [97]. If the concentration of two-dimensional colloidal particles is high, one has to take multiple scattering of capillary waves into account [98].

Fig. 1 shows a scheme of capillary wave scattering by a two-dimensional circular particle of the radius a. Eqs. (1), (2) describe the liquid motion and Eqs. (5), (6), (11) give the velocity potential, the pressure and the stream function of the incident wave. The difference between the wave numbers for the particle and the ambient medium causes the formation of a transitional region with the thickness δ at the perimeter of the scatterer where the liquid motion becomes more complicated (Fig. 1). The main assumption is the neglectiong of the excess of energy dissipation in the transitional region in comparison with the whole energy dissipation in the liquid under the scatterer.

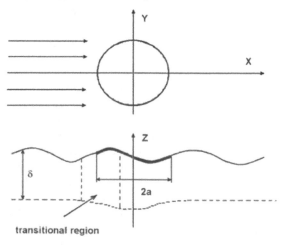

Fig. 1. Scattering of surface waves by a single two-dimensional particle

In the outer region the wave field is a superposition of the incident and scattered wave. The latter is described by Hankel functions. The Bessel functions describe also the wave motion in the inner region. The conditions at the boundary line between two-dimensional phases allow us to "sew together" the solutions of the hydrodynamic equations in the inner and outer regions. The wave motion in the transitional region can be rather complicated. However, if we are not interested in the details of the liquid dynamics in this region, the solutions obtained at a certain distance can be continue up to the boundary line.

The conditions at the boundary line can be reduced to the following four equations:

1. The balance of the surface forces normal to the boundary line (γ_{rr} components of the surface tension tensor in polar coordinates r, ϑ):

$$\gamma_{rr} - \gamma_{rr}^0 = \frac{\chi}{R},$$

$$(23)$$

where χ is the line tension, R is the radius of curvature of the contact line, and the superscript "0" indicates the inner region.

2. The balance of tangential forces.

3. The continuity equation for the normal components of the bulk pressure tensor in the liquid at the boundary line.

4. The continuity of the velocity vector at the boundary line.

According to the Helmholtz theorem, the liquid velocity is a sum of two components $\vec{v} = \vec{grad}\varphi + \vec{rot}\vec{\psi}$. In cylindrical coordinates r, ϑ, z the scalar potential of the scattered wave φ_2 takes the form of

$$\varphi_2 = e^{kz + i\omega t} \sum_{n=0}^{\infty} A_{2n} H_n^1(kr) \cos n\vartheta, r > a,$$

$$(24)$$

where H_n^l is the Hankel function of the first kind and A_{2n} are the amplitudes of the corresponding wave harmonics.

The vector potential of the scattered wave $\vec{\Psi}_2$ can also be represented as a sum of two components:

$$\vec{\Psi}_2 = \Psi_{21} \vec{e}_z + \vec{rot}(\Psi_{22} \vec{e}_z), r > a,$$

$$(25)$$

where

$$\Psi_{2j} = e^{m_j z + i\omega t} \sum_{n=0}^{\infty} B_{nj} f_{nj} H_n(k_j r); \ f_{1n} = \sin n\vartheta; f_{2n} = \cos n\vartheta; j = 1,2$$

$$(26)$$

and B_{nj} are the expansion coefficients.

One can find the wave numbers m_j from the conditions at the interface. Note that the potential ψ_{21} relates to the surface shear waves which rapidly fades away at a short distance from the boundary between the two-dimensional phases. One can find the scattering amplitudes (A_{2n}, B_{n2}) from the boundary conditions. For long waves ($ka \ll 1$) it is sufficient to consider only the first two harmonics. For transverse surface waves ($m \gg k$, $A \gg B$) the corresponding expressions for A_{20} and A_{21} take the form [95] of

$$A_{20} = -\frac{i\pi a^2}{4} A \left\{ \left[(k^0)^2 \frac{\gamma^0}{\gamma} - k^2 \right] - \frac{k^3}{m} \left(1 - \frac{\varepsilon}{\varepsilon^0 - \kappa/2a} \right) \right\} (1 - \frac{k}{m})^{-1}$$
(27)

$$A_{21} \approx \frac{\pi a^2 k^2}{2} \frac{\gamma^0 - \gamma}{\gamma + \gamma^0} A$$
(28)

For transverse waves the amplitudes of the vortical flow can be neglected: $A_{20} \gg B_{02}$ and $A_{21} \gg B_{12}$. This means that the waves are scattered mainly by surface tension inhomogeneities.

From the obtained results the scattering cross-section and the corresponding effective damping coefficient can be calculated if the concentration of scatterers is low. In the general case the effects of multiple scattering have to be taken into account.

If $A \gg B$ (scattering of transversal surface waves), one can approximately consider the liquid motion to be potential so that the scalar potential φ satisfies the following equation [98]:

$$(\Delta + k^2)\varphi(r) = 0.$$
(29)

The exciting field acting upon the scatterer with the centre in the point r_1 $\varphi^E(r/r_1)$ is a sum of the incident wave φ^{inc} and the waves scattered by other particles. The contribution of the given particle depends on the distribution of the other scatterers. If the scattering is weak ($n_0 Q_s/k \ll 1$, where Q_s is the total scattering length and n_0 is the number density of the particles), the configurationally averaged exciting field $\rangle\varphi^E\langle$ is the solution of the following integral equation [105]:

$$\rangle\varphi^E(r/r_1)\langle = \varphi^{inc}(r) + \int dr' n(r'/r_1)T(r')\rangle\varphi^E(r/r')\langle - R(r/r')$$
(30)

where $T(r')$ is the scattering operator, $n(r''/r_1)$ is the conditional density of the scatterers, and $R(r/r')$ is a small correction taking correlations between the scatterers into account.

The approximate solution of Eq. (30) takes the form [98] of

$$\rangle\varphi^E(r/r_1\langle = e^{kz+i\alpha t}\sum_{n-0}^{\infty}C_n(x_1)J_n(k|r-r_1|)\cos n\vartheta(r-r_1) \qquad (31)$$

where $\vartheta(r-r_1)$ is the angle between vector $r-r_1$ and the x-axis.

The scattering amplitude at $r \to \infty$ reduces to the following expression:

$$\frac{e^{ikr}}{\sqrt{r}}\frac{1-i}{\sqrt{\pi k}}\sum_{n=0}^{\infty}\varepsilon_n\tilde{A}_n\cos n\vartheta \qquad (32)$$

where $\varepsilon_0=1$, $\varepsilon_n=2$ at $n=1,2,...$; $A_{2n} = A\varepsilon_n i^n \tilde{A}_{2n}$.

The phase velocity of capillary waves under the conditions of scattering deviates from the value of the medium without scatterers. Therefore it is natural to introduce an effective wave number K for the former case and choose the following trial functions for the coefficients C_n: $C_n(x_1) \propto e^{iKx_1}$ (33)

Then Eqs. (31)-(32) lead to the following expression for the complex effective wave number [98]:

$$K^2 = [k+(1-i)\sqrt{\frac{\pi}{k}}n_0 f(0)]^2 -[(1-i)\sqrt{\frac{\pi}{k}}n_0 f(\pi)]^2 \qquad (34)$$

where $f(0)$ and $f(\pi)$ are the forward and the back scattered amplitudes:

$$f(0) = \frac{1-i}{\sqrt{\pi k}}\sum_{j=0}^{\infty}\varepsilon_j\tilde{A}_{2j}; \quad f(\pi) = \frac{1-i}{\sqrt{\pi k}}\sum_{j=0}^{\infty}(-1)^j\varepsilon_j\tilde{A}_{2j} \qquad (35)$$

Relation (34) allows determination of the effective wave number at the propagation of surface waves along the interface with two-dimensional scatterers on the basis of characteristics of scattering on a single particle.

If the condition of weak scattering is satisfied and scatterers are statistically independent, Eq. (34) holds for two-dimensional particles of any sizes. The condition of "two-dimensionality" of scatterers is not too rigid. It is important only that the scattered wave should satisfy approximately Eq. (29) and that it should be possible to define the scattered amplitude. Probably a rather large group of particles meets these conditions, for example, three-dimensional particles, which are held off at the interface by capillary forces, if the wavelength exceeds their size.

In the general case transverse surface waves have a vortical component and Eq. (29) is only an approximation. This means that Eq. (34) does not allow calculating the damping coefficient. However, the real part of the complex wave number K for slightly damped capillary waves ($Re\,K \gg Im\,K$) can be estimated. Both the inhomogeneities of the surface tension and dynamic surface elasticity scatter capillary waves. In the former case one can estimate the effects of multiple scattering if the line tension and the particle size are known. The existing experimental estimates of the line tension for the boundary between two-dimensional fluid-like phases give too low values, which cannot lead to any measurable scattering effects. Larger numbers can be expected for the boundary between solid-like phases, especially under non-equilibrium conditions. In this case a noticeable influence of the multiple scattering was predicted [98]. Although fluctuations of the surface elasticity usually almost do not influence the scattering of transversal waves, a special situation arises when the continuous surface film contains regions of a gaseous two-dimensional phase (two-dimensional bubbles). If the dynamic surface elasticity of the scatterer tends to zero, large multiple scattering effects can be observed [98].

Although Eqs. (34), (35) do not allow to estimate the contribution of scattering to the effective damping coefficient, this can be done via the total scattering length Q_s [105]

$$Im\,K = Im\,k + nQ_s/2 \tag{36}$$

The total scattering length was calculated in [95]. Noticeable changes of the effective damping of surface waves arise simultaneously with corresponding changes of the wavelength. In this case the scattering leads to the increase of the effective damping coefficient and probably explains the experimental results for some insoluble monolayers [91].

C. EXPERIMENTAL STUDIES

Capillary waves attracted attention of experimentalists not only in connection with measurements of surface rheological properties. Spontaneous capillary waves play an important role in the modern theory of interfaces [106-109]. Therefore, many efforts have been directed to the determination of the intrinsic roughness of a liquid surface by means of static light [110-112] and X-ray [113, 114] scattering. Recently, spontaneous capillary waves have been observed directly by optical microscopy in a system with ultralow interfacial tension [115]. The development of capillary wave studies was also caused by the impact of waves on various processes in technology and nature, for example, mass exchange in chemical reactors [116], and between atmosphere and ocean [117], remote sensing [80, 83, 117, 118], disruption of surface [119] and free [120-122] films. The discussion of these studies and experimental techniques of the surface rheology is beyond the scope of this work. The reader can find a description of SQELS and electrocapillary wave methods in Chapter 6. Some other set-ups to measure the characteristics of transverse and longitudinal capillary waves are described in [123], and only some surface rheological capillary wave studies are briefly discussed here.

1. Insoluble monolayers

The first studies of capillary wave damping at low frequencies (≤ 1 kHz) were mainly dedicated to insoluble monolayers on aqueous substrates as model systems to examine the dispersion relation of surface waves [4, 5, 7, 9, 10, 124-128]. The experimental results confirmed the main theoretical conclusions but for some systems there was no quantitative agreement [7]. The dependencies of the damping coefficient on surface concentrations of some spread polymer films displayed even a number of local maxima, which were initially connected with conformational transitions in the film [125, 126]. The shape of the dependence of the damping coefficient on concentration in general proved to be extremely sensitive to the details of the surface equation of state [129].

The first studies by SQELS were mainly devoted to the surface rheological parameters of insoluble monolayers [66-68, 130-140], and main attention was paid to the transverse surface viscosity [66-68, 132, 134, 137]. Unfortunately, as was shown by Buzza [54], these results may be rather rough due to erroneous assumptions on the parameterisation of surface visco-elasticity (cf. Section B.2.). For most of the systems the real part of the dynamic surface elasticity was much larger than the imaginary one and the surface films were mainly elastic.

The damping coefficient of capillary waves is sensitive to surface rheological properties in the range of relatively low surface concentrations when both the surface pressure and the surface elasticity modulus are low (section B.3). Therefore, one can expect the best agreement between experimental data on the capillary wave damping and the results of calculations by Eq. (20) just in this range. However, this is not the case. Moreover, the scatter of the damping coefficient obtained at low surface pressures often largely exceeds the instrumental error. More careful measurements showed that the deviations from theoretical results were mainly caused by the heterogeneity of the monolayer [92, 93, 102, 104, 139, 141-143]. In a solid-like two-dimensional phase the surface pressure can be inhomogeneous and change along the surface. Miyano showed that even if the surface pressure was approximately homogeneous, the surface elasticity was not thereby leading to fluctuations of the capillary wave characteristics [102]. Other effects arise when a gaseous coexists with a liquid-like phase. The behaviour of the system can change abruptly when separate islands of a condensed phase form an immobile rigid structure [93, 141, 142]. For fatty acid monolayers (myristic, pentadecanoic, palmitic, stearic) the damping coefficient at close to zero surface pressures displays chaotic fluctuations indicating the inhomogeneity of the film [142]. If α is measured as a function of time, one can observe both an almost constancy of damping and chaotic fluctuations. Most often a sequence of jumps between definite values can be noted. The effects of this kind can be explained by the convective movement of macroscopic islands with relatively high surface elasticity. At initial stages of monolayer compression the scatter of the α values almost does not change, and then, in the region of initial surface pressure rise, a sharp transition to a much smaller scatter occurs. This behaviour is connected with the establishment of the connectivity of the islands in the heterogeneous monolayer [142]. The islands are probably associates of smaller size aggregates. Up to the beginning of the surface pressure rise the monolayer has structured (islands) and unstructured (two-dimensional dispersion of small aggregates) regions with different surface elasticities. In this case, surface convection leads to fluctuations of the damping coefficient. When the whole monolayer becomes structured a sharp transition to small fluctuations occurs. The monolayer becomes heterogeneous but the corresponding scale becomes smaller. The dependence of the mean damping coefficient on surface concentration in such a system with a few scales of heterogeneity differs from the results for a continuous compact film. Although the mean damping goes through a small maximum with increasing surface concentration, a sharp maximum, as predicted for the latter case, has never been observed for fatty acid monolayers [141].

A different behaviour can be observed if the expanded monolayer consists not of separate islands but forms a continuous two-dimensional foam even at very low surface concentrations. In this case the compression of the monolayer leads to a continuous increase of the dynamic surface elasticity and, consequently, to a sharp maximum of the damping coefficient. The gradual transition from a two-dimensional foam to a two-dimensional gas emulsion can lead to increased capillary wave scattering according to the theory discussed in Section B.4 and, consequently, to the increase of the effective damping coefficient. A similar behaviour of the damping coefficient was really observed for spread monolayers of alkyldimethylphosphine oxides [91].

The heterogeneity of insoluble monolayers can explain deviations from predictions of the simple theory for most of the systems at low frequencies ($\leq$ 1 kHz). However, the large differences between the dynamic surface elasticity measured by the electrocapillary wave method on one hand, and the static surface elasticity, and the results of SQELS on the other hand, for spread monolayers of poly(vinyl acetate) [143] and hydroxypropylcellulose [144] probably requires special explanation.

Recently SQELS has been frequently applied to spread polymer films on liquid substrate in spite of the unsolved problem of negative surface viscosity and probable inadequacy of the dispersion equation [44, 46-49, 51, 52, 145-151]. A successful application of SQELS obviously means that calculations by the dispersion relation lead to the correct order of the surface elasticity. The problems mentioned above concern only a relatively small correction to the complex dynamic surface elasticity. The real part of the surface elasticity is usually much higher than the imaginary part and can be determined with a reasonable accuracy. Frequently the real part of the surface elasticity almost coincides with the static surface elasticity. In some systems the static elasticity is lower thus indicating that the main relaxation processes correspond to lower frequencies [47, 148, 149]. At the same time, any analysis of small dilational surface viscosities is dubious at the current level of knowledge. This conclusion does not relate to some important cases when both components of the surface elasticity are comparable and strong relaxation processes proceed in the surface film [149]. The surface visco-elasticity of polymer films is discussed in detail in Chapter 6. Here we only mention an important application of the capillary wave method to polyelectrolyte multilayers [152]. The surface elasticity proved to be an indicator of the core zone formation in multilayers.

2. Solutions of soluble surfactant

2.1. Non-micellar solutions

The propagation of capillary waves in liquids with an adsorbed surfactant film has some peculiarities in comparison with insoluble film. If the adsorption kinetics is controlled by diffusion and the characteristic time of the surfactant diffusion from the bulk phase to the surface is much less than the oscillation period, the waves propagate like in pure liquids. If the oscillation period decreases, some intermediate cases are possible. These features are connected with the peculiarities of the dilational surface rheology of surfactant solutions. The shear rheological properties are negligible in this case and the dynamic surface elasticity is determined by the surfactant exchange between the bulk phase and the surface layer [153, 154]:

$$\varepsilon = -\frac{\partial \gamma}{\partial \ln \Gamma} \frac{i\omega\tau + (1+i)\sqrt{\dfrac{\omega}{2D}}\dfrac{\partial \Gamma}{\partial c}}{1 + i\omega\tau + (1+i)\sqrt{\dfrac{\omega}{2D}}\dfrac{\partial \Gamma}{\partial c}} \tag{37}$$

where c is the surfactant concentration, D is the surfactant diffusion coefficient in the bulk phase, and τ is the adsorption relaxation time, which is connected with the adsorption barrier.

If the adsorption kinetics is controlled by diffusion, $\tau = 0$ and Eq. (37) reduces to the well-known expression of Lucassen and Hansen [8]. Note that although ε is a surface property, it depends on the kinetic coefficient of the bulk phase (D). This is a consequence of the non-autonomy of the surface layer as a separate phase [155].

Eq. (37) together with the dispersion relation (20) has to describe the characteristics of capillary waves in surfactant solutions. On the other hand, this equation shows that capillary waves can give information on the surfactant adsorption kinetics. This idea was used first by Lucassen and Hansen, who investigated the adsorption kinetics of fatty acids [7], and many authors applied later the method of capillary waves to kinetic studies of surfactant adsorption [70, 75, 78, 79, 156-168]. The possibility to determine kinetic adsorption coefficients depends on the relation between three characteristic times: ω^{-1}, τ and the characteristic time of diffusion to the surface $\tau_d \equiv (\partial \Gamma / \partial c)^2 / 2D$. If $\omega^{-1} \ll \tau_d$, the adsorption film behaves as an insoluble monolayer and the wave characteristics do not depend on the adsorption kinetics. This condition holds for many conventional surfactants at frequencies typical for SQELS ($\geq$ 10 kHz) [40, 169]. However, this method can be successfully used to study the adsorption kinetics of surfactants with low surface activity like pentanoic acid when $\omega^{-1} \sim \tau_d$ [165]. The increase of the surfactant's surface activity leads to a decrease of the SQELS sensitivity to the adsorption kinetics. This is probably the reason why a 10-times higher diffusion coefficient was obtained from surface dynamic scattering for alkylglucoside surfactants [168].

The method of externally excited capillary waves allows measurements at lower frequencies and, can consequently be applied to a larger number of surfactants [158-162, 167]. This technique has some important advantages in comparison with traditional methods of adsorption kinetics. SQELS excludes in principle any external perturbations, but even in the case of excited capillary waves the perturbations are very small. Measurements are possible when the wave amplitude is significantly less than 1% of the wavelength. The conventional methods of surface rheology and adsorption kinetics are usually based on much stronger perturbations [64, 170-172]. Besides, the complicated phenomena at the three phase contact line do not influence the results obtained by the capillary waves unlike, for example, for methods based on t bubbles or drops. As a result the hydrodynamics of linear capillary waves is simpler in comparison with the latter case [172].

The application of the capillary wave method allowed, for example, proving the diffusion mechanism of adsorption kinetics of normal alcohols. The mechanism of adsorption in these systems has been discussed in literature for about sixty years and some authors discovered adsorption barriers [167]. The traditional methods of adsorption kinetics based on the application of bubbles and liquid jets led to a low accuracy for the fast adsorption of alcohols with small hydrocarbon tails. Capillary waves proved to be more useful [167]. Fig. 2 shows as an example the concentration dependence of the damping coefficient at a frequency of 200 Hz for pentanol solutions. The lines present the results of calculations by Eqs. (20) and (37) at different τ. The derivatives $\partial\gamma/\partial\ln\Gamma$ and $\partial\Gamma/\partial c$ were calculated from equilibrium data.

The curves corresponding to the barrier and mixed adsorption mechanisms do not fit the experimental data unlike the curve for the pure diffusion adsorption mechanism.

The results for surfactants with higher surface activity are not so convincing. The adsorption kinetics of dodecyltrimethylammonium bromide is close to a diffusion mechanism only if some corrections to the Frumkin equation of state are used [158]. An approximate correspondence to a diffusion mechanism was also obtained for sodium dodecylsulfate (SDS) [161], sodium decylsulfate (SDeS) [166] and dodecylammonium chloride [162]. The accuracy proved to be insufficient for substances with higher surface activity like for dodecyl dimethyl phosphine oxide (C_{12}DMPO) [173]. A diffusion controlled adsorption kinetics of this surfactant was established by means of longitudinal capillary waves at frequencies of the order of 1 Hz [79]. In this case $\omega^{-1} \sim \tau_d$ and the experimental method is precise enough. For ionic surfactants the longitudinal wave method indicates the importance of an electrostatic adsorption barrier at the oil-water interface [75, 163].

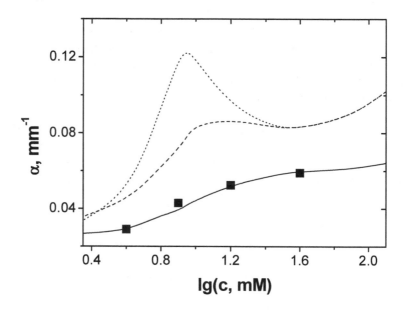

Fig. 2. The damping coefficient of capillary waves as a function of pentanol concentration at frequency 200 Hz. The curves are calculated according to Eqs. (20) and (37) at $\tau = 0$ (solid line), $\tau = 0.001$ s (dashed line) and $\tau = 1$ s (dotted line).

2.2. Micellar solutions

The formation of micelles usually leads to a kink in the concentration dependencies of properties of surfactant solutions. The same is true for the wavelength of capillary waves which is almost constant beyond the critical micellar concentration (CMC) like the surface tension [174]. The behaviour of the damping coefficient of micellar solutions depends on the relation between the period of oscillations, and the fast τ_1 and slow τ_2 micellar relaxation times. If $\omega^{-1} \ll \tau_2$, micelles influence only slightly the dynamic surface properties and, consequently, the damping coefficient is almost constant beyond the CMC. This behaviour was observed, for example, for SDS and $C_{12}DMPO$ solutions at a frequency of 200 Hz [161, 173]. If $\omega^{-1} \geq \tau_2$, micelles contribute significantly to the surfactant exchange between the surface layer and the bulk phase. As a result the dynamic surface elasticity and the damping coefficient of capillary waves decrease with the surfactant concentration beyond the CMC. As an example, the concentration dependence of the damping coefficient for decylpyridinium bromide solutions at a frequency of 200 Hz is shown in Fig. 3 [175]. The lines present the results of calculations by Eqs. (20), and (37) at $\tau = 0$ (diffusion controlled adsorption kinetics) and $\tau = 0.01$ s (barrier controlled adsorption kinetics) for non-micellar solutions. The local maximum corresponds to the CMC. Beyond the CMC the damping decreases due to the influence of micelles. The subsequent increase at higher concentrations is caused by the increase of the solution's bulk viscosity. Similar results were obtained for SDeS [166], $C_{10}DMPO$ [173] and decylpyridinium bromide [176].

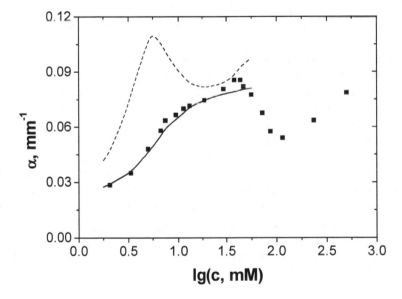

Fig. 3. The damping coefficient of capillary waves as a function of concentration of decylpyridinium bromide at a frequency of 200 Hz. The curves are calculated according to Eqs. (20) and (37) at $\tau = 0$ (solid line) and $\tau = 0.01$ s (dashed line).

Lucassen made a first mathematical description of the dilational surface rheology of micellar solutions [177]. However, this theory was created before the development of the detailed description of micellisation kinetics by Anniansson and Wall [178, 179], and, consequently, did not take into account the two-step mechanism of this process. More elaborated theories of the dynamic surface properties of micellar solutions were published by several authors [180-184]. Unlike the case of non-micellar solutions, no general equation can be derived and one should consider various different approximations. In the most important case, when the wave period is comparable with the slow micellar relaxation time, the adsorption kinetics is determined by diffusion and most surfactant molecules belong to micelles, the dynamic surface elasticity takes the following form [180]:

$$\varepsilon = -\frac{\partial \gamma}{\partial \ln \Gamma} \frac{1 + p(q-1)^{1/2} - ip(q+1)}{\left[1 + p(q-1)^{1/2}\right]^2 + p^2(q-1)} \tag{38}$$

where $p = (\partial \gamma / \partial c_1)^{-1} (2\omega^2 \tau_2 / D_1)^{-1/2}$, $q = (1 + \omega^2 \tau_2^2)^{1/2}$, c_1 is the concentration of monomers and D_1 is the diffusion coefficient of monomers.

The application of Eq. (38) together with the dispersion relation of capillary waves (20) to experimental data on the damping coefficient (Fig. 3) allows describing the dependence of α on concentration using τ_2 as the only fitting parameter [166, 173-176]. The micellar relaxation times obtained in this way agree with values obtained from studies of the bulk properties of micellar solutions.

At low frequencies corresponding to longitudinal capillary waves, the surface elasticity of conventional surfactant solutions drops to almost zero within a narrow concentration range close to the CMC [78, 79]. In this case Eq. (38) is inapplicable and another approximate relation must be used [180]. The dynamic surface elasticity close to the CMC depends strongly on the mean aggregation number, which can be used as fitting parameter to describe the experimental data on the wave characteristics. The fitting for $C_{12}DMPO$ solutions gives an expected value of 35 [79].

2.3. Solutions of polymeric surfactants

The dilational rheology of polymer solutions is discussed in Chapter 6 of this book. Here we only limit ourselves to two short remarks on the peculiarities of capillary wave propagation in the liquids with polymer adsorption film.

The rheological properties of bulk polymer solutions can change already at very low concentrations leading to a higher damping of capillary waves. If the only polymer effect in the bulk phase is the increase of the Newtonian bulk shear viscosity, the impact on the capillary wave characteristics is trivial and reduces to the change of a single parameter. However, if the bulk phase becomes visco-elastic, new effects arise. The theory predicts a crossover from capillary waves to Rayleigh elastic waves when the bulk rheological properties change [185-187]. Significant efforts were directed to discover the crossover by SQELS [188-190]. From the point of view of surface rheological studies, it is important that a new wave mode can appear even at rather low concentrations. Munoz et al. have shown recently that the classic capillary wave spectrum can describe experimental data of solutions of the triblock copolymer poly(ethylene oxide-b-propylene oxide-b-ethylene oxide) (PEO_{76}-PPO_{29}-PEO_{76}) only for concentrations below 0.04 mM [190].

The surface elasticity of non-ionic homopolymer solutions close to equilibrium is usually very low (≤ 5 mN/m) and can be easily determined by the capillary wave technique [123, 191-193]. It is just this method that allows measurements of the non-monotonic dependencies of the surface elasticity on concentration and surface age [191, 193]. On the contrary, the dynamic surface elasticity of polyelectrolyte solutions is at least one order of magnitude higher and only the initial steps of adsorption film formation can be studied by this technique [194, 195]. At the same time, the dynamic surface elasticity of polyelectrolyte/surfactant [196-198] and protein/surfactant solutions [199] is usually high at extremely low concentrations but drops to low values close the critical aggregate concentration or the CMC.

It is also worth to mention the recent application of standing capillary waves (cylindrical meniscus waves) to the investigation of DNA interaction with a lipid monolayer at the air-water interface [200-201].

D. CONCLUSION

Since ancient times studies of capillary waves involve a touch of mystery [11]. Nowadays it is still impossible to deny this effect. The studies of non-linear waves are only at the beginning and some experimental findings still require explanation [58, 69]. Even in the linear case the discovery of a negative surface viscosity for some systems indicates deficiencies in the existing theory. Nevertheless, the existing problems relate only to small effects and to a significant extent are the problems of accuracy. If the processes at the liquid surface influence strongly the capillary wave characteristics, the corresponding methods can give reliable kinetic information. The reliable determination of the surface elasticity modulus is somehow restricted by the condition $|E|/\gamma \leq 0.16$ for transverse capillary waves. The longitudinal wave method has probably no principal limitations and is more promising at low frequencies (~ 1 Hz).

Acknowledgement. This work was financially supported by the Russian Foundation of Basic Research (RFFI № 08-03-00207_a).

E. REFERENCES

1.   L.D. Landau and E.M. Lifshitz, Fluid Mechanics. Pergamon Press, Oxford, 1987.
2.   M. Eigen and L. De Mayere, in Investigation of Rates and Mechanisms of Reactions, G.G. Hammes Ed. John Wiley, New York, 1974 (Chap. 3).
3.   J. Stuehr, in Investigation of Rates and Mechanisms of Reactions, G.G. Hammes Ed. John Wiley, New York, 1974 (Chap. 7).
4.   W.D. Garrett, J.D. Bultman, J. Colloid Sci., 18 (1963) 798.
5.   J.A. Mann, R.S. Hansen, J. Colloid Sci., 18 (1963) 805.
6.   J.T. Davies, R.W. Vose, Proc. R. Soc. (London), A286 (1965) 218.
7.   J. Lucassen, R.S. Hansen, J. Colloid Interface Sci., 22 (1966) 32.
8.   J. Lucassen, R.S. Hansen, J. Colloid Interface Sci., 23 (1967) 319.
9.   R.L. Bendure, R.S. Hansen, J. Phys. Chem., 71 (1967) 2889.
10.   D. Thiessen, A. Scheludko, Kolloid-Z. Z. Polym., 218 (1967) 139.
11.   D. Tabor, J. Colloid Interface Sci., 75 (1980) 240.
12.   Aristotle, Problemata Physica, Book 23, n. 38.
13.   Plutarch, Moralia, L. Pearson and F.H. Sandbach trans. Harvard U. Press, Cambridge Ma, 1965, p. 77.
14.   C. Plinius Secundus, Historia Naturalis, Book 2, Chaps. 103 and 106.
15.   B. Franklin, Philos. Trans. R. Soc., 64 (1774) 445.
16.   N.Ya. Oserezkovskiy, The Journey along Ladoga and Onega, Imperial Acad. Sci., St.Petersburg, 1792 (Chap. 3).
17.   C.H. Giles, Chem. Ind., 8 (1969) 1616.
18.   J. Mertens, Phys. Today, 59 (2006) 36.
19.   Lord Rayleigh, Phil. Mag., 30 (1890) 386.
20.   L.A. Kalen, Ann. Phys., 7 (1902) 440.
21.   L. Kolovrat-Chervinskiy, J. Russian Phys.Chem. Soc., 36(1904)265.
22.   R.C. Brown, Trans. Faraday Soc., 31 (1935) 205.
23.   R.C. Brown, Proc. Phys. Soc. 48 (1936) 312.
24.   H. Lamb, Hydrodynamics, Dover, New York, 1932.
25.   V.G. Levich, Soviet Phys. JETP, 10 (1940) 1296; 11 (1941) 340.
26.   R. Dorrestein, Proc. Koninkl. Ned. Akad. Wet., B54 (1951) 260; 350.
27.   F.C. Goodrich, J. Phys. Chem., 66 (1962) 1858.
28.   J.A. Mann, R.S. Hansen, J. Appl. Phys., 35 (1964) 151.

29. M. van den Temple, R.P. van de Riet, J. Chem. Phys., 42 (1965) 2769.
30. E.H. Lucassen-Reynders, J. Lucassen, Adv. Colloid Interface Sci., 2 (1969) 347.
31. M.G. Hegde, J.C. Slattery, J. Colloid Interface Sci., 35 (1971) 183.
32. L. Kramer, J. Chem. Phys., 55 (1971) 2097.
33. J. Lucassen, Trans. Faraday Soc., 64 (1968) 2221.
34. J. Lucassen, Trans. Faraday Soc., 64 (1968) 2230.
35. R.H. Katyl, U. Ingard, Phys. Rev. Lett., 19 (1967) 64; 20 (1968) 248.
36. Light Scattering by Liquid Surfaces and Complimentary Techniques, D. Langevin Ed. Marcel Dekker, New York, 1992.
37. K. Sakai, P.K. Choi, H. Tanaka, K. Takagi, Rev. Sci. Instrum., 62 (1991) 1192.
38. R.S. Hansen, J. Ahmad, in Progress in Surface and Membrane Sci. J. Danielli, M. Rosenberg, D. Cadenhead. Academic Press, New York, 1971, vol. 4, p. 1.
39. J.A. Mann, in Surface and Colloid Sci. E. Matijevic, R.J. Good Eds. Plenum Press, New York, 1984, vol. 13, p. 145.
40. J.C. Earnshaw, E. McCoo, Langmuir, 11 (1995) 1087.
41. J.C. Earnshaw, D. Sharpe, J. Chem. Soc., Faraday Trans., 92 (1996) 611.
42. D. Sharpe, J. Eastoe, Langmuir, 12 (1996) 2303.
43 .R.W. Richards, M.R. Taylor, J. Chem. Soc., Faraday Trans, 92 (1996) 601.
44. S.K. Peace, R.W. Richards, Polymer, 37 (1996) 4945.
45. R.W. Richards, M.R. Taylor, Macromolecules, 30 (1997) 3892.
46. S.K. Peace, R.W. Richards, N. Williams, Langmuir, 14 (1998) 667.
47. C. Booth, R.W. Richards, M.R. Taylor, G.E. Yu, J. Phys. Chem. B 102 (1998) 2001.
48. A.J. Milling, R.W. Richards, R.C. Hiorns, R.G. Jones, Macromolecules, 33 (2000) 2651.
49. A.J. Milling, L.R. Hutchings, R.W. Richards, Langmuir, 17 (2001) 5297.
50. F. Monroy, M.G. Munoz, J.E.F. Rubio, F. Ortega, R.G. Rubio, J. Phys. Chem. B 106 (2002) 5636.
51. M.G. Muñoz, F. Monroy, P. Hernández, F. Ortega, R.G. Rubio, D. Langevin, Langmuir 19 (2003) 2147.
52. S. Rivillon, M.G. Munoz, F. Monroy, F. Ortega, R.G. Rubio, Macromolecules, 36 (2003) 4068.
53. B.A. Noskov, G. Loglio, Colloids Surf. A, 143 (1998)167.
54. D.M.A. Buzza, Langmuir, 18 (2002) 8418.
55. M. Hennenberg, X.L. Chu, S. Slavtchev, B. Weyssow, J.C. Legros, J. Colloid Interface Sci., 230 (2000) 216.

56. D.M.A. Buzza, J.L. Jones, T.C.B. McLeisch, R.W. Richards, J. Chem. Phys., 109 (1998) 5008.

57.J. Giermanska-Kahn, F.Monroy, and D. Langevin Phys. Rev. E, 60 (1999) 7163.

58:S. Joo, A. Messiter, W. Schultz, J. Fluid Mech., 229 (1991) 135.

59.G.S. Lapham, D.R. Dowling, W.W. Schultz, Exp. Fluids, 30 (2001) 448.

60. V.G. Levich, Physicochemical hydrodynamics. Engelwood Cliffs, New York, 1962.

61. L.E. Scriven, Chem. Eng. Sci., 12 (1960) 98.

62. F.C. Goodrich, Proc. R. Soc. (London), A260 (1961) 490, 503.

63. F.C. Goodrich, Proc. R. Soc. (London), A374 (1981) 341.

64. A.I. Rusanov and V.A. Prokhorov, Interfacial Tensiometry, Elsevier, 1996.

65. E. Mayer, J.D. Elliassen, J. Colloid Interface Sci., 37 (1971) 228.

66. D. Langevin, J. Colloid Interface Sci., 80 (1981) 412.

67. S. Hard, R.D. Neuman, J. Colloid Interface Sci., 83 (1981) 315.

68. J.C. Earnshaw, P.J. Winch, J. Phys.: Condens. Matter, 2 (1990) 8499.

69. J. Lucassen, M. Van den Tempel, J. Colloid Interface Sci., 42 (1972) 491.

70. H.C. Maru, D.T. Wasan, Chem. Eng. Sci., 34 (1979) 1295.

71. N.F. Djabbarach, D.T. Wasan, Chem. Eng. Sci., 37 (1982) 175.

72. K. Miyano, B.M. Abraham, L. Ting, D.T. Wasan, J. Colloid Interface Sci., 92 (1983) 297.

73. C. Lemaire, D. Langevin, Colloids Surf., 65 (1992) 101.

74. Q. Jiang, Y.C. Chiew, J.E Valentini, J. Colloid Interface Sci., 159 (1993) 477.

75. A. Bonfillon, D. Langevin, Langmuir, 10 (1994) 2965.

76. A. Dussaud, M. Vignes Adler, J. Colloid Interface Sci. 167 (1994) 256.

77. J.E. Valentini, J.T. Chandler, Q. Jiang, Y.C. Chiew, L.J. Fina, Ind. Eng. Chem. Res. 35 (1996) 434.

78. B.A. Noskov, D.A. Alexandrov, E.V. Gumennik, V.V. Krotov, R. Miller, Colloid J., 60 (1998) 204.

79. B.A. Noskov, D.A. Alexandrov, R. Miller, J. Colloid Interface Sci., 219 (1999) 250.

80. H. Huehnerfuss, P.A. Lange, W. Walter, J. Colloid Interface Sci., 108 (1985) 430.

81. R. Cini, P.P. Lombardini, C. Manfredi, E. Cini, J. Colloid Interface Sci., 119 (1987) 74.

82. S.J. Brown, M.S. Triantafyllou, D.K.P. Yue, Proc. R. Soc. (London) A, 458 (2002) 1167.

83. S.A. Ermakov, Izvestya - Atmospheric and Ocean Physics, 39 (2003) 624.
84. J.C. Earnshaw, A.C. Mc Laughlin, Proc. R. Soc. (London) A, 433 (1991) 663.
85. J.A. Stone, W.J. Rice, J. Colloid Interface Sci. 61 (1977) 160.
86. M. Hennenberg, P.M. Bisch, M. Vignes Adler, A. Sanfeld, J. Colloid Interface Sci., 74 (1980) 495.
87. M. Hennenberg, A. Sanfeld, P.M. Bisch, AIChE J., 27(1981)1002.
88. B.A. Noskov, Colloid J., 45 (1983) 689; 912.
89. M. Hennenberg, X.L. Chu, A. Sanfeld, M.G. Velarde, J. Colloid Interface Sci., 150(1992)7.
90. P.M. Bisch, A. Steichen, A. Sanfeld, J. Colloid Interface Sci., 95 (1983) 561.
91. B.A. Noskov, D.O. Grigoriev, R. Miller, Langmuir, 13(1997) 295.
92. K. Miyano, K. Tamada, Langmuir, 8 (1992) 160.
93. K. Miyano, K. Tamada, Langmuir, 9 (1993) 508.
94. B. A. Noskov, T. U. Zubkova, J. Colloid Interface Sci., 170 (1995) 1.
95. B.A. Noskov, Fluid Dyn. (USSR), 26 (1991) 106.
96. T. Chou, D.R. Nelson, J. Chem. Phys., 101 (1994) 9022.
97. T. Chou, S. K. Lukas, H. A. Stone, Phys. Fluids., 7 (1995) 1872.
98. B.A. Noskov, J. Chem. Phys., 108 (1998) 807.
99. H.M. McConnel, Annu. Rev. Phys. Chem., 42 (1991) 171.
100. C.M. Knobler, R.C. Desai, Annu. Rev. Phys. Chem., 43(1992)207.
101. S. Henon, J. Meunier, Thin Solid Films, 234 (1993) 471.
102. K. Miyano, Langmuir, 6 (1990) 1254.
103. E.K. Mann, D. Langevin, S. Henon, J. Meunier, L.T. Lee, Ber. Bunsenges. Phys. Chem., 98 (1994) 519.
104. Q. Wang, A. Feder, E. Mazur, J. Chem. Phys., 98 (1994) 1272.
105. P.C. Waterman, R. Truell, J. Math. Phys., 2 (1961) 512.
106. J.C. Rowlinson, B. Widom, Molecular Theory of Capillarity, Clarendon Press, Oxford, 1982.
107. F.H. Stillinger, J.D. Weeks, J. Chem. Phys., 99 (1995) 2807.
108. K.R. Mercke, S. Dietrich, Phys. Rev. E, 59 (1999) 6766.
109. A. Milchev, K. Binder, Europhys. Lett., 59 (2002) 81.
110. L. Mandelstam, Ann. Phys. 41 (1913) 609.
111. P.S. Hariharan, Proc. Indian Acad. Sci., A16 (1942) 290.
112. A. Vrij, Adv. Colloid Interface Sci., 2 (1968) 39.
113. M.K. Sanyal, S.K. Sinha, K.G. Huang, B.M. Ocko, Phys. Rev. Lett., 66 (1991) 628.
114. C. Fradin, A. Braslau, D. Luzet, D. Smilgles, M. Alba, N. Boudet, K. Mecke, J. Daillant, Nature, 403 (2000) 871.

115. D.G.A.L. Aarts, M. Schmidt, H.N.W. Lekkerkerker, Science, 304 (2004) 847.
116. H. Savistowski, Ber. Bunsenges. Phys. Chem., 85 (1981) 905.
117. Marine Surface Films - Chemical Characteristics, Influence on Air-Sea Interactions, and Remote Sensing, M. Gade, H. Hühnerfuss, and G. Korenowski Eds. Springer, Berlin, 2006.
118. M. Gade, W. Alpers, H. Huehnerfuss, P. A. Lange, J. Geophys. Res., 103 (1998) 3167.
119. K. Kargupta, A.Sharma, R. Khanna, Langmuir, 20 (2004) 244.
120. A. Scheludko, Adv. Colloid Interface Sci., 44 (1967) 391.
121. R. Tsekov, B. Radoev, Adv. Colloid Interface Sci., 38 (1992) 353.
122. J.E. Coons, P.J. Halley, S.A. McGlashan, T. Tran-Cong, Colloids Surf. A, 263 (2005) 258.
123. B.A. Noskov, A.V. Akentiev, A.Yu. Bilibin, I.M. Zorin, R. Miller, Adv. Colloid Interface Sci., 104 (2003) 245.
124. R.A. Avetisyan, A.A. Trapesnikov, Russian J. Phys. Chem., 38 (1964) 3036.
125. W.D. Garrett, W.A. Zisman, J. Phys. Chem., 74 (1970) 1796.
126. R.L. Shuler, W.A. Zisman, J. Phys. Chem., 74 (1970) 1523.
127. R. Cini, P.P. Lombardini, J. Colloid Interface Sci., 65 (1978) 387.
128. R. Cini, P.P. Lombardini, J. Colloid Interface Sci., 81 (1981) 125.
129. J. Ahmad, Bull. Chem. Soc. Japan, 56 (1983) 3142.
130. S. Hard, H. Lofgren, J. Colloid Interface Sci., 60 (1977) 529.
131. D. Byrn, J.C. Earnshaw, J. Phys. D, 12 (1979) 1145.
132. D. Langevin, C. Griesmar, J. Phys. D, 13 (1980) 1189.
132. Y.L. Chen, M. Sano, M. Kawaguchi, H. Yu, G. Zografi, Langmuir, 2 (1986) 349.
134. M. Kawaguchi, M. Sano, Y.L. Chen, G. Zografi, H. Yu, Macromolecules, 19 (1986) 2606.
135. B.B. Sauer, M. Kawaguchi, H. Yu, Macromolecules, 20(1987)2732.
136. J.C. Earnshaw, R.C. McGivern, P.J. Winch, J. Phys., 49(1988)1271.
137. R.G. McGivern, J.C. Earnshaw, Langmuir, 4 (1989) 545.
138. F.E. Runge, H. Yu, Langmuir, 9 (1993) 3191.
139. K. Sakai, K. Takagi, Langmuir, 10 (1994) 802.
140. W. Lee, A.R. Esker, H. Yu, 102 (1995) 191.
141. K. Miyano, K.Tamada, Jpn. J. Appl. Phys., 33 (1994) 5012.
142. B.A. Noskov, T.U. Zubkova, J. Colloid Interface Sci., 170(1995)1.
143. R. Skarlupka, Y. Seo, H. Yu, Polymer, 39 (1998) 387.
144. C.S. Gau, H. Yu, G. Zografi, Macromolecules, 26 (1993) 2524.
145. A.R. Esker, L.H. Zhang, B.B. Sauer, W. Lee, H. Yu, Collods Surfaces A, 171 (2000) 131.
146. P. Cicuta, I. Hopkinson, J. Chem. Phys., 114 (2001) 8659.

147. J.T.E. Cook, R.W. Richards, 8 (2002) 111.
148. C. Kim, H. Yu, Langmuir, 19 (2003) 4460.
149. Y. Seo, A.R. Esker, D. Sohn, H.J. Kim, S. Park, H. Yu, Langmuir, 19 (2003) 3312.
150. P. Cicuta, I. Hopkinson, Colloids Surfaces A, 233 (2004) 97.
151. C. Kim, A.R. Esker, F.E. Runge, H. Yu, Macromolecules, 39 (2006) 4889.
152. M. Safouane, R. Miller, H. Möhwald, J. Colloid Interface Sci., 292 (2005) 86.
153. B.A. Noskov, Colloid J. 44 (1982) 492.
154. M. van den Tempel and E. Lucassen-Reynders, Adv. Colloid Interface Sci. 18 (1983) 281.
155. R. Defay, I. Prigogine, A. Bellemans, Surface Tension and Adsorption, John Wiley and Sons, New York, 1966.
156. M. Sasaki, T. Yasunaga, N. Tatsumoto, Bull. Chem. Soc. Japan, 50 (1977) 858.
157. C.I. Christov, L. Ting, D.T. Wasan, J. Colloid Interface Sci., 85 (1982) 363.
158. C. Stenvot, D. Langevin, Langmuir, 4 (1988) 1179.
159. B.A. Noskov, A.A. Vasiliev, Colloid J., 50 (1988) 909.
160. B.A. Noskov, M.A. Schinova, Colloid J., 51 (1989) 69.
161. B.A. Noskov, O.A. Anikieva, N.V. Makarova, Colloid J., 52 (1990) 909.
162. B.A. Noskov, Colloids Surf. A, 71 (1993) 99.
163. A. Bonfillon, D. Langevin, Langmuir, 9 (1993) 2172.
164. B.A. Noskov, D.O. Grigoriev, Progr. Colloid Interface Sci., 97 (1994) 1.
165. K. Sakai, K. Takagi, Langmuir, 10 (1994) 257.
166. B.A. Noskov, D.O. Grigoriev, Langmuir, 10 (1996) 257.
167. B.A. Noskov, Adv. Colloid Interface Sci., 69 (1996) 63.
168. O.J. Rojas, R.D. Neuman, P.M. Claesson, J. Phys. Chem., 109 (2005) 22440.
169. V. Thominet, C. Stenvot, D. Langevin, J. Colloid Interface, 126 (1988) 54.
170. S.S. Dukhin, G. Kretzschmar and R. Miller, Dynamics of Adsorption at Liquid Interfaces, in "Studies in Interface Sci.", Vol. 1, D. Möbius and R. Miller (Eds.), Elsevier, Amsterdam, 1995.
171. K.D. Wantke, H. Fruhner, in Studies in Interface Sci., D. Möbius and R. Miller (Eds.), Elsevier, Amsterdam, Vol. 6, 1998, 327.
172. V.I. Kovalchuk, J. Krägel, E.V. Aksenenko, G. Loglio, L. Liggieri, in Studies in Interface Sci., D. Möbius and R. Miller (Eds.), Elsevier, Amsterdam, Vol. 11, 2001, 327.

173. B.A. Noskov, D.O. Grigoriev, R. Miller, J. Colloid Interface Sci., 188 (1997) 9.
174. B.A. Noskov, Adv. Colloid Interface Sci., 95 (2002) 237.
175. D.O. Grigoriev, V.V. Krotov, B.A. Noskov, Colloid J., 56(1994)637.
176. D.O. Grigoriev, B.A. Noskov, S.I. Semchenko, Colloid J., 55 (1993) 45.
177. J. Lucassen, Faraday Discuss. Chem. Soc., 52 (1975) 76.
178. G.E.A. Aniansson, S.N. Wall, J. Phys. Chem., 78 (1974) 1024.
179. G.E.A. Aniansson, S.N. Wall, J. Phys. Chem., 79 (1975) 857.
180. B.A. Noskov, Fluid Dyn. (USSR), 25 (1990) 105.
181. B.A. Noskov, Colloid J., 52 (1990) 509.
182. A. Johner, J.F. Joanny, Macromolecules, 23 (1990) 5299.
183. C.D. Duschkin, I.B. Ivanov, Colloids. Surf. 60 (1991) 213.
184. C.D. Duschkin, I.B. Ivanov, P.A. Kralchevsky, Colloids. Surf. 60 (1991) 213.
185. J.L. Harden, H. Pleiner, P.A. Pincus, J. Chem. Phys., 94(1991)5208.
186. C.H. Wang, Q.R. Huang, J. Chem. Phys., 107 (1997) 5898.
187. H. Nakanishi, S. Kubota, Phys. Rev. E, 58 (1998) 5898.
188. R.B. Dorshow, L. Turkevich, Phys. Rev. Lett., 70 (1993) 2439.
189. F. Monroy, D. Langevin, Phys. Rev. Lett., 81 (1998) 3167.
190. M.G. Munoz, M. Encinar, L.J. Bonales, F. Ortega, F. Monroy, R.G. Rubio, J. Phys. Chem. B, 109 (2005) 4694.
191. B.A. Noskov, A.V. Akentiev, G. Loglio, R. Miller, J. Phys. Chem. B, 104 (2000) 7923.
192. B.A. Noskov, A.V. Akentiev, R. Miller, J. Phys. Chem. B, 255 (2002) 417.
193. B.A. Noskov, A.V. Akentiev, A.Yu. Bilibin, D.O. Grigoriev, G. Loglio, I.M. Zorin, R. Miller, Langmuir, 20 (2004) 9669.
194. B.A. Noskov, S.N. Nuzhnov, G. Loglio, R. Miller, Macromolecules, 37 (2004) 2519.
195. B.A. Noskov, A.Yu. Bilibin, A.V. Lezov, G. Loglio, S.K. Filippov, I.M. Zorin, R. Miller, Colloids Surf. A, 298 (2007) 115.
196. B.A. Noskov, G. Loglio, R. Miller R. J. Phys. Chem. B, 108 (2004) 18615.
197. B.A. Noskov, D.O. Grigoriev, S.-Y. Lin, G. Loglio, R. Miller, Langmuir, 23 (2007) 9641.
198. B.A. Noskov, A.G. Bykov, S.-Y. Lin, G. Loglio, R. Miller, Colloids Surf. A, 322 (2008) 71.
199. A.V. Latnikova, S.-Y. Lin, G. Loglio, R. Miller, B.A. Noskov, J. Phys. Chem. C, 112 (2008) 112.
200. C. Picard, L. Davoust, Rheol. Acta. 45 (2006) 497.
201. C. Picard, L. Davoust, Langmuir, 23 (2007) 1394.

DILATIONAL RHEOLOGY OF ADSORBED LAYERS BY OSCILLATING DROPS AND BUBBLES

F. Ravera[1], L. Liggieri[1] and G. Loglio[2]

[1] CNR - Istituto per la Energetica e le Interfasi, 16149 Genoa, Italy
[2] University of Florence, 50019 Sesto Fiorentino (Firenze), Italy

Contents

A. INTRODUCTION

The mechanical behaviour of liquid interfaces has a large relevance for many technological and natural processes involving multiphase systems subjected to dynamic conditions. This behaviour is primarily linked to the value of the interfacial tension and to its instantaneous variation under the effect of changes of the available interfacial area, i.e, to the dilational rheology of the interfacial layer.

In particular for systems characterised by a high specific area, such as liquid films, emulsions or foams, these aspects may become the main driving force for the system evolution and the key-feature for stability [1, 2, 3, 4].

Moreover, a close link exists between dynamic interfacial tension and the adsorption mechanisms, which, in turns, reflect some molecular characteristics of surfactants and of the transport and kinetic processes involving them. Thus the interpretation of dilational rheology data with the help of suitable models provides a unique opportunity to access basic physicochemical mechanisms concerned with adsorption and to derive quantitative and qualitative information.

Deepening and quantify the above relations requires accurate and effective methodologies for the investigation of interfacial rheology and the measurement of the related parameters. During the last decade several progresses have been made on this path, either by upgrading and adapting existing tensiometric techniques or developing new methodologies. Among them, the methods utilising oscillating bubble and drop are particularly suitable due to their versatility.

In its general meaning, interfacial rheology is the study of surface modification induced by mechanical forces, like shear or dilational stresses [5, 6]. In case of dilational, or compression/expansion, rheology, the stress is the variation of the interfacial tension while the corresponding surface modification is the area change.

The dilational aspect of surface rheology is particularly important when surfactant or, more generally, composite interfacial layers are concerned. In this case, in fact, a viscous elastic modulus, or dilational visco-elasticity, can be attributed to the interface due to the mass transfer or to the modification of the surface composition.

It is then clear that appropriate experimental techniques for dilational rheology must provide the monitoring of interfacial tension as response to controlled variations of the surface area. Accordingly most classical surface rheological studies [7], dealt with techniques like Langmuir Blodgett balances, where the area change of the liquid/air interface was controlled by means of mobile barriers and the monitoring of the surface tension performed by pulling devices like a du Nöuy ring or a Wilhelmy plate. These traditional techniques are still used in upgraded versions to study surfactants, polymers and proteins [8, 9, 10] at water/air interfaces.

On the other side, bubble/drop techniques are today widely employed for dilational rheology investigations, mainly for their always better applicability to both liquid-liquid and liquid-air interfaces [11, 12].

Most bubble/drop methods derive from classical methods, originally conceived for interfacial tension measurements, that, own to the technological development of the last decades, have enlarged their potentialities. In particular, the implementation of modern control techniques allows for a better dynamic performance in measuring the response of interfacial tension to surface area variations, improving the accuracy of dilational visco-elasticity measurements.

Two main types of bubble/drop techniques are available for monitoring the interfacial tension: one is based on the analysis of the drop profile and the other one on the capillary pressure measurement. Both techniques are effective for liquid-liquid and liquid air interfaces and can be used according to different methodologies for investigating different aspect of the rheological properties.

In particular this chapter focuses on those methodologies utilising harmonic variations of the interfacial area, which are then particularly suitable to investigate dilational rheology in the frequency domain.

Some experimental studies based on the utilisation of such methodologies are also reviewed, dealing not only with surfactant adsorbed layer but also with different kind of composite interfacial layers containing polymers, proteins or nanoparticles.

B. DILATIONAL RHEOLOGY OF ADSORBED LAYERS

1. Generalities on the dilational visco-elasticity

The rheological properties of adsorption layers are expressed by the relationship between the variation of the interfacial tension, γ, from its initial value, γ^0, and the expansion (or contraction) of the surface area, A.

The dilational stress of a system subjected to expansion (or contraction) of the surface area A. can be written as the sum of two terms

$$\Delta\gamma = E_0\alpha + \eta\dot{\alpha} \tag{1}$$

The first one, purely elastic, is proportional to the relative variation of the area, $\alpha = \Delta A/A_0 = (A(t)-A_0)/A_0$. The second one is instead a viscous term, proportional to the rate of the area variation, $\dot{\alpha} = d\alpha / dt$ [5]. This viscous character arises from the relaxation processes in the adsorbed layer or from the adsorption re-equilibration

The coefficient E_0 and η, are then termed as the dilational surface elasticity and viscosity respectively.

Eq. (1) leads to a definition of the complex dilational visco-elastic modulus or the dilational visco-elasticity, E. In fact for a low amplitude harmonic perturbation of angular frequency ω, $\Delta A = \tilde{A}e^{i\omega t}$, from Eq. (1) one obtains

$$E = \frac{\Delta\gamma}{\Delta A / A_0} = E_0 + i\omega\eta \tag{2}$$

The visco-elasticity E is then a frequency dependent complex quantity, where the real part, $E' = E_0$, is the dilational elasticity and the imaginary part, $E'' = \omega\eta_d$, is directly linked to the dilational viscosity.

Considering only small amplitude perturbation, the system behaves linearly. This means that when a purely harmonic perturbation (for example, a sinusoidal perturbation) at a given frequency $v = \omega/2\pi$ is applied to the surface, the interfacial tension and the other quantities influenced by this perturbation vary according to the same harmonic function at the same frequency.

Because under these conditions, the variation of a time dependent quantity can be written as a superposition of harmonic functions, using the Fourier formalism, from Eq. (1), it follows that the response of the interfacial tension to an arbitrary area variation of the adsorbed layer is given by [13],

$$\Delta\gamma(t) = \int_0^t \hat{\varepsilon}(\tau)\alpha(t-\tau)d\tau \tag{3}$$

where $\hat{\varepsilon}$ is the inverse Fourier transform of E. Thus the interfacial layer can be considered as a linear system characterised by the transfer function $E(\omega)$. Eq. (3) evidences that for a any small area perturbation, the interfacial tension response can be assessed in a way as accurate as larger is the frequency range where $E(\omega)$ is known.

2. Theoretical models

Because the frequency dependence of E is due to the relaxation processes occurring in the interfacial layer, dilational rheology is often used to investigate the dynamic aspects of the surfactant adsorption and the processes involved in it like diffusion or other rearrangement in the adsorbed layer. To this aim and to access and quantify these features, it is however necessary to develop suitable models.

According to its definition the dilational visco-elasticity is determined both by the thermodynamic properties of the adsorbed layer and by its kinetic characteristics.

In fact, while the interfacial tension dependence on the surface area expresses the two-dimensional equation of state, as very clear for insoluble surfactants, the frequency dependence of E is strictly related to the kinetic of re-equilibration.

For small amplitude harmonic perturbations, eq. (2), reads

$$E = \frac{d\gamma}{d\ln A} \tag{4}$$

which is the expression of ε originally proposed by Gibbs [14, 15].

Eq. (4) has been widely used as key equation to find the dependence of E on the frequency, applying different thermodynamic and kinetic model for the adsorbed layer.

Considering as an example the case of one soluble surfactant Eq. (4) can be written as

$$E = -E_0 \frac{d\ln \Gamma}{d\ln A} \tag{5}$$

where $E_0 = \left(\frac{d\gamma}{d\ln\Gamma}\right)_{eq}$ is the so-called Gibbs elasticity, which is a function

solely of the thermodynamics, being essentially equivalent to the surface equation of state. On that basis it can be trivially calculated in the framework of any given adsorption model assumed. The other factor in the right-hand side of Eq. (5) contains instead all the information about the dynamic adsorption mechanism. In the hypothesis of small area perturbation it can be expressed as a function of the frequency by solving the set of transport and mass conservation equations describing the kinetic model, in the framework of a perturbation scheme. For diffusion controlled adsorption Eq. (5) leads to the well known Lucassen–van den Tempel equation [16],

$$\varepsilon = \varepsilon_0 \frac{1 + \xi + i\xi}{1 + 2\xi + 2\xi^2} \tag{6}$$

where $\xi = \sqrt{\dfrac{\omega_D}{2\omega}}$ and $\omega_D = D\left(\dfrac{dc}{d\Gamma}\right)^2_{eq}$ is the characteristic angular frequency of the diffusion mechanism. The thermodynamic adsorption model enters into the expressions of ω_D and E_0.

Eq. (6) has been widely utilised by many authors, for liquid/air and liquid / liquid interfaces [17,18], coupled with classical adsorption models like Langmuir or Frumkin isotherms. More recently, other thermodynamic models, taking into account different possible states for the adsorbed molecules have been utilized with Eq. (6). Such models consider for example the molecular re-orientation or surface aggregation [19, 20]. Moreover, the effect of a two-dimensional compressibility of the adsorbed layer has been considered to explain the behaviour of some surfactant layers at increasing surface coverage [21, 22, 23] In this latter approach described in details in ref, [24, 25], the molar surface area in the adsorbed layer is assumed to decrease linearly with the surface pressure. This provides a limited Gibbs elasticity at large surfactant concentration, which better agrees with most experimental observation.

Though extensions of Eq. (6) to investigate mixed surfactant layers at liquid interfaces can be found already in classical literature [18, 26], in more recent works advances in this problem are presented for different multicomponent systems like ionic-non ionic surfactants, surfactant-polymers and surfactants-proteins [27, 28]. This is in fact an important topic for practical applications, because of the wide utilisation in modern technologies of these mixtures to obtain effects that cannot be obtained by single components.

Nevertheless, it is important to underline that, using the Lucassen – van den Tempel equation is correct only if the equilibrium can be assumed for the surface relaxation processes with respect both to the diffusion exchange with the bulk and to the perturbation of the surface.

This assumption is not always appropriate especially for systems containing polymers or proteins for which the kinetic transformations of the adsorbed layer have often very long characteristic times .

Moreover, even for some common non-ionic surfactants, especially at water oil interfaces, Eq. (6) and in general a diffusion controlled model is sometime not suitable to describe the experimental dilational results [29]. It is then necessary to take into account not only new surface isotherms but also multiple kinetic processes.

3. Surface rheology and relaxation phenomena

When further relaxation processes are present besides surfactant diffusion, like internal transformation of the adsorbed layer, an extension of the Lucassen – van den Tempel approach is required, to obtain an expression of $E(\omega)$ in terms of the characteristic times, or frequencies, of such kinetic processes. Thus, by applying appropriate kinetic models, expressions of ε are found, for example, for surface molecular re-orientation [30], aggregation [31], or chemical reactions [6, 32].

The principal results of these studies summarized and analyzed in ref. [33], have shown that the characteristic time of a given kinetic process corresponds to a characteristic frequency around which the dilational visco-elasticity is mostly significant.

This is obviously true also for the Lucassen - van den Tempel model where, being the diffusion the only dynamic process involved in the adsorption equilibrium, the key parameter is the characteristic frequency ν_D related to the diffusion coefficient D. The typical feature of $\varepsilon(\nu)$ according to this model and reported in Fig. 1, evidences the fact that at $\omega=\omega_D$ the imaginary part of E presents a maximum and the real part an inflection point. Also the role of the Gibbs elasticity is shown in Fig. 1, which is, in this case, the high frequency limit of the elasticity.

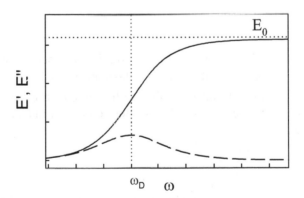

Fig. 1 Real (solid line), and imaginary (dashed line) part of E versus frequency according to the Lucassen - van der Tempel model, Eq.(6). At $\omega = \omega_D$, E' has an inflection point and E'' a maximum.

In ref. [34] a general approach is proposed to calculate the expression of $E(\omega)$, where diffusion is considered together with other kinetic surface processes. The key-point of this approach is the assumption that, to describe the state of the system out of equilibrium, an additional variable is needed for each kinetic process in addition to those necessary to describe the equilibrium state. In practice this means that out of equilibrium, the surface tension (or the surface pressure) is dependent, not only on the total adsorption, Γ, but also on a further variable for each process. Such variable may be for example the average surface area per molecule when re-orientably adsorbed molecules are considered in the interfacial layer.

Thus, assuming a generic interfacial process described by a generic variable X, the dependence of the surface pressure on X has to be considered as well as on the total adsorption Γ, i.e. $\Pi = \Pi(\Gamma,X)$,

According to eq. (4) the dilational visco-elasticity can then be written as,

$$E = -E_{0\Gamma} \frac{d\ln\Gamma}{d\ln A} - E_{0X} \frac{d\ln X}{d\ln A} \tag{7}$$

where, $E_{0\Gamma} = \dfrac{\partial\Pi}{\partial\ln\Gamma}$ and $E_{0X} = \dfrac{\partial\Pi}{\partial\ln X}$, are thermodynamic quantities calculated from the surface equation of state.

Notice that, by definition, $E_{0\Gamma}$ is related to the Gibbs elasticity through

$$E_{oG} = E_\Gamma + E_{0X} \frac{d\ln X^0}{d\ln\Gamma} \tag{8}$$

As explained in details in [34], assuming a linear rate equation, $dX/dt = -K(X - X^0)$, for the surface relaxation process and solving the mass balance at the interface together with such equation according to a harmonic perturbative scheme, leads after some re-arrangements to the following expression of E:

$$E = \frac{E_{0\Gamma} - i\lambda E_{0G}}{-(1+\xi - i\xi)i\lambda + \dfrac{1+G+\xi - i\xi}{1+G}} \tag{9}$$

where $\lambda = \omega_k/\omega$, being ω_k, is the characteristic frequency of the surface kinetic process, while G is a thermodynamic characteristics, $G = \left(1 - \dfrac{dc_s}{d\Gamma} \Big/ \dfrac{\partial c_s}{\partial\Gamma}\right)$, where c_S is the sublayer concentration. In most realistic cases c_S is not dependent on the variable describing the surface relaxation process, that implies G=0.

In the limit case of very fast equilibration of the relaxation process (λ very large), Eq. (9) reduces to Eq. (6). Notice that the high frequency limit of the elasticity given by Eq. (9) is $E_{0\Gamma}$, which differs from the Gibbs elasticity. This means that the presence of other relaxation process besides diffusion has the effect of changing the behaviour of ε at high frequency as shown in Fig. 2, where E(ω) calculated by Eq. (9) is compared to the Lucassen-van den Tempel model. Moreover the further relaxation process produces an additional maximum in the imaginary part of ε.

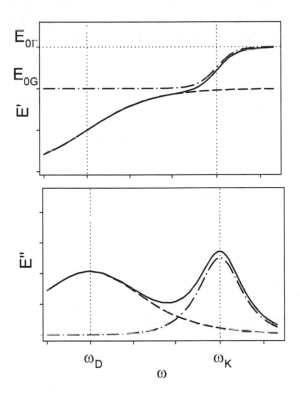

Fig. 2 Real, E′, and imaginary, E″ part of ε versus frequency: according to the general mixed kinetic model, Eq. (9) (solid line), the diffusion controlled model, Eq. (6) (dashed line), and the insoluble monolayer with kinetic process model, Eq. (10) (dashed-dot line). At ω= ω_D and ω= ω_k, E′ has an inflection point and E″ a maximum.

In Fig. 2 it is also reported the other limit case obtained for vanishing ξ, corresponding to an insoluble layer, for which Eq. (9) becomes

$$E = \frac{E_{0\Gamma} - i\lambda E_{0G}}{1 - i\lambda} \tag{10}$$

With some rearrangement Eq. (10) can be recast as

$$E = E_{0G} + \left(E_{0\Gamma} - E_{0G}\right)\frac{i\omega/\omega_k}{1 + i\omega/\omega_k} \tag{11}$$

which has the same form of the Maxwell visco-elasticity modulus often utilised to describe dilational rheology of insoluble layers [35]. In Fig. 3 $E(v)$ calculated by Eq. (9) is reported for different values of v_D.

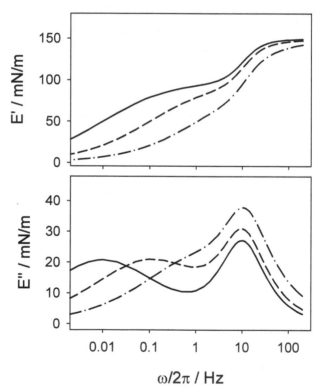

Fig. 3 Real, E', and imaginary, E'' part of E versus frequency according to the mixed kinetic model, Eq. (9), calculated for $E_{0\Gamma} = 150$ mN/m, $E_{0G} = 100$ mN/m, $\omega_K/2\pi = 10$ Hz and for different values of $\omega_D/2\pi$: 0.01 Hz (solid line), 0.1 Hz (dashed line), and 1 Hz (dashed-dot line).

The approach described above is quite general because can be applied to any relaxation process described by a linear rate equation, In fact for its application the only requirement is that the system has linear behaviour. This is usually true for surfactant adsorbed layers and in general for interfacial layer not far from the equilibrium. Eq. (9) has been applied so far to soluble and insoluble surfactant systems and also, more recently, for interpreting rheological data obtained with mixed surfactant nanoparticle systems, in order to better investigate the relaxation process inside the mixed interfacial layer [36, 37]

There are other theoretical works available in literature concerning dilational rheology and surface relaxation processes. Among these Noskov et al [38, 39] developed theoretical approaches for polymers at liquid interfaces taking into account different surface kinetic processes besides diffusion. Though in these works the used thermodynamic approach differs from the present one, the resulting frequency behaviour of the dilational visco-elasticity has the same feature as that calculated by Eq. (9). and in particular the limit cases of very fast surface processes and insoluble monolayer lead to the same equations.

The importance of appropriate theoretical models for systems undergoing mixed diffusion and kinetic processes is clear from several recent works concerning multicomponent adsorbed layers or more complicated mixtures of surfactant and polymers [40, 41, 42, 43]

Other systems important for practical application are those undergoing phase transition or surface aggregation [44, 45, 46] whose dynamics characteristics cannot be properly described by classical theoretical models.

C. THE OSCILLATING DROP/BUBBLE METHODS

From the above considerations it follows that to investigate the dynamic behaviour of an adsorbed layer, it is important measuring dilational visco-elasticity in suitable frequency ranges especially around those characteristic frequencies corresponding to the time scale of the relaxation processes involved in adsorption dynamics. To this aim drop/bubble tensiometers are particularly suitable when they are used according to the oscillating drop methodology to measure the visco-elasticity versus frequency.

In order to enlarge the frequency range, the oscillating drop technique is often applied coupling two different tensiometers, one based on the drop profile acquisition which is appropriate at low frequency and the other one based on the pressure measurement providing the visco-elasticity in a larger frequency range which can arrive to few hundreds Hz, depending on the fluid properties and on the measuring cell geometry.

Whatever is the tensiometer employed for the oscillating drop experiments, the surface dilational modulus $E(\omega)$ is obtained by imposing a frequency sweep to the surface area A. For each angular frequency ω it holds:

$$A = A^0 + \tilde{A} \sin(\omega t) \tag{12}$$

where A^0 is the reference surface area and $\tilde{A}$ the amplitude of the area oscillations[*].

The perturbation produces an harmonic response of the surface tension γ with the same frequency, *i.e.*

$$\gamma = \gamma^0 + \tilde{\gamma} \sin(\omega t + \phi) \tag{13}$$

where γ^0 is the equilibrium reference surface tension and $\tilde{\gamma}$ the amplitude of the surface tension oscillations. The phase shift ϕ between the area perturbation and the response of the interfacial tension, is the phase of the complex dilational modulus, which is related to the viscous aspect of the interfacial layer. In fact, according to its definition, Eq, (2), E at a given frequency reads,

$$E = A^0 \frac{\tilde{\gamma}}{\tilde{A}} \cos\phi + i\, A^0 \frac{\tilde{\gamma}}{\tilde{A}} \sin\phi \tag{14}$$

Eq. (14) provides an expression for E in terms of quantities that can be determined experimentally, either in a direct way or by an appropriate calculation procedure, depending on the utilized tensiometer.

1. Drop/ Bubble Shape Tensiometry

A bubble or a drop of one liquid inside another fluid, under the gravity effect, assumes a shape which minimises the total energy of the system and depends on the surface tension and on the density difference between the two fluids. For axis-symmetric drops, the shape profile is provided by the Bashforth-Adams equation [47], which is an integration of the Laplace equation, expressing the mechanical equilibrium at each point of the surface. In the Drop/Bubble Shape tensiometer (DST), the surface tension is obtained by fitting such theoretical drop profile to the acquired one.

The DST is commonly used for measurements of equilibrium interfacial tension, for example of solutions at different surfactant concentration for adsorption isotherm investigations, and also for

---

[*] Here and in the whole text ~ indicates the amplitude of the oscillating quantities while the apex 0 means that the quantity is calculated at the reference state

measuring the dynamic interfacial tension during the ageing of a fresh surface at constant area (adsorption kinetics) or again during slow variations of the surface area to investigate interfacial rheology at low frequency.

According to its working principle, there are two fundamental requirements for the application of this technique: the two fluids involved must have an appreciable density difference and the interface must be not far from the mechanical equilibrium.

In fact if the liquids are isodense the drop is spherical whatever is the interfacial tension which then results undetermined. The dimensionless Shape Factor $\beta = \Delta\rho \, gb^2/\gamma$, where g is the gravity acceleration, $\Delta\rho$ is the density difference and b the curvature radius at the drop apex, gives an estimation of the level of drop deformation due to gravity effect. In many practical cases, with the most common acquisition systems, accurate data are obtained for $|\beta| > 0.1$.

The requirement of mechanical equilibrium of the interface implies that only slow variation of the surface area can be applied to obtain accurate measurements. For rheology studies by oscillating drop experiments, this sets an upper limit in the frequency range. Specific investigations have shown [48] that, for amplitude of the area oscillations below 10%, the drop can be considered at mechanical equilibrium for frequencies below 1 Hz. This condition holds for water-air systems while for more viscous liquids or liquid-liquid interfaces, the limit frequency reduces to about 0.1 Hz.

A typical DST is composed by a cell where a drop or a bubble is formed inside the other fluid, continuously monitored by a video-camera coupled to a computer. The drop profile is then acquired in an automatic way in order to calculate the surface tension by means of a numerical fitting procedure. Different algorithms and procedures have been proposed and are available for these two tasks [11, 49, 50, 51. 52].

In modern version of DST a controlled variation of the surface area is ensured by a feed back control algorithm utilising the synchronous acquisition of the surface tension and of all the geometrical characteristics of the drop.

The implementation of this control task led to a large improvement of the technique, allowing for example to investigate adsorption kinetics under really constant surface area conditions. At the same it has been a key step for the utilisation of DST for dilational rheology investigation. Indeed, for dilational visco-elasticity measurements, γ is acquired while a controlled harmonic perturbation of the surface area is applied. At each frequency the complex dilational visco-elasticity is calculated by the acquired area and surface tension signals according to Eq. (14). In practice, assuming the area and interfacial tension response to vary according to Eqs. (12) and (13), respectively, the amplitudes $\tilde{A}$ and $\tilde{\gamma}$ and the phase shift ϕ are obtained as amplitudes and phase of the components of frequency ν, extracted by the experimental signals via DFT (Discrete Fourier Transform) algorithms as explained in details in refs. [53, 54].

In Fig. 4 data acquired during the surface oscillation are reported as an example with the extracted theoretical harmonics utilised for the calculation of E.

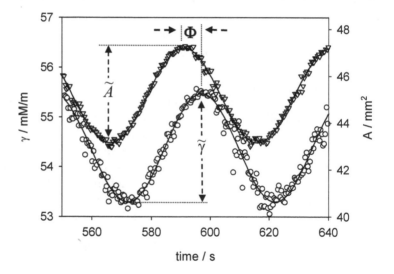

Fig. 4 Example of acquired surface tension (o) and surface area (∇) by the Oscillating Drop method in the DST, at $\omega/2\pi=0.02$ Hz.

2. Drop / Bubble Pressure Tensiometry

The Capillary Pressure Tensiometer (CPT) is based on the direct measurement of the pressure difference between two phases separated by a spherical (or nearly-spherical) interface of a drop/bubble. Indeed, according to the Laplace equation, such pressure, P, is directly linked to the interfacial tension

$$P = \frac{2\gamma}{R} + P^0 \tag{15}$$

where R is the drop radius (or the curvature radius at the apex of the drop) and P^0 can be either an hydrostatic constant or a fluid-dynamic term, depending on the adopted experimental methodology. The dynamic interfacial tension can be then inferred from the above relationship by knowing P, R and P^0 versus the time.

A typical CPT is composed by two chambers connected by a capillary tube, as schematically sketched in Fig 5. One of the chambers is closed and contains both the pressure sensor and a piezoelectric rod. The latter is utilised to control the volume of the drop (or bubble) formed at a tip of a glass capillary. A video camera allows for a continuous monitoring of the drop. Submillimetric droplets are typically utilised to obtain measurable capillary pressure values, which are typically of the order of a hundred Pascal. The drop radius is either measured by direct imaging or calculated from the injected liquid volume, if the compressibility of the closed phase is known. The other cell either is open to atmospheric pressure or, if closed, contains another pressure sensor. Technical details about this kind of tensiometer can be found elsewhere [55, 56, 57]. An important characteristics of this technique is that it does not require gravity deformed droplet but it works preferably with spherical interfaces, which makes it suitable both for liquid-liquid and liquid-air systems, using either small drops or small density difference.

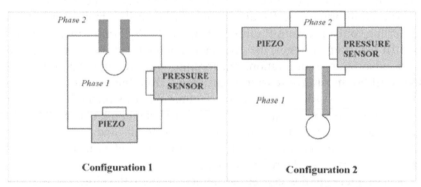

Fig. 5 Sketch of the measurement cell of the Capillary Pressure Tensiometer according to two possible configurations. In both cases the closed part of the cell is equipped by a pressure sensor and a piezoelectric driver.

The CPT has been widely utilised in many experimental investigations of adsorbed layers based on the measurement of equilibrium and dynamic interfacial tension and of the dilational visco-elasticity [56], both at water/air and water/oil interfaces. Moreover, this tensiometer is also the core of an apparatus (FAST - Facility for

Adsorption and Surface Tension) utilised for studies about surfactant adsorption at liquid-liquid and liquid-air system under weightlessness conditions (microgravity) [58, 59] on board the Space Shuttle. More details about aims and results of these experiments are given in section D.4.

The CPT is a versatile technique which can be used according to different experimental methodologies [55], to investigate dynamic and equilibrium aspects of pure and composite surfactant systems. For example, according to the growing drop experiment, the drop volume is continuously increased while the pressure is measured. If the surface tension is constant, the Laplace equation provides a linear relationship between the pressure and the curvature, and the interfacial tension can be obtained from the slope of the pressure-curvature straight line [60]. By this method it is possible to measure both the interfacial tension of pure systems and equilibrium interfacial tension of surfactant systems presenting adsorption kinetics fast with respect to the drop growing rate. The method has also been proposed to investigate adsorption kinetics [61]. This method does not require the direct drop imaging, being possible to calculate the drop curvature from the volume rate.

The growing drop experiments are also used for side measurements in dilational studies [56] to evaluate the compressibility of the system.

The equilibrium interfacial tension can be easily obtained in a CPT according to the Pressure-Radius Step method [56]. For a certain number of different drop volumes, the drop radius and the pressure are measured after adsorption equilibration. Interpreting the pressure-curvature data the equilibrium surface tension is calculated from the slope of the fitting straight-line, according to the Laplace equation. This method is used for equilibrium adsorption studies versus the surfactant bulk concentration to access the adsorption isotherm of the system, and also for side measurements in the framework of rheological studies to evaluate the interfacial tension of the reference state.

Rheological studies can be performed in the CPT either by the low amplitude stress/relaxation method [62] or by the Oscillating Drop/Bubble [63, 64]. In the first one the layer is perturbed with a low amplitude step function of the surface area and, because the interface behaves as a linear system, the response of the surface tension due to the surface relaxation process is given by eq. (3).

The Oscillating Drop/Bubble technique in the CPT is based on the acquisition of the pressure response to harmonic perturbations of the interfacial area of a drop/bubble. This method was introduced several years ago [65] and used by many authors with different experimental set-ups, more and more upgraded and with different approaches for the interpretation of acquired experimental data [21, 56, 63, 66, 67, 68].

In this technique, a sinusoidal volume variation of piezoelectric rod, is applied,

$$V_{pz} = V_{pz}^0 + \tilde{V}_{pz} \, sin(\omega t) \qquad (16)$$

This provides, in response, a harmonic oscillation of the interfacial area and, consequently, of the interfacial tension. The measured pressure is an oscillating quantity, with a phase shift, φ, in respect to the forcing V_{pz}, i.e. (39)

$$P = P^0 + \tilde{P} \, sin(\omega t + \varphi) \qquad (17)$$

$\tilde{P}$ and φ are obtained as amplitude and phase of the component of frequency ν, extracted by the experimental signal via DFT algorithms. Considering the sampling of the pressure signal at the times tj, the j-th element of such harmonics is,

$$\Delta P_j = A \, cos(\omega t_j) + B \, sin(\omega t_j) \qquad (18)$$

where $A = \dfrac{1}{N} \sum\limits_{k=1}^{N} P_k 2 \, cos(\omega t_k)$, $B = \dfrac{1}{N} \sum\limits_{k=1}^{N} P_k 2 \, sin(\omega t_k)$, N is the number of experimental points and P_k is the measured pressure value at the time t_k. Thus one obtains,

$$\tilde{P} = \sqrt{A^2 + B^2} \qquad (19)$$

$$\varphi = tan\left(\frac{B}{A}\right) \qquad (20)$$

In a typical oscillating drop experiment the pressure signal is acquired for a number of cycles sufficient to warrant the achievement of stationary oscillations and a significant statistics. For each frequency, a couple of $\tilde{P}, \varphi$, is obtained from which E can be calculated.

In fact a specific experiment theory is needed to calculate the dilational visco-elasticity from the acquired pressure signal because the measured pressure during the oscillating drop experiment is the superposition of different fluid dynamic contributions.

To this aim several methods have been used, depending on the experimental conditions and on the studied system. Some of them need either reference measurements on the pure systems, without surfactants

[63], or the knowledge of specific characteristics of the studied sample and of the experimental cell [66], which are hardly obtained with suitable accuracy.

One of the most important problems connected with the data interpretation is originated by the compressibility of the closed phase (cell plus liquid) which, in general, cannot be neglected because of the large ratio between the droplet and the cell volume.

Recently [33] a method has been proposed to calculate ε from the acquired pressure and piezo volume signals. The measured pressure is theoretically calculated considering the contributions not only of the capillary pressure but also of the Poiseuille pressure due to the fluid motion through the capillary and of hydrodynamic inertial terms. In this way an expression of E can be found without the need of reference measurements on the pure system.

The main characteristics of this method is the modelling of the closed part of the measurement cell as equivalent to an incompressible liquid containing an amount of gas of volume V_g , related to the pressure P through the perfect gas law. A relationship between this equivalent volume and the compressibility C can be easily found according to the definition of this latter, i.e.

$$C = -\frac{1}{V}\frac{\partial V}{\partial P} \cong \frac{1}{P_{atm}}\frac{V_g^0}{V^0} \tag{21}$$

where V^0 is the volume of the closed phase (phase 2 in the configuration 1 and phase 1 in the configuration 2). Thus using the present model, the compressibility can be expressed through the ratio between the effective gas volume and the total volume of the closed phase.

As an example, considering a drop/bubble radially varying its volume inside a liquid for the configuration 1 of the CPT, the measured pressure variation is obtained as

$$\Delta P = \frac{2\gamma\Delta R}{R^{0^2}} - 2\frac{\varepsilon}{R^0}\frac{\Delta A}{A} - 4i\omega(\eta_1 - \eta_2)\frac{\Delta R}{R^0} + \rho_1\omega^2 R^0\Delta R + (\omega^2 K_2 - i\omega K_1)\Delta V_d \tag{22}$$

where θ is the phase shift of the drop volume V_d, which also varies harmonically, K_1 and K_2 only depend on the fluid characteristics density ρ and the viscosity η - and on the geometry of the capillary tube - length L and internal radius a -

$$K_1 = \frac{8\eta_2 L}{\pi a^4} \tag{23}$$

$$K_2 = \frac{4\rho_2 L}{3\pi a^2} \tag{24}$$

Using the expressions of the oscillating functions and taking only the first order terms, after some re-arrangements one obtains the complex $E(\omega)$

$$E = -\frac{\pi h^0 R^{0^3}}{2} \frac{\tilde{P}}{\tilde{V}_d} e^{i(\varphi-\theta)} +$$

$$\frac{(h^0 - R^0)R^0}{2h^0}\left[\frac{2\gamma^0}{R^0} - 4i\omega(\eta_1 - \eta_2) + \rho_1\omega^2 R^{0^2}\right] + \frac{\pi h^0 R^{0^3}}{2}\left(\omega^2 K_2 - i\omega K_1\right) \quad (25)$$

$\tilde{V}_d$ and θ are also functions of the measured quantities and can be found through the following relationship (valid for configuration 1) in terms of the effective gas volume Vg,

$$\frac{dV_d}{dt} = -\frac{dV_{pz}}{dt} - \frac{dV_g}{dt} = -\frac{dV_{pz}}{dt} + \frac{V_g^0}{P^0}\frac{dP}{dt} \quad (26)$$

Using Eq. (25) requires also the side evaluation of γ^0 and V_g^0. While γ^0 is obtained by a pressure-radius step experiment right before the oscillating drop experiment, the effective gas volume at the reference state can be determined executing a growing drop experiment, where the piezo is made decrease its volume at continuous rate, w, while the drop volume and the pressure are acquired. Assuming the drop volume V_d to reach its reference value V_d^0 at the time t_0 and integrating Eq. (26), between t_0 and a generic time t, after some arrangements one obtains

$$V_g^0 = (P + P_{atm})\frac{V_d - V_d^0 - w(t - t^0)}{(P - P^0)} \cong P_{atm}\frac{V_d - V_d^0 - w(t - t^0)}{(P - P^0)} \quad (27)$$

Eq. (25) has been derived on the assumptions of incompressible and quasi stationary conditions of the fluid inside the capillary [68]. This sets a frequency limitation in the applicability of this approach, depending on the value of the capillary internal radius. For a typical value of a = 0.3 mm, this limit frequency is of the order of 100 Hz, for liquid-gas (gas inside the capillary) and of 10Hz for liquid-liquid.

In Fig. 6, an example of experimental pressure amplitude and phase shift data is reported with the corresponding calculated visco-elasticity data.

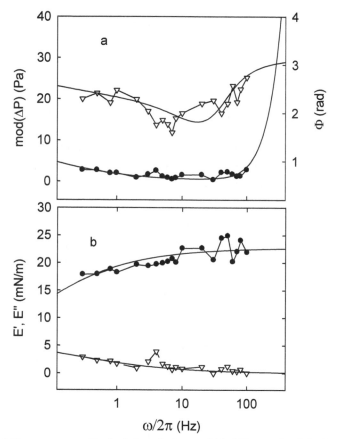

Fig. 6 Pressure amplitude (•) and phase shift (∇) acquired during an oscillating drop experiment in the CPT, (a), for $C_{10}E_4$ 2×10^{-5} M solution/air, and corresponding calculated values of the real (•) and imaginary (∇) part of E, (b).

D. EXPERIMENTAL RESULTS IN DILATIONAL RHEOLOGY

1. Non ionic surfactant at water/air and water/oil interface

There are several experimental studies in literature concerning the dilational rheology of surfactant adsorption layer and most of them are based on experimental techniques exploiting drop/bubble tensiometers.

Investigating dilational rheology of surfactant systems is important for different reasons. Firstly, these experimental studies can provide a better understanding of the processes occurring in the adsorbed layers, because of the strict link existing between the characteristics of macroscopic quantities, like dynamic interfacial tension and dilational visco-elasticity, and the adsorption mechanisms which are often driven by the molecular characteristics of surfactants. This approach is particularly effective if appropriate theoretical models are utilised for interpreting the experimental data and, as better explained in previous sections, these latter have to be acquired in a range of frequency adequately wide.

Many experimental results are available dealing with rheological properties of surfactant systems in relation with the adsorption mechanisms. Among them, those reported in refs. [33, 29, 69] concern polyoxyethylene glycol ethers (C_iE_j) investigated at water/air and water/hexane interfaces. Due to their molecular structures these surfactants can present different possible equilibrium states when they are adsorbed at the fluid interface. These states are characterized by different molecular orientations with respect to the interface. The thermodynamic characteristics of these systems can be accessed already by equilibrium measurements, providing the adsorption isotherm proper of the system [57,70]. Nevertheless the possibility of different states for the adsorbed molecules implies the existence of a surface re-orientation process during the adsorption dynamics, which can be investigated by measuring the dynamic interfacial tension and the dilational visco-elasticity as a function of the frequency. By means of these last measurements it was possible, not only to evidence the presence of a surface relaxation process driving the adsorption together with the diffusion, but also, by interpreting the data with a suitable theoretical model, to evaluate the re-orientation characteristic time.

As an example, in Fig. 7, the equilibrium interfacial tensions of $C_{10}E_4$ at water/hexane interface are reported, from which it was found that this system can be suitably described by the two-state adsorption model. For the same system, the dilational visco-elasticity (Fig, 8) was measured by the oscillating drop techniques in a frequency range from 0.01 to 100 Hz, coupling the Drop Shape and the Capillary Pressure tensiometers. From Fig. 8 it is clear that the re-orientation approach describes better the experimental data with respect to diffusion controlled approach in the studied frequency range.

This is a case in which the dilational rheology provides an effective tool for accessing the kinetic features of a surface process. From the visco-elasticity data here obtained, in fact, even if the frequency range is not yet adequate to evidence the second maximum in the imaginary part, it is possible to give an estimation of the orientation rate, k_{or}.

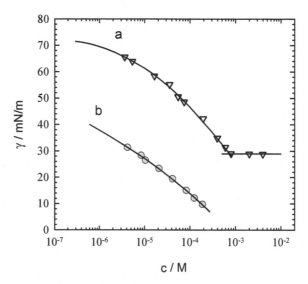

Figure 7: Equilibrium surface tension versus surfactant concentration in water for $C_{10}E_4$ at water/air (a) and water/hexane (b), measured by DST at T =20 °C. The theoretical adsorption isotherm are according the two-state model.

All the results reported show that a better understanding of adsorption properties is possible only by a contemporaneous investigation of equilibrium properties, adsorption kinetics and surface rheology.

Another very important aspect of dilational rheology of surfactant adsorbed layers concerns the application to technological fields where high specific surface systems are involved, like foams and emulsions, where ionic and non-ionic surfactants are employed as stabiliser additives [1,4].

In fact, being by definition the response of the interfacial layer to dilational stresses, the dilational rheology is expected to have an important role in damping the external area disturbances. Thus the behaviour of the visco-elasticity can be relevant for the stability conditions of liquid films and consequently for the drop or bubble coalescence [71, 72, 73,74,75, 76] in emulsions and foams.

Experimental studies based on measurements of interfacial tension and dilational visco-elasticity can then be useful to investigate the relation between the single interfacial properties and the stability of emulsions and foams, in order to improve their formulation by the choice of suitable stabilising additives.

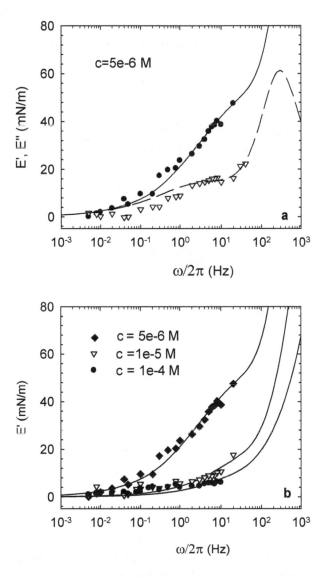

Figure 8: Real •) and imaginary (∇) part of E versus frequency for $C_{10}E_4$ at water/hexane, a, and real part of ε versus frequency for different concentration of $C_{10}E_4$ in water, b. Experimental data obtained by DST, for $\omega/2\pi < 0.2$ Hz, and by CPT for higher frequency and theoretical curves by the re-orientational model according to Eq. (9).

In ref. [77] the dilational visco-elasticity for two non-ionic surfactants n-dodecyl-β-D-maltoside   (β-$C_{12}G_2$)   and   tetraethyleneglycol-monodecylether ($C_{10}E_4$) was measured as a function of the concentration in a frequency range from 0.005 to 100 Hz, in order to investigate the correlation with the liquid film stability properties.

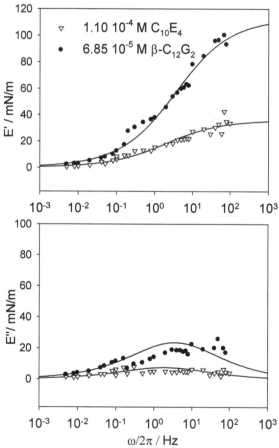

Figure 9: Real (E') and imaginary (E') part of the E versus frequency for β-$C_{12}G_2$ at c = 6.85×10^{-5} M ($\bullet$) and for $C_{10}E_4$ at c = 1.1×10^{-4} M (∇). Theoretical curves according the diffusion controlled model Eq. (6).

The same surfactants had been used as film stabilizers in a previous study [78] based on disjoining pressure measurements. The stability conditions for these systems were not completely explained accounting only for the surface forces. In fact it was found, for example, that films presenting similar values of the surface charge had completely different stability conditions.

As expected a correlation was found in ref. [77] between the film stability and the values of the dilational elasticity. This is clear from the data reported in Fig. 9, considering that the films produced by β-$C_{12}G_2$ solutions result more stable than those by $C_{10}E_4$. Moreover the measurements performed as a function of the concentration for the $C_{10}E_4$ solutions (Fig. 10) evidence a correlation between the elasticity at high frequency and the film stability. In fact the stability of films obtained with the same $C_{10}E_4$ solutions increase with increasing concentration.

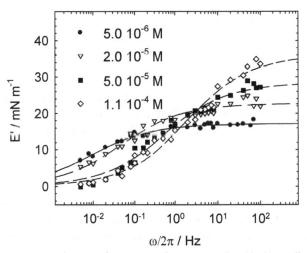

Figure 10: Real part of E versus frequency for $C_{10}E_4$ at four different concentrations, Experimental data obtained by DST for $\omega/2\pi < 0.2$ Hz and by CPT for higher frequency. Theoretical curves calculated according to the diffusion controlled model Eq. (6).

The same correlation with the high frequency values of the dilational elasticity has been found for a completely different systems consisting of an oil soluble surfactants used to stabilize water in oil emulsions [79]. Also in this study the dilational visco-elasticity has been measured for different surfactant concentration according to the oscillating drop coupling the DST and CPT techniques. A set of data is reported in Fig. 11 obtained for a micellar solution of Span- 80 in paraffin oil at the water interface. One of the conclusion drawn from this study was that higher values of high frequency limit of the elasticity of the water/oil interface corresponds to more stable emulsions.

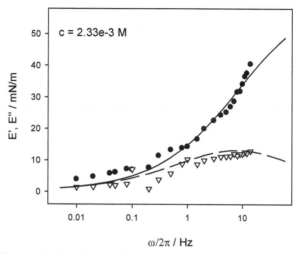

Figure 11: Real (•) and imaginary (∇) part of dilational visco-elasticity, for 0.1 wt% of Span80 in paraffin oil. Experimental data obtained by DST for $\omega/2\pi < 0.2$ Hz and by CPT for higher frequency.

2. Polymers, proteins and mixed systems

Monolayers of adsorbed polymers with amphiphilic character have been often investigated mainly because of their application to fields of practical interest such as cleaning, lubrication, coating, foaming, and emulsification.

The principal characteristics of these systems concern the conformational changes or phase transitions occurring at the air/solution interface during the adsorption kinetics, even while the surface tension remains almost constant. Thus these relaxation processes, which may have important consequences in the dynamic mechanical properties of the interface, can be more effectively investigated by dilational rheology, compared to simple dynamic surface tension studies. Like for common surfactants in fact the dilational visco-elasticity seems to be more strictly related to the microscopic state of the monolayer (molecular conformations etc.).

Another characteristics of these systems is the very long time required for reaching the equilibrium state, due to diffusion and slow reorganization processes.

Oscillating drops and bubbles methods are utilized to investigate the visco-elastic behaviour of adsorbed and spread polymer films often in association with other surfactants at the air-solution interface [80. 81, 82, 83, 84]. In some of these studies the drop technique is used to measure dilational visco-elasticity versus frequency and often it is combined with other methods to ensure a broad frequency range, like oscillating barriers and electrocapillary waves which work between 100 Hz and 400 Hz [80, 81].

Oscillating drop experiments utilising drop shape tensiometers are also used to investigate the evolution of the visco-elasticity during the equilibration of the adsorbed layer at fixed frequency of the area perturbation. In this kind of experiment an initially fresh surface is monitored during the adsorption re-equilibration acquiring the interfacial tension while a harmonic oscillation with a period much shorter than the time of the adsorption re-equilibration is imposed to the surface area. The resulting dynamic interfacial tension is the superposition of a decreasing behaviour, characteristic of the adsorption kinetics, plus an oscillation at the same frequency of the surface area. At each period a value of ε can be then calculated according to Eq. (14), after calculating the interfacial tension and surface area oscillations amplitudes and phases via DFT. The method provide then E vs. time during adsorption kinetics. Associating to each of these ε the value of the average surface tension during the oscillation, it is possible from a single experiment to derive information about the dependence of E on the surface status: either on the surface pressure or on the adsorption, if the equation of state is known. This kind of studies have been utilised especially to investigate polyelectrolyte/surfactant mixed interfacial layers and aggregation processes at the water/air interfaces [82 ,83].

Oscillating drop methods based on drop shape tensiometers are also broadly utilised to investigate dilational rheology of protein and protein-surfactant adsorption layers. In fact also these systems are characterized by long time evolution of the adsorbed layer with slow unfolding and rearrangement within the interfacial layer. These investigations are important in a broad field of applications that spans from human physiology and bio-medical science to food technology, where protein adsorption underlie important biological processes or proteins are employed as emulsion and foams stabilizer.

Studies on dilational rheology of phospholipids at water/air interfaces, in association with natural pulmonary surfactant proteins, have been utilised to investigate respiratory physiology [85]. Oscillations of the interfacial area of bubbles simulate in this case the behaviour of alveolar surfaces covered by this kind of proteins, during breathing cycles [86].

Concerning proteins important for food science and technology, many investigations on dilational rheology are available, both for liquid-air and liquid-liquid interfaces [28, 41, 87, 88, 89], where the visco-elasticity change is acquired during the adsorption by an oscillating drop technique. In other studies, protein-surfactant mixtures have been investigated, allowing for a better comprehension of the phenomenon of the competitive adsorption with ionic and non-ionic surfactants and for the development of suitable theoretical models for the interpretation of the visco-elasticity data [88, 89].

3. Nanoparticles at fluid interface

The effect of nanoparticles and more in general of surfactant/particle structures at fluid interfaces is an important scientific topic which also finds applications in emulsions and foams technology and still presents many open questions.

Depending on their wettability, nanoparticles tend to transfer from the bulk of the dispersion to the surface. Such surface accumulation provides a change in the mechanical properties of liquid/air or liquid/liquid interfacial layers, that can be either the construction of quasi solid interfacial barriers, like particle networks, or, for lower surface coverage, a variation of the interfacial tension and of the dilational rheology. For these reasons nanoparticles can play a role in preventing coalescence of drops and bubbles and have a stabilizing effect for emulsions and foams.

Understanding the particle/fluid interface interactions and their effect on the macroscopic interfacial properties, with particular attention to the low particle surface coverage, is the subject of recent experimental works [36, 37] where mixed silica nanoparticle - cationic surfactant layers have been investigated at the dispersion/air and dispersion/oil interfaces, using drop/bubble tensiometers and performing rheological investigation according to oscillating drop/bubble techniques.

In these studies the solid surface of the nanoparticles was modified by the adsorption of the cationic surfactant, hexadecyltrimethylammonium bromide, (CTAB). This makes them vary their characteristics from completely hydrophilic to partially hydrophobic so that the affinity with the water/air or water/oil interface increases with the increasing of surfactant concentration.

In the case of partial wetting of particles, at nanometric scale, the interfacial layers is a multiphase zone for which a rigorous definition of the interfacial tension is a critical item. In refs. [36, 37] this problem has been overcome by the definition of an *effective interfacial tension*, as the quantity entering in the mechanical equilibrium conditions of a macroscopic interface. Owing to its definition this effective quantity can be unambiguously measured by tensiometric methods exploiting the Laplace equation, like bubble/drop shape and capillary pressure methods.

In the following, for sake of brevity, the terms "interfacial tension" is utilised, standing for the above defined effective quantity.

The principal results obtained in these studies concern the variation of the effective interfacial tension due to the transfer of nanoparticle into the interface and the occurring of relaxation process in the composite surfactant/nanoparticle layer.

In ref. [36], a decrease of the equilibrium interfacial tension with the CTAB concentration was observed due to the increasing nanoparticle concentration in the interfacial layer, both for dispersion/air and dispersion/oil interfaces. (Fig. 12).

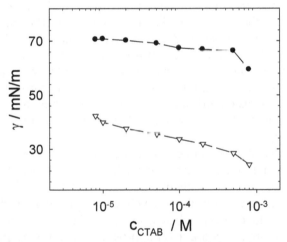

Figure 12: Equilibrium effective interfacial tension versus CTAB concentration.of silica dispersion / air, (●), and silica dispersion / hexane, (∇), systems The silica nanoparticle concentration was of 1 wt %. Data obtained by DST.

Moreover the dynamic interfacial tension was found to vary according to a multiple kinetic process, leading to two major conclusions: the incorporation of nanoparticles into the interfacial layer modifies the equilibrium interfacial tension and different kinetic processes are involved in the equilibration of such composite system. To corroborate such conclusions the low frequency surface dilational rheology was also investigated and it was found a dependence of the dilational visco-elasticity on the surfactant concentration, indicating the presence of an interfacial relaxation process.

In fact, the observed dynamic behaviour, which typically occur when the particle density at the interface is not so high to saturate it, is expected to be due to the bulk particle diffusion with consequence attachement to the fluid interface, but also to other mechanisms involving particle – particle interactions, surface aggregation or other re-organisation interfacial processes.

To advance in the comprehension of the dynamic properties of the dispersion interfacial layer, in ref. [37] the same composite system was further investigated by measuring the dilational visco-elasticity of the dispersion/hexane at higher frequency.

In these rheological studies, measuring the dilational visco-elasticity versus frequency and interpreting the data with the theoretical model according to the approach presented in section B.3, provided further information on the transport processes and on the dynamics of the interfacial layer. In fact it was found that, when the CTAB concentration is low, i.e. at low nanoparticle hydrophobicity, the visco-elasticity varies with the frequency according to a diffusion controlled model (Eq (6)), see Fig. 13. Increasing the surfactant concentration a second kinetic process - probably a reorganisation of the interfacial particle layer – appears, sided to diffusion, driving long time equilibration, as proven by the better agreement of the experimental data with theoretical curves derived by Eq. (9).

Another important aspect concerns the dependence of the $\varepsilon(v)$ on the age of the interfacial layer. In fact, it has been observed that, during the ageing of the layer, even without any appreciable variation of the interfacial tension, $\varepsilon(v)$ shift from a behaviour typical of a diffusion controlled system to one typical of a mixed kinetics. Increasing time diffusion seems more and more less important, so that at very long time the system behaves almost as an insoluble layer (Fig. 14). This observed shift from diffusion-controlled to kinetic-controlled regime during the ageing of the layer calls for an irreversible aspect of the nanoparticle transfer into the fluid interface, appearing above a certain degree of hydrophobicity.

This phenomenon has afterwards been confirmed by other observations of the oil droplet behaviour in dispersion phases and related to the corresponding emulsion stability. However the important point to underline in the present context is that dilational rheology and in particular the experimental studies based on oscillating drop/bubble techniques can provide a very effective tool to investigate the dynamic aspect of this complex systems. In fact the presence of surface relaxation processes besides the diffusion and the evaluation of their characteristic times are provided by these experimental studies even without a deep knowledge of the mechanisms at the basis of them.

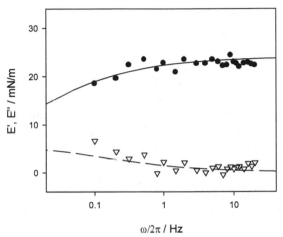

Fig. 13 Real (•) and imaginary (∇) part E versus frequency for dispersion/heaxane system with 1% of silica nanopartices and c_{CTAB}=8×10^{-6} M. Experimental data obtained by CPT. Theoretical curves according to Eq. (9) with E_{0G}=24.0 mN/m and $\omega_D/2\pi$=0.01 Hz: values found fitting Eq.(6) which provided $E_{0\Gamma}$ = E_G and ω_k undetermined.

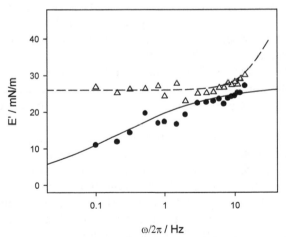

Fig. 14 Real part of E versus frequency for silica dispersion / heaxane system with 1% of silica nanopartices and c_{CTAB}=1×10^{-5} M. Data obtained by CPT after 600 s (•) and after 2000 s (Δ) from the formation of the drop. Theoretical E' according to Eq. (6) (solid line) and with Eq. (10) (dashed line).

4. Microgravity experiments

Microgravity conditions aboard of orbital vehicles present a unique experimental environment for the study of interfacial and transport phenomena in fluid systems. As the hydrostatic pressure is vanishing to zero, the shape of the interfaces is exclusively determined by the surface tension and by the boundary conditions. Moreover all fluid processes driven by capillary forces and by differences in chemical potentials are not affected by gravity-driven convection. These conditions make the microgravity environment particularly suitable for studying the time evolution of the adsorption process at fluid interfaces, for which a simple geometry of the interface is preferably required together with the absence of gravity-driven convection, which could modify to a large extent the bulk diffusion profile near the interface.

Experiments about the dynamic adsorption properties at fluid interfaces were performed by the FAST instrument (Facility for Adsorption and Surface Tension), aboard the Space Shuttle Columbia during the STS-107 scientific mission [90]. FAST implements two capillary pressure tensiometers used according to different methodologies: oscillating drop/bubble and transient relaxation experiments, to investigate both the liquid/liquid and liquid/air interfaces.

During this scientific mission, measurements of $\varepsilon(\nu)$ were performed according to the Oscillating Drop/Bubble methods, at different temperatures and surfactant concentrations. The principal results of these studies are reported in Refs. [91, 92, 93] .

The low frequency oscillating drop/bubble techniques was applied, determining the surface/interfacial tension with great accuracy through the direct measurement of the drop/bubble radius and the corresponding capillary pressure.

Values of the modulus of the dilational visco-elasticity as a function of the frequency obtained for $C_{10}E_8$ (octaoxyethylenmonodecylether) at water-hexane [92] for different initial bulk concentrations in water are reported as example in Fig. 15. The concentrations in water were chosen in order to span from very dilute monomeric to micellar solutions. It is worth to note that an accurate measurement of E is possible in microgravity even at the latter large concentrations.

Under these conditions, due to the very small interfacial tension (a few mN/m), these measurements are nearly impossible on ground, in particular when increasing the frequency, since the interface is extremely deformed and unstable because of the drop weight. The curves reported in Fig. 15 are the predictions of the model accounting for the surface reorientation process previously developed and already applied to other C_iE_j surfactants [30]. These experiments provided then a further check of the validity of such model. As underlined in previous section, at given frequency, E is determined by the equilibrium isotherm parameters, as well as by the kinetic parameters, which are in this case the bulk diffusion coefficients and the rate, K_{or}, of the re-orientation process. Such kinetic parameters have been determined as best fit values, while the isotherm parameters were taken from side equilibrium measurements.

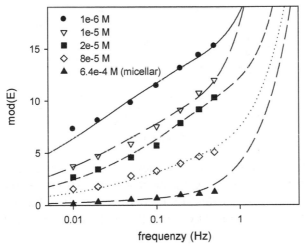

Fig. 15 Modulus of E versus frequency at T= 15 °C for $C_{10}E_8$ at water/hexane for different surfactant concentrations. The theoretical curves are the best fit according to the two-state model. Data obtained by the FAST modulus during the flight experiment.

The low frequency dilational rheology of $C_{12}DMPO$ (dodecyl-dimethy-phosphine-oxide) has been investigated with similar oscillating bubble experiments at the water/air interface, by varying concentrations and temperatures.

Figure 16 shows the frequency dependence of the modulus of E, as obtained using Fourier series analysis on the acquired $\gamma(t)$, $A(t)$ data [53,54] for each concentration.

As shown in Fig. 17, the Lucassen-van den Tempel model provides an excellent description of the elasticity data. That means that the adsorption is controlled by diffusion at least within the time scale corresponding to the investigated low frequency range. The best fit of

Eq. (6) equation provided the thermodynamic parameter E_0, (which is in this case also the limiting elasticity) and the characteristic frequency v_D.

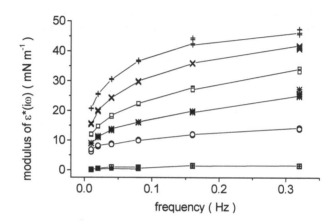

Fig. 16 Modulus of E versus frequency at T= 25°C, for $C_{12}DMPO$ at water / air for increasing surfactant concentrations: (from top to down) 2.95×10^{-5}, $4.22\ 10$-5, 6.78×10^{-5}, 1.19×10^{-4}, 2.20×10^{-4}, 4.21×10^{-4}, 8.17×10^{-4} M. Data obtained by the FAST modulus during the flight experiment.

Trapezoidal small amplitude perturbations have also been utilised as an alternative way to investigate the dilational properties of the same system in the time domain. An example of such investigations is reported in Fig.18.

The theoretical curve is obtained on the bases of Eq. (3), assuming a trapezoidal variation of the surface area and diffusion controlled model, as described in ref. [62]. It is important to notice that, the best fit values of ε_0 and v_D, obtained in this way are substantially coincident with those determined via Eq. (6) from the data obtained in the frequency domain by the oscillating bubble experiments. This is a direct proof that the investigated adsorption layer system behaves as a linear system over a wide perturbation amplitude range.

The rheological properties of the same non ionic surfactant at the water/air interface have been investigated in the FAST at higher frequency by a pressure based oscillating bubble methodology [93].

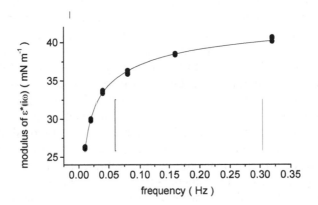

Fig. 17: Modulus of E versus frequency at T= 25°C, for $C_{12}DMPO$ at water / air for surfactant concentration of 1.67×10^{-5} M. Best fit theoretical curve by Lucassen - van den Tempel model with E_0 = 46.6 mN/m and $\omega_D/2\pi$ = 0.0148 Hz. Data obtained by the FAST modulus during the flight experiment.

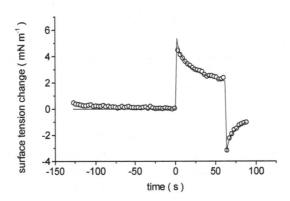

Fig. 18: Surface tension response vs. time, consequent to a trapezoidal pulse perturbation of interfacial area. Best fit theoretical curve according to Eq. 3 with E_0 = 46.1 mN/m and $\omega_D/2\pi$ = 0.0157 Hz Data obtained by the FAST module during the flight experiment.

The procedure adopted for the calculation is based on the same assumptions of that exposed in section C.1. apart for the effective compressibility which in this case is estimated from the eigenfrequency of the experimental cell. In fact this approach provides an accurate interpretation of the data and satisfactory results especially in microgravity conditions due to the sphericity of the drop and in general to a better performing of the geometry control. The characteristics of the adsorption layers obtained from the results for the complex visco-elasticity are in good agreement with adsorption isotherms and equations of state accounting for the intrinsic (2D) monolayer compressibility [24].

An important conclusion that could be drawn from the FAST experiment results was that, among the multiple advantages offered by the microgravity conditions, there is also the fact that the normal modes of drop or bubble oscillation were exclusively radial (i.e., prolate / oblate oscillation modes were not triggered) which provides a wider frequency range for accurate acquisition of the dilational visco-elasticity.

E. CONCLUSIONS

Dilatational rheology represents a powerful tool to investigate equilibrium and dynamic properties of interfacial layer of surfactants, proteins, polymers and particles on a macroscopic and microscopic scale. The development of oscillating bubble and drop techniques during the last ten years responded to the need to obtain more and more accurate measurements of the parameters charactering dilational rheology, mainly dilational visco-elasticity, on a wide frequency range.

These methods provide today measurements in the range between 10^{-3} to about 10^3 Hz, filling the previously existing gap between the methods based on Langmuir troughs, utilising either slowly moving barriers or based on the surface waves damping.

At present Oscillating Bubble/Drop techniques are limited at the large frequencies by the poor knowledge of mass and momentum transport in droplet/bubbles subject to highly dynamic conditions. In spite simplified dynamic conditions, such as those set by microgravity, can be utilised to partially overcome these limitations, focussed investigations, theoretical analysis and simulations of these fluid-dynamic aspects could be very useful to extend the potentialities of these methods in the standard laboratory utilisation.

During the last years different commercial dilational surface rheometers appeared on the market, either as dedicated instruments or as accessories of tensiometers. Nevertheless, to boost the research in the field and to promote the utilisation of dilational rheology methods in research and industrial applications, it is today necessary an effort towards a larger standardisation of the methodologies and of the procedures.

ACKNOWLEDGEMENTS

The work was financially supported by a project of the European Space Agency (FASES MAP AO-99-052)

F. REFERENCE

1. P.M. Kruglyakov and D.R. Exerowa, Foams and Foam Films, in Studies of Interface Science, Vol. 5, D. Möbius and R. Miller (Editors), Elsevier, Amsterdam, 1997
2. I.B. Ivanov (Ed.); Thin liquid films: Fundamentals and applications, In: Surfactant Science Series Vol. 29, Marcel Dekker Inc., New York, 1988
3. A.J. Sjöblom (Ed.); " Emulsion and emulsion stability" Marcel Dekker, 1996, Vol.61.
4. I.B. Ivanov, P.A. Kralchevsky, Colloids and Surfaces A, 128 (1997) 155
5. D.A. Edwards, H. Brenner, D.T. Wasan, Interfacial transport process and rheology, Butterworth-Heinemann, Boston 1991.
6. I.B. Ivanov, K.D. Danov, K.P. Ananthapadmanabhan, A. Lips, Adv. Colloid Interface Sci., 114 (2005), 61.
7. G. Kretzschmar, Prog. Colloid Polym. Sci. 77 (1988) 72.
8. R. Miller, J.B. Li., M.. Bree, G. Loglio, A.W. Neumann, H. Möhwald, Thin Solid Films, 327-329 (1998) 275.
9. P. Cicuta, E. M. Terentjev, Eur. Phys. J., 16 (2005) 147.
10. R. Wüstneck, N. Wüstneck, D.O. Grigoriev, U. Pison, R. Miller, Colloids and Surfaces B, 15 (1999) 275.
11. G. Loglio, P. Pandolfini, R. Miller, A.V. Makievski, F. Ravera, M. Ferrari, L. Liggieri, in "Novel Methods to Study Interfacial Layers", Möbius, D.; Miller, R.; (Eds.), Studies in Interface Science Series, Elsevier-Amsterdam 2001; Vol. 11, pp 439-483.
12. V.I.; Kovalchuk, J. Krägel, E.V Aksenenko, G. Loglio, L. Liggieri, in "Novel Methods to Study Interfacial Layers", Möbius D., Miller R. Eds., Studies in Interface Science Series, Elsevier-Amsterdam 2001; Vol.11, pp 485-516.
13. G. Loglio, U. Tesei, R. Cini, J. Colloid Interface Sci., 71 (1979), 316,
14. J.W. Gibbs, Collected Works, vol. 1, Dover Publishing Co. Inc, New York, 1961, p. 301.

15. G. Serrien, G. Geeraerts, L. Ghosh, P. Joos, Colloids Surf. 68 (1992), 219.
16. J. Lucassen, M. van den Tempel, Chem. Eng. Sci., 27 (1972) 1283.
17. R. Miller, G. Loglio, U. Tesei, Colloid Polymer Sci., 270 (1992), 598.
18. P.R. Garret, P.J. Joos, Chem. Soc. Faraday Trans., 69 (1976), 2161.
19. V.B. Fainerman, R. Miller, V.I. Kovalchuk, Langmuir 18 (2002) 7748.
20. V.B. Fainerman, V.I. Kovalchuk, E.V. Aksenenko, M. Michel, E. Leser, R. Miller, J. Phys. Chem. B, 108 (2004), 13700.
21. K.D. Wantke, H. Fruhner, J. Fang, K. Lunkenheimer, J. Colloid and Interface Sci., 208 (1998), 34.
22. K.D. Wantke, H. Fruhner, J. Colloid and Interface Sci., 237 (2001), 185.
23. V. I. Kovalchuk, J. Krägel, A. V. Makievski, G. Loglio, F. Ravera, L. Liggieri, R. Miller, J. Colloid and Interface Sci., 252 (2002), 433.
24. V.I. Kovalchuk, R. Miller, V.B. Fainerman, G. Loglio, Advances in Colloid and Interface Science 114–115 (2005) 303– 312
25. V. I. Kovalchuk, G. Loglio, V.B. Fainerman, , R. Miller, J. Colloid and Interface Sci., 270 (2004), 475
26. E.H. Lucassen-Reynders, J Colloid Interface Sci., 42 (1973), 573.
27. E.V. Aksenenko, V.I. Kovalchuk, V.B. Fainerman c, R. Miller, Advances in Colloid and Interface Science 122 (2006) 57
28. E. H. Lucassen-Reynders, V. B. Fainerman and R. Miller, J. Phys. Chem. B , 108 (2004), 9173
29. L.Liggieri, M. Ferrari, D. Mondelli, F. Ravera, Fararday Discussions, 129 (2005) 125.
30. F. Ravera , M. Ferrari.,R. Miller R., L. Liggieri, J. Physical Chemistry B, 105 (2000) 195.
31. R. Palazzolo, F. Ravera, M. Ferrari, L. Liggieri, Langmuir, 18 (2002) 3592.
32. L. Liggieri, M. Ferrari, F. Ravera, In: Colloid Stability : The Role of Surface Forces - Part II. Colloids and Interface Science Series vol. 2, Th. Tadros. (Ed); Wiley-VCH, 2007. pp. 313-344
33. F. Ravera, M. Ferrari, E. Santini, L. Liggieri, Advances in Colloid and Interface Sci. 117 (2005), 75.
34. F. Ravera, M. Ferrari, L. Liggieri, Colloids and Surfaces A, 282-283 (2006), , 210.
35. A. Klebanau, N. Kliabanova, F. Ortega, F. Monroy, R. G. Rubio and V. Starov J. Phys. Chem. B, 109 (2005), 18316.
36. F. Ravera, E. Santini, G. Loglio, M. Ferrari, L. Liggieri, J. of Physical Chemistry B, 110 (2006), 19543

37. F. Ravera, M. Ferrari a, L. Liggieri , G. Loglio, E. Santini, A. Zanobini, Colloids Surf. A, 323 (2008), 99.
38. B.A. Noskov, A.V. Akentiev, A.Yu. Biblin, I.M. Zorin, R. Miller, Adv. Colloid Interface Sci., 104 (2003) 245.
39. B.A. Noskov, G. Loglio, Colloids and Surfaces A , 143 (1998) 167
40. R.G. Muñoz, L. Luna, F. Monroy, R. G. Rubio, F. Ortega, Langmuir, 16 (2000) 6657.
41. J. Maldonado-Valderrama, V.B. Fainerman, E. Aksenenko, M. J. Galvez-Ruiz, M.A. Cabrerizo-Vilchez, R. Miller, Colloids and Surfaces A, 261 (2005), 85.
42. C. Stubenrauch, V.B. Fainerman, E.V. Aksenenko, R. Miller, J. Phys. Chem. B, 109 (2005), 1505.
43. D. Wu, Y. Feng, G. Xua, Y. Chen, X. Cao, Y. Li, Colloids and Surfaces A,, 299 (2007) 117,
44. L. Liggieri, F. Ravera, M. Ferrari, Langmuir, 19 (2003) 10233.
45. B. A. Noskov, S.-Y. Lin, G. Loglio, R. G. Rubio, and R. Miller, Langmuir, 22 (6) (2006), 2647
46. B.A. Noskov, DO Grigoriev , SY Lin, G. Loglio, R. Miller, Langmuir, 23 (2007), 9641
47. F. Bashforth and C. Adams, An attempt to test the Theories of Capillary Action, Cambridge Univ. Press, 1883.
48. M.E. Leser , S. Acquistapace, A. Cagna , A.V. Makievski, R. Miller, Colloids and Surfaces A, 261 (2005) 25–28
49. C. Maze, G. Burnet, Surface Sci. 13 (1969), 451.
50. Y. Rotenberg, L. Boruvka, A.W. Neumann, J. Colloid Interface Sci. 93 (1983), 169.
51. N.R. Pallas, Y. Harrison, Colloids Surfaces 43 (1990), 169
52. L. Liggieri, A. Passerone, High Temperature Tech. 7 (1989), 80.
53. G. Loglio, P. Pandolfini, R. Miller, A.V. Makievski, J. Krägel, F. Ravera, B.A. Noskov, Colloids and Surfaces A, 261 (2005) 57.
54. G. Loglio, P. Pandolfini, R. Miller, A.V. Makievski, J. Krägel, F. Ravera, Phys. Chem. Chem. Phys. 6 (2004) 1375.
55. L. Liggieri, F. Ravera, in "Drops and Bubbles in Interfacial Research" (Möbius, D., Miller, R., Eds.), Elsevier, Amsterdam 1998; Vol. 6, pp 239-278.
56. L. Liggieri, V. Attolini, M. Ferrari, F. Ravera, J. Colloid and Interface Sci., 252 (2002) 225.
57. M. Ferrari, L. Liggieri, F. Ravera, J. Phys. Chem. B, 102 (1998) 10521.
58. J. Krägel, V.I. Kovalchuk, A.V. Makievski, M. Simoncini, F. Ravera, L. Liggieri, G. Loglio and R. Miller, Microgravity Science and Technology Journal, 16 (2005) 186

59. L. Liggieri , F. Ravera, M. Ferrari, A. Passerone Microgravity Science and Technology Journal, 16 (2005) 201
60. A. Passerone, L. Liggieri, N. Rando, F. Ravera, E. Ricci, J. Colloid and Interface Sci., 146 (1991), 152
61. C.A. MacLeod, C.J. Radke, J. Colloid and Interface Sci., 160 (1993), 435.
62. G. Loglio, R. Miller, A. Stortini, U. Tesei, N. Degli Innocenti, R. Cini, Colloid Surfaces A, 90 (1994), 251.
63. H. Fruhner, K.-D. Wantke, Colloids and Surfaces A, 114 (1996), 53.
64. Y-H Kim, K.Koczo, D.T Wasan, J. Colloid and Interface Sci., 187 (1997) 29.
65. K. Lunkenheimer, G. Kretzschmar, Z. Phys. Chem. (Leipzig), 256 (1975), 593.
66. V.I. Kovalchuk, E.K. Zholkovskij, J., R. Miller, V.B Fainerman, R Wüstneck, G. Loglio, S.S. Dukhin, J. Colloid and Interface Sci., 224 (2000) 245.
67. V.I.; Kovalchuk, J. Krägel, E.V Aksenenko, G. Loglio, L. Liggieri, in "Novel Methods to Study Interfacial Layers", Möbius D., Miller R. Eds., Studies in Interface Science Series, Elsevier-Amsterdam 2001; Vol.11, pp 485-516.
68. V.I. Kovalchuk, J. Krägel, R. Miller, V.B. Fainerman, N.M. Kovalchuk, E.K. Zholkovskij, R. Wüstneck, S.S. Dukhin, J. Colloid and Interface Sci., 235 (2000), 232.
69. L. Liggieri, V. Attolini, M. Ferrari, F. Ravera, J. Colloid and Interface Sci., 252 (2002) 225.
70. L. Liggieri, M. Ferrari, A. Massa, F. Ravera, Colloids and Surfaces A, 156 (1999) 455.
71. J. Lucassen in: Anionic surfactants. Physical chemistry of surfactant action. Surf. Sci. Series, v.11, Ed. by E.H. Lucassen-Reynders, NY: Dekker, 1981, 217-265.
72. D. Langevin, Adv. Colloid Interface Sci., 88 (2000) 209.
73. D.S. Valkovska, K.D. Danov, I.B. Ivanov, Adv. Colloid Interface Sci., 96 (2002)101
74. D.S. Valkovska, K.D. Danov, I.B. Ivanov, Colloids and Surfaces A, 175 (2000) 179
75. C. Stubenrauch, R. Miller, J. Phys. Chem B, 108 (2004), 6412.
76. N.A. Mishchuk, A. Sanfeld, A. Steinchen, Adv. Colloid and Interface Sci., 112 (2004), 129.
77. E. Santini; F. Ravera, M. Ferrari, C. Stubenrauch, A. Makievski, J. Krägel, Colloids and Surfaces A., 298 (2007), 12.
78. J. Schlarmann, C. Stubenrauch, R. Strey, Phys. Chem. Chem. Phys., 5 (2003), 184

79. E. Santini, L. Liggieri, L. Sacca , D. Clausse, F. Ravera, Colloids and Surfaces A, 309 (2007), 270
80. A. M. Díez-Pascual, F. Monroy, F. Ortega, R.G. Rubio, R. Miller and B.A. Noskov, Langmuir, 23 (2007), 3802
81. B. A. Noskov, G. Loglio and R. Miller, J. Phys. Chem. B 108 (2004), 18615
82. B. A. Noskov, D. O. Grigoriev, S.Y. Lin, G. Loglio and R. Miller, Langmuir 23 (2007), 9641
83. B.A. Noskov, A.Yu. Bilibin, A.V. Lezov, G. Loglio, S.K. Filippov, I.M. Zorin, R. Miller, Colloids and Surfaces A, 298 (2007) 115
84. V. G. Babak, J. Desbrieres, V. E. Tikhonov, Colloids and Surfaces A, 255 (2005) 119
85. N.Wüstneck, R.Wüstneck, V.B. Fainerman, R. Miller, U. Pison, Colloids Surf. B 21 (2001) 191
86. M. Vranceanu, K. Winkler, H. Nirschl, G. Leneweit, Colloids and Surfaces A, 311 (2007) 140
87. J. Benjamins, J. Lyklema, and E. H. Lucassen-Reynders, Langmuir 22 (2006), 6181.
88. R. Miller, M. E. Leser, M. Michel and V. B. Fainerman, J. Phys. Chem. B, 109 (2005), 13327
89. V. S. Alahverdjieva, D. O. Grigoriev, V. B. Fainerman, E. V. Aksenenko, R. Miller and H. Molhwald, J. Phys. Chem. B, 112 (2008), 2136
90. L. Liggieri, F. Ravera, M. Ferrari, A. Passerone, G. Loglio, R. Miller, J. Krägel, A. V. Makievski, Microgravity Sci. and Technology, XVI (I) (2005), 196.
91. L. Liggieri, F. Ravera, M. Ferrari, A. Passerone, G. Loglio, R. Miller, A. Makievski, J. Krägel, Microgravity sci. technol. XVIII-3/4 (2006) 112
92. L, Liggieri, F. Ravera, M. Ferrari, A. Passerone, Microgravity Sci. and Technology, XVI (I) (2005), 201
93. V.I. Kovalchuk, J. Krägel, A.V. Makievski, F. Ravera, L. Liggieri, G. Loglio, V.B. Fainerman, and R. Miller, J. Colloid Interface Sci., 280 (2004) 498

SURFACE RHEOLOGY STUDIES OF SPREAD AND ADSORBED POLYMER LAYERS

Francisco Monroy[1], Francisco Ortega[1], Ramón G. Rubio[1] and Boris A. Noskov[2]

[1] Dept. Química Física I, Fac. Química, Univ. Complutense, 28040-Madrid, Spain.
[2] St. Petersburg State University, Chemical Faculty, 198904 St. Petersburg, Russia

Contents

A. INTRODUCTION

A great number of problems on surface and interface science deal with molecular motions and relaxation at fluid surfaces containing polymer molecules [1]. In fact, modern soft-matter physics has opened a great number of questions dealing with the dynamics of soft surfaces, and particularly, with polymer dynamics at interfaces or at reduced dimensionality, *e.g.* polymer chains in pores or confined between solid walls [2,3]. Among these systems, adsorbed polymer films represent a paradigmatic example; in fact, polymer films are frequently considered as model systems to explore this new physics (see [4] for a recent review). Neutron and x-ray techniques, in combination with classical spectroscopy, have revealed in this context as very powerful microscopic probes of structure and dynamics [5-7]. It is now clear that for polymer at interfaces there are at least two controlling length scales which may be rather different in magnitude: the range of interactions between the segments of the polymer chains, or of the segments with the interface, and the size of the polymer chain. The reduction of the number of available coil conformations imposed by the presence of the interface makes the interfacial region thicker than for non-polymeric systems. It is the size of the adsorbed polymer coil (e.g. the 2D radius of gyration, R_g) that sets the size of the interface [1,4].

The most traditional characterization of polymers at the fluid/fluid interface consists on the measurement of the surface pressure $\Pi = \gamma\text{-}\gamma_0$, where γ is the surface tension of the interface with the polymer, and γ_0 that of the bare interface [8,9]. However, it has been shown that the surface pressure may be insensitive to the existence of some structural or conformational transitions within the interfacial region [10,11]. Other experimental techniques, such as surface potential [12], ellipsometry [13], infrared or UV reflectivity [14, 15], surface plasmon resonance [16], Maxwell currents [17], and non-linear optical techniques [18], as well as computer simulations [19-23], have allowed to understand some aspects of such processes.

Time-resolved versions of the scattering techniques, *i.e.* x-ray photon correlation spectroscopy (XPCS) and neutron spin-echo (NSE), have emerged as powerful probes of the microscopic dynamics in these soft and fragile systems [23]. However, long-time relaxation phenomena, like collective diffusion, or more sophisticated mechanisms of relaxation like reptation motions, arm retractions, etc. show in some cases too slow dynamics to accomplish their coherence requirements. As a recent example, XPCS has been successfully used to follow slow collective dynamics in solid-supported thin polymer films close to the glass transition [24], however high-brilliance synchrotron sources are needed to carry out these experiments.

Surface rheology is one of the best options to study the dynamics of slow processes at fluid interfaces. The experimental advances in the last decades allow the researcher to explore a rather broad frequency range: 1 mHz – 1 MHz by combining different experimental techniques. Moreover, both the dilational and the shear rheology can be routinely studied in the laboratory [25-28]. In the case of polymer films at fluid interfaces, this tools have revealed the existence of several dynamic processes, both in insoluble [29-32] and soluble polymers [11,33-36]. In spite of the advances reported in the last years, several fundamental problems still remain: a) Which is the physical mechanism underlying the dynamics of polymer coils at fluid interfaces (quasi-2D geometry)? Does the reptation mechanism apply to quasi-2D? b) Is the usual phase-behaviour phenomenology found in 3-D polymer systems (e.g. solid-fluid or fluid-glass transitions, plastic behaviour, etc.) also be observable at fluid interfaces? c) The physically unsound results (e.g. negative apparent dilational viscosities) suggest that the hydrodynamic description of the fuid interface most frequently used is incomplete. In this chapter we will review some of the recent work on the behaviour of polymer monolayers at fluid interfaces (mainly at the liquid/air interface) in relation to the above issues. In this chapter we will focuss on the dilational viscoelastic moduli because for polymer monolayers the dilational component is two or three orders of magnitude larger than the shear one [37-39]

B. SURFACE RHEOLOGY AT FLUID INTERFACES

It is well known that many surface active substances, polymers with hydrophilic groups among them, are able to adsorb at water interfaces. They can form either Gibbs monolayers, when adsorbed from a bulk polymer solution, or Langmuir films, when insoluble polymers are formed by spreading at the interface. From the equilibrium point of view, the surface tension, γ, is in both cases decreased with respect to the bare interface γ_0. The adsorbed film is characterized by the surface pressure Π, that represents the decrease of surface free energy per unit area, resulting of the spontaneous adsorption of the film. In order to discuss the system under dynamical conditions, let us consider an infinitesimal change of the surface area, $\delta A(t)$. The area perturbation induces a change in the adsorption state of the film, thus leading to a change in the surface pressure $\delta\Pi(t)$ which evolves in time until a new equilibrium state is reached. The change in surface pressure depends on the time scale probed in the experiment, and, to first order, it can be expressed as

$$\partial\Pi(t) = \Pi(t) - \Pi_0 = \frac{\partial\Pi}{\partial A}\delta A = -E(t)u(t) \qquad (1)$$

where $E(t) = -A_0\frac{\partial\Pi}{\partial A}$ is the time-dependent dilational modulus, that accounts for the elastic energy storage on dilating the film. It usually contains both dilational and shear components: $E(t) = E_K + E_S$. For fluid systems, the shear component is vanishing. Therefore, for fluid films at equilibrium, the dynamic modulus equals the Gibbs elasticity ε_0

$$E(t) \rightarrow E_0 = \Gamma\left(\frac{\partial\Pi}{\partial\Gamma}\right)_{eq} \qquad (2)$$

where $\Gamma = 1/A$ is the surface concentration.
In the limit of constant elasticity modulus, Eq. (1) is formally equivalent to Hooke's law for a pure elastic 2D-body. If friction is present some delay may exist in the response function, that can be written in terms of a visco-elastic modulus including a dissipation operator

$$-d\Pi(t) = E(t)\cdot u(t) = \left[E(t) + \zeta(t)\frac{\partial}{\partial t}\right]\cdot u(t) \qquad (3)$$

The second dissipative term in Eq. (3) accounts for the viscous losses, that are proportional to the dilation rate (Newton's law). The proportionality factors are respectively the elastic modulus $E(t)$ and the dilational viscosity $\zeta(t)$, that also depends on time, t. A linear visco-elastic modulus can be defined as:

$$\tilde{E}(t) = -\left(\frac{\partial\Pi}{(\partial A/A)}\right)_T \qquad (4)$$

For an small-amplitude oscillatory motion $[\delta A(t) \sim e^{i\omega t}]$ of frequency ω, this complex modulus is a complex quantity

$$E(i\omega) = E(\omega) + i\omega\zeta(\omega) \qquad (5)$$

The visco-elastic dilational modulus contains information about how the surface pressure changes on dilating the film, and thus, about the changes on the adsorption state and the molecular conformation stressed by the external deformation. These changes are, in general, dependent of the time scale probed in the particular rheological experiment (or of the frequency ω of the applied deformation in surface waves experiments). The constitutive visco-elastic parameters $\varepsilon(\omega)$ and $\omega\zeta(\omega)$ obtained as a function of ω contain the time-dependent response of the system subject to a small disturbance, *i.e.*, they allow one to probe the surface dynamics of the adsorbed film.

For insoluble films at equilibrium, or when the film is disturbed along a *quasi*-static path ($\omega \to 0$), the surface dilation causes an instantaneous change in surface concentration, $\delta A/A = -\delta\Gamma/\Gamma$. In this case, the elasticity modulus equals the compression modulus, which can be obtained as the relative slope of the equilibrium isotherm:

$$E(\omega \to 0) = E_0 = \Gamma \left(\frac{\partial \Pi}{\partial \Gamma} \right)_T \tag{6}$$

and the dilational viscosity equals its frequency-independent Newtonian limit:

$$\zeta(\omega \to 0) = \zeta_0 \tag{7}$$

In general, the linear visco-elastic response of the adsorbed film must obey the Kramers-Krönning relationships. They state that the elastic response increases monotonously with the frequency of the deformation; contrarily, the viscosity κ decreases with frequency [40]. Moreover, the following inequalities hold:

$$E(\omega) \geq E_0$$
$$\omega\zeta(\omega) \leq E(\omega) \tag{8}$$

Figure 1 summarizes the material response expected for a visco-elastic polymer film exhibiting a relaxation process due to chain diffusion. Once the relevancy of the dilational modulus to get insight on surface dynamics has been pointed out, it is necessary to discuss how to measure it.

Two different, but complementary, approaches can be followed to measure $\tilde{\varepsilon}(t)$: a) mechanical relaxation, and b) surface waves experiments. The direct rheological methods, and in particular, mechanical stress relaxation, allow one to obtain the dilational modulus from the stress response function of the film strained by a fast external dilation of the interfacial plane. The stress relaxation is measured by registering the changes in surface pressure as a function of time, $\delta\Pi(t)$. Eventually, in a compliance experiment the film could be stressed upon a pressure gradient imposed by the barriers so monitoring the resulting change of strain necessary to maintain the imposed stress [21]. Obviously, these methods are restricted to a time scale longer than a few seconds, the time necessary to strain the film. It must be stressed that the available phenomenological theory that allows one to obtain the constitutive parameters from the relaxation (or fluency) experiments is currently restricted to the linear visco-elastic regime [41,42].

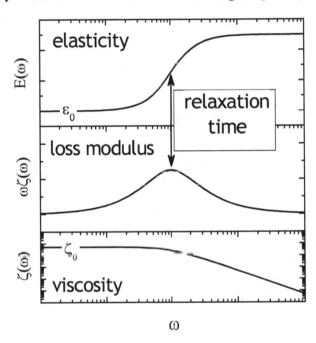

Figure 1. Typical frequency dependence of the elasticity, loss modulus and viscosity of a visco-elastic material undergoing a dynamic process.

On the other hand, the study of the propagation of small-amplitude surface waves, excited either, by thermal agitation or by an externally disturbance (electrical or mechanical), represents the second category of methods of surface rheology. This approach does not allow the direct determination of the response functions, surface hydrodynamics being necessary to calculate the visco-elastic parameters from the dispersion curve of the surface modes. When compared with the direct relaxation experiments, the surface waves methods are more intricate, not only from a conceptual and interpretative point of view, but also from the experimental point of view [28, 43-47]. Furthermore, the underlying hydrodynamic theory seems to be incomplete for some systems, leading to physically unsound results, such as negative apparent viscosities [48-51].

1. Surface wave experiments

Either under thermal agitation or mechanically applied stresses, a fluid interface at thermal equilibrium shows some surface roughness. This surface motion can be conceived as the restoring response of the interface for re-establishing its flat shape characteristic of the mechanical equilibrium.

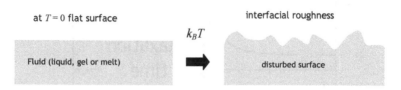

Figure 2. A fluid interface is flat at 0 K. Thermal energy generates a finite roughness.

For thermal random motion, the interfacial shape can be described by a Fourier series of independent plane waves propagating within the *xy*-plane, z being perpendicular to the surface, and slightly penetrating the bulk phases adjacent to the interface. If the equilibrium position of the interfacial plane is placed at $z = 0$, displacements of the surface, u(r,t), can be written as follows:

$$u(r,t) = \sum_q u_q(q) e^{i[q \cdot r - \omega(q)t]} \qquad (9)$$

where the sum extends over all wave vectors q_j and the associated frequencies $\omega(q)$, that are related through the dispersion equation.

For short wave-length and low amplitude waves, the surface tension is the restoring force for transversal or capillary motions of the surface: any local curvature causes a γ-governed Laplace stress that restores the planar equilibrium shape of the interface. For linear waves it takes the following form

$$f_z = \gamma \frac{\partial^2 u_z}{\partial x^2} \qquad (10)$$

Gravity is also a restoring force, but its role is less important except for very high amplitude and long wave-length waves displacing large amounts of liquid, as the sea wind and tidal waves.

As stated above, there are two main types of surface visco-elastic modes, shear (*S*) and compression (*K*), that appear coupled together. However, for polymer films the compression modulus is several orders of magnitude larger than the shear one [52]. The dilational modulus governs the longitudinal interfacial motion, or Marangoni flux. This flux is driven by the surface tension gradient originated by dilating the surface with area A. The longitudinal restoring force is then proportional to dilational modulus, that contains both shear and dilational contributions, through [53]:

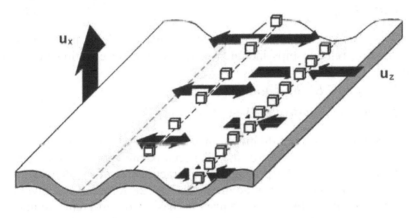

Figure 3. The rheology of a fluid interface can be described as a sum of hydrodynamic modes. When surface active molecules are adsorbed at the interface, in addition to the transverse motion in the z direction (capillary mode), there is an in-plane mode associated to the concentration gradients (dilational mode).

$$f_x = \frac{d\gamma}{dx} = \tilde{\varepsilon}(\omega)\frac{\partial^2 u_x}{\partial x^2} \qquad (11)$$

The response functions in Eqs. (10)-(11) are included in the boundary conditions for the transversal and longitudinal motions of the interface. When the Navier-Stokes equations are solved for oscillatory motion Eq. (9) the following dispersion equation is obtained [26,28]

$$D = L\left[E(i\omega);q,\omega\right]\cdot T\left(\gamma;q,\omega\right) - C\left(\eta,\rho\right) \equiv 0 \qquad (12)$$

Here, the constant C couples the longitudinal relationship $L[E(i\omega)]$, which only depends on the dilational modulus, with the transversal one $T(\gamma)$, governed by surface tension and gravity. This coupling constant depends only on the densities ρ and the viscosities η of the adjacent bulk phases. C is near-zero for *mechanically symmetric* interfaces, *e.g.* the oil/water interface or a suspended membrane, for which both modes, capillary and dilational, remains almost decoupled. The coupling constant takes non-zero values for the air/fluid interfaces, which show kinematic asymmetry:

$$C(\eta,\rho) = \left[\omega\eta(q-m)\right]^2 \neq 0 \tag{13}$$

Here, the penetration depth m^{-1} of the surface velocity field depends on density ρ and viscosity η of the fluid beneath:

$$m^2 = q^2 - i\omega\rho/\eta \tag{14}$$

For visco-elastic films adsorbed at the air/fluid interface the propagation relationships included in the dispersion equation Eq. (10) are:

$$L[E(i\omega);q,\omega] = E(i\omega)q^2 + i\omega\eta(q+m)$$
$$T(\gamma;q,\omega) = \gamma q^2 + i\omega\eta(q+m) - \omega^2\rho/q \tag{15}$$

Hence, capillary and dilational modes are found coupled together at the free surface of a fluid, so, the dilational parameters can be obtained from the analysis of the propagation characteristics of capillary waves probed by transversal waves devices [see Ref. 28 for a general review]. In general, this dispersion equation has two different solutions corresponding to two propagative modes, the capillary and dilational ones; at a given q two different complex roots are obtained ($\omega = \omega_q + i \Gamma_q$, the propagation frequency is ω_q and Γ_q the mode damping). For bare interfaces ($E = \zeta = 0$) a limiting solution can be obtained for pure capillary transversal modes: the so-called Kelvin's law for the frequency of capillary waves propagating on the free surface of a liquid:

$$\omega_q^2 = \frac{\gamma}{\rho}q^3 \tag{16}$$

and the Stokes' law accounting for the capillary damping due to the viscous friction with the bulk fluid:

$$\Gamma_q = 2\frac{\eta}{\rho}q^2 \tag{17}$$

If visco-elastic dilational effects do exist within the interface the capillary damping exceeds this minimum value expected for bare interfaces.

Methods based on excited capillary waves are the most frequently used, in particular, light scattering by thermally excited capillary waves (SQELS), or propagation of monochromatic transversal waves excited at a given frequency by mechanical or electrical actuators (ECW). Since transversal motion causes strong fluctuations of refractive index (between the value corresponding to the fluid $n \sim 1.3$ typically and that of air, $n \sim 1$), methods based on capillary waves are more usual and popular than those probing longitudinal modes, which are directly governed by the relevant dilational modulus.

1.1. Surface Quasi-Elastic Light Scattering (SQELS)

This method probes the light scattered by thermally excited capillary waves. The *quasi*-elastic scattering arises from fluctuations of refractive index due to capillary roughness. Because of the small amplitude of these fluctuations, $<u_z>^2 \sim k_B T/\gamma q^2 \sim 5$ Å for non-critical interfaces at room temperature, thus leading to a very weak light scattering, the heterodyne detection is preferred in the modern set-up's [28].

In a typical configuration, monochromatic and vertically polarized light coming from a continuous very stable laser, working typically in the mW rage is directed to the interface with a reflectivity geometry (incident angles of $\theta_i = 45\text{-}50$ degrees work well at air/water interfaces, but grazing incidence is usually necessary at oil/water or critical interfaces, where optical contrast is vanishing).

Before being reflected by the interface, the light beam passes through a transmission diffraction grating with constant spacing ($d = 275$ μm in our set-up). Diffraction orders travel at well-defined directions θ_n given by the Bragg condition, $d \sin \theta_n = n\lambda$. Using a long-distance converging lens (~ 1 m) placed at a distance twice the focal length, the diffracted beams and the transmitted main-beam (which contains more than a 95% of the laser power) are focused together at the interface plane, forming here a one-to-one image of the diffraction grating. At the photomultiplier the reflected diffracted beam (local oscillator) at a wave vector q_n given by:

$$q_n = \frac{2\pi}{\lambda}\sin\theta_n \cos\theta_i = \frac{2\pi n}{d}\cos\theta_i \qquad (18)$$

mixed with the scattered light giving a heterodyne beating,

The use of diffraction a grating is essential because the heterodyne beating allows not only improving the signal to noise ratio by orders of magnitude, but also it eliminates spurious scattering contributions arising from low-frequency mechanical disturbances of the surface. In fact, the diffraction grating is the more efficient anti-vibration element since only the light scattered by the surface waves with equal length than the "optical spatial template" created by the grating on the surface is coherently mixed at the photomultiplier with the light reflected in this direction.

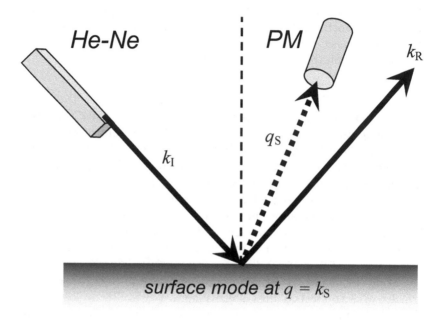

Figure 4. Typical geometry of a quasi-elastic surface light scattering (SQELS).

The heterodyne signal $I_q(t) \sim u_q^2$ can be analysed, either by autocorrelation in the time domain, or by measuring the power spectrum of the detected intensity $P_q(\omega)$ by means of a Fourier-transform spectrum analyzer. Since thermal energy is distributed over all the excited modes, equipartion imposes that the scattering intensity decreases with wavevector as $I_q \sim k_B T / \gamma q^2$.

Figure 5 shows some examples of autocorrelation functions for the air-water interface. They can be properly fitted to a dependence of the type:

$$G(t) = \langle I(\tau) I(t+\tau) \rangle = \langle u_z^2(\tau) u_z^2(t+\tau) \rangle =$$
$$= A \cos(\omega_q t) e^{-\Gamma_q t} e^{-\beta t^2/2} + G(\infty) \tag{19}$$

where ω_q is the propagation frequency of the capillary mode, and Γ_q the capillary damping. The baseline $G(\infty)$ is proportional to the time-averaged intensity allowing one to analyze the optical reflectivity of the surface. Due to the finite size of the detected spot Δq, the autocorrelation function is affected of some extra instrumental damping β. This instrumental contribution is properly accounted for by a Gaussian function of width β which increases with the product $q^{1/2} \Delta q$ as:

$$\beta = \Delta\omega_q \approx \frac{2}{3}\omega_q \frac{\Delta q}{q} \approx \frac{2}{3}\left(\frac{\gamma}{\rho}\right)^{1/2} q^{1/2}\Delta q \qquad (20)$$

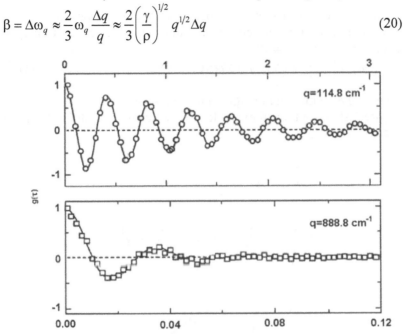

Figure 5. Autocorrelation functions of the air-water interface at 25 °C, and at two values of the wave vector q. The lines are the fits to Eq. (19)

In the frequency domain the spectrum of light scattered by a q-mode is nearly Lorentzian-distributed around the propagation frequency ω_q; here, the band-width $\Delta\omega_q = 2\,\Gamma_q$ is due to the dissipation existing within the capillary mode. If the instrumental broadening is properly accounted for by a Gaussian profile, the experimental spectra can be fitted by the convolution of a Lorentzian function with the Gaussian instrumental function, more precisely the Voigt profile:

$$P_z(q,\omega) = FT\left[G(t)\right] \approx \int\limits_0^\omega \frac{\Gamma_q^2}{4\Gamma_q^2 + \left(\omega' - \omega_q\right)^2} e^{-(\omega'/2\beta)^2} d\omega' \qquad (21)$$

However, in most cases the exact expression for the power spectrum must be used. It can be obtained from the fluctuation-dissipation formalism; so, for capillary modes:

$$P_z\left[E(i\omega),\gamma;q,\omega\right]=\frac{k_B T}{\pi\omega}\,\mathrm{Im}\left[\frac{L\left[E(i\omega);q,\omega\right]}{D\left[E(i\omega),\gamma;q,\omega\right]}\right] \tag{22}$$

Actually, the detected power spectrum $P_{Exp}(\omega)$ is the convolution between the theoretical spectrum and the Gaussian-like instrumental function $I(\beta) \sim \exp(-\omega^2/4\beta^2)$:

$$P_{Exp}\left[E(i\omega),\gamma\right]=P_z\left[E(i\omega),\gamma\right]\otimes I(\beta) \tag{23}$$

β being the width at half height of the Gaussian.

Figure 6 shows a typical example of experimental spectra obtained at different wavevector, q.

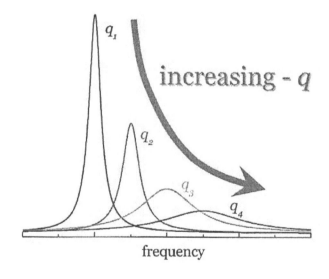

Fig. 6 Wave vector dependence of the spectrum of the light scattered by a fluid surface.

1.2. Electrocapillary Waves (ECW)

Although the scattering of light by thermal surface waves increases at low-q 's ($q < 100$ cm^{-1}, $\lambda > 0.1$ mm), the width of the instrumental function also increases, becoming similar to the width of the spectrum of the scattered light. In this case, external excitation (either mechanical or electrocapillary) of larger amplitude ($u_z \sim 1$ μm) is a better option [54]. Because of its non-invasive character, electrical excitation is the most frequently used. In a typical electrocapillary waves set-up one obtains the propagation characteristics (wavelength λ and spatial damping α) of a capillary wave excited by an external electric AC field oscillating at a frequency $\omega/2$ (see Figures 7 and 8).

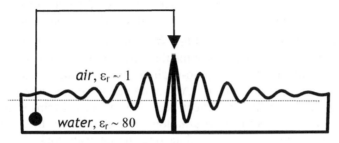

Figure 7. Surface waves generated by an oscillating electrical field due to the difference of the permittivity of the two fluids.

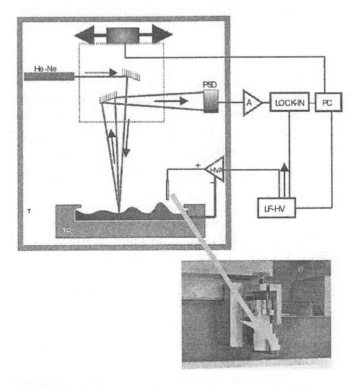

Figure 8. Overall design of an electrocapillary waves (ECW) technique.

The deformation is locally applied by a cutter-shaped electrode placed at air near to the interface ($\sim$ 1 mm). A 500 V AC field with frequency ω is applied between this the electrode and the fluid. Because of the dielectric constant difference between both electrodes, the deformable interface follows instantaneously the direction of the applied electric field. This dielectric ponderomotive deformation is proportional to the square of the applied field $u_z \sim \Xi^2(\omega_0) \sim (\ e^{i\omega_0 t}\ )^2 \sim \cos(\ e^{i2\omega_0 t}\ )$, thus the response has a frequency double than the applied field, $\omega = 2\omega_0$ [28].

The spatial profile of the so-excited electrocapillary wave is then scanned by laser reflectometry. A low intensity He-Ne laser is focused on and reflected by the interface; here, the reflected beam oscillates as the surface does. Surface oscillations are monitored by a position sensitive photodiode (PSD), by detecting the intensity locked-within the excitation signal in a lock-in amplifier. This allows one to record both the amplitude and the phase difference of the excited capillary wave as a function of the distanced to the excitation point, x. The spatial profile of the ECW can be described by

$$u_z\left(x\right) \approx \cos\left(\frac{2\pi}{\lambda}x + \phi\right)e^{-\alpha x} \tag{24}$$

where λ is the wavelength ($\lambda = 2\pi/q$), and α is the spatial damping for a given excitation frequency ω of the ECW. Both, α and the temporal damping Γ_q obtained in the SQELS experiments are mutually related through the group velocity of the surface wave, $U = d\omega/dq$, obtained from the dispersion relationship:

$$\Gamma_q = U\left(\omega\right)\alpha = \frac{d\omega}{dq}\alpha \tag{25}$$

Data analysis must be performed through the numerical solution of the dispersion equation Eqs. (10)-(13), that must be solved for a given set of propagation parameters, ω and $q = (2\pi/\lambda) - i\ \alpha$.

2. Mechanical relaxation experiments

2.1. Stress relaxation

Experiments of mechanical rheology can be easily performed in commercial Langmuir troughs. Particularly, the stress relaxation $\sigma(t)$ can be recorded as a function of time t after a sudden uniaxial in-plane compression of the Langmuir film is performed with the barriers of the trough. The surface stress is defined as $\sigma(t) \equiv \Delta\Pi(t) = \Pi(t) - \Pi(t\rightarrow\infty)$

[44-49]. As a consequence of the surface compression the surface pressure deviates from the equilibrium value, and a surface pressure change $\Delta\Pi$ appears, the so-named dilational stress σ_D, and acts as restoring force for recovering the initial state of the film when strain ceases. Here, after compression of the film the surface pressure, $\Pi(t) = \gamma_0 - \gamma(t)$, relaxes with respect to its value in the final equilibrium state $\Pi(t\rightarrow\infty)$. The surface dilation, $\theta = -\Delta A/A_0 = -(A_0 - A)/A_0$, is typically adjusted in order to accommodate a strain within the linear elastic regime (typ. θ is kept well below 5%). In a typical experiment the film area A is rapidly reduced by the compression factor θ with respect to the initial area A_0. A typical experimental record is plotted in Figure 9.

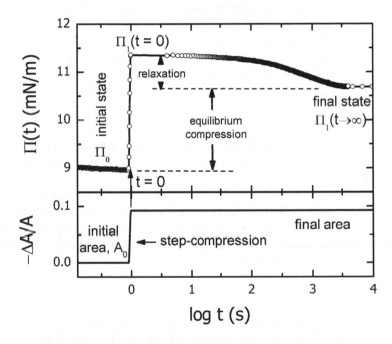

Figure 9. Behaviour of the time-dependence of the area and of the surface pressure in a typical step-relaxation experiment.

Routinely, it must be checked that the visco-elastic response is linear at the strain imposed, and that the signal-to-noise ratio of the recorded signal high enough. At times long enough the system must recover always a final equilibrium state compatible with the equilibrium isotherm.

2.2. Oscillatory barriers

In this case, the barriers of the trough are subjected to a sinusoidal motion at a constant frequency ω [$\delta A(t) \sim \exp(i\omega t)$]. It is extremely important to produce a true sinusoidal strain perturbation. Fourier transform analysis (FT) of the strain perturbation has to show the absence of multiple ω components superimposed to the fundamental frequency. To discard any influence of the time dependence of the pressure sensor response to a change of pressure, blank experiments at different sensor to barrier distances have to be performed. It has also to be checked that the material parameters are not biased by the propagation time of the compression wave, which must be at least one order of magnitude lower than the experimental period (typically higher than 1s). Films are strained in the Langmuir trough by an uniaxial in-plane compression of amplitude u. Since the Langmuir trough has a fixed width, the strain ratio is actually an elongation/compression ratio; if the trough barriers are placed at a distance L:

$$u(t) \equiv \frac{A_0 - A(t)}{A_0} = \frac{L_0 - L(t)}{L_0} \tag{26}$$

In order to perform experiments at constant frequency but variable strain amplitudes the barrier speed v_b is in practice conveniently changed.

It must be noticed that because of the uniaxial character of the compressive deformation, both the applied strain u and the response stress σ_D have dilational (xx) and in-plane shear (xy) components. In terms of a generalized stress function, the dilational response function can be then written as follows:

$$\sigma(\omega; u) \equiv E(i\omega; u) \cdot u \tag{27}$$

where $\varepsilon(\omega; u)$ is in general a frequency-strain-dependent dilational modulus, containing two components coupled together, the dilational (K) and the shear (S) components [27,53-55]. Only at strains small enough the elastic response becomes linear, i.e. at $u \to 0$ the dilational modulus is a strain-independent constant taking values close to the equilibrium compression modulus, ε_0.

Since the applied strain is sinusoidal in shape, its time dependence has the form:

$$u(t) = \frac{u_0}{2} e^{i\omega t} \tag{28}$$

where u_0 is the strain amplitude. Only in the linear regime the elastic response closely follows the sinusoidal shape of frequency ω imposed by the strain:

$$\Pi(t) = \Pi_0 + \sigma(t)$$

$$\sigma(t) = \frac{\sigma_0}{2} e^{i(\omega t + \phi_\sigma)}$$

(29)

Here, σ_0 is the amplitude of the stress response, measured with respect to the pressure background Π_0, and ϕ_σ a phase factor accounting for the viscous delay in the response. In fact, a pure elastic response leads to a zero phase difference, which reaches a maximum value of $\pi/2$ for a pure viscous surface. In the linear regime, the elasticity modulus ε and the dilational viscosity ζ can be obtained as follows:

$$E = |E| \cos \phi_\sigma \quad \omega \zeta = |\varepsilon| \sin \phi_\sigma$$

$$\text{where} \quad |E| = \frac{\sigma_0}{u_0}$$

(30)

At high strains (the limit strongly depends on each system! [56]), the visco-elastic response becomes non-linear, and the sinusoidal model in Eq. (30) no longer is a good description of the stress response. Thus, the usual methods of direct Fourier-transform rheology have to be used when the oscillation of the barriers drives the system beyond the linear regime, and there are no interactions between the harmonics [57]. For a given strain, the non-linear stress can be expanded as:

$$\sigma(t) = \sigma_0 e^{i\omega t} + \sigma_1 e^{2i\omega t} + \sigma_2 e^{3i\omega t} + \dots$$

(31)

where σ_1, σ_2, σ_3, are the amplitudes of the harmonic components of the non-linear response. For 3-D systems, obeying time reversibility $\sigma(-t) = \sigma(t)$, only odd harmonics must be present in Eq. (31), as extensively observed in 3-D polymer systems [58]. However, even harmonics have been observed for some systems, e.g. concentrated agarose gels [59]. In order to analyze the behaviour of the monolayers beyond the linear regime, the experimental $\Pi(t)-\Pi_0$ curves have to be Fourier transformed, and the corresponding spectrum is obtained [60]. It is worth to notice that a reasonable noise/signal ratio is only achieved if the time resolution is maximised and if the experimental data contain a large number of periods (typically at least 50 periods).

D. EQUILIBRIUM BEHAVIOUR OF LANGMUIR MONOLAYERS
OF INSOLUBLE POLYMERS

1. General behaviour

Langmuir polymer films are usually considered as two-dimensional objects since their thickness changes only slightly during dilation [1,4,5,61-65]. The measurement of the surface pressure Π vs. surface concentration Γ at constant temperature (hereinafter called an isotherm) is the very first piece of information that has to be known for the study of Langmuir monolayers. Π provides the basic equilibrium thermodynamic information about the system, and, as we shall see, the isotherm is a reference framework for some of the rheological experiments [29].

Equilibrium isotherms have been measured for many insoluble neutral homopolymers [9,25,29,32,46,52,53,55,66,67]. As far as a homopolymer film is insoluble and neutral, in all the cases the isotherms have the shape shown in Figure 11. In the dilute regime (usually $\Pi < 1$ mN·m^{-1}), Π slightly increases with the surface concentration of the polymer in the film, Γ, closely follows the ideal gas law at high dilution, $\Pi \sim k_B T \Gamma$. As Γ is increased in this regime, chain interactions emerge imposing deviations from the ideal behaviour well quantified as virial contributions to the surface pressure:

$$\frac{\Pi}{k_B T \Gamma} = 1 + B_2(T)\Gamma + \cdots \tag{32}$$

$B_2(T)$ being the second virial coefficient [8,25,68]. At higher concentrations it is possible to describe the surface pressure by a power law $\Pi \sim \Gamma^y$, y being a constant that depends on the polymer, and the nature of the fluids in equilibrium. For reasons to be explained below, the value of Γ where the two type of behaviours merge is called the *overlapping concentration* Γ^*. At still higher concentrations (above Γ^{**}, see Figure 11), the power law does not describe the experimental behaviour any longer. Figure 10 shows the two limiting values of the exponent y found for different systems.

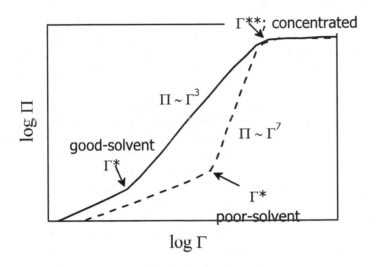

Figure 10. General behaviour of insoluble polymer monolayers for good and poor solvent conditions.

Contrary to what happens in the monolayers of many insoluble surfactants [8,69,70], those of insoluble neutral homopolymers do not show regions in which $\Pi \sim \Gamma^0$. These regions are frequently interpreted as surface phase transitions. However, they exist for monolayers of copolymers with a block soluble in one of the adjacent phases [71-78], or in the case of polyelectrolytes [79,80]. They are also observed when there is a transition from monolayer to multilayer [21].

2. Scaling in quasi-2D polymer systems

2.1. The nature of the interfacial solvent

Taking into account that for a quasi 2-D system like a Langmuir monolayer, the surface pressure plays the role of the osmotic pressure in 3-D systems [81-83], it seems natural trying to explain the power-law behaviour shown in Figure 11 in terms of the theory of polymer solutions. In this case the power-law behaviour of the osmotic pressure vs. concentration has been successfully described in terms of Flory's critical exponent ν, and the size of the polymer chain through $R_F \sim bN^\nu$, N being the polymerization degree, and b the size of the monomer [b ≈ O(0.1 nm)] [84,85]. The proportionality between Π and Γ is strongly influenced by the strength of interactions with the surrounding medium through the exponent v. So, $v = \frac{1}{2}$ at poor-solvent conditions, $v = 0.58$ in the melt or the Θ-state, $v = \frac{3}{4}$ for chains conformed in 2D at good-solvent-like conditions and, $v = 1$ for rigid rods, *i.e.* the maximum contour length of a rigid chain in the highest extension limit is $R_F = L = b\,N$.

The nature of the solvent is not obvious in the case of polymer films adsorbed at interfaces. In adsorbed two-dimensional films, the polymer chains lye at the interface between two bulk media, as a consequence they lose one degree of freedom with respect to the 3D conformation. Thus, in general there exists an entropic penalty on adsorption, the decrease in interfacial tension being the driving-force responsible for spontaneous adsorption. One can therefore conceive an adsorbed polymer chain as the result of the interplay between different forces: monomer-monomer, chain-chain and all the interactions of the polymer with the interface. In this case, the interface itself plays the role of the solvent, and it is in this sense that one must define the thermodynamic quality of the solvent. This concept, usual in bulk, refers to the strength of the adsorption interaction in the case of interfacial films. Then, good-solvent conditions refer to net cohesive adsorption interactions between the polymer and the interface. Weak adsorption as compared with the monomer-to-monomer interaction leads to adsorbed polymers at poor-solvent conditions. Chain conformation at adsorbed films is then governed by the strength and directionality of the adsorption forces compared with the intra- and inter-chain interactions. From a structural point of view good-solvent conditions are equivalent to high excluded-volume conformations like the extended coil or the flexible worm-like conformations. The poorer the adsorption interaction is, the smaller the excluded volume is thus leading to compressed coil conformations in order to minimize the number of contacts with the interface, and to increase those between monomers inside the chain.

2.2. Scaling theory

Once the "solvent" has been identified, the description of the power-law behaviour faces one of the most discussed issues in polymer monolayers: do the chains interpenetrate, i.e. are there entanglements in the quasi-2D systems? It is well known that chain overlapping is not possible in strictly two dimensions [84]. Based on this idea, de Gennes concluded that the polymer chains form segregated disks of density close to the 2D bulk density even at high surface concentrations [84]. This picture allows a simple interpretation for the overlapping concentration shown in Figure 11 as the point at which the individual patches start to contact each other. At this special concentration, the polymer concentration in the film become similar to the monomer concentration within a single chain, *i.e.* Γ^* and R_F are mutually related by:

$$\Gamma^* = \frac{N}{\pi R_F^2} \tag{33}$$

Then, one can easily obtain the chain size R_F from the Π–Γ isotherms. Within this segregated chains description, R_F is predicted to follow a simple dependence on the chain size $R_F \sim N^{0.5}$.

However, the domain occupied by a chain might be interpenetrated by the domains of other neighbour chains, as suggested by computer simulations [19,20,22]. Also recent AFM experiments on monolayers deposited on mica have shown that in some cases the individual chains do not form condensed patches [67,85]. Also the experiments performed by Maier and Rädler on DNA chains marked with fluorescent probes, and supported on lipid bilayers have found $R_F \sim N^{0.74}$ instead of $R_F \sim N^{0.5}$ [86]. Furthermore, the two-photon correlation fluorescence experiments of Granick's group seem to suggest that the polymer coils do not always form condensed patches in monolayers onto mica [87-89]. The existence of entanglements in very thin PMMA films (up to 10 nm thick) has been demonstrated by Itagaki et al. [90] from the measurement of the Försters distance. Probably the most clear demonstration that long chains do not necessarily form independent patches when they form a monolayer onto a solid substrate is given by the experiments of Wang and Foltz [91] using AFM experiments on nanorope monolayers.

In order to describe the power-law behaviour observed between Γ^* and Γ^{**} in the equilibrium isotherms, and to interpret the different values of the power exponent y, it is necessary to recall the equivalence of Π in quasi-2D systems to the osmotic compressibility in 3D ones. In this way it is possible to use the same arguments of des Cloizeaux and Jannik to describe the equilibrium properties of polymer solutions [92]. Within this theoretical framework, the adequacy of the power-law description of the Γ^*-Γ^{**} branch of the Π–Γ isotherms might underlay the existence of a re-scalable characteristic distance, ξ, inversely decreasing with the polymer concentration, $\xi \sim \Gamma^{-v/(2v-1)}$. In this way, the scaling law of the surface pressure at the semidilute regime can be written as [92].

$$\frac{\Pi}{\Pi^*} = \left(\frac{\xi}{\xi^*}\right)^m \tag{34}$$

Since any local property, as the osmotic susceptibility, must be N-independent and $\xi^* \sim R_F \sim N^\nu$, one may conclude that $\Pi^* \sim k_B T / R_F^d \sim N^{-d\nu}$, and therefore $m + d = 0$. As a consequence, particularized to the case of Langmuir polymer films, $m = -d = -2$. A power law like that in Eq. (32) is then recovered, $\Pi \sim k_B T / \xi^2 \sim N0$ $\Gamma^{2\nu/(2\nu-1)}$, from which one may identify $y = 2\nu/(2\nu-1)$. Now it is possible to interpret the values of the exponent y found for different monolayers in terms of the "solvent quality" concept used in polymer solutions. If $\nu = \frac{3}{4}$ the surface pressure increases as $\Pi \sim \Gamma^3$. Stepper dependencies are observed as the solvent-quality becomes poorer, e.g. $\Pi \sim \Gamma^7$ at the Θ-state, $\nu = 0.58$. Finally, in the hypothetic limit $\nu = \frac{1}{2}$, a hard-sphere-like divergence is expected. At a surface concentration exceeding a critical value Γ^{**} the concentrated regime is then entered; the surface solvent is almost excluded and the film becomes pure polymeric. At $\Gamma \gg \Gamma^{**}$ the surface pressure reaches a plateau value corresponding to the saturation of the film. Let us remark that the above *scaling* description relies on the existence of a characteristic length ξ, i.e. on the existence of entanglements. This impicitly implies that N has to be high enough to allow for entanglements, and probably that the monolayer is not strictly a 2D system, but has a finite thickness that allows those entanglements to take place. Based on this theoretical description, and by analogy to the 3D polymer solutions, one may consider that the transition from the diluted to concentrated regimes can be described as shown schematically in Figure 11. Here de Gennes' idea of segregated patches is recovered for near-Θ conditions.

An important test for the scaling description is to check one of the implicit consequences of Eq. (34): the plot of an equilibrium property rescaled by its value at Γ^* vs. the rescaled concentration Γ/Γ^* should lead to a master curve for N > Nc, Nc being the critical length above which the chains can entangle. Figure 12 collects the experimental isotherms obtained for monolayers of PtBAc on water at 298.15 K [37]. The molecular weights of the samples range from 102 to 106 Da, and the polydispersity index of the samples is below 1.1.

It is clear that for polymers long enough the results collapse in a master curve. The slope of the straight line is y = 3, i.e. $\nu = 0.75$, that means that the air/water interface is a good solvent for PtBAc. Similar results were obtained by Vilanove and Rondelez for poly(vinylacetate), PVAc, using two samples of different molecular weight [83]. They reached similar conclusions for monolayers of poly(methyl methacrylate), PMMA, (with $v = 0.56$, i.e. near Θ conditions) using three different samples. More recently Maestro et al. [93] have confirmed the same behaviour using six samples of PMMA with molecular weights in the range 14-400 kDa. This is a very strong piece of evidence of the validity of the scaling concept behind Eq. (33), which, as it has been already said, relies on the existence of an internal scale length ζ.

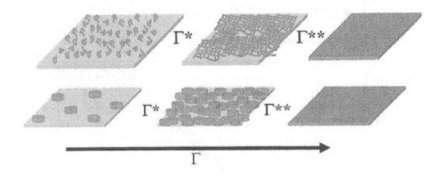

Figure 11. Schematic change of the polymer coils configuration in a monolayer under good solvent conditions (top), and under bad solvent conditions (bottom).

The above data allow one to discuss the N-dependence of the radius of gyration through the use of Eq.(32). Figure 14 shows that Flory's scaling $R_G \sim bN^v$ holds with $v = 0.73$ above $N_c = 100$ (this corresponds to $M_w \approx 10.000$ Da) and $b = 0.46$ nm. This value of v is the same that the one obtained from the Π vs. Γ results (see Figure 13). An exponent $v = 0.74$ was also reported by Maier and Rädler for DNA monolayers onto mica [86]. For $N < N_c$ the slope corresponds to $v = 1$. This plot is similar to the well known behaviour of R_G in three dimensional polymer systems [104]. The analysis of the Π vs. Γ isotherms $\Pi \sim \Gamma^{2v/2v-1}$ has been the most frequently used method to determine v. Miscible polymer blends of PVAc ($v = 0.75$) and P4HS ($v = 0.51$) have been found to show a continuous but sharp change of v with blend composition [95,96].

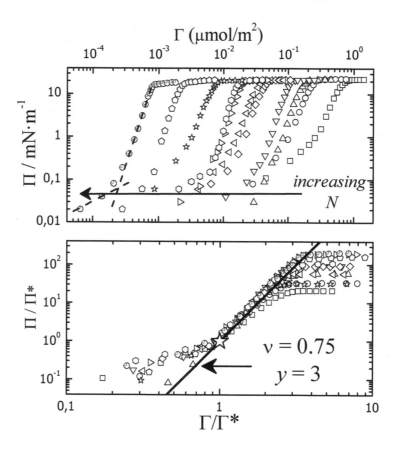

Fig. 12. Top) Molecular weight dependence of the isotherm for monolayers of poly(t-butyl acrylate). The range of molecular weights is $10^2 – 10^6$ Da. Bottom) Scaling behaviour of the experimental data shown at the top of the figure.

The existence of high density patches under poor solvent conditions ($\nu = 0.5$) (see Figure 1) allows one to predict that as Γ increases and the monolayer crosses from the dilute to the semidilute regime, the characteristic behaviour of the percolation transition should be observed. Such a behaviour was confirmed by Monroy et al. [97]. Moreover, the cartoon of the poor solvent conditions system (Figure 11) suggests a strong similarity with a 2D colloid. A glass transition has been found in colloids of a given composition as the temperature is decreased, and at constant temperature as the concentration is increased [98]. The temperature dependence of the Π vs. Γ monolayer may be use to confirm the existence of such a transition in a monolayer.

Figure 14 shows the expansivity plot for monolayers of poly(vinyl stearate) (poor-solvent conditions) at a fixed surface pressure. In 3D systems, the break of the isobaric expansivity has been identified as the signature of the glass transition [94]. The existence of the glass transition in this 2D system will be further confirmed by rheological measurements (see below).

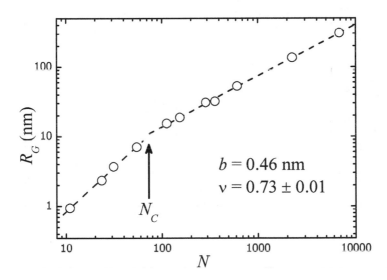

Figure 13. Chain length dependence of the radius of gyration calculated from the isotherms shown in Figure 12.

The precision of the data does not allow one to discuss the density dependence of the glass transition. In spite of the above evidences, the existence of entanglements is still actively discussed in the literature [31,32,88,91]. The analysis of the rheology of monolayers will shed some light into this problem.

E. THE RHEOLOGY OF POLYMER MONOLAYERS

Although many experiments have been reported on the shear viscosity of monolayers [99-109], in this section we will only review the dilational visco-elastic moduli.

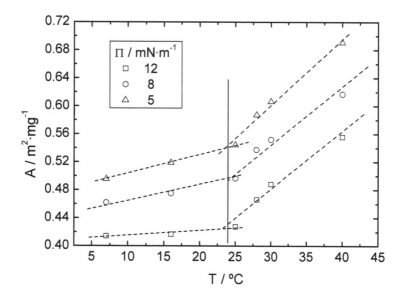

Figure 14. Temperature dependence of the area occupied by the poly(vinyl stearate) monolayers at three different values of surface pressure.

4.1. Good-solvent conditions

4.1.1. The low frequency range

The dynamic properties of the polymer monolayers can also be described by power-laws when they are studied as a function of Γ and of N [32,11,111]. Also the dynamics of polymer monolayers on solid substrates has been found to present similar power-law behaviour [87-89].

The most frequently used scaling description of the rheology of 3D semidilute polymer solutions is based on the concept of reptation of a polymer chain within a tube defined by the entanglements with the neighbour chains [112]. If excluded volume interactions are screened, the diffusion of a self-avoiding walk (SAW) chain through a distance comparable to its own size R_G takes place during the so-called Rouse time, $\tau_R = R_G^2 / D_R$, independently of the polymer concentration Γ. In the semidilute regime and if chain entanglements are present, the characteristic time depends on both N and Γ through

$$\tau \sim R_G^d \left(\frac{\Gamma}{\Gamma^*} \right)^m \sim N^{vd+m(vd-1)} \Gamma^m \tag{35}$$

For good solvent conditions, it has been quite usual to assume that the dependency of τ on N is $\tau \sim N^3$, and therefore, one can write m = 3, and $\tau \sim N^3 \Gamma^3$. This N^3 law represents a much stronger dependence than the simpler Rouse dynamics expected for individual chains, $\tau \sim N$. It must be stressed that the N^3 law implicitly assumes that the polymer chain is formed by blobs within which the excluded volume interactions are screened. As a consequence, the scaling law for the dilational viscosity can be written as:

$$\zeta \sim E\tau \sim N^3 \frac{k_B T}{\xi^2} \tau \sim N^3 \Gamma^6 \tag{36}$$

The strong Γ dependence of τ and of ζ have been tested for several polymer monolayers [82,83,110,113]. Figure 15 shows such a behaviour for monolayers of PVAc [105], as obtained from low-frequency mechanical relaxation experiments. It is worth noticing that one obtains the same exponent v by fitting the dynamic variables τ or ζ to Eqs. 34 and 35, than by fitting the equilibrium variable Π to $\Pi \sim \Gamma^{2v/2v-1}$. Similar conclusions were obtained for miscible polymer blends [111]. It must be remarked that the results shown are in marked contrast with the predictions of SAW chains moving without hydrodynamic interactions: $\tau \sim \Gamma$.

However, the N^3 law is not free from controversy [114]. Therefore, it is important to check its validity for monolayers. Figure 16 shows the results obtained by step relaxation experiments on monolayers of PtBA, and for molecular weights in the range $10^2 - 10$ It can be observed that the N^3 law is satisfied for $N > N_c$, with $N_c = 100$ ($M_w \approx 10^4$ Da), the same value that was obtained from the analysis of the equilibrium isotherms (see Figure 13).

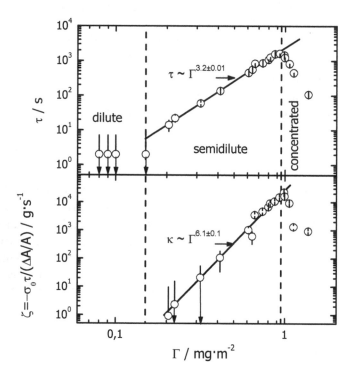

Figure 15. Surface concentration dependence of the relaxation time and of the dilational viscosity of poly(vinyl acetate) monolayers at 25 °C. The data were obtained by step relaxation experiments. The straight lines are the fits to the scaling laws.

These results represent a strong support for reptation as the mechanism for motion of the polymer chains in the monolayer. Such a mechanism assumes the existence of entanglements that can give rise to the tube concept.

The existence of entanglements at polymer interfaces has been discussed by Brown and Russell [115]. who found that the density of entanglements at the interface was lower than in the bulk.

Previous results were not as conclusive about the validity of the predictions of reptation theory on quasi-2D systems. In effect, the surface diffusion measurements of poly(ethylene glycol) were found to be compatible within their experimental error with several possible mechanisms: reptation, ameboid motion, sticky reptation, and overcraft motion [89]. Ameboid motion predicts $\tau \sim N^{5/2}$, that is far from the results shown in Figure 17. Sticky reptation leads to $\tau \sim N^{2.75}$ for good solvent conditions; this again is not compatible with the value $\nu = 3.0\pm0.1$ obtained for PtBA. Finally, Figure 16 shows that $\tau \sim N$ for $N < N_c$, in agreement with the Rouse dynamics expected for short non-entangled chains.

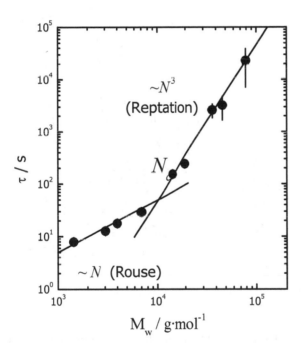

Figure 16 Molecular weight dependence of the relaxation time of poly(t-butyl acrylate) monolayers. The data were obtained by step-relaxation experiments at 25 °C. Above $N = 100$ the data follow the predictions of reptation theory, while below it the Rouse dynamics is obeyed.

Even though Figure 16 confirms the validity of the N^3 power-law, previous works reporting surface shear viscosity η_s studies have lead to different conclusions. Using a canal viscometer, Sacchetti et al found $\eta_s \sim N^1$ for poly(t-butyl methacrylate) (poor-solvent conditions!) [103]. Using a different type of rheometer, Gravanovic et al. found similar results for low and medium values of Π [21]. Sato et al. found $\eta_s \sim N^{1.3}$ for poly(vinyl octanal acetal) using a canal viscometer [102]. In both cases the interface behaves as a poor solvent, thus one may expect the coils to adopt a more compact (pancake-like) structure than in the case of PtBA. Finally, in a study of end-tethered polymer monolayers, Luap and Goedel [104] suggested that arm-retraction, instead of reptation, seemed to be the most adequate mechanism for their system. More recently Maestro et al. [37] have measured the shear viscosity of PtBa using two different oscillatory rheometers and found the same power laws shown in Figure 16 for the dilational viscosity.

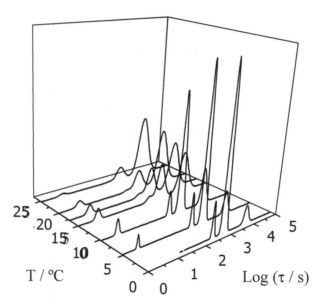

Figure 17. Relaxation spectra of the poly(vinyl acetate) monolayers calculated from step-relaxation experiments.

A further test of the mechanism of chain dynamics can be done by calculating the relaxation spectrum from $\Pi(t)$ vs. t curves of the step-compression experiments [116]. In effect, the elasticity can be calculated from

$$E(i\omega) = \frac{i\omega}{\theta} \int_0^\infty \Delta\Pi(t)\exp[-i\omega t]dt \qquad (37)$$

where θ is the relative area change. Usually, $\Delta\Pi(t)$ has been modelled as a sum of exponentials [63]. However, this is an arbitrary and empirical method, and it is difficult to assign a physical meaning to the different terms. It is more rigorous to estimate the relaxation spectrum H(lnτ) from the experimental data using

$$\frac{\Delta\Pi(t)}{\Delta\Pi(t=0)} = \int_0^\infty H(\ln\tau)\exp\left[-\frac{t}{\tau}\right]d\ln \tag{38}$$

The calculation of H(lnτ) is an ill-defined problem, for which the Tikhonov's regularization method can be used [117]. This method is the same used to analyze the autocorrelation functions of the quasi-elastic light scattering experiments in polymer solutions.

Figure 17 shows the results obtained for PVAc monolayers on water [29,96]. Once H(lnτ) is known, it is possible to calculate the elasticity and the loss modulus as

$$E(\omega) = \frac{1}{\theta}\int_0^\infty H(\ln\tau)\frac{\omega^2\tau^2}{1+\omega^2\tau^2}d\ln\tau$$

$$ \tag{39}$$

$$\omega\zeta(\omega) = \frac{1}{\theta}\int_0^\infty H(\ln\tau)\frac{\omega\tau}{1+\omega^2\tau^2}d\ln\tau$$

Although the form of the relaxation spectrum shown in Figure 18 looks rather complex, one must not forget that some theories about the rheology of polymer systems also predict the existence of several relaxation modes [94,112,114,116]. The same ideas can be also applied to surface polymer films. The theory describes the dynamics of the quasi-2D polymer coils in terms of two main mechanisms [35]: reptation of the chains on the plane of the interface, and a diffusive dynamics of the adsorption-desorption of loops and tails. The reptation takes place within a 2D tube defined by the points at which the loops and tails protrude into the subphase. According to the theory [118],

$$E(i\omega) = \left[\frac{\partial\gamma}{\partial\ln A}\right]_{\omega=0}\left\{\frac{16\Gamma_3}{\pi^3\Gamma_s}\sum_{p\,odd}\frac{i\omega\tau_p}{1+i\omega\tau_p}+\left(1-\frac{2\Gamma_3}{\pi\Gamma_s}\right)\frac{i\omega\tau_2}{1+i\omega\tau_2}\right\}$$

$$\tau_p = \frac{\tau_B}{p^2}\qquad p=1,3,5,7,\ldots \tag{40}$$

where Γ_3 is related to the surface concentration of loops protruding into the subphase. The relaxation time τ_p has the usual definition of the reptation theory, and τ_2 is the characteristic time for the adsorption-desorption of the loops and tails. Eq.(40) includes the relationship of the relaxation times of the different dynamic modes with the total disengagement time of the chain constrained within the tube τ_2. Figure 18 shows that the reptation prediction describes the relaxation times of the modes shown in Figure 18. Similar results were found for other polymers [96]. Furthermore, the amplitudes of the different modes shown in Figure 17 also follow a relationship $A_p = A_B/p^2$, as predicted by reptation theory. The amplitude of the relaxation mode corresponding to the shortest time in Figure 17 τ_2 is inversely proportional to the elasticity of the monolayer [96] as predicted by theory [35].

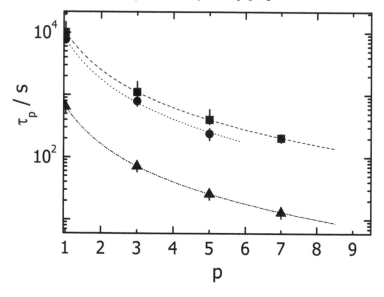

Figure 18. Test of the reptation prediction for the characteristic times of the relaxation spectrum of monolayers. Symbols: ■, PVAc; ●, 0.83 PVAc+0.17P4HS; ▲, 0.53 PVAc+0.47P4HS.

These results, together with the scaling laws discussed previously represent a strong support for reptation in quasi-2D systems, even though it is still unclear the physical nature of ξ and of the reptation tube. A further, though indirect, support comes from the plastic behaviour of monolayers under high strains. In effect, if entanglements exist, and the monolayer is perturbed at a frequency higher than the inverse of the terminal relaxation time, one can imagine that the entanglements play the role of physical crosslinks in the quasi-2D system. As a consequence one may expect the monolayer to behave as quasi-2D rubbers; this is exactly what has been reported by Hilles et al. [60].

For intermediate strains, not too deep into the non-linear regime, the stress-strain curve is described by a 2D affine model for rubbers, while a transition to plastic behaviour is found for higher strains. The strong similarity with the behaviour of 3-D systems can be taken as an indirect support for the existence of entanglements. However, a direct experimental evidence of their existence is still lacking.

1.2. The high frequency range

From the analysis of the high-frequency rheological data obtained from capillary waves experiments, internal or segmental relaxation times can be probed. Figure 19 shows, as a representative example, the Maxwell-like visco-elastic relaxation observed in PVAc films at SQELS-frequencies [29], which is characterized by a rather large strength $\triangle E/E_0 \approx 0.5$ and a relaxation time τ in the microsecond window.

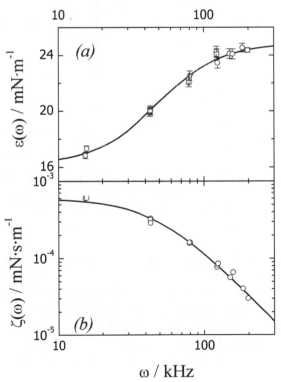

Figure 19.- Frequency dependence of (a) the dilational elasticity, and (b) the dilational viscosity of poly(vinyl acetate) monolayers measured by SQELS in the high-frequency regime. The data correspond to $\Gamma \sim \Gamma^{**}$. A viscoelastic relaxation is observed at $\omega \approx 60$ kHz.

A high value of the activation energy $E_A \sim 30\ k_B T$, pointing out the collective character of the probed motion, has been derived from the Arrhenius-like T-dependence of the relaxation times. Internal motions ranging the sub-millisecond window are typically found in these experiments.

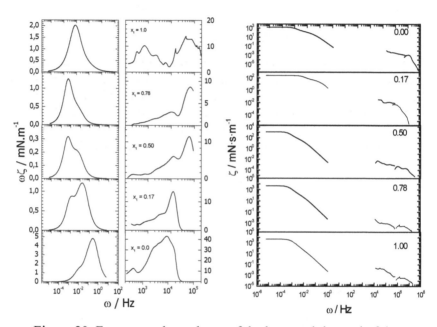

Figure 20. Frequency dependence of the loss modulus and of the dilational viscosity for monolayers of a miscible polymer blend: x_1 PVAc + $(1-x_1)$ P4HS. Step-relaxation, ECW, and SQELS data have been combined to sweep the whole frequency range.

The spatial scale involved here $\xi_0 \sim (D\tau_0)^{1/2} \leq 1$ nm, obviously corresponds to highly correlated segmental motions. Maybe, the particular dynamics observed here arises from the internal motions of the blobs comprised between entanglement points, but it remains yet a matter of debate. In fact, since diffusion within a given blob with n monomers and size, $R_{blob} \sim \xi \sim n^\nu$, governed by a Rouse-like diffusion coefficient $D_R \sim D_0/n \sim D_0\, \xi^{-1/\nu}$, this internal relaxation time might scale as:

$$\tau_0 \approx \frac{\xi^2}{D_R} \sim \frac{1}{D_0}\xi^{2+1/\nu}$$

(41)

i.e. $\tau_0 \sim \xi^{3.3} \sim \Gamma^{-5} \sim \Pi^{-1.6}$ at good-solvent conditions, while stronger dependencies are expected at poor-solvent conditions ($v = 0.78$); $\tau_0 \sim \xi^4 \sim \Gamma^{-13.5} \sim \Pi^{-1.9}$. In fact, Π–decreasing relaxation times are always found from the rheological information achieved on surface waves experiments (SQELS), as experimentally observed [29,55,96]. The validity of the scaling laws that describe the Γ dependence of ϵ and of ζ have been tested in this high frequency range. It has been found that they describe the experimental results quite satisfactory. Moreover, the value of v obtained coincides with that obtained from the power law that describes the dependence of Π [111,113]. No test of the N^3 law has been carried out so far using dynamic results in this high frequency range.

The combination of the step-relaxation, oscillatory barrier (or oscillatory drop), ECW and SQELS techniques allow one to explore the rheology of polymer monolayers over a broad frequency range. Figure 20 shows the results obtained for the PVAc (1) + P4HS (2) miscible blend as a function of ω, and of the mole fraction of PVAc [96].

This points out the complex dynamic behaviour of insoluble polymers in quasi-2D systems. It also indicates that the solvent-quality of the interface ($v = 0.75$ for PVAc, and $v = 0.51$ for P4HS) has an important effect on the dynamics, specially in the high frequency range. This is in agreement with the conclusions of Yu's group (for a review see [30]).

It is worth noticing the huge change in κ found in the low-frequency range. In this range $\zeta \sim \omega^{-\alpha}$ with $\alpha \approx -1$, a result close to that found in tribology experiments for thin polymer films between mica walls [3]. A further conclusion from Figure 20 is that polymers in quasi-2D films show a rather complex dynamic behaviour. Its theoretical prediction, though necessary for modelling many processes of technological interest, is still far from satisfactory.

2. Poor-solvent conditions

As it was discussed in Section 3.2.2, for near-Θ conditions, one may expect the coils to form a dense pancake, thus for the dilute and semidilute regimes the monolayer looks like a 2D colloid (see Figure 11). In Figure 14 we have shown that the equilibrium isotherms indicate that as the temperature is decreased the monolayer undergoes a transition that mimics the glass-transition in 3D polymers. The glass-transition has a dynamic character [119-121], and therefore rheology measurements are best suited for studying it. Figure 21 shows the surface pressure dependence of the step-relaxation experiments on monolayers of P4HS at fixed temperature [122].

It is obvious that the relaxation is not a single exponential in most of cases. The experimental results of the stress $\sigma(t)$ have been described by means of a stretched exponential function

$$\sigma(t) = \sigma_0(t)\exp\left[-\left(\frac{t}{\tau}\right)^\beta\right] \tag{42}$$

where $\sigma_0 = \sigma(t=0) = \Pi(t=0) - \Pi(t=\infty)$, and the stretching parameter $0 < \beta \le 1$ is related to the coupling parameter n of Ngai's Coupling Model n by $\beta = 1 - n$.

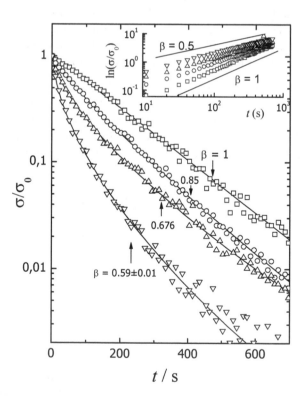

Figure 21. Relaxation curves for monolayers of P4HS at different surface pressures and 25 °C. From top to bottom, the data correspond to the following surface pressures: 0.2, 1.0, 2.0, and 3.0 mN·m^{-1}, respectively.

The higher n the higher is the cooperative character of the dynamics [123]. The results show that the relaxation is almost exponential ($\beta = 1$) for low density states ($\Gamma \sim \Gamma^*$). However, β decreases (n increases) as the density of coils increases, indicating a strong increase of the degree of cooperativity. As it can be observed in Figure 22, similar results are obtained for fixed Γ and varying T. β strongly decreases as T is decreased for $\Gamma = \Gamma^*$.

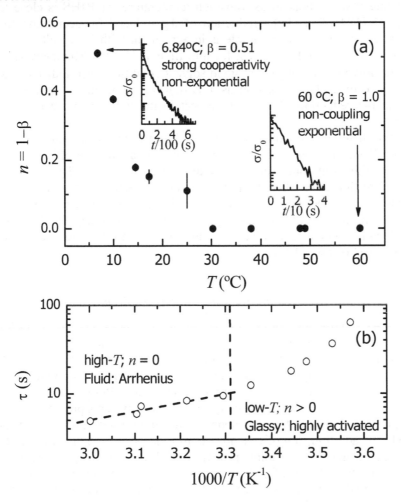

Figure 22. Top) Temperature dependence of the coupling parameter n for monolayers of P4HS at $\Gamma = \Gamma^*$. The inset shows the relaxation curves for typical non-cooperative and a cooperative dynamics. Bottom) Temperature dependence of the mean relaxation time for the same system.

The Arrhenius-like plot of the relaxation time indicates that the apparent activation energy of the dynamics increases below $T \approx 30\ °C$, i.e. the same temperature at which n starts to increase. Moreover, a plot of the expansivity of P4HS monolayers, similar to Figure 14, shows a break at 30 °C, and the temperature dependence of the shear viscosity shows a sharp change at the same temperature [93]. This further confirms the glasslike character of the dynamics observed. It is worth noticing that the bulk glass transition temperature for P4HS is close to 115 °C [124]. This huge reduction of Tg induced by decreasing the thickness of the polymer sample is in agreement with the results of the groups of Jones and of Forrest for thin films [125,126]. The analysis of τ_B for polymer monolayers at good solvent conditions also indicates the existence of a glass transition as T is reduced, although in the case of PVAc the reduction of Tg with respect to the bulk sample was much smaller [29].

The different reduction of Tg with respect to the bulk value may be rationalized in terms of the chain-interface interactions. In effect, in the case of thin films onto solid substrates, it was found that Tg decreased with film thickness when the polymer-wall interaction was weak (as for monolayers at poor solvent conditions!). However, when such interaction was strong (as for monolayers at good solvent conditions!), the decrease of Tg was smaller; in fact Tg was found to increase with respect to the bulk in some cases [127].

3. The interface as a selective solvent

Block copolymers formed by two insoluble blocks may be useful for testing the influence of solvent-quality on the dynamics of monolayers. Even though many authors have carried out rheological studies on block copolymer monolayers [71-78], most of them deal with copolymers that contain a soluble block, mainly poly(oxyethylene) (PEO) [63,7178,128-130]. In addition to the in-plane dynamics, these systems present surface phase transitions and the corresponding adsorption-desorption dynamics [131]. In addition, long-lived non-equilibrium structures are frequently formed during the spreading process [130,132]. Changing the relative length of the blocks of polystyrene-b-PtBa it is possible to tune the solvent-quality of the interface. Table I shows the characteristics of the copolymers studied in [133].

Table I. Characteristics of the PS-b-PtBa copolymers used for studying the rheology . Flory's exponent ν has been calculated from the Π vs. Γ isotherm (25°C), from step-relaxation experiments (low frequency), and from SQELS (high frequency)

Polymer	$M_{w,PtBA}$	$M_{w,PS}$	ν		
			Equilibrium	Low frequency	High frequency
P4HS	-	30000*	0.59±0.04	≈ 0	0.59±0.06
B-1	201800	206200	0.63±0.03	0.44±0.04	0.59±0.07
B-2	312800	309800	0.68±0.03	0.48±0.06	0.57±0.06
B-3	489000	236600	0.73±0.02	0.61±0.04	0.70±0.05
PtBA	287000	-	0.76±0.03	0.75±0.03	0.70±0.06

* Molecular weight of P4HS

The first point to notice is that there is a good agreement between the values of ν obtained from equilibrium isotherms and from SQELS at high frequencies. However, the values obtained at low frequencies by step relaxation experiments show clear discrepancies for poor-solvent conditions. Figure 23 shows the Π-dependence of the relaxation times obtained from relaxation experiments. It points out that the behaviour of the copolymer monolayers is different from that of PtBA, even for the B3 sample, for which the equilibrium and SQELS experiments indicate that the interface is a good solvent. Even though these results seem to contradict the conclusions of Esker et al. [30], and those discussed above, it is necessary to take into account that non-equilibrium structures such as micelles, cylinders, etc. have been found in copolymer monolayers [134-137]. Therefore, the analysis in terms of scaling, and even the solvent-quality concept, should be taken with care for this type of monolayers.

Kent et al. have reported studies on block copolymers for which the interface is a poor solvent [134,135], or good solvent for both blocks [136,137]. Also Seo et al. have reported on the block copolymers under bad solvent conditions [138,139].

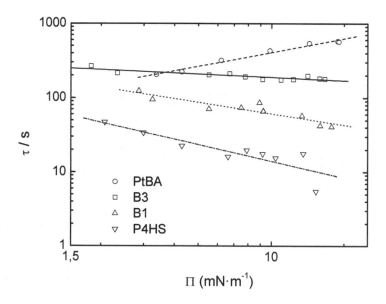

Figure 23. Surface pressure dependence of the relaxation time of polystyrene-b-poly(t-butyl acrylate) monolayers at 25 °C. The results were obtained by step-relaxation experiments.

F. POLYMER SOLUTIONS

Unlike spread polymer monolayers, the adsorbed polymer films at the liquid – air interface are investigated to a less extent. In particular, this also relates to the surface dilational rheology of polymer films. More or less systematic studies of the surface visco-elasticity of polymer solutions have been started only in the last years. The reason of this lag as compared with the studies of spread films is probably connected with experimental difficulties at first. The interpretation of surface rheological data is much facilitated, for example, if the adsorbed amount is known in advance as in the case of spread insoluble films. On the contrary, the determination of the adsorption of soluble polymers at the liquid – gas interface by traditional methods is a serous problem.

For dilute solutions of monomeric surfactants, the adsorbed amount at the liquid-gas interface can be calculated by the Gibbs adsorption equation:

$$\Gamma = -\frac{1}{RT}\left[\frac{\partial \gamma}{\partial \ln c}\right] \tag{43}$$

where R is the gas constant and c is the bulk molar concentration.

Unlike the solutions of conventional surfactants, the molar concentration of polymers of high molecular weight is very small even in concentrated solutions. Therefore, the surface tension almost does not depend on the concentration, and the application of Eq. (43) leads to large errors when used to determine polymer adsorption [140-142].

Reliable data on the adsorbed amount of main soluble polymers have appeared only within about last ten years due to the application of neutron reflection, which allows also the determination of the monomer distribution normal to the liquid surface [140, 142-149]. The progress in the understanding of the equilibrium structure of adsorbed polymer films influenced surface rheological studies as well.

The long time non-equilibrium surface properties of polymer solutions were mainly investigated via the measurements of dynamic surface tension [150-163] and by comparison of the results with the solution of the corresponding boundary problem for the diffusion equation [151,153,157,158,163] or with empirical equations, such as that given by Hua and Rosen [164]. This approach, when applied to conventional surfactant solutions, allowed establishing the diffusion-controlled adsorption for a large number of systems [165, 166]. For most of polymer solutions the diffusion from the bulk phase to the surface influenced strongly only the initial steps of adsorption [154-156, 158, 160, 162, 163]. The authors usually connected the subsequent very slow decrease of the surface tension with the conformational changes in the surface layer and described these processes only in general terms [158, 160, 162]. The application of the more sophisticated kinetic theory [173], though allowed Zhang and Pelton to describe the adsorption of poly(N-isopropylacrylamide) (PNIPAM), led to insignificant progress in the elucidation of the real mechanism of slow changes of surface properties [164]. The number of parameters to be extracted from the fitting to monotonic kinetic dependencies increased together with the complexity of the model.

The application of the methods of surface dilational rheology in a broad frequency range gives a general possibility to separate the contributions of different processes in the surface layer, and to study the details of the conformational transitions of macromolecules at the liquid surface. Similar ideas proved to be very useful in the interpretation of the dilational rheology of the bulk phase, and led to the development of the ultrasound relaxation spectrometry [168].

Unfortunately, this program has not been executed yet in the case of surface films because of both theoretical and experimental hindrances. The range of accessible frequencies is seriously limited in this field. Although the SQELS technique gives a possibility to work at frequencies up to a few megahertz, the interpretation of the results becomes a serious problem when the apparent surface viscosity takes negative values (Section 2). At lower frequencies (10 – 3000 Hz) the data of the electrocapillary wave method (Section 2.1.2) usually agree better with the theory [165, 166], although some systems also exhibit negative apparent surface viscosity [49].

At relatively high frequencies (≥ 10 Hz) the surface visco-elastic behaviour of polymer solutions usually coincides with that of insoluble monolayers [50, 169-172]. This is because bulk-to-surface polymer exchange is limited by slow diffusion. The spread and adsorbed films of PEO [169, 176], hydroxypropylcellulose and hydroxyethylcellulose [177] were compared using the surface light scattering technique and found to exhibit almost identical surface dilational rheological properties. This indicates that the characteristic times of the main relaxation processes in adsorbed polymer films should correspond to lower frequencies.

Low-frequency surface rheological methods have been applied to the solutions of most of soluble polymers only recently [11, 33, 50, 173-182]. In this section we review these results after a short account of the theory related to the surface dilational rheological properties of polymer solutions. The main attention will be paid to the surface rheology of solutions of non-ionic homopolymers [11, 33, 173-175]. The experimental data on the solutions of block copolymers [46, 178, 180, 181] and polyelectrolytes [176, 177] are more scarce and will be discussed briefly. The surface rheology of natural polymer solutions [57, 64, 178, 183-186] and mixed polymer/surfactant solutions [187-194] is beyond the scope of this chapter.

1. Mechanical relaxation in adsorbed polymer films

The first expression of the complex dilational surface elasticity was derived by Lucassen and Hansen about 40 years ago in the case of solution of a single surfactant [195]:

$$E(i\omega) = -\left(\frac{\partial \gamma}{\partial \ln \Gamma}\right) \frac{(1+i)\sqrt{\dfrac{\omega}{2D}}\dfrac{\partial \Gamma}{\partial c}}{1+(1+i)\sqrt{\dfrac{\omega}{2D}}\dfrac{\partial \Gamma}{\partial c}} \tag{44}$$

where D is the diffusion coefficient of the surfactant in the bulk phase.

Relation (43) assumes that the only relaxation process in the system is the surfactant adsorption/desorption and it is controlled by the diffusion in the bulk phase. The methods of non-equilibrium thermodynamics allow one to derive more general expressions for the solutions with an arbitrary number of surfactants, with an arbitrary adsorption mechanism and other relaxation processes in the system [196]. The general expressions are rather cumbersome. However, if apart from the diffusion in the bulk phase only the second relaxation process is possible, for example, the transition through the adsorption/desorption barrier, the complex surface elasticity takes the compact form [196]

$$E(i\omega) = \left(\frac{\partial \gamma}{\partial \ln A}\right)_{\xi,c} \frac{i\omega\tau + (1+i)\sqrt{\dfrac{\omega}{2D}}\dfrac{\partial \Gamma}{\partial c}}{1 + i\omega\tau + (1+i)\sqrt{\dfrac{\omega}{2D}}\dfrac{\partial \Gamma}{\partial c}} \qquad (45)$$

where τ is the relaxation time, ξ is the chemical variable (structural parameter for the processes of the structural rearrangement), the subscript ζ means that the derivative corresponds to a non-equilibrium process of the surface dilation when the relaxation process is forbidden. If this process is adsorption/desorption, $(\partial\gamma/\partial\ln A)_{\xi,c}$ coincides with - $\partial\gamma/\partial\ln\Gamma$.

If the molecular weight of the surfactant increases, the derivative $\partial\Gamma/\partial c$ increases much faster than the decrease of the diffusion coefficient, and the characteristic time of the diffusion to and from the surface $\tau_d = (\partial\Gamma/\partial c)^2/D$ increases too. Therefore the diffusion exchange of macromolecules between the surface layer and the bulk phase is usually negligible for solutions of polymers of high molecular weight and can only influence the dynamic surface properties of oligomer solutions. Application of relations (43), (44) to polymer systems [187-189] has no strict substantiations up to now, and in this case it is necessary to consider them as semi-empirical equations.

If the diffusional exchange between the surface layer and the bulk phase is negligible, the relaxation processes in the adsorption film can be reduced to N linear-independent normal processes and the thermodynamic approach leads to the following expression of the dynamic dilational surface elasticity [196]:

$$E(i\omega) \equiv \frac{\delta\gamma}{\delta \ln A} = \left(\frac{\partial \gamma}{\partial \ln A}\right)_{\xi} - \sum_{i=1}^{N} \frac{\left(\dfrac{\partial \gamma}{\partial \ln A}\right)_{\xi} - \left(\dfrac{\partial \gamma}{\partial \ln A}\right)_{A_i,\xi_j}}{1 + i\omega\tau_i} \qquad (46)$$

where τ_i is the isothermal relaxation time of the normal process i in the surface layer, ξ_i is the corresponding chemical variable, A_i is the chemical affinity of the process i. The subscript A_i at the derivative indicates the chemical equilibrium for the process i.

One can apply in principle relation (44) to polymer adsorption films. However, two questions arise. Is it possible to reduce the complicated motion of polymer chains in the film to a discrete number of relaxation processes? What is the meaning of these processes? It is hardly possible to study these problems in the framework of thermodynamics. Therefore, more detailed investigation requires a molecular model of the polymer adsorption film [35].

Although the exchange of macromolecules between the surface layer and the bulk phase is negligible, the exchange of monomers between different regions of the surface layer has to be taken into account. Two main parts of the surface film are considered: a relatively narrow concentrated region (I) contiguous to the gas phase (the proximal region), and region (II) of «tails» and «loops» protruding into the bulk of the liquid where the global concentration of monomers is essentially lower (the middle and distal regions). It is essential that the film thickness in region (II) exceeds that of region (I) and the monomer concentration in region (I) is significantly higher. In this case one can describe the motion of a polymer chain as a whole in region (I) using conventional ideas developed for semi-dilute and concentrated polymer solutions, in particular, the concept of chain reptation in the tube made of entanglements [84, 112]. It is quite probable indeed that the chains containing hydrophilic groups are entangled in region (I) due to small loops consisting of few monomers (cf. the discussion of the role of entanglements in insoluble polymer films in Sections 3.2.2. and 4.1.1).

One can also consider that the surface tension depends first of all on the concentration in region (I) [197]. Then the relaxation of surface dilational stresses can proceed at the expense of drawing chains up to the surface (transition of monomers from (II) to (I)) or squeezing chains out of the surface (the reverse transition). If the surface deformations are small, the departure of a polymer coil from the surface (disentanglement of the coil), or the reverse process of attaching of a free coil to the surface (entering region (I)), turn out to be relatively rare events and can be neglected. In this case the exchange of monomers between regions (I) and (II) under expansion (compression) of the surface can take place as a result of two processes: the relaxation of inner strains of a polymer chain or the drawing up (squeezing in) of the chain as a whole. The first process can be described mathematically by means of a solution of the corresponding boundary problem for the Rouse equation [112]:

$$\frac{\partial r_j}{\partial t} = B\left(f_j + \frac{3k_BT}{b^2}\frac{\partial^2 r_j}{\partial j^2}\right) \tag{47}$$

Here the real polymer chain is assumed to consist of a large number of frictional units (Rouse segments). These segments have the mobility constant B, and are connected by harmonic springs with the mean square separation b^2. r_j is the position of the j-th Rouse segment at time t, f_j is an external force.

Eq. (47) corresponds to ideal chains and does not take into account excluded volume effects and hydrodynamic interactions. However, in the proximal region, where the local concentration is high and the chains are unfolded, this assumption can be satisfactory.

The second process of the motion of a polymer chain as a whole in region (I) can be considered in the framework of the two-dimensional reptation model by de Gennes [84, 112]. The high monomer concentration at the surface leads to interactions between the adsorbed polymer trains and the neighbouring adsorbed polymer chains form a two-dimensional network of entanglements. In the model, the real nature of the entanglements is not important, and the adsorbed chain wriggles (reptates) inside a quasi-2D network of entanglements, which form a two-dimensional tube.

In the general case there are three kinds of polymer chains in region (I): the chains completely lying on the surface, the parts of the chains with a single transitional point region (I) – region (II) and the trains with two transitional points (Fig. 24). The velocity of a train with one transitional region from (I) to (II) as a whole dL_2/dt is proportional to the deviation of the surface from the equilibrium value $\Delta\tilde{\Pi}$:

$$\Delta\Pi \cdot a = \frac{N_2}{B}\frac{d L_2}{d t} \tag{48}$$

where a is the mean distance between the walls of the two-dimensional tube of entanglements, N_2 is the number of monomers in the train under consideration, L_2 is the length of the train, B is the mobility constant.

These simple assumptions allow one to describe the motion of polymer trains in the surface layer under local surface dilation. If the number of loops and tails (in region II) per chain n is large (n » 1), one can obtain relation (40) for the complex dynamic surface elasticity [35]. In the case of soluble adsorption films it is necessary to exchange the first coefficient in the square brackets by $-d\gamma/d\ln\Gamma$ and τ_B is the relaxation time of inner stresses of a polymer chain:

$$\tau_B = \frac{b^2 N_0^2}{3\pi^2 k_B T B}$$

(49)

The relaxation time τ_2 corresponds to the motion of a train in region (I) as a whole

$$\tau_2 = \frac{N_2 L_2}{B\Gamma_2 a\left(\partial \tilde{\Pi}/\partial \Gamma_s\right)}$$

(50)

where Γ_2 is the number of moles of monomers per a unit area in region (I), which belong to trains confined by one transitional region from (I) to (II) (Fig. 24b), N_0 is the number of monomers inside the two-dimensional tube of entanglements.

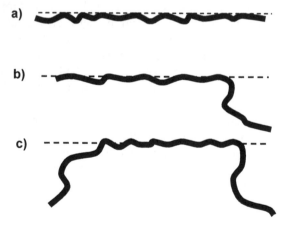

Fig. 24. Three important cases in region (I): a) a chain entirely lying in the region, b) a polymer train with only one end in (I) and contiguous to a loop or a tail, c) a polymer train without ends in (I) and contiguous to two loops or tails.

The relaxation times τ_B and τ_2 decrease with the decrease of N_0 and N_2, and consequently with the increase of the number of loops and tails. The additional loops and tails create new paths for the relaxation of stresses in region (I) of the surface layer and therefore decrease the resistance to this relaxation and the dynamic surface elasticity according to relation (40).

If the chains are short enough and the number of adsorbed chains with more than one loop or tail in region (II) is small (n < 1, $\Gamma_3 \ll \Gamma_2 < \Gamma_s$), it follows from relation (40) that the relaxation of surface stresses at the frequencies comparable with $1/\tau_2$ will mainly take place at the expense of squeezing separate trains out of region (I) (drawing into the region (I)). In this case the relaxation of local surface dilational stresses can proceed mainly at the expense of reptation of trains with a tail in the bulk phase. Then

$$E(i\omega) = -\left(\frac{\partial \gamma}{\partial \ln \Gamma_s}\right)\frac{i\omega\tau_2}{1+i\omega\tau_2} \tag{51}$$

and the rheological Maxwell model describes the surface dilational visco-elasticity.

Relation (40) describes the dynamic surface elasticity of the both adsorbed and spread polymer films. In the latter case the prediction of the relationship between the relaxation times amplitudes is in accordance with relations (40), (49), (50) (Section 4.1.1.). In the former case the quantitative comparison has not been performed yet. However, qualitative conclusions of the theory [35] agree well with the experimental results on the surface visco-elasticity of polymer solutions, and of Langmuir films of insoluble polymers (see Figures 17 and 18).

The conformations of polymer chains at the liquid – gas interface are determined mainly by the energy of the segment interactions with the surface and between themselves, and also by the entropic penalty due to restrictions of the conformations at the surface. When an amphiphilic polymer adsorbs at the liquid surface at very low concentrations the first contribution into the excess free energy is usually the main one, and the macromolecules take extended, almost flat, conformation without long loops and tails ("pancake" conformation). At the increase of the concentration the repulsion between the adsorbed segments compensate partially the adsorption energy, the entropic contribution becomes significant and the adsorbed macromolecules can form loops and tails protruding into the bulk phase. In the case of spread films of amphiphilic homopolymers [175, 198, 199] or block copolymers with amphiphilic blocks [128, 180, 181, 200] this can lead to the "pancake" – "mushroom" conformational transition. The increase of the number of loops and tails results in the decrease of the relaxation times in accordance with relations (48), (49) and, consequently, to the decrease of the dynamic surface elasticity at low frequencies. If the frequency is less about 100 Hz, the dynamic surface elasticity of spread films of amphiphilic polymers as a function of the surface concentration, indeed, has a local maximum in the range of conformational transition [50, 71, 175, 180, 181, 199, 200]. The maximum can disappear if the frequency exceeds a few kilohertz [201].

The neutron reflectivity shows that the adsorbed amphiphilic homopolymers also form loops and tails at the liquid-gas interface [140]. However, the observation of the "pancake" conformation is difficult for polymers of high molecular weight because of the very slow changes of surface properties with the bulk polymer concentration (cf. discussion at the beginning of this section). For example, the surface pressure of PEO solutions is about 9 mN·m^{-1} in a broad concentration range [174], which is much higher than the surface pressure of the conformational transition in PEO spread films (about 5 mN·m^{-1} [198]). The "pancake" conformation probably corresponds to the case of very dilute solutions when the slow adsorption kinetics does not allow measurements of the equilibrium surface properties and only the surface elasticity of the system far from equilibrium can be measured. The decrease of the polymer molecular weight leads to the shift of the range of flat conformations to the direction of more concentrated solutions. The probability of long loops and tails is lower indeed for shorter chains. One can expect that the observation of the "pancake" – "mushroom" conformational transition and, consequently, the local maximum of the surface elasticity are possible if the polymer chains are short enough.

It is impossible to observe the surface elasticity maximum for the equilibrium system and to check up the theoretical prediction directly at high polymer molecular weights. However, one can examine the states of local equilibrium between the surface and the subsurface layers when the system slowly approaches the adsorption equilibrium. Indeed the adsorption kinetics of most non-ionic homopolymers is controlled by diffusion in the bulk phase [33, 157, 163] and the system meets the conditions of local equilibrium. In this case the kinetic dependence of the dynamic surface elasticity is an image of its concentration dependence. Every point of the former dependence must correspond to a definite equilibrium subsurface concentration and the transition from $\varepsilon(t)$ to $\varepsilon(c)$ requires only a (non-linear) transformation of the abscissa. Thus it follows from the theory of surface visco-elasticity of polymer solutions that the dynamic surface elasticity must be a non-monotonic function of the surface age.

2. Solutions of homopolymers

Scott and Stephens were the first who measured the characteristics of capillary waves and estimated the surface elasticity for solutions of poly(ethylene glycol) (PEG) but at a single concentration and a single frequency [202]. Myrvold and Hansen used the oscillating bubble method to determine the dynamic surface elasticity of ethylhydrooxyethyl cellulose solutions at a few concentrations in the frequency range from 0.2 to 2 Hz [203]. Zhang and Pelton applied the oscillating drop method to PNIPAM solutions but the surface deformations were large and they did not determine the dynamic surface elasticity [162]. The limited concentration and frequency ranges in these studies did not allow the authors to observe the main peculiarities of the surface visco-elasticity of polymer solution described in the preceding section.

The local maximum of the surface elasticity concentration dependence was discovered first by the capillary wave method, which is highly sensitive to low values of surface rheological properties [173, 174]. The damping coefficient of capillary waves α increases abruptly when the absolute value of the dynamic surface elasticity increases from zero up to about 0.15γ but changes only slightly at higher values (Chapter 4 of this book). Unlike the surface tension of PEG and PEO solutions, the damping coefficient for these systems proved to be a non-monotonic function of concentration and depended on the polymer molecular weight (Fig. 25).

At a first sight, the dependence $\alpha(c)$ for solutions of PEG4 of molecular weight 4000 resembles the corresponding curves for conventional surfactants where a local maximum is also followed by a local minimum [165, 166]. However, a careful inspection allows one to detect significant differences. Firstly, the damping coefficient of the polymer solutions changes slowly with concentration. The concentration range in Fig. 26 spans over 5 orders of magnitude while the corresponding range is much shorter for solutions of conventional surfactants. Secondly, in the latter case the damping coefficient in the region of the local minimum corresponds to high surface elasticity (approximation of the infinite surface elasticity in the theory of capillary waves) and equals to about $3\alpha_0$, where α_0 is the damping coefficient at zero surface elasticity. For polymers the obtained α values in the region of the local minimum are significantly smaller.

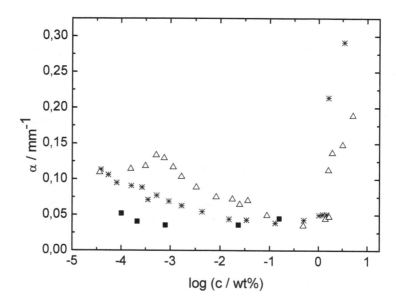

Figure 25. Concentration dependency of the damping coefficient of capillary waves at the frequency 200 Hz for solutions of PEG4 (triangles), PEG20 (asterisks) and PEO (squares).

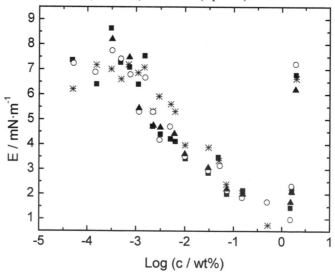

Figure 26 Real part of the dynamic surface elasticity of PEG4 solutions vs. concentration at the frequencies 120 Hz (squares), 180 Hz (circles), 220 Hz (triangles pointing up) and 270 Hz (asterisks).

For solutions of PEG20 and PEO of molecular weights 20 kDa and 100 kDa, respectively, the decrease in damping for dilute solutions proceeds more slowly, and the local maximum disappears. One can assume that this maximum shifts towards lower concentrations where measurements of the surface properties are difficult because of the slow equilibration. For PEO solutions the damping coefficient is almost constant. Similar results have been obtained also for solutions of PNIPAM [179] and poly(vinylpyrrolidone) (PVP) [33].

Fig. 27 shows the real part of the dynamic surface elasticity of PEG4 solutions calculated according to dispersion equation (10) from the characteristics of capillary waves at different frequencies. The imaginary part of the dynamic elasticity was close to zero.

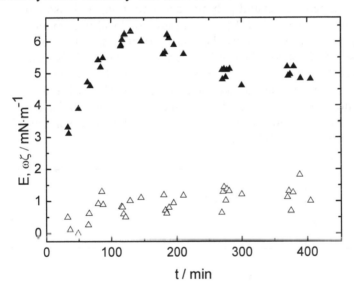

Figure 27. Kinetic dependence of the real (solid symbols) and imaginary (open symbols) components of the dynamic surface elasticity of PVP55 solutions at the concentration 0.00003 wt.%.

One can see in Fig. 26 a local maximum of the surface elasticity real part at low concentrations in agreement with the considerations above. When the polymer molecular weight increases, the local maximum probably shifts to low concentrations where measurements close to the equilibrium are impossible. The decrease of the surface elasticity with concentration becomes more gradual in this case. For PEO solutions the elasticity is almost independent of concentration. Similar results have been obtained at other frequencies using different experimental methods and other polymers [11, 33, 174, 179]. Note that the strong increase of the damping coefficient for concentrated solutions is accompanied by a decrease of the surface tension and can be connected with the change of interactions between PEG and water in this concentration range.

The observed surface visco-elastic behaviour is entirely different from that of conventional surfactant solutions where the dynamic surface elasticity can be one or two orders of magnitude higher [165, 166] and they cannot be described by relations (43) and (44). At the same time the results are in a qualitative agreement with the theory of Section 5.1.

Figs. 27 and 28 show that the kinetic dependencies of the surface elasticity of PVP and PNIPAM solutions are also non-monotonical in agreement with conclusions of the preceding section. Similar results were obtained for several polymer concentrations and also corroborated these conclusions [33, 179]. In the range of local maximum of the real part of the surface elasticity the imaginary part deviates from zero as in the case of PEG4 solutions at low frequencies [174]. This means that the adsorbed film becomes visco-elastic there, and the characteristic time of the processes in the surface layer is comparable with the period of oscillations. For solutions of PEG4 this gives a possibility to determine the frequency dependence of the both dynamic surface elasticity components (Fig. 30) [174].

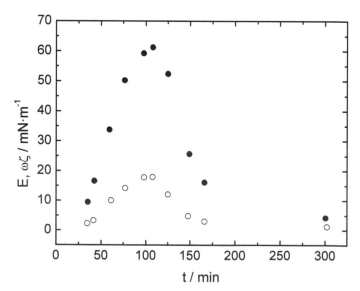

Figure 28. Kinetic dependence of the real (solid symbols) and imaginary (open symbols) components of the dynamic surface elasticity of PNIPAM solutions at the concentration 0.0001 wt.%.

In this case the polyoxyethylene chain is relatively short and we can assume that the number of loops and tails per adsorbed chain is small (n < 1). Therefore, one can apply the Maxwell rheological model (50) to describe the frequency dependencies in Fig. 29. The parameters $\partial\gamma/\partial\ln\Gamma$ and τ_2 have been determined by a non-linear regression analysis and the curves in Fig. 29 represent the results of calculations according to Eq. (50). The deviation of the real part of the surface elasticity from the curve at low frequencies is probably caused by other relaxation processes.

Note that Eq. (50) describes only the contribution of the main relaxation process to the surface elasticity. $\partial\gamma/\ln\partial\Gamma$ equals to 7.2 mN·m^{-1} and indicate a looser surface structure than in the case of surface films of conventional surfactants. The relaxation time ($\tau_2 = 0.06$ s for 0.001 wt. % PEG solutions) probably corresponds to the chain reptation inside the tube of entanglements.

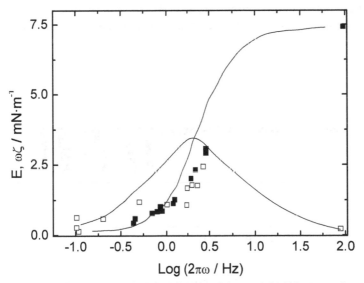

Figure 29. Frequency dependence of the real (solid squares) and imaginary (empty squares) parts of the dynamic surface elasticity for solutions of PEG4 at the concentration 0.001 wt. %. The lines are the best fit to equation (50) for the real (solid line) and imaginary (dotted line) parts of the surface elasticity.

The obtained experimental results agree qualitatively with the conclusions of Section 5.1. and lead to the following sketch of the adsorption process (Fig. 30). At very low bulk concentrations the adsorbed macromolecules have an almost flat two-dimensional conformation and do not form long loops and tails. The film is elastic and the elasticity increases with concentration at the expense of increasing repulsion between neighbouring segments in the surface layer. At further increase of the bulk and, consequently, surface concentrations the segments in the region (I) hinder complete unfolding of the adsorbing polymer coils in the surface layer. Gradually some loops and tails dangling in the bulk phase appear. This leads to a relaxation of surface stresses at the expense of monomer exchange between the regions (I) and (II) of the surface layer. The imaginary part of the surface elasticity increases and the film becomes visco-elastic. Subsequent increase of the number of loops and tails leads to faster exchange between the two regions of the surface layer and, consequently, to a decrease of the dynamic surface elasticity. The elasticity goes through a maximum. At higher concentrations, beyond the elasticity maximum, the number of loops and tails increases further. This leads to a fast exchange of monomers between the regions (I) and (II), and the reverse characteristic time of this process exceeds the frequencies accessible to experimental techniques.

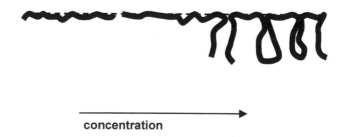

concentration

Figure 30. Schematic illustration of the structural changes with concentration in the surface layer of polymer solutions.

At the same time, a slow co-operative process involving the reorganization or the whole surface layer can accompany the fast monomer exchange between the two regions of the surface layer. The latter process can be responsible for the low but non-zero values of the real part of the dynamic surface elasticity for most of the studied systems. Ellipsometric studies confirm this picture and allow observation of the two steps of the adsorption process [179, 204]. The first one is the formation of region (I) and proceeds at an almost constant adsorption layer film thickness. The second step, which is connected with the formation of loops and tails, leads to a significant increase of the thickness and only small changes of the adsorbed amount [179].

3. Solutions of block copolymers

The rheological properties of insoluble spread films of block copolymers on an aqueous subphase are discussed in Section 4.3. The rheological surface properties of block copolymer solutions are also a subject of a number of studies. Most of the investigated soluble block copolymers contain PEO blocks and among them only the surface properties of triblock copolymers of poly(ethylene oxide)-poly(propylene oxide)-poly(ethylene oxide) (PEO-PPO-PEO) have been studied in detail [50, 148, 149, 178, 180, 181, 200, 205-215]

If the macromolecules contain blocks with different degrees of hydrophobicity, the surface properties of their solutions differ significantly from the properties of corresponding homopolymer solutions. For example, unlike the case of PEO solutions, the surface tension isotherms of PEO-PPO-PEO solutions display abrupt changes in the slope, even at the concentrations below the critical micellar concentration [206-208, 210]. To explain these results Alexandridis et al. assumed conformational changes in the surface layer of block copolymer solutions, in particular, protrusion of PEO segments into the solution bulk beyond a critical concentration [207]. The PPO segments are deemed to remain at the interface in the whole concentration range. These ideas were further developed by other authors who used ellipsometry and neutron reflectivity to study conformation changes of the polymer at the surface [210-212]. Some authors proposed a brush structure at surface pressures higher than about 15 mN·m^{-1} [211, 212]. In this case the macromolecules are tethered by PPO segments to the surface and the repulsion between neighbouring PEO chains leads to their partial stretching perpendicular to the liquid surface (Fig. 31). However, more careful measurements of the neutron reflectivity [148, 149] and mean field calculations [214] indicated at a more complex structure with mixing PEO and PPO segments in the sublayer (region (II) in terms of the previous sections).

The surface rheological properties of PEO-PPO-PEO solutions are also more complicated than those of PEO solutions and can be explained at least partly by the successive protrusion in the bulk phase of the both PEO and PPO blocks at high surface pressures [50, 181]. Fig. 32 shows as an example the real part of the dynamic surface elasticity as a function of surface pressure for PEO$_{76}$-PPO$_{29}$-PEO$_{78}$ spread and adsorbed surface films. The imaginary part is about an order of magnitude less than the real one and the film is elastic. Both sets of experimental data almost coincide at $\Pi < 19$ mN·m^{-1}. Measurements of the surface properties close to the equilibrium are difficult at $\Pi < 14$ mN·m^{-1} for copolymer solutions (cf. Section 5.2), but the non-monotonic kinetic dependencies of the surface elasticity indicate that the surface rheological properties of spread and adsorbed films are indistinguishable also in this region [181]. At $\Pi < 10$ mN·m^{-1} the results coincide with the PEO surface elasticity isotherm. In this case the surface rheological properties are determined by PEO blocks and are explained in the preceding section. One can connect the increase of the surface elasticity beyond the local minimum (Fig. 32) by the interactions between neighbouring PPO blocks.

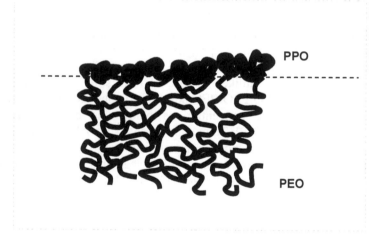

Figure 31. Brushlike structure of PEO-PPO-PEO copolymers at the liquid-gas interface.

Although the dynamic surface elasticity of spread films at the frequency 0.14 Hz increases monotonically beyond the region of the local minimum, the static elasticity does not. At $\Pi > 13$ mN·m^{-1} the static elasticity begins to decrease to almost zero (Fig. 33) thus indicating that PPO chains can protrude into the bulk phase. At $\Pi > 19$ mN·m^{-1} the equilibrium spread films do not exist. The contraction of the film at $\Pi \approx 19$ mN·m^{-1} leads to its partial dissolution in the subphase. At the same time the adsorbed films display the surface pressure much higher than 19 mN·m^{-1} but in this case the dynamic surface elasticity is much lower.

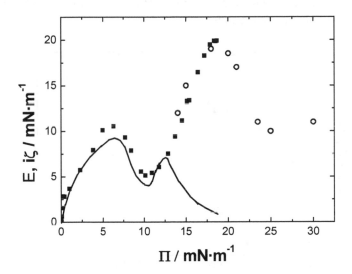

Figure 32. Real part of dynamic surface elasticity vs. surface pressure for PEO$_{76}$-PPO$_{29}$-PEO$_{78}$ spread (squares) and adsorbed (circles) films at the frequency 0.14 Hz. The line represents the static surface elasticity of spread films.

The maximum of the dynamic surface elasticity of PEO$_{76}$-PPO$_{29}$-PEO$_{78}$ solutions at $\Pi \approx 19$ mN·m^{-1} can be connected also with the mixing of PPO and PEO segments in the subphase and formation of a looser surface structure. The results in Fig. 32 demonstrate that the brush model can give only very rough description of the adsorbed films of block copolymers in agreement with conclusions [148, 153, 214]. Ah high surface pressures the both blocks can protrude into the bulk phase for both the spread and adsorbed PEO-PPO-PEO surface films. Another argument in favour of the mixing of PEO and PPO segments in the subphase stems out of the comparison between the elasticity of PEO$_{76}$-PPO$_{29}$-PEO$_{78}$ surface films and that of the films of a diblock copolymer of polystyrene and PEO (PS-PEO) [181]. The brush formation in spread films of PS$_{38}$-PEO$_{250}$ was confirmed by the neutron reflectivity [217] and surface dilational rheology methods [86, 216], and one can consider these films as physical models of polymer brushes. The surface elasticity increases much steeper for PS$_{38}$-PEO$_{250}$ surface films with increasing surface concentration in the range of high surface pressures in comparison with the PEO$_{76}$-PPO$_{29}$-PEO$_{78}$ surface films where the

SPREAD AND ADSORBED POLYMER LAYERS

elasticity does not exceed 20 mN·m^{-1}. The different behaviour is obviously connected with the deviations of the real PEO-PPO-PEO film structure from a classic brush model.

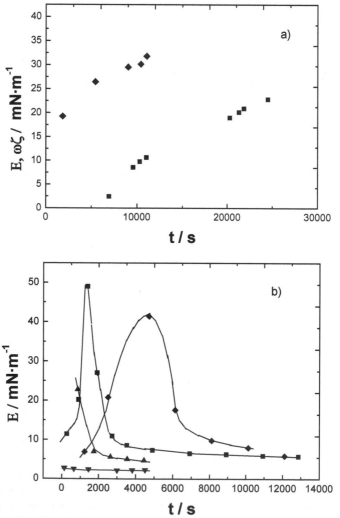

Figure 33. Kinetic dependencies of the real part of the dynamic surface elasticity of AMPS - NIPAM solutions in water (a) at polyelectrolyte concentrations of 0.1 (squares), 0.2 wt. % (diamonds) and in 1 M NaCl (b) at concentrations 0.0002 (diamonds), 0.0005 (squares), 0.001 (triangles up) and 0.1 wt. % (triangles down). Lines are guides for the eye.

4. Solutions of synthetic polyelectrolytes

In spite of numerous applications of polyelectrolyte aqueous solutions any information on their surface properties is scarce. The first studies of the dilational surface rheology of solutions of synthetic polyelectrolytes were published in the last few years [176, 198].

Only adsorbed and spread films of natural polyelectrolytes, mainly polyampholytes, have been intensively investigated [57, 67, 178, 183-188]. The situation for conventional synthetic polyelectrolytes is different and some authors even mention that these polymers do not display any surface activity [195, 218]. However, this is only true if the polyelectrolyte concentration is sufficiently low, usually less than 0.1 wt.% or sometimes 1 wt.%, and the solution does not contain excess of inorganic salts. The surface tension of polyelectrolyte solutions begins to decrease only when the ionic strength is high enough and the lateral repulsion of charged groups is shielded to a great extent [177, 219, 220].

The surface properties of synthetic polyelectrolyte solutions were studied mainly by the tensiometry. The neutron reflectivity was applied only to solutions of sodium poly(styrenesulfonate) (PSS) [146, 147]. While qualitative changes of the adsorbed amount and film thickness with the concentrations of the polymer and inorganic salt proved to be consistent with predictions of the self-consistent-field theory (SCF) [1, 221-223], quantitative agreement was not achieved. Unlike the predicted monotonic segment profile, the experimentally observed one consists of two distinct layers with different segment concentrations [146]. The concentration dependencies of the adsorption at different salt contents had maximums at concentrations correlated with the overlap polymer concentrations [149].

Recent studies of the surface dilational rheological properties of PSS solution also discovered some distinctions from the theoretical conclusions [176, 177]. At the same time, the dilational surface visco-elasticity for solutions of other polyelectrolytes, for example the copolymer of sodium 2-acrylamido-2-methyl-1-propansulfonate with N-isopropylacrylamide (AMPS-NIPAM), was in qualitative agreement with the theory [224]. The dynamic surface elasticity of AMPS-NIPAM solutions changed very slowly with the surface age probably due to a strong electrostatic adsorption barrier and reached high values, much higher than for nonionic polymer solutions (Fig. 33). According to SCF calculations the polyelectrolytes form a narrow adsorption film without long loops and tails and with strong repulsion between the charged segments [223]. This structure must lead to the high dynamic surface elasticity. The increase of the solution ionic strength must lead to shielding of the electrostatic interactions and, consequently, to the decrease of the surface elasticity. One can observe this effect for solutions of AMPS-NIPAM [224].

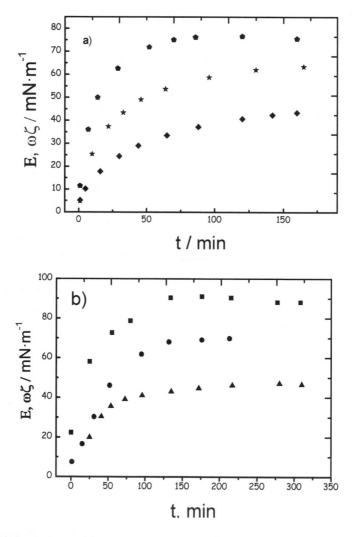

Figure 34. Kinetic dependencies of the real part of the dynamic surface elasticity of PSS solutions in water (a) and 1 M NaCl (b) at the frequency 0.12 Hz for concentrations 0.01 wt.% (triangles); 0.05 wt.% (circles); 0.1 wt.% (diamonds); 0.11 wt.% (squares); 0.5 wt.% (asterisks); 0.8 wt.% (pentagons).

The addition of sodium chloride leads to an increase of the polyelectrolyte surface activity and of the rate of surface property changes with time. The surface visco-elastic behaviour of AMPS-NIPAM solutions approaches that of non-ionic PNIPAM solutions (Fig. 33).

In the case of PSS solutions the surface visco-elastic behaviour is different [176, 177]. Firstly, the dynamic surface elasticity is higher thus indicating at stronger interactions in the surface layer. Secondly, the addition of sodium chloride leads not to a decrease but even to a slight increase of the dynamic surface elasticity (Fig. 34).

Thirdly, the influence of the increase of solution ionic strength on the rate of the adsorption film formation was also in the disagreement with the theoretical predictions. Calculations according to the SCF and Kramers rate theories under the assumption of only electrostatic interactions between polymer segments predict that the adsorption rate increases by several orders of magnitude at high salt content [225]. However, measurements of the dynamic surface tension and elasticity showed that addition of sodium chloride to PSS solutions induced only an acceleration of the first step of adsorption while the characteristic time of the whole equilibration process remained changes only a little (Fig. 34).

The electrostatic interactions are obviously insufficient to explain the peculiarities of the surface properties of PSS solutions. It was assumed that the hydrophobic attraction between the neutralized PSS segments led to aggregate formation in the surface layer and strong cohesion between the polymer chains at the high ionic strength [175, 177].

G. THE NON-LINEAR REGIME

In Section 2 it has been stressed that the equations relating the visco-elastic moduli with the stress-relaxation or the oscillatory barrier experiments were valid in the linear regime (LR). The range of strains for which linear behaviour is found strongly depends on the polymer used. In this section we will present results on two polymers for which maximum strain within the LR is quite small. This fact not only complicates the analysis of the rheological experiments, but also may pose some constrains on the way the measurements of the equilibrium isotherms can be carried out.

Poly(octadecyl acrylate) PODA and poly(vinyl stearate) PVS molecules differ only in the reversed position of the ester group within each monomer. PODA has been studied by Gabrielli et al. [226] and Mumby et al. [227]. PVS monolayers have been studied by Gabrielli et al. [226], Peng et al. [228], by Chen et al. [229], and by O'Brien and

Lando [230]. Noticeable discrepancies are found in the results obtained by the different authors, and also by using continuous compression or sequential addition methods in the same laboratory [230].

Figure 35 shows the stress-strain plot for the two polymers obtained by the oscillating barrier technique in a Langmuir trough. Even though one might think that the LR holds up to strains of c.a. 10%, the slope of the straight lines are much larger than the equilibrium compressibility of the isotherms for PODA monolayers. The inset shows that the maximum strain compatible with the LR u_m is in fact much smaller for both polymers. The slope of the straight line equals the equilibrium compressibility of PODA. The low value of u_m imposes severe difficulties to the rheological experiments, because it lowers the signal to noise ratio.

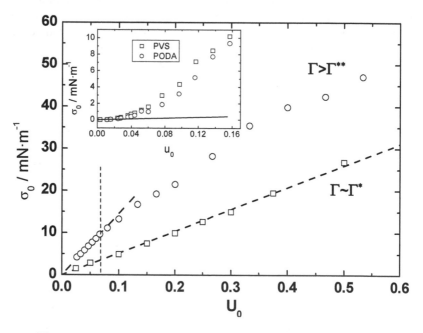

Figure 35. Stress-strain plot for poly(octadecyl acrylate), PODA, and poly(vinyl stearate), PVS, monolayers at 25 °C. The data were obtained by oscillatory barrier experiments. The inset shows the low-strain range. The slope of straight line of the inset equals the equilibrium compressibility of the PODA monolayer.

As it was mentioned in Section 2.2.2, the methods of Fourier transform rheology are useful outside the LR [65]. Figure 36 shows part

of the stress function of an oscillatory experiment for PODA at 25 °C and
$\Gamma \approx \Gamma^{**}$.

Despite that the area perturbation was true-sinusoidal, it can be
observed that the stress response is not. The corresponding FT-spectrum
shows a fundamental peak at the frequency of the area perturbation ω_0,
and two harmonics at $3\omega_0$, and $5\omega_0$, respectively. Notice that the strain
imposed, 9% is still within the apparent range of validity of the LR in
Figure 35.

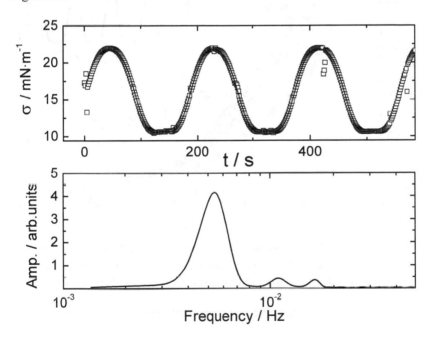

Figure 36 Top) Part of the response function in an oscillatory barrier
experiments beyond the linear regime for a PODA monolayer at 25 °C.
Notice that the stress function is not a sinusoidal. Bottom) FT-spectrum of
the stress function shown at the top.

Figure 37 shows the system response for oscillatory barrier
experiments of the same frequency but different amplitudes. It can be
observed that the increase of the perturbation amplitude strongly
increases the area of the hysteresis cycle. Within the LR, the area of the
cycle is negligible. At least for monolayers of PODA and PVS,
hysteresis seems to be associated to experiments performed outside the
LR. It is worth noticing that in these experiments, the increase of the
perturbation amplitude at constant frequency implies the increase of the

barriers rate. It must be stressed that the barriers at which hysteresis cycles appear correspond to rates at which the Π vs. Γ isotherms obtained by continuous compression lead to values of Π higher than isotherms obtained by directly spreading polymer solution.

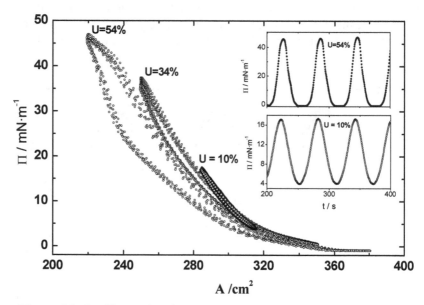

Figure 37. Oscillatory barrier experiments for a PODA monolayer at constant frequency, and for different values of the strain amplitude. The inset show part of the oscillatory response function for two strains.

ACKNOWLEDGEMENTS

The work done in Madrid was supported in part by MEC (grants: VEM2003-20574-C03-03, CTQ2006-6208/BQU, FIS2006-01305 and FIS2006-12281-C02-01) and by Comunidad de Madrid (grants: INTERFASES S05-MAT-227 and NANOBIO-M S05-MAT-283). The work done in St. Petersburg was supported in part by the Russian Foundation of Basic Research (RFFI № 08-03-00207_a).

H. REFERENCES

1. G. J. Fleer, M.A. Cohen Stuart, J.M.H.M. Scheutjens, T. Cosgrove and B. Vincent, Polymers at Interfaces, Chapman-Hall, Cambridge, 1993
2. H.-W. Hu, S. Granick, and K.S. Schweizer, J. Non-Crystalline Solids, 172-174 (1994) 721

3. B. Bhushan, and JN Israelachvili, in Handbook of Micro/Nanotribology, B. Bhushan, Ed. 2nd Ed. CRC Press, Boca Raton, CA, 1999 (Chap. 9)
4. R.A.L. Jones and R.W. Richards, Polymers at Surfaces and Interfaces, Cambridge Univ. Press, Cambridge, 1999
5. .P. Russell, Annu. Rev. Mater. Sci., 21 (1991)249
6. J.A. Henderson, R.W. Richards, J. Penfold, R.K. Thomas, Macromolecules, 26 (1993) 65
7. F. Rondelez, D. Ausserré, and H. Hervé, Annu. Rev. Phys. Chem., 38 (1987) 317
8. L. Gaines Jr. Insoluble Monolayers at Liquid-Gas Interfaces, Interscience, New York, 1966
9. A-F. Mingotaud, Ch. Mingotaud, L.K. Patterson, Handbook of Monolayers, Academic Press, San Diego, 1993
10. Liggieri, M. Ferrari, D. Mondelli, and F. Ravera, Faraday Discuss., 129 (2005) 125
11. B.A. Noskov, A.V. Akentiev, A. Yu. Bilibin, I.M. Zorin, R. Miller, Adv. Colloid Interface Sci., 104 (2003) 245
12. D.M. Taylor, Adv. Colloid Interface Sci., 87 (2000) 183
131 A. Kapilashrami, M. Malmsten, K. Eskilsson, J-W. Benjamins, T. Nylander, Colloids Surfaces A, 225 (2003) 181
14. R. Mendelsohn, J.W. Brauner and A. Gericke, Annu. Rev. Phys. Chem., 46 (1995) 305
15. P. Dynarowicz-Latka, A. Dhanabalan, O.N. Oliveira Jr. Adv. Colloid Interface Sci., 91 (2001) 221
16. D.K. Kambhampati, W. Knoll, Curr. Opinion Colloid Interface Sci., 4 (1999) 273
17. H. Fujimaki, T. Manaka, H. Ohtake, and A. Tojima, J. Chem. Phys., 119 (2003) 7427
18. V. Vogel, Y.R. Shen, Annu. Rev. Mater. Sci. 21 (1991) 515
19. A. Yethiraj, Macromolecules, 36 (2003) 5854
20. B. Ostrovsky, M.A. Smith, Y. Bar-Yam, Int. J. Mod. Phys., C 8 (1997) 931
21. G.T. Gavranovic, J.M. Deutsch, G.G. Fuller, Macromolecules, 38 (2005) 6672
22. I. Carmesin, K. Kremer, J. Phys. (Paris), 51 (1990) 915.
23. I. Sikharulidze, B. Farago, I.P. Dolbnya, A. Madsen, and W.H. de Jeu, Phys. Rev. Lett., 91 (2003) 65504
24. H. Kim, A. Rúhm, L.B. Lurio, J.K. Basu, J. Lal, D. Lumma, S.G.J. Mochrie, and S.K. Sihna, Phys. Rev. Lett., 90 (2003) 068302
25. M. Kawaguchi, Prog. Polym. Sci., 18 (1993) 341

26. V.G. Levich, Physicochemical Hydrodynamics, Prentice Hall, New York, 1962
27. J.C. Earnshaw, Adv. Colloid Interface Sci., 68 (1996) 1
28. D. Langevin, Ed. Light Scattering by Liquid Surfaces and Complementary Techniques, Marcel Dekker, New York, 1992
29. F. Monroy, F. Ortega, and R.G. Rubio, Phys. Rev. E, 58 (1998) 7629
30. A.R. Esker, L. H. Zhang, B.B. Sauer, W. Lee, and H. Yu, Colloids Surfaces B, 171 (2000) 131.
31. S.K. Peace, and R.W. Richards, Polymer, 37 (1996) 4945
32. P. Cicuta, and I. Hopkinson, J. Chem. Phys., 114 (2001) 8659
33. B.A. Noskov, A.V. Akentiev and R. Miller, J. Colloid Interface Sci., 255 (2002) 417
34. Q.R. Huang, and C.H. Wang, Langmuir, 12 (1996) 2679
35. B.A. Noskov, Colloid Polym. Sci., 273 (1995) 263
36. S.S. Dukhin, G. Kretzschmar and R. Miller, Dynamics of Adsorption at Liquid Interfaces, Studies in Interface Sci., Vol. 1, D. Möbius and R. Miller (Eds.), Elsevier, 1995
37. A. Maestro, H. Hilles, F. Monroy, F. Ortega, D. Langevin, R.G. Rubio, submitted to J. Chem. Phys.
38. L.J. Bonales, H. Ritacco, J.E.F. Rubio, R.G. Rubio, F. Monroy, R.G. Rubio, F. Ortega, Open J. Phys. Chem., 1 (2007) 25
39. M. Sickert, F. Rondelez, H.A. Stone, Europhys. Lett., 79 (2007) 66005
40. N.W. Tschoegl, The Phenomenological Theory of Linear Viscoelastic Behavior, Springer-Verlag, Berlin, 1989.
41. G. Loglio, U. Teseo, and R. Cini, J. Colloid Interface Sci., 71 (1979) 316
42. B.A. Noskov, G. Loglio, Colloids Surfaces A, 143 (1998)167
43. J.C. Earnshaw, R.C. McGivern, and P.J. Winch, J. Phys. France, 49 (1988) 1271
44. D.M.A. Buzza, J.L. Jones, T. McLeish, and R.W. Richards, J. Chem. Phys., 109 (1998) 5008
45. D.M.A. Buzza, Langmuir, 18 (2002) 8418
46. F.E. Runge, M.S. Kent, and H. Yu, Langmuir, 10 (1994) 1962
47. F.Monroy, F. Ortega, R.G. Rubio, M.G. Velarde, Adv. Colloid Interface Sci., 134-135 (2007) 175
48. D. Sharpe, and J. Eastoe, Langmuir, 12 (1996) 2303
49. J. Giermanska-Kahn, F.Monroy and D. Langevin, Phys. Rev. E, 60 (1999) 7163
50. M.G. Muñoz, F. Monroy, P. Hernández, F. Ortega, R.G. Rubio, D. Langevin, Langmuir, 19 (2003) 2147

51. B.A. Noskov and T.U. Zubkova, J. Colloid Interface Sci., 170 (1995) 1
52. H.C. Maru and D.T. Wasan, Chem. Eng. Sci., 34 (1979) 1295
53. M. van den Tempel and E. Lucassen-Reynders, Adv. Colloid Interface Sci., 18 (1983) 281
54. R. Skarlupka, Y.Seo and H. Yu, Polymer, 39 (1998) 387
55. F. Monroy, S. Rivillon, F. Ortega and R.G. Rubio, J. Chem. Phys., 115 (2001) 115
56. H.M. Hilles, A. Maestro, F. Monroy, F. Ortega, R.G. Rubio, M.G. Velarde, J. Chem. Phys., 126 (2007) 124904
57. J.T. Petkov, T.D. Gurkov, B.E. Campbell and P. Borwankar, Langmuir, 16 (2000) 16
58. W.K. Findley, J.S. Lai and K. Onaran, Creep and Relaxation on Nonlinear Viscoelastic Materials, Dover Pub., New York, 1989
59. L.M.C. Sagis, M. Ramaekers and E. van der Linden, Phys. Rev. E, 63 (2001) 051504-1
60. H.M. Hilles, F. Monroy, L.J. Bonales, F. Ortega and F.G. Rubio, Adv. Colloid Interface Sci., 122 (2006) 122
61. K.A. Dill, J. Naghizadeh and J.A. Marqusee, Ann. Rev. Phys. Chem., 39 (1988) 425
62. P. Joos and M. Van Uffelen, Colloids Surfaces A, 75 (1993) 273
63. A.E. Cárdenas and A.I. Valera, Colloids Surfaces A, 79 (1993) 115.
64. P. Cicuta and E.M. Terentjev, Eur. Phys. J. E, 16 (2005) 147
65. M. Wilhelm, Macromol. Mater. Eng., 287 (2002) 287
66. J.P. Coffman and C.A. Naumann, Macromolecules, 35 (2002) 1835
67. J. Kumaki, T. Kawauchi and E. Yashima, J. Amer. Chem. Soc., 127 (2005) 5788
68. R. Vilanove and F. Rondelez, Phys. Rev. Lett., 45 (1980) 1502
69. D. Vollhardt, Adv. Colloid Interface Sci., 64 (1996) 143
70. Ch.M. Knobler and R.C. Desai, Annu. Rev. Phys. Chem., 43 (1992) 207
71. S. Rivillon, M.G. Muñoz, F.Monroy, F. Ortega and R.G. Rubio, Macromolecules, 36 (2003) 4068
72. S.K. Peace, R.W. Richards and N. Williams, Langmuir, 14 (1998) 667
73. R.W. Richards, B.R. Rochford and M.R. Taylor, Macromolecules, 29 (1996) 1980
74. Q.R. Huang and C.H. Wang, Langmuir, 12 (1996) 2679
75. A.J. Milling, L.R. Hutchings and R.W. Richards, Langmuir, 17 (2001) 5305; 5297
76. M.C. Fauré, P. Bassereau, M.A. Carignano, I. Szleifer, Y. Gallot and D. Andelman, Eur. Phys. J. E, 3 (1998) 365

77. M.C. Fauré, P. Bassereau, L.T. Lee, A. Menelle and C. Lhevender, Macromolecules, 32 (1999) 8538
78. J.K. Cox, K. Yu, A. Eisenberg and R.G. Lennox, Phys. Chem. Chem. Phys., 1 (1999) 4417
79. M. Kawaguchi, S. Itoh and A. Takahashi, Macromolecules, 20 (1987) 1052
80. L. Gargallo, B. Miranda, H. Ríos, F. González-Nilo and D. Radic, Polymer Intl., 50 (2001) 858
81. M. Kawaguchi, A. Yoshida and A. Takahashi, Macromolecules, 16 (1983) 956
82. R. Vilanove, D. Poupinet and F. Rondelez, Macromolecules, 21 (1988) 2880
83. D. Poupinet, R. Vilanove and F. Rondelez, Macromolecules, 22 (1989) 2491
84. P.G. de Gennes Scaling Concepts in Polymer Physics, Cornell Univ. Press, Ithaca, N.Y., 1993
85. J. Kumaki, T. Kawauchi and E. Yashima, Macromolecules, 39 (2006) 1209
86. B. Maier and J.O. Radler, Macromolecules, 34 (2001) 5723
87. S. Sukhishvili, Y. Chen, J.D. Müller, E. Gratton, K.S. Schweizer and S. Granick, Nature, 406 (2000) 146
88. S.C. Bae, F. Xie, S. Jeon and S. Granick, Current Opinion Solid State Mater. Sci., 5 (2001) 327
89. S.A. Sukhishvili, Y. Chen, J.D. Müller, E. Gratton, K.S. Schweizer and S. Granick, Macromolecules, 35 (2002) 1776
90. H. Itagaki, Y. Nishimura, E. Sagisaka, Langmuir 22 (2006) 742.
91. X. Wang, V.J. Flotz, J. Chem. Phys. 121 (2004) 8158.
92. J. des Cloizeaux and G. Jannick, Polymers in Solution, Clarendon Press, Oxford, 1990
93. A. Maestro, F. Ortega, F. Monroy, R. Miller submitted to Langmuir. February 2009.
94. J.D. Ferry, Viscoelastic Properties of Polymers, 3rd Ed., John Wiley & Sons, New York, 1980
95. F. Monroy, M.J. Esquinas, F. Ortega and R.G. Rubio, Colloid Polym. Sci. 276 (1998) 960
96. S. Rivillon, F. Monroy, F. Ortega and R.G. Rubio, Eur. Phys. J. E, 9 (2002) 375
97. F. Monroy, F. Ortega, R.G. Rubio, H. Ritacco, D. Langevin, Phys. Rev. Lett., 95 (2005) 056103-1
98. P.N. Pusey and W. van Megen, Phys. Rev. Lett., 59 (1987) 2083
99. H.E. Gaub and H.M. McConnell, J. Phys. Chem., 90 (1986) 6830

100. B.M. Abraham, K. Miyano, S.Q. Su, J.B. Ketterson, Phys. Rev. Lett., 49 (1982) 1643

101. R.S. Ghaskadvi, T.M. Bohanon, P. Dutta and J.B. Ketterson, Phys. Rev. E, 54 (1996) 1770

102. N.Sato, S. Ito and M. Yamamoto Polymer J. 28 (1996) 784; Macromolecules, 31 (1998) 2673

103. M. Sachetti, H. Yu and G. Zografi, Rev. Sci. Instrum. 64 (1993) 1941; Langmuir, 9 (1993) 2168

104. C. Luap and W.A. Goedel, Macromolecules, 34 (2001) 1343

105. K. Balashev, A. Bois, J.E. Proust, Tz. Ivanova, I. Petkov, S. Masuda and I. Panaiotov, Langmuir, 13 (1997) 5362

106. C. Barentin, P. Muller, C. Ybert, J. F. Joanny and J. M. di Meglio, Eur. Phys. J. E, 2 (2000) 153

107. C.A. Naumann, C.F. Brooks, W. Wiyatno, W. Knoll, G.G. Fuller and C.W. Franck, Macromolecules, 34 (2001) 3024

108. E.M. Freer, K.S. Yim, G.G. Fuller and C.J. Radke, J. Phys. Chem. B, 108 (2004) 3835

109. C. Monteux, R. Mangeret, G. Laible, E. Freyssingeas, V. Bergeron and G. Fuller, Macromolecules, 39 (2006) 3408

110. F. Monroy, H. M. Hilles, F. Ortega and R.G. Rubio, Phys. Rev. Lett., 26 (2003) 268302-1

111. F. Monroy, F. Ortega and R.G. Rubio, J. Phys. Chem. B, 103 (1999) 2061

112. M. Doi and S.F. Edwards, The Theory of Polymer Dynamics, Oxford Univ. Press, Oxford, 1986

113. P. Cicuta, E.J. Stanvik and G.G. Fuller, Phys. Rev. Lett., 90 (2003) 236101

114. H. Watanabe, Prog. Polym. Sci., 24 (1999) 1253

115. H.R. Brown and T.P. Russell, Macromolecules, 29 (1996) 798

116. E. Riande, R. Díaz-Calleja, M.G. Prolongo, R.M. Masegosa and C. Salom, Polymer Viscoelasticity. Stress and Strain in Practice, Marcel Dekker, New York, 2000

117. J. Jakes, Coll. Czech. Chem. Commun., 60 (1995) 1781

118. B.A. Noskov, Colloid Polym. Sci., 273 (1995) 263

119. C.A. Angell, J. Phys. Condens. Matter, 9 (2000) 6463

120. E.J. Donth, Relaxation and Thermodynamics in Polymers, Akademie Verlag, Berlin, 1992

121. P.G. Debenedetti, Metastable Liquids: Concepts and Principles, Princeton Univ. Press, Princeton, N.J., 1996

122. H.M. Hilles, F. Ortega, R.G. Rubio and F. Monroy, Phys. Rev. Lett., 92 (2004) 255503-1

123. K.L. Ngai, Eur. Phys. J. E, 8 (2002) 225

124. J. Brandrup, E.H. Immergut, E.A. Grulke, Eds. Polymer Handbook, Wiley Interscience, New Jersey, 1999
125. J.L. Keddie, R.A.L. Jones and R.A. Cory, Europhys. Lett., 27 (1994) 59
126. J.A. Forrest and J. Mattsson, Phys. Rev. E, 61 (2000) R54
127. J. Rault, J. Macromol. Sci. B: Physics, 42 (2003) 1235
128. A.M. Gonçalves da Silva, A.L. Simoes Gamboa and J.M.G. Martinho, Langmuir, 14 (1998) 5327
129. A.M. Gonçalves da Silva, E.J.M. Filipe, J.M.R. d'Oliveira and J.M.G. Martinho, Langmuir, 12 (1996) 6547
130. C.A. Devereaux and S.M. Baker, Macromolecules, 35 (2002) 1921
131. S. Rivillon, M.G. Muñoz, F.Monroy, F. Ortega and R.G. Rubio, Macromolecules, 36 (2003) 4068
132. A.E.Hosoi, D. Kogan, C.E. Devereaux, A.J. Bernoff and S.M. Baker, Phys. Rev. Lett., 95 (2005) 037801.
133. H.M. Hilles, M. Sferrazza, F. Monroy, F. Ortega and R.G. Rubio, J. Chem. Phys., 125 (2006) 074706-1
134. M.S. Kent, J. Majewski, G.S. Smith, L.T. Lee and S. Satija, J. Chem. Phys., 110 (1999) 3553
135. M.S. Kent, J. Majewski, G.S. Smith, L.T. Lee and S. Satija, J. Chem. Phys., 110 (1999) 5635
136. M.S. Kent, L.T. Lee, B.J. Factor, F. Rondelez and G.S. Smith, J. Chem. Phys., 103 (1995) 2320
137. F.E. Runge, M.S. Kent and H. Yu, Langmuir, 10 (1994) 1962
138. Y. Seo, J. H. Im, J. S. Lee and J. H. Kim, Macromolecules, 34 (2001) 4842
139. Y. Seo, A.R. Esker, D. Sohn, H. J. Kim, S. Park and H. Yu, Langmuir, 19 (2003) 3313
140. J.R. Lu, T.J. Su, R.K. Thomas, J. Penfold and R.W. Richards, Polymer, 37 (1996) 109.
141. Q.R. Huang, C.H. Wang, J. Chem. Phys., 105 (1996) 654
142. S.W. An, R.K. Thomas, C. Forder, N.C. Billingham, S.P. Armes and J. Penfold, Langmuir, 18 (2002) 5064.
143. L.T. Lee, B. Jean, A. Menelle, Langmuir, 15 (1999) 3267.
144. R.M. Richardson, R. Pelton, T. Cosgrove and Ju Zhang, Macromolecules, 33 (2000) 6269.
145. J. Penfold, Curr. Opinion Colloid Interface Sci. 7 (2002) 139.
146. H. Yim, M.S. Kent, A. Mathewson, R. Ivkov, S. Satija, J. Majewski and G.S. Smith, Macromolecules, 33 (2000) 612
147. H. Yim, M.S. Kent, A. Mathewson, M.J. Stevens, R. Ivkov, J. Majewski and G.S. Smith, Macromolecules, 35 (2002) 9737.

148. J.B. Vieira, Z.X. Li and R.K. Thomas, J. Phys. Chem. B ,106 (2002) 5400.

149. J.B. Vieira, Z.X. Li, R.K. Thomas and J. Penfold, J. Phys. Chem. B, 106 (2002) 10641.

150. A. Couper and D.D. Elley, J. Polym. Sci., 3 (1948) 345.

151. H.L. Frish and S. Al Madfai, J. Amer. Chem. Soc., 80 (1958) 3561

152. M.N. Gottlieb, J. Chem. Phys., 63 (1959) 1687

153. J.E. Glass, J. Phys. Chem., 72 (1968) 4459.

154. K. Uebereiter, S. Morimoto and R. Steulman, Colloid. Polymer Sci., 252 (1974) 273.

155. S. Matsuzawa and K. Uebereiter, Colloid Polymer Sci., 256 (1978) 490.

156. K. Uebereiter and M. Okubo, Colloid Polymer Sci., 256 (1978) 1030.

157. B.B. Sauer and H. Yu, Macromolecules, 22 (1989) 78

158. I. Nahringbauer, J. Colloid Interface Sci., 176 (1995) 318.

159. Ju Zhang, R. Pelton, Langmuir, 12 (1996) 2611.

160. S. U. Um, E. Poptoschev and R. Pugh, J. Colloid Interface Sci., 193 (1997) 41.

161. Ju Zhang and R. Pelton, Langmuir, 15 (1999) 8032.

162. Ju Zhang and R. Pelton, Colloids Surfaces A, 156 (1999) 111.

163. Ju Zhang and R. Pelton, Langmuir, 15 (1999) 5662.

164. X. Y. Hua and M.J. Rosen, J. Colloid Interface Sci., 124 (1988) 652.

165. B.A. Noskov, Adv Colloid Interface Sci., 69 (1996) 63.

166. B.A. Noskov, Adv. Colloid Interface Sci., 95 (2002) 237.

167. M.C.P. Van Eijk, M.A. Cohen Stuart, Langmuir, 13 (1997) 5447.

168. M. Eigen, L. de Maeyer, in Investigation of Rates and Mechanisms of Reactions, G.G. Hammes Ed, Jonh Wiley, New York, 1974 (Chapter 3).

169. B.B. Sauer, M. Kawaguchi and H. Yu, Macromolecules, 20 (1987) 2732.

170. B.B. Sauer and H. Yu, Macromolecules, 22 (1989) 78

171. E.J. McNally and G. Zografi, J. Colloid Interface Sci., 138 (1990) 61.

172. Q.R. Huang and C.H. Wang, Langmuir, 105 (1996) 654

173. B.A. Noskov, A.V, Akentiev, G. Loglio and R. Miller, Mendeleev Commun., 8 (1998) 190.

174. B.A. Noskov, A.V. Akentiev, G. Loglio and R. Miller, J. Phys. Chem. B, 104 (2000) 7923.

175. B.A. Noskov, A.V. Akentiev, D.A. Alexandrov, G. Loglio and R. Miller, In: Food Colloids. Fundamentals of Formulation,

E. Dickinson, R. Miller (Eds.), Royal Society of Chemistry, Cambridge, 2001, P. 191.

176. S. N. Nuzhnov, R. Miller and B. A. Noskov, Mendeleev Commun., 13 (2003) 25

177. B.A. Noskov, S. N. Nuzhnov, G. Loglio and R. Miller, Macromolecules, 37 (2004) 2519.

178. A. Hambardzumyan, V. Aguie-Beghin, M. Daoud and R. Douillard, Langmuir, 20 (2004) 75

179. B.A. Noskov, A.V. Akentiev, A.Yu. Bilibin, D.O. Grigoriev, G. Loglio, I.M. Zorin and R. Miller, Langmuir, 20 (2004) 9669.

180. B. Rippner Blomqvist, T. Wärnheim and P. Claesson, Langmuir, 21 (2005) 6373.

181. B.A. Noskov, S. Y. Lin, G. Loglio, R.G. Rubio and R. Miller, Langmuir, 22 (2006) 2647.

182. A.M. Díez-Pascual, F. Monroy, F. Ortega, R.G. Rubio, R. Miller, B.A. Noskov, Langmuir, 23 (2007) 3802.

183. J. Benjamins and E.H. Lucassen-Reynders, in Proteins at Liquid Interfaces, Studies in Interface Sci., Vol. 7, D. Möbius and R. Miller (Eds.), Elsevier, Amsterdam, 1998, p. 341

184. J. Benjamins, J. Lyklema and E.H. Lucassen-Reynders, Langmuir, 22 (2006) 6187.

185. B.A. Noskov, A.V. Latnikova, S.-Y. Lin, G. Loglio, R. Miller, J. Phys. Chem. C, 111 (2007) 16895.

186. A.V. Latnikova, S.-Y. Lin, G. Loglio, R. Miller, B.A. Noskov, J. Phys. Chem. C, 112 (2008) 112.

187. A. Bhattacharya, F. Monroy, D. Langevin, J. F. Argillier, Langmuir, 16 (2000) 8727.

188. N.J. Jain, P. A. Albouy and D. Langevin, Langmuir, 19 (2003) 5680.

189. N.J. Jain, P. A. Albouy, R. v. Klitzing and D. Langevin, Langmuir, 19 (2003) 8371.

190. B.A. Noskov, G. Loglio and R. Miller, J. Phys. Chem. B, 108 (2004) 18615.

192. B.A. Noskov, G. Loglio, S. Y. Lin and R. Miller, J. Colloid Interface Sci., 301 (2006) 386

193. B.A. Noskov, D.O. Grigoriev, S.-Y. Lin, G. Loglio, R. Miller, Langmuir, 23 (2007) 9641.

194 B.A. Noskov, A.G. Bykov, S.-Y. Lin, G. Loglio, R. Miller, Colloids Surf. A, 322 (2008) 71.

195. J. Lucassen and R.S. Hansen, J. Colloid Interface Sci,. 23 (1967) 319.

196. B.A. Noskov, Colloid J., 44 (1982) 492.

197. C. Barentin, P. Muller and J.F. Joanny, Macromolecules, 31 (1998) 2198

198. D.J. Kuzmenka and S. Granick, Macromolecules 21 (1988) 779.

199. A.V. Akentiev and B.A. Noskov, Colloid J., 64 (2002) 129.

200. A.V. Akentiev and R. Miller, B.A. Noskov, Colloid J., 64 (2002) 653.

201. C. Kim and H. Yu, Langmuir, 19 (2003) 4460.

202. J.C. Scott, R.W.B. Stephens, J. Acoust. Soc. Am., 52 (1972) 871.

203. R. Myrvold and F.K. Hansen, J. Colloid Interface Sci., 207 (1998) 97.

204. B.A. Noskov, A.V. Akentiev, D.O. Grigoriev, G. Loglio and R. Miller, J. Colloid Interface Sci., 282 (2005) 38.

205. K.N. Prasad, T.T. Luong, A.T. Florence, J. Paris, C. Vaution, M. Seiller and F. Puisieux, J. Colloid Interface Sci., 69 (1979) 225.

206. G. Wanka, H. Hoffman and W. Ulbricht, Colloid Polymer Sci., 266 (1990) 101.

207. P. Alexandridis, V. Athanasiou, S. Fukuda and T.A. Hatton, Langmuir, 10 (1994) 2604.

208. E.P.K. Currie, F.A.M. Leermakers, M.A. Cohen Stuart and C.J. Fleer, Macromolecules, 32 (1999) 487.

209. L. C. Chang, C. Y. Lin, M. W. Kuo and C. S. Gau, J. Colloid Interface Sci., 285 (2005) 640.

210. J.K. Ferri, R. Miller and A.V. Makievski, Colloids Surfaces A 261 (2005) 39.

211. M.G. Munoz, F. Monroy, F. Ortega, R.G. Rubio and D. Langevin, Langmuir, 16 (2000) 1083.

212. R. Sedev, D. Exerowa and G.H. Findenegg, Colloid Polym. Sci., 278 (2000) 119.

213. R. Sedev, R. Steitz and G.H. Findenegg, Physica B, 278 (2002) 119.

214. M.G. Munoz, M. Encinar, L.J. Bonales, F. Ortega, F. Monroy and R.G. Rubio, J. Phys. Chem. B, 109 (2005) 4694.

215. P. Linse and T.A. Hatton, Langmuir, 13 (1997) 406

216. B.A. Noskov, A.V. Akentiev and R. Miller, J. Colloid Interface Sci., 247 (2002) 117.

217. H.D. Bijsterbosch, V.O. de Haan, A.W. de Graaf, M. Mellema, F.A.M. Leermarkers, M.A. Cohen Stuart and A.A. van Well, Langmuir, 11 (1995) 4467.

218. D.J.F. Taylor, R.K. Thomas and J. Penfold, Langmuir, 18 (2002) 4748.

219. T. Okubo, J. Colloid Interface Sci., 125 (1988) 38

220. O. Theodoly, R. Ober and C.E. Williams, Eur. Phys. J. E, 5 (2001) 51.

221. J. Papenhuijzen, H.A. Van der Schee and G.J. Fleer, J. Colloid Interface Sci., 104 (1985) 540.
222. M.R. Bohmer, O.A. Evers and J.M.H.M. Scheutjens, Macromolecules, 23 (1990) 2288.
223. G.J. Fleer, Ber. Bunsenges. Phys. Chem., 100 (1996) 93
224. B.A. Noskov, A.Yu. Bilibin, A.V. Lezov, G. Loglio, S.K. Filippov, I.M. Zorin and R. Miller, Colloids Surfaces A, 298 (2007) 115
225. M.A. Cohen-Stuart, C.W. Hoogendam and A. de Keizer, J. Phys. Cond. Matt., 9 (1997) 7767.
226. G. Gabrielli, M. Puggelli and P. Baglioni, J. Colloid Interface Sci., 86 (1982) 485
227. S.J. Mumby, J.D. Swalen and J.F. Rabolt, Macromolecules, 19 (1986) 1054
228. J.B. Peng, G.T. Barnes and B.M. Abraham, Langmuir, 9 (1993) 3574
229. Y. L. Chen, M. Kawaguchi, H. Yu and G. Zografi Langmuir, 3 (1987) 31
230. K.C. O'Brien and J.B. Lando, Langmuir, 1 (1985) 453

INTERFACIAL RHEOLOGY OF ADSORBED PROTEIN LAYERS

J. Benjamins[1] and E.H. Lucassen-Reynder[2]

[1] Doornakkers 4, 9467PR Anloo, the Netherlands
[2] Mathenesselaan 11, 2343HA Oegstgeest, the Netherlands

Contents

A. INTRODUCTION.

Proteins can be extremely effective in producing and stabilizing foams and emulsions. To a degree, their role in this is similar to that of low molecular weight surface active molecules: both types of molecules adsorb at air/water (foams) and oil/water (emulsions) interfaces. The primary effect of such molecular adsorption is that it reduces the tension of the interface. However, the reduction of the interfacial tension cannot in itself explain the formation of emulsions and foams with more than transient stability. If this were the case, it should be possible to prepare emulsions and foams in the absence of surface active solutes, from pure low tension liquids. In practice it is impossible to obtain emulsions with any degree of stability in this way.

The essential role of surface active molecules, including proteins, during emulsification and foaming is that, by forming an adsorbed interfacial layer, they retard or even prevent recoalescence of newly formed droplets or bubbles. This adsorbed layer provides the interface with properties like elasticity and viscosity. Such properties enable the interface to resist tangential stresses from the adjoining flowing liquids [1]. This liquid flow will induce tension gradients in the interface that will in turn oppose further deformation. Interfacial tension gradients may arise if liquid flow causes two emulsion droplets to collide. The gradient is caused by the drainage of the liquid out of the film between the droplets. In turn, the gradient will retard further drainage and so the droplets will survive the collision without coalescence. In a similar way these gradients can undo local thinning of a liquid film separating two emulsion droplets or foam bubbles.

Dynamic interfacial properties or interfacial rheology can be measured in two types of deformation:

(i) surface compression/dilation, which measures the response to changes in area at constant shape of a surface element.

(ii) surface shear, which measures the response of the surface to changes in shape at constant area.

Both types of measurement can be performed at small periodic deformations as well as under continuous expansion or shear [2-6]. Surface shear viscosity has enjoyed the greater attention of experimentalists working with small-molecule surfactants [7, 8]. Macromolecules, e.g., food proteins, have been studied in surface shear by Graham and Phillips [9], Martinez-Mendoza and Sherman [10], Kiosseoglou [11], Benjamins and Van Voorst Vader [12], Murray and Dickinson [13], Roberts [14], Wierenga [15] and Martin [16]. Dilational rheological properties of adsorbed protein layers have been investigated by many authors. For reviews see references [17-19]. Most papers deal with experiments performed with the "barrier-and-plate" method [18, 20, 21] while the more recent work was performed using the experimentally very convenient "drop" method [22-28] which especially facilitates experiments at oil/water interfaces.

These studies have shown that surface shear viscosity/elasticity of protein layers are far higher than those of small molecules, and that they keep increasing considerably with increasing age of the interface. While such viscosity may contribute appreciably to the long-term stability of emulsions and foams, it cannot be expected to have much relevance to their short-term stability for two reasons. First, the type of deformation that interfaces undergo during emulsification and foaming is expansion and, to a lesser extent, compression rather than shear. Second, the new interface which is continuously formed during the dispersion process, in time scales which may be as low as 1 ms or less, cannot build up the high shear viscosity/elasticity found for aged interfaces [29]. For short-term stability, therefore, interfacial rheology in compression/expansion is considered to be far more relevant.

B. DILATIONAL MODULUS: PRINCIPLES, PROBLEMS AND METHODS.

1. Principles

The surface dilational modulus in compression and expansion, E, is defined by the expression originally proposed by Gibbs [30] for the surface elasticity of a soap-stabilised liquid film as the increase in surface tension for a small increase in area of a surface element:

$$E = \frac{d\gamma}{d\ln A}$$

(1)

where γ is the surface tension and A the area of the surface element. In the simplest case, the modulus is a pure elasticity with a limiting value, E_0, to be deduced from the surface equation of state, i.e., from the equilibrium relationship between surface tension and surfactant adsorption, Γ:

$$E_0 = \left(\frac{-d\gamma}{d\ln\Gamma} \right)_{eq}$$

(2)

This limiting value is reached only if there is no exchange of surfactant with the adjoining bulk solution (i.e., $\Gamma \times A$ is constant), and if, moreover, the surface tension adjusts instantaneously to the equilibrium value for the new adsorption. Deviations from this simple limit occur when relaxation processes in or near the surface affect either γ or Γ within the timescale of the measurement. In such cases, the modulus E is a surface visco-elasticity, with an elastic part accounting for the recoverable energy stored in the interface and a viscous contribution reflecting the loss of energy through any relaxation processes occurring at or near the surface. Elastic and viscous contributions can be measured separately by subjecting the surface to small periodic compressions and expansions at a given frequency. In such experiments the visco-elastic modulus E is a complex number, with a real part E' (the storage modulus) equal to the elasticity and the imaginary part E'' (the loss modulus) given by the product of the viscosity, η_d, and the imposed angular frequency, ω, of the area variations:

$$E = E' + i E'' \equiv E' + i\omega\eta_d$$

(3)

Experimentally, the imaginary contribution to the modulus E is reflected in a phase difference ϕ between stress ($d\gamma$) and strain (dA), which means that the elastic and viscous contributions are given by:

$$E' = |E| \cos\phi \quad ; \quad E'' = |E| \sin\phi$$

(4)

where $|E|$ is the absolute value of the complex modulus and ϕ is the viscous phase angle.

2. Problems

2.1. Non-uniform area deformation due to liquid viscosity (oil/water interfaces)

In the conventional barrier-and-plate technique for measuring surface dilational moduli, the surface is periodically expanded and compressed, by a barrier which oscillates in the plane of the surface, and the response of the surface tension is monitored by a probe, e.g., a Wilhelmy plate, some distance away from the barrier. This technique was derived from the longitudinal wave method developed by Lucassen and van den Tempel [2]. The area variation generated by the barrier travels over the surface as a longitudinal wave. In Chapter 2 it is discussed that the most convenient experimental condition is the limiting case of low wave damping. In that case the area deformation is uniform through multiple reflections of the wave between the barrier and the vessel wall. Wave damping and hindrance of wave propagation due to hydrodynamic friction between interfacial layer and the adjoining liquids, increase with increasing viscosity of these liquids (e.g., triacylglycerol (TAG) oils are highly viscous). In that case the condition for uniform deformation requires a smaller effective length of the trough. Such small effective lengths L are more easily achieved in the Dynamic Drop Tensiometer [22]. This instrument (see Section B.3.2) subjects a small drop or bubble, which either hangs from or sits on the tip of a capillary, to sinusoidal oscillations of its volume and, therefore, also of its area. Here the role of the oscillating barrier is played by the rim of the capillary against which the interface is being alternately compressed and expanded. For an external diameter of the capillary of 2 mm, the distance over which the wave can travel over the drop is only a few mm, i.e., much smaller than the effective trough length in the conventional set-up. The drop method, therefore, facilitates rheological measurements at viscous-oil/water interfaces.

The conventional barrier-and-plate method has proven to be very suitable to investigate the dilational properties and relaxation mechanisms of many surfactants adsorbed at the air/water interface [2, 31-33]. However, with proteinaceous surfaces, surface shear resistance can be a complicating factor. The effect of surface shear resistance on the area deformation in the barrier-and-plate method can be visualized by monitoring the displacement of small paper particles floating on the surface. Fig. 1 shows slightly idealized pictures of the displacement patterns, found for low molecular weight surfactants and proteins. For low molecular weight surfactants (Fig 1a) usually the displacement is uniform across nearly the entire width of the trough and decreases with increasing distance from the barrier. Provided the modulus is sufficiently high and/or frequency is sufficiently low, this results in an isotropic and uniform deformation over the trough (Chapter 2) [34]. For proteins the displacement pattern (Fig 1b) can be completely different, depending on adsorbed amount and protein type. The displacement of the particles increases with increasing distance from the side walls. Across the width of the trough a parabolic displacement profile was observed, with maximum halfway the width of the trough. As with low molecular weight surfactants, the displacement decreases with increasing distance from the barrier. However, this decrease is not linear, consequently the deformation over the length of the through is not uniform.

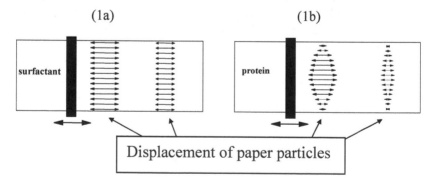

Fig. 1 The effect of strong shear resistance on local area deformation.

2.2. Non-uniform area deformation due to shear effects.

The non-uniform deformation over the length and width of the trough is in line with the non-uniformity of the surface tension variations, which was found to depend not only on distance from the oscillating barrier, but also on distance from the side walls [18]. The results indicate that shear effects are not negligible here and point to a high shear modulus of the protein-covered surface. Similar effects were observed with polymers [35] and polypeptides [36]. The parabolic flow profile of the surface is caused by a high shear resistance and by the fact that the surface sticks to the sidewalls of the trough (zero displacement at the sidewalls). Thus, the non-uniform area deformation can be eliminated when stick of the surface layer to the sidewalls would be prevented. For ovalbumin this could be achieved by using sidewalls with an electrical charge of the same sign as that of the protein [37].

Instruments designed to obviate interference of shear effects include the square-band barrier of Benjamins et al. [38] (described in Section B.3.1.), the ring trough of Kokelaar et al. [39], the oscillating bubble method of Wantke et al. [34] and the dynamic drop tensiometer of Benjamins et al. [22]. This last instrument, described in Section B.3.2., is suitable for both air/water and oil/water interfaces and has the advantage of a small effective trough length. Thus, it can generate a regime of compression/expansion which is both isotropic and uniform.

3. Methods for dilational rheology without interference of shear deformation.

3.1 Modified barrier-and-plate method

In the conventional barrier-and-plate set-up (Fig. 1), the compression/expansion effected by the barrier is unidirectional rather than isotropic. The surface deformation, therefore, contains a shearing component which can interfere with the measurement of the dilational modulus in cases of interfacial layers with appreciable resistance against shear. A modified version [17], shown in Fig. 2, avoids such complications by producing isotropic dilational deformation of the surface. The area to be subjected to compression/expansion is isolated from the rest of the surface by a square of elastic bands placed vertically in the surface. The sinusoidal and isotropic area change is effected by moving the corners synchronously along the square's diagonals. In this set-up the cycle frequency ω can be varied between 10^{-3} and 1 rad/sec. The amplitude $\Delta \ln A$ of the sinusoidal change of the surface area A can be varied between 0 and 0.25 by adjusting the eccentricity of the driver system.

The sinusoidal area oscillation as well as the resulting interfacial tension oscillation are recorded. The absolute value of the dilational modulus is simply related to the amplitudes of these oscillations $\Delta \gamma$ and ΔA by $|E| = A \, \Delta \gamma / \Delta A$. Simultaneous measurement of the phase angle ϕ results in values of the elastic and viscous components of the modulus through Eq. (4).

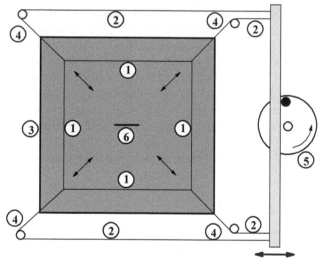

Fig. 2. Modified barrier-and-plate apparatus ensuring isotropic compression/expansion (top view). (1) square of rubber bands; (2) metal wires; (3) glass vessel; (4) wheels; (5) eccentric driver system; (6) Wilhelmy plate.

3.2. Dynamic Drop Tensiometer

The Dynamic Drop Tensiometer, available since 1995 [22], is a modified version of the Automatic Drop Tensiometer developed by Cagna et al. [40]. In traditional drop tensiometers [41-45], axisymmetric drop shape analysis is applied to an undisturbed drop or bubble to study the time dependence of the interfacial tension during adsorption or desorption. The modified instrument, manufactured by IT Concept (Longessaigne, France) differs from these in that it allows us to impose sinusoidal oscillations of the drop volume, and thereby of the interfacial area. The resulting sinusoidal oscillations of the interfacial tension can be evaluated by applying the Young-Laplace equation to the shape of the oscillating drop. As in the case of square of elastic bands set-up, the absolute value of the dilational modulus is simply related to the amplitudes of the oscillations $\Delta\gamma$ and ΔA by $|E| = A\ \Delta\gamma/\Delta A$. Simultaneous measurement of the phase angle ϕ results in values of the elastic and viscous components of the modulus through Eq (4). The main features of the instrument are shown in Fig. 3; for more details see [22].

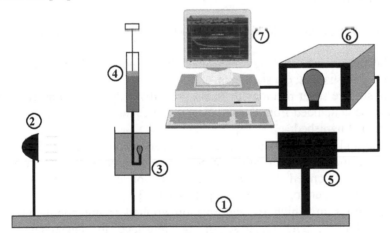

Fig. 3. Dynamic Drop Tensiometer (side view). (1) optical bench; (2) light source; (3) cuvette in which drop is formed; (4) syringe with drop phase and DC motor driving piston; (5) CCD camera; (6) video monitor; (7) PC.

The set-up differs from the barrier-and-plate technique in three respects: (i) the oscillations in area and tension are measured on one and the same small interfacial area, (ii) homogeneity of deformation of this area is far more easily ensured and (iii) the tension changes are evaluated by means of the Young-Laplace equation from measurements of the fluctuating drop shape rather than by means of a Wilhelmy plate. The set-up permits us to determine the interfacial rheological properties in compression/expansion at a chosen amplitude and frequency.

In Section B.2. it was argued that this method alleviates two problems, (i) interference of shear effects and (ii) non-homogeneous deformation with viscous oil. However, especially with viscous oil this method suffers from a drawback that is related to measuring interfacial tension by drop shape analysis. At too high frequencies and high viscosity, the viscous drag affects the drop shape by retarding the shape adaptation .[46, 47]. It has been shown that for a TAG-oil/water interface the frequency limit is between 0.1 and 1 Hz.[46]. An advantage of the set-up is that it prevents leakage of adsorbed material, such as can occur along the barrier.

C. SURFACE SHEAR RHEOLOGY PRINCIPLES AND METHODS.

Surface shear rheology measures the response of the surface to changes in shape at constant area.

In surface shear rheology there is a great diversity of methods available as can be concluded from the survey given in Chapter 10. Four main types can be distinguished:

- methods where in a Couette type set-up the layer is deformed continuously and the stress response is measured on the inner ring or disk suspended from a torsion wire or connected to a more sophisticated torque measuring device [48]
- methods in which the deformation applied is not continuous but oscillatory. The response of the interface to sinusoidal, low-amplitude, shear waves at an externally imposed frequency is measured, and the elastic and viscous properties are evaluated from the wave dispersion equation [3, 49]

- a torsion pendulum with damped oscillation, here the shear properties of the layer are derived from time-damping characteristics of the oscillation. [50]
- 2D analogues of Couette type viscometers with either an oscillating inner ring or disk or an oscillating outer ring and measurement of the resulting torque on the inner ring . [12, 18, 51]

Surface shear waves (ii) have not been in wide use, but they are interesting because their background is very similar to that of the longitudinal wave method that is common in dilational rheology (see Chapter 1). A liquid surface or an interface between two liquids can carry four different types of waves: compression/expansion waves, which can be transverse or longitudinal, and shear waves. Transverse waves, where the surface has a velocity component normal to the direction of the propagation and to the plane of the undisturbed surface, can be divided into capillary waves or ripples (small wavelength, surface tension-dependent) and gravity waves (long wavelength). Longitudinal waves are characterized by the fact that the motion of the surface is in the direction of the propagation of the wave. The properties of this wave depend mainly on the ability of the surface to support surface tension gradients[31]. In the case of shear waves the motion of the surface is in the plane of the interface, but in contrast to longitudinal waves, perpendicular to the wave propagation. The properties of this wave are determined by surface shear elasticity and viscosity. The absence of vertical motion in both waves results in a great similarity of their dispersion relations [2, 49].

Surface shear waves can be generated by e.g. an oscillatory motion of a ring or disk as used in many shear rheological methods. Due to hydrodynamic coupling between surface and solution in contact with it [3, 49], the oscillatory motion of the outer ring generates a damped transverse shear wave in the surface. This has consequences for optimal dimensions of the set up (distance between oscillating ring and measuring body). If the distance is too long no signal is measured because the wave has damped completely. In the most convenient set-up, the distance is short enough for multiple reflections (as in the case of longitudinal waves) between the two rings to occur, which results in uniform deformation across the gap. Limiting conditions with respect to dimension, modulus and frequency have been given in references [3, 12].

The main advantage of shear methods that use sufficiently small oscillatory, sinusoidal, shear-deformations (≈0.05 rad) with an externally imposed frequency, is that the network remains intact [12]. In this way information about intra- and intermolecular interaction of the adsorbed molecules forming a 2D network can be obtained. Combining these results with the information on dilational properties, surface concentration and layer thickness [17, 18] will help to reach a more complete understanding of the inter- and intramolecular interactions occurring in adsorbed protein layers.

The surface shear rheometer shown in Fig. 4 consists of two concentric glass rings which lie flat in the interface. The opposing faces of the rings were roughened to suppress slip. The outer ring oscillates periodically around its axis with a small amplitude (≈ 0.05 rad). The inner ring, which is stationary, is connected to a torque-measuring device consisting of an air bearing and a capacitive rotation transducer. A small rotation (≈ 10^{-4} rad) of this transducer is coupled back automatically by a torque motor. The voltage on the torque motor, required to prevent motion of the inner ring, is a measure for the torque exerted on this ring. The sinusoidal outer ring (deformation) oscillation as well as the resulting stress oscillation on the inner ring are recorded The absolute value of the shear modulus is simply related to the amplitudes of the oscillating shear stress S and deformation tanα by:

$$|G| = S/\tan \alpha. \tag{5}$$

For a given ring configuration the stress on the inner ring, S, and tanα can be calculated as follows:

$$S = \frac{T_M(R_i)}{2\pi R_i^2}, \quad tan\,\alpha = \frac{2A_oR_o}{(R_o^2 - R_i^2)} \tag{6}$$

Here $T_M(R_i)$ is the maximum torque at the inner ring, R_i = outer radius of the inner ring, R_o = inner radius of the outer ring and A_o = amplitude of the outer ring.

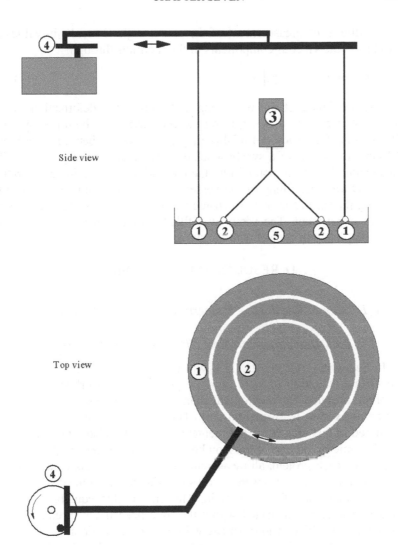

Side view

Top view

Fig 4 Concentric ring surface shear rheometer; 1 - outer ring, 2 - inner ring, 3 - torque measuring device, 4 - eccentric driver system, 5 - solution

Simultaneous measurement of the phase angle Θ results in values of the elastic and viscous components of the modulus through:

$$G' = |G| \cos \Theta, \ G'' = |G| \sin \Theta \tag{7}$$

The amplitude of the outer ring and, hence, the deformation of the surface, can be varied by adjusting the eccentricity of the driver system.. The angular frequency (ω) of this ring can be varied between 0.008 and 0.4 rad/s by means of a synchronous electric motor with a gear box. The protein solutions were poured into the measuring vessel using a separating funnel. During the filling of the vessel the tip of the funnel was kept just touching the surface in order to ensure that the surface was fresh at the start of each experiment. The vessel was filled until the two concentric rings were completely in contact with the surface.

D. RESULTS AND DISCUSSION

1. Dilational Rheological Properties of Adsorbed Protein Layers

1.1. Induction period and time effects. Relationship $\Pi(\Gamma)$ curve

In Sections D.1.2 and D.1.3 it will be shown that, from the shape of the surface equation of state (surface pressure vs adsorbed amount) of adsorbed proteins, the rheological characteristics of these adsorbed layers can be derived at least qualitatively. Therefore, for most of the proteins discussed in this chapter, experimental values of the adsorptions, measured by ellipsometry, and the corresponding surface pressures are presented in Fig. 5. For each protein, all measurements at different bulk concentrations and different surface ages were found approximately to coincide on a single curve characteristic for the protein. This means that equilibrium in the surface is largely established within the time needed for an ellipsometric measurement, which is approximately 5 min. Thus, equilibration within the surface is very much faster than between surface and bulk solution, and each curve in Fig. 5 can be considered to represent the equilibrium surface equation of state to a fair approximation. From the shape of these curves, for each protein three characteristic regions can be distinguished (i) a low adsorption region where the surface pressure does not measurably deviates from zero, (ii) a region, starting at $\Gamma_{\Pi>0}$, where the surface pressure steeply increases with increasing adsorption and (iii) a third region where this steep increase flattens off. For proteins with flexible, random coil-like molecular structure, the shape of the curves can be characterised by a low $\Gamma_{\Pi>0}$ and a gradual increasing surface pressure above $\Gamma_{\Pi>0}$. With globular proteins like Ovalbumin, BSA and Lysozyme $\Gamma_{\Pi>0}$ is higher and beyond $\Gamma_{\Pi>0}$ the surface pressure increase is steeper.

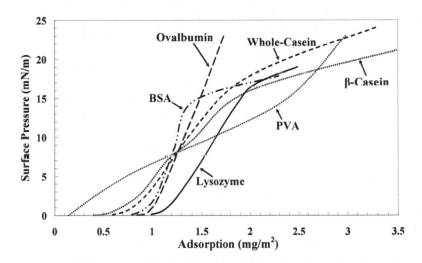

Fig. 5 Surface pressure as a function of adsorption Bovine Serum Albumin, Ovalbumin, Lysozyme, b-Casein, Whole Casein, PVA205.

The region where the increase of the surface pressure with increasing Γ flattens off is somewhat more pronounced with the flexible proteins than in the case of the globular proteins. In the region of flattening multilayer formation is supposed to occur. The idea that a low $\Gamma_{\Pi>0}$ and a gradual increase of the curve beyond this value are related to a flexible molecular structure is supported by the shape of the curve for the synthetic, random-coil, polymer PVA. With this molecule $\Gamma_{\Pi>0}$ and the steepness of the slope are even lower than with β-casein and there is no sign of levelling off at high adsorptions.

Fig. 6 illustrates the slow equilibration of a 1 mg/l BSA solution, reflected in a steady increase of the modulus and the surface pressure as a function of the adsorption time. For such a low protein concentration, a time lag is observed during which the modulus is too small to be measured with sufficient accuracy. This time lag corresponds with the time lag ("induction" period) in the surface pressure, which is a consequence of the fact, as illustrated in Fig. 5, that a certain value of the adsorption ($\Gamma_{\Pi>0}$), characteristic for each protein, is required to produce a measurable non-zero value of the surface pressure and, in consequence, also of the dilational modulus. Only at low concentrations this induction time is noticed experimentally. At higher concentration, due to faster adsorption, $\Gamma_{\Pi>0}$ is reached within the time the first surface tension and modulus measurement is finished. After take-off, both modulus and surface pressure, in most cases gradually, increase with time, until almost constant values are reached when the system is close to adsorption equilibrium.

During the equilibration process, the modulus E depends on the adsorption time and protein concentration in the solution. Over a large range of surface pressures, however, these two variables combined, merely serve to determine the changing values of the really important variable, i.e., the protein adsorption. As a result, modulus *vs* time curves at different protein concentrations tend to correspond to a single modulus *vs* adsorption curve and consequently, a modulus *vs* pressure curve as illustrated in Fig. 7 for BSA [17, 18]. At the highest rate of compression/expansion, ω=0.8 rad/s, surface behaviour is, within experimental error, purely elastic over the whole adsorption/surface pressure range. Except in the case of Ovalbumin, we observed for all proteins in the higher adsorption/surface pressure range that the modulus becomes frequency dependent (visco-elastic behaviour; see Fig.7). This is an indication that in this region relaxation processes in close-packed protein surfaces take place. In the lower adsorption/surface pressure region the elasticity is equal to the limiting value E_0, derived from the $\Pi(\Gamma)$ curve, as shown in Fig. 7.

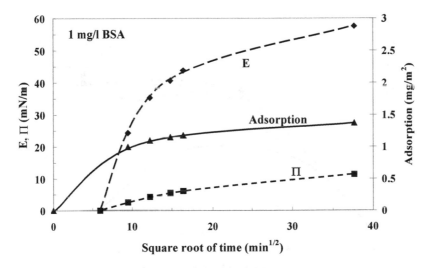

Fig. 6 Surface dilational modulus, adsorption and surface pressure as a function of the adsorption time of Bovine Serum Albumin at 1 mg/l. Frequency: 0.84 rad/s; pH=6.7; $\Delta A/A$=0.07

Ovalbumin, like many other proteins, shows a maximum surface concentration near the I.E.P., at pH=4. This maximum is consistent with the more extended structure of the protein molecule outside the I.E.P. region, due to the repulsive effects of equally charged parts of the molecule [52] or to lateral repulsive forces between adsorbed molecules. The modulus at equilibrium was found to increase if the pH decreases from high pH to the iso-electric region [18]. This is in line with the generally found increase of the modulus with increasing surface concentration. It was also found that at pH< I.E.P. the modulus remained at the level that was reached at the I.E.P. This is not surprising, because at pH 3.6 the surface concentration remains at a level where the modulus no longer increases upon further increase of the surface concentration; see Fig. 8.

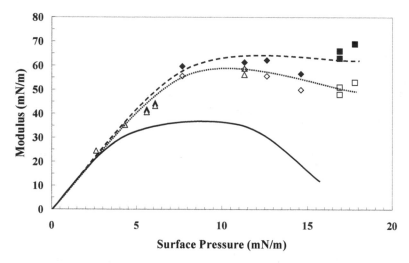

Fig. Surface dilational modulus as a function of surface pressure for Bovine Serum Albumin. pH=6.7; $\Delta A/A$=0.0 Closed symbols and dashed line: ω=0.84 rad/s; open symbols and dotted line: ω=0.084 rad/s. Triangles: c=1 mg/l; diamonds: c=5 mg/l; Squares: c=10 mg/l. Drawn line: limiting modulus, E_0, from Π vs Γ curve in Fig. 5

As illustrated above (Section D.1.1.) interfacial data are best compared at equal surface pressure or equal surface concentration. Therefore we have plotted in Fig. 8 the modulus vs surface concentration at different pH values. Because of the fast adsorption near the I.E.P at high protein concentrations, only moduli at high surface pressures are available in this pH region. Consequently there is only little overlap between data at high and low pH. However, all measured points in this plot, once again, more or less collapse into one single $E(\Pi)$ curve. This further illustrates that the major part of the effect of pH on the modulus can be related to the effect of pH on the surface concentration. The absence of an effect of pH on E(Π) can also be deduced from the finding that at pH values not too far from the I.E.P the Π(Γ) curves do not differ significantly [18].

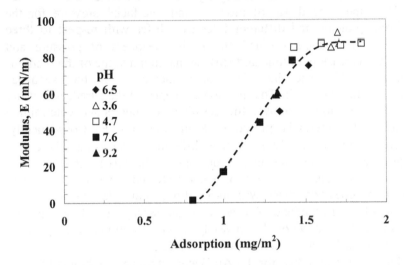

Fig. 8 Modulus *vs* adsorbed amount for Ovalbumin at different pH values; frequency= 0.1 Hz

Near the I.E.P. the interface becomes more viscous, which is in line with the general trend that the viscous component of the modulus increases with a further increase of the surface concentration. It is not very likely that in this case the viscous component (φ<9°) is related to multilayer adsorption, because the surface concentration did not exceed 2 mg/m^2.

However if the charge of the ovalbumin molecule is not affected by pH but by succinilation [15] it was found that the Π(Γ) curve shifts to lower Γ with increasing negative charge. This also results in an effect on E(Π).

1.2. Effect of type of interface on the dilational rheology of adsorbed protein layers.

In this section we will compare the dilational modulus of adsorbed protein layers at three liquid interfaces viz., Air/Water, Tetradecane(TD)/Water and Triacylglycerol (TAG)/Water.

The time dependence of modulus and interfacial pressure for the different proteins and different interfaces differ with respect to three aspects: (i) "induction time" ($t_{\Pi>0}$), (ii) steepness of pressure and modulus increase with time and (iii) the maximum value of the modulus [18, 24, 25]. The induction time increases in the sequence TAG < tetradecane< air. The increase is steepest at the tetradecane/water interface. As argued above, the adsorbed amount is the determining parameter for interfacial pressure (Π) and modulus (E). Consequently, differences in $\Pi(t)$ and $E(t)$ curves at different interfaces must be due to differences in either the adsorption rate or the $\Pi(\Gamma)$ curve. For the air/water interface, Figs.5 and 6 demonstrate that considerable protein adsorption does take place without resulting in any detectable value of Π. Therefore, the "induction" time is not necessarily caused by a specific barrier against adsorption, but possibly also to strong non-ideality of the $\Pi(\Gamma)$ relationship, as argued before [18, 53-55].

Fig. 7 illustrates that, for the Air/Water interface, combining the $\Pi(t)$ and $E(t)$ data for different protein concentrations results in one single $E(\Pi)$ plot, at least in the lower Π range where E is purely elastic. At both oil/water interfaces the behaviour is similar. Figs.9a-d show $E(\Pi)$ plots for the proteins β-casein, β-lactoglobulin (BLG), Bovine Serum Albumin (BSA) and Ovalbumin at the three different interfaces. At the three interfaces we find qualitatively the same difference between the globular proteins (BLG, BSA and Ovalbumin) and β-casein. The curves of the globular proteins can be characterized by, initially, a linear increase followed by a flattening off of the increase or even a decrease. With β-casein we also observe initially a linear increase, however only up to lower pressure. With further increasing pressure a significant decrease is observed followed by a second increase at higher pressures, especially at

Air/Water and TD/Water. This second decrease is most likely due to multilayer formation [17, 18].

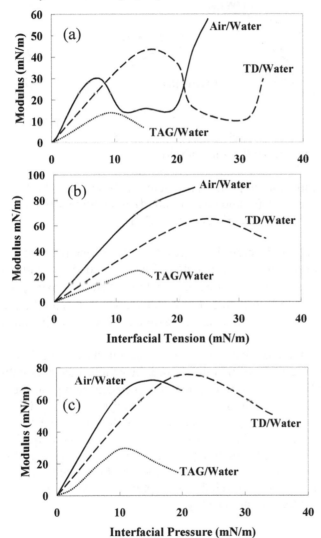

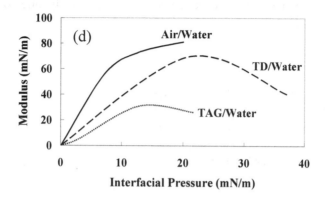

Fig. 9. E(Π) plots for the proteins β-casein (a), β-lactoglobulin (b), Bovine Serum Albumin (c) and Ovalbumin (d) at the three different interfaces. Frequency= 0.1 Hz

When comparing the different interfaces, the main differences are the initial slope, the extent of the linear range (Table 1) and the maximum modulus value (Fig. 9). For all proteins, at both oil/water interfaces, the modulus decreases significantly after this maximum (Fig. 9). At all three interfaces globular proteins show a higher maximum value for the modulus than β-casein.

The following trend is observed at the three interfaces: the initial slope and the pressure range of this steep linear behaviour increases with decreasing flexibility of the protein molecule, β-casein being the most flexible molecule and ovalbumin the most compact one. In this linear range, the modulus is almost purely elastic, and should be equal to the value of E_0 evaluated from the equilibrium $\Pi(\Gamma)$ curve. Such equality was confirmed for adsorbed protein layers at the air/water interface [17, 18]. At present, however, it is not possible to verify this equality for the two oil/water interfaces because reliable adsorption data over a sufficient range of interfacial pressures are lacking.

Table 1 Initial slope, dE/dΠ, and extent of linear range of the proteins at the three interfaces

Protein	air/water		tetradecane/water		TAG-oil/water	
	dE/dΠ	Extent mN/m	dE/dΠ	Extent mN/m	dE/dΠ	Extent mN/m
β-casein	5	5	3.4	10	1.5	8
BLG	6.2	9	3.8	11	2.0	8
BSA	8.5	7	5.5	12	3.5	8
Ovalbumin	9.4	7	4.4	14	3.0	10

1.3. The surface equation of state

Fig. 5 presents the $\Pi(\Gamma)$ relationship for various proteins at the Air/Water interface . The adsorbed amount at this interface was determined by ellipsometry [17, 18]. For all proteins the $\Pi(\Gamma)$ relationship is clearly non-linear, which usually indicates a strongly non-ideal surface equation of state [54, 55]. At both oil/water interfaces experimentally determined adsorption data are lacking. However, the 'induction time' for Π and E found at these interfaces indicates that here the $\Pi(\Gamma)$ relationship is also non-linear.

In the elastic range, where the dilational modulus is dominated by the equation of state, the modulus $|E|$ can be equated to its limiting value E0, if equilibrium between Π and Γ is established within the time scale of the modulus measurement. Such equilibrium is observed in all cases where $E(\Pi)$ curves measured at different times collapse into a single curve. A simple two-dimensional solution treatment [17, 18, 25, 56] can be used to interpret some main features of the measured moduli in the elastic range. In order to describe the distribution of solvent (component 1) and solute (component 2) between bulk solution and surface, this thermodynamic model defines the interface as a Gibbs dividing surface not by the convention $\Gamma 1 \equiv 0$, but by the relationship $\omega_1\Gamma_1 + \omega_2\Gamma_2 + ... \equiv 1$, where the parameters ωi are partial molar surface areas. In this convention, the surface excess quantities Γi combined with ωi make up the contents of the interface in the same way as volume concentrations combined with partial molar volumes do in three dimensions. An analytical equation can be obtained only by adopting a non-thermodynamic approximation to account for surface non-ideality. First-order approximations are used for both enthalpy and non-ideal entropy of mixing, in a Frumkin-type correction and an adaptation [57] of Flory's equation for the entropy of 3-D solutions to the surface, respectively. In the resulting equation of state for a binary mixture:

$$\Pi = -\frac{RT}{\omega_1}\left[ln(1-\Theta_2) + (1-\omega_1/\omega_2)\Theta_2 + \frac{H^\sigma}{RT}\Theta_2^2 \right] \tag{8}$$

Θ_2 ($=\omega_2\Gamma_2$) is the degree of surface coverage and the interaction constant H^σ is the partial molar surface excess enthalpy of mixing in the limit of $\Pi=0$. Positive values of H^σ above a critical value must cause phase separation into a protein-rich surface phase and a protein-depleted one (for detailed information see [58]).

Summarising, Eq (8) is a phenomenological relationship between Π and Γ, based on thermodynamic equations without *a priori* assumptions about molecular geometry and interactions at the surface, combined with first-order approximations for non-ideal mixing at the surface. Specifically, protein molecules are not assumed to be adsorbed as either flat disks or flexible polymer chains. In its simplest form, the equilibrium relation between Π and Γ depends on three constant phenomenological parameters ω_1, ω_2 and H^σ, with numerical values to be determined from the experimental results. The value of ω_2 can be chosen equal to the 'excluded' area, i.e., the inverse of the protein's saturation adsorption. For more details about applications of this model so far see references [17, 25, 56, 59, 60].

Theories derived from statistical-thermodynamic models [61-64], like the self-consistent field theory [63], developed originally for flexible-chain polymers are less suitable for proteins because of the strong specific intra- and intermolecular interactions that distinguish proteins from flexible-chain polymers. Theories based on scaling laws [65-67] provide a simple description of the steep linear ascent of the $E(\Pi)$ curve for values of $\Pi < 10$ mN/m. However, including the pre-exponential constants for all regimes, the number of adjustable parameters is fairly large, and a single analytical formula for the entire range of adsorptions cannot be given.

According to Eq (8) the limiting elastic modulus is given by

$$E_0 = \frac{RT}{\omega_1}\left[\frac{\Theta_2}{1-\Theta_2} - (1-\omega_1/\omega_2)\Theta_2 - \frac{2H^\sigma}{RT}\Theta_2^2\right] \tag{9}$$

Examples of applying Eqs (8) and (9) to measured $E_0(\Pi)$ data are presented in Fig. 10, for ovalbumin at the three fluid interfaces. In the absence of experimental information on the molecular area at the two oil/water interfaces, the same values of ω_i as derived for air/water, were applied at the three interfaces. The differences between the three curves then depend only on the difference in the interaction enthalpy parameter, H^σ. High values of the enthalpy combined with high non-ideal entropy produce near-zero surface pressures in the initial range of adsorptions (see Fig. 5) followed by a steep increase of $dE_0/d\Pi$ (see the top curve in Fig. 10) [17]. At air/water, where the highest value of H^σ applies, the model curve undergoes an abrupt change in gradient at very low pressure (in this case, at $\Pi < 0.07$ mN/m), indicating the occurrence of phase separation in the surface. In the phase-separation range, considerable adsorption takes place without resulting in any measurable surface pressure [58]. Therefore, a pronounced 'induction time' can result from a strongly non-ideal equation of state. Values of the interaction parameter H^σ (Table 2) increase in the order triacylglycerol < hydrocarbon < air. This can be interpreted as a decrease in solvent quality of the apolar phase for the more hydrophobic amino acids in the same sequence. We note that the same sequence is also followed by the interfacial tensions of the clean interfaces, γ^0, as illustrated in Table 2.

Comparing the moduli for the different interfaces at a given, fairly high, surface pressure is somewhat arbitrary. Nevertheless, it does show the quantitative effect of differences in H^σ and γ_0, and it is representative for the entire curves of both ovalbumin and β-lactoglobulin. For bovine serum albumin, the differences between tetradecane and air are only small, and the curve for β-casein at air/water is too complex for an easy comparison. In all cases, however, Figs.9a-d clearly show that the interface with the lowest value of γ_0 yields the smallest moduli over the entire range of pressures.

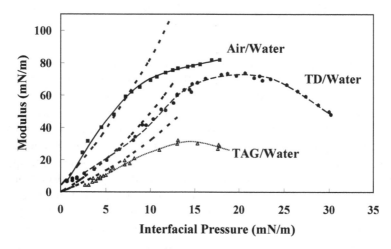

Fig. 10. E(Π) for ovalbumin at three interfaces; comparison between experimental data and the simple model; frequency= 0.1 Hz Dotted lines: theory of Eq (9), with values of H^σ given in Table 2

Values of γ^0 reflect the differences in nature of the interfaces formed by water and non-aqueous liquids with various degrees of polarity, defined as the contribution of polar forces to the surface free energy and judged by their dielectric constants. Through its effect on adsorption levels, γ_0 appears to be an important factor in determining the interfacial pressure - and thereby the dilational modulus - at different interfaces.

Table 2. Correlation of measured modulus E_0 (at $\Pi=15$ mN/m) with enthalpy parameter, H^σ, clean-interface tension, γ_0, and dielectric constant. Example for ovalbumin (see Fig. 10)

Apolar Phase	Dielectric Constant	Tension γ_0 (mN/m)	Enthalpy H^σ/RT	Modulus E_0 (mN/m)
Air	1.007	71.2	0.84	80
n-Tetra-decane	2.0	51.3	0.45	60
TriAcyl-Glycerol	3.2	31.2	0	30

Two aspects that limit the applicability of the simple three-parameter model are:

(i) The model of Eqs (8) and (9) is very simplified, not only because of its first-order approximation of entropy and enthalpy, but also because the protein's molecular area, ω_2, was assumed to be single-valued. This last assumption is the reason why Eq (8), like all current similar equations of state, predicts an ever steeper increase of Π with increasing Γ, in conflict with the flattening-off of the experimental $\Pi(\Gamma)$ curves for proteins (see Fig. 5).

(ii) Quantitative information on protein adsorption – and, therefore, on protein molecular area – is available only at air/water, not at the two oil/water interfaces.

In both respects, progress can be made with the aid of two promising recent developments:

(i) The model on the basis of Eq (8) has been improved by accounting for multiple conformations with different molecular areas of adsorbed protein molecules [68]. The extended model assumes that (1) the protein's molar area can vary in small steps from a maximum value (ω_{max}) at low adsorption to a minimum value (ω_{min}) at high adsorption, (2) the standard free adsorption energy is the same for all states, and (3) the different states are in equilibrium with each other but otherwise behave independently. Thus the extended model adds one phenomenological parameter (ω_{min}) to the simple model of Eqs (8) and (9). The result is a quantitative and internally consistent theory for molecularly 'compressible' adsorbates, for which the term 'soft' particles was coined by De Feijter and Benjamins [69]. In this context, 'compressible' refers to the surface area of the molecule, not to its volume.

(ii) Novel methods have been developed for the measurement of protein adsorption at oil/water [70, 71] which may remove the need for arbitrary assumptions about protein molecular area.

Application of this extended theory to BSA and β-casein at air/water [68, 72] revealed a much better fit of the experimental data, including the flattening-off of the modulus with increasing Π. Moreover, the measured values of E0 for the rigid molecule of bovine serum albumin can be described by a moderate degree of molecular surface compressibility ($\omega max/\omega min$ = 2.5), but the flexible molecule of β-casein requires a much larger compressibility ($\omega max/\omega min$ = 10). The higher compressibility of the flexible molecule directly leads to a much lower modulus at all but the lowest pressures [72].

1.4. Surface tension relaxation --- viscous phase angle

In the low-pressure region E is almost purely elastic (tan φ <0.1) at a frequency of 0.1 Hz. Depending on protein type and interface the extent of this region is up to 10 – 15 mN/m. In this region, the results at different concentrations and different interfacial ages collapse into single E(Π) curves. In the higher-pressure region where this is not the case any more, phase angles are no longer negligible; therefore, the measured modulus |E| no longer equals the pure elasticity E₀. As an example the pressure dependence of the modulus and the phase angle for β-LG at the TD/Water interface is presented in Figs.11A and 11B, respectively. In this high pressure region, the phase angles in most cases increased up to 20°, at all three interfaces. This indicates that within adsorbed protein layers relaxation processes take place with a characteristic timescale of about 10 s. Diffusional interchange with the bulk solution can be ruled out as a possible mechanism in such a short time [17, 18].

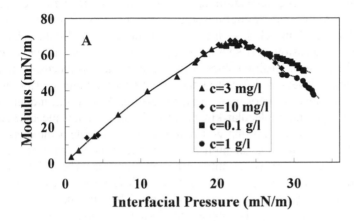

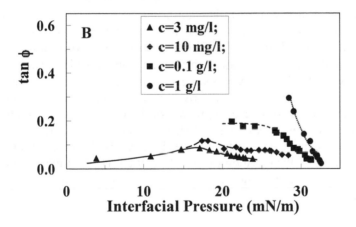

Fig. 11. Pressure dependence of the modulus (A) and the phase angle (B) for
β-LG at the TD/Water interface for different protein concentrations in
solution; frequency= 0.1 Hz.

Possible in-surface relaxation processes listed in the literature [55] include slow re-orientation after adsorption, internal reconformation, such as unfolding of long chains, involving changes in molecular shape, molecular compression into smaller areas, formation of multilayers and collapse phenomena. Distinguishing between these various mechanisms in densely packed layers is difficult as they may occur simultaneously, and their rate constants can vary with surface pressure. As a result, modelling in terms of specific molecular-kinetic models has not advanced much beyond the stage of curve fitting with a number of adjustable parameters.

Fig. 11B illustrates that at high pressures the viscous phase angle at different concentrations splits into different branches. The differences between the branches are probably caused by differences in the ratio of the rates of adsorption and re-conformation. The branches invariably appear to revert to zero phase angle with further increase of pressure, at the given frequency. This frequency leaves a time span of the order of only 10 s, which is far too short for very slow in-surface relaxations to make noticeable progress towards equilibrium between Π and Γ. Thus, our near-zero phase angles are in line with recurrent suggestions in the literature on proteins [20, 26, 73-77] that relaxations proceed at a much slower rate in dense layers at high Π than in the more dilute layers at lower Π. As early as 1972, Bull reported very slow relaxation of ovalbumin layers over a time span of 105 s at fairly high surface pressures [74]. Such a long time span would indeed imply that in the frequency range studied here the slow relaxation has barely started. Therefore, elasticities at such high pressures cannot be equated to the limiting elasticity E0, although they still depend on the equation of state. Specific relaxation mechanisms for a quantitative interpretation at high surface coverages have yet to be developed.

2. Shear Rheological Properties of Adsorbed Protein Layers

2.1 Reproducibility

Shear modulus measurements of adsorbed protein layers suffer from bad reproducibility, especially when measured at separately prepared surfaces [12, 16, 18, 51, 78]. The differences are far beyond the accuracy of duplicates when measured at the same surface [12, 18, 51] and are apparently due to irreproducibility of surface formation. Other factors probably affecting the reproducibility are (i) very small amounts of adventitious surface active admixtures [79-81], (ii) inhomogeneous layer/structure formation by unintentional disturbances in the surface layer during the preparation stage or (iii) irreproducible adherence of surface layer to the glass rings [12, 18] or the torque measuring ring [51]. A major structure-disturbing action is creating the interface and establishing contact between the interface and the rings. Consequently, irreproducible disruption of the structure is likely to occur already at the start of an experiment. Roughening those parts of the rings that are in contact with the surface layer in order to promote adherence of the surface layer to the glass rings did not solve the reproducibility problem [12, 18].

Despite this poor reproducibility the differences of G with protein type and surface age are sufficiently large to draw conclusions as shown below.

2.2. Effect of magnitude of the deformation on shear modulus G.

In order to obtain the surface shear modulus for the undisturbed structure of the adsorbed protein film, the deformation must be so small that $|G|$ no longer depends on the magnitude of the deformation. It was verified whether our measurements satisfied this condition by determining $|G|$ and Θ for different values of the deformation amplitude ranging from 0.012 to 0.25. For both BSA and ovalbumin, at all protein concentrations, surface ages and angular frequencies, we found $|G|$ to increase and Θ to decrease linearly with decreasing deformation. Only for $|G| < 5$ mN/m, which is the case at low protein concentration and at the initial stage of the adsorption process (low surface ages), $|G|$ was found to be constant over the whole deformation range (Fig. 12).

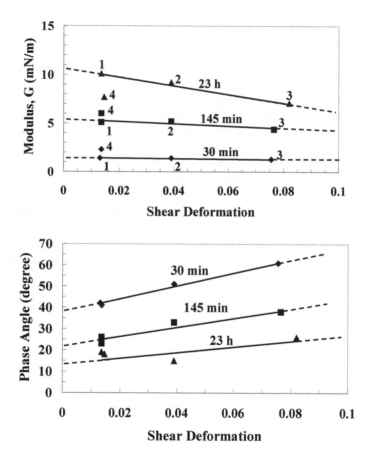

Fig. 12. Shear modulus and phase angle vs. deformation at increasing adsorption time; BSA : 0.005 g/l, $\omega = 0.42$ s^{-1}. The numbers indicate the sequence of the measurements.

By extrapolation to zero deformation the modulus for the unperturbed surface layer is approached. Unless stated otherwise, all values of G' and G'' reported here have been calculated by linear extrapolation as in Fig. 12. The clear deformation dependency of the modulus suggests breaking of bonds that contribute to the structure. The experiments were performed at increasing deformation followed by an experiment at the initial first small deformation. For $G > 5$ mN/m, this latter modulus is always lower than on the way up, which also points to breaking of bonds. It further indicates that restoring these bonds takes at least 0.5 hour, being the time needed for a set of experiments at the different frequencies at one deformation.

From bulk rheology of protein gels a linear region up to deformations ≥ 0.025 was found [82]. Considering the adsorbed protein layer as a thin gel layer of protein, a similar linear region would be expected for surface shear rheology. A considerably smaller linear region points to inhomogeneities in the adsorbed layer. Due to these inhomogeneities the local deformation can be much larger than the mean applied deformation.

It was found [18] that only G' and not G'' showed this deformation dependency. Consequently, the bonds that are broken at large deformations and slowly restore, do not contribute to G''.

2.3. Comparison between literature data

As mentioned before, numerous experimental techniques are in use in surface shear rheology. Each instrument has its own geometry, way and extent of deformation and method of preparation of the surface layer. Table 3 is an illustration of the state of affairs at this moment. Here a comparison has been made between results obtained with various proteins where we limited ourselves to the most commonly used techniques. Considering substantial differences in experimental conditions and preparation of the monolayer, the agreement with respect to the modulus (Table 3) is satisfactory. This is in the first place the case for the ranking order of the maximum values of the surface shear modulus (Na-caseinate< BSA< Ovalbumin). Burgess [83] reported considerably higher values for the shear modulus of BSA. However, these high values were only found for high protein concentrations. There are indications that with high protein bulk concentrations the thickness of the adsorbed layer increases considerably, probably by the formation of multilayers (see Table 4).

Table 3 Comparison between shear moduli obtained with different techniques. The data in this table were determined after adsorption equilibrium was established.

protein	shear modulus (mN/m)	method (ref)	angular freq.(s^{-1})
β-casein	<0.1	creep [9]	
	<1	osc.ring [84]	
Na-caseinate	0.5 (0.3 g/l)	stress-strain [18]	0.001 (def. rate)
	0.2 (0.3 g/l)	osc. ring [18]	0.084
κ-casein	6 (0.3 g/l)	osc. ring [18]	0.42
β-lactoglobulin	5-10 (spread layer)	creep [85] stress-strain [86]	
	2-3 (0.1g/l)		
BSA	8 (spread layer)	creep [87]	
	5 (0.05 g/l)	creep [9]	
	12 (1 g/l), 237 (10 g/l)	osc.ring [83] osc.disk [88]	66
	30 (1 g/l)	stress-strain [18]	0.001 (def. rate)
	12 (0.3 g/l)	osc. ring [18]	0.42
	14.5 (0.1 g/l)	osc. ring [18]	0.42
HSA	17 (0.3 g/l)	osc. disc [89]	
	3 (0.005 g/l), 12 (1 g/l)	osc. disc [50]	
	0.01 (0.1 g/l)		
Lysozyme	5 (spread layer)	osc. disc [89]	
	10 (0.1 g/l)	osc. ring [14]	5
Ovalbumin	27-60 (1 g/l)	osc.disc [88]	66
	20 (0.3 g/l)	stress-strain [18]	0.001 (def. rate)
	19 (0.1 g/l)	osc. ring [18]	0.42
	40 (0.3 g/l)	osc. ring [18]	0.42
	2 (0.1 g/l)	stress-strain [86]	

2.4. Effect of surface age and adsorbed amount on shear modulus.

In literature [12, 14, 90] the increase of G with time is often interpreted as caused by a structure formation process, where bond formation is assumed between neighbouring macromolecules according to a first order reaction. In those studies the possibility of an increase of G caused by an increase of adsorption was ignored. In Fig. 13 this increase of the shear modulus with time is compared with the time dependence of the adsorption and the dilational modulus, E. This comparison indicates that the shear modulus continues to increase when the dilational modulus is already constant (or even decreasing, depending on frequency, not indicated in this plot). In this range of surface ages, the adsorption increases still somewhat. Consequently, the increase of $|G|$ with time may be due to at least two different time-dependent contributions: (i) increase of the adsorbed amount and (ii) increase of number of bonds between adsorbed molecules.

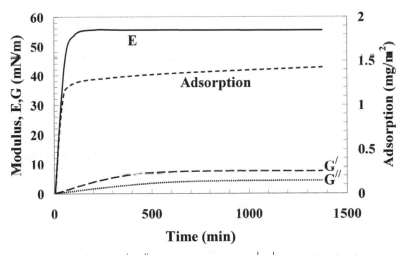

Fig. 13 Shear modulus G', G'', dilational modulus $|E|$, and adsorbed amount as a function of the age of the surface. BSA conc. 0.005 g/l, frequency = 0.084 rad s^{-1}.

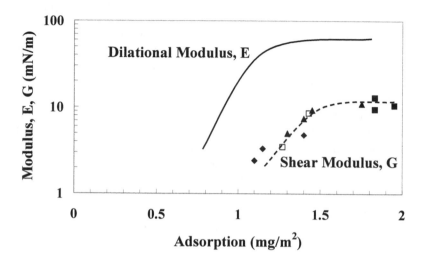

Fig. 14 The shear modulus $|G|$ and the dilational modulus $|E|$ as a function of the adsorbed amount. Both curves are the combined result of various BSA concentrations at increasing surface age. BSA concentrations: ♦ 0.001 g/l, □ 0.005 g/l, ▲ 0.03 g/l, ■ 0.1 g/l; $\omega = 0.42$ s^{-1}

In Fig. 14 $|G|$ is plotted as a function of the surface concentration for different BSA concentrations. For the sake of comparison the dilational modulus vs. surface concentration curve ($|E|(\Gamma)$) is given as well. The adsorption data were obtained by ellipsometry, not at the same surfaces, but under identical experimental conditions. The adsorption time was used to link adsorbed amount to the shear modulus. The data in this plot were obtained by measuring the shear modulus in the course of the adsorption process, consequently at increasing surface ages. So equality of adsorptions for different concentrations means higher surface ages for the lower concentration. If the formation of intermolecular bonds is slow compared to the rate of adsorption, moduli determined at equal adsorptions, but at different protein concentrations, would be higher for the lower concentration. However, the results with the different concentrations seem to merge into one master curve of $|G|(\Gamma)$. Consequently we can conclude that, (i) not the intermolecular bond formation step, but the adsorption rate controls the increase of the shear modulus with time and (ii) the adsorbed amount dominates the value of the shear modulus.

This conclusion also implies that the formation of bonds between adsorbed molecules (the structure formation process) is fast compared to the adsorption process. Whether this is also true for much higher concentrations (initially much faster adsorption) is questionable.

Due to the poor reproducibility of the shear modulus measurements, the accuracy of the shape of $|G|(\Gamma)$ curve is less than that of the $|E|(\Gamma)$ curve. The results, however, justify the conclusion that, compared to $|E|$, a higher adsorption is required to produce a non-zero value of the shear modulus. The resistance against shear deformation becomes measurable at an adsorbed amount where $|E|$ is already at half its maximum value. With a further increase of the adsorption the shear modulus increases rather steeply.

The shear modulus was found to deviate from zero above a critical value of the surface pressure, $2 \leq \Pi_{crit} \leq 5$ mN/m [18]. This in contrast to the dilational modulus, which linearly increases with increasing surface pressure as soon as Π deviates from zero. A critical value for the surface pressure was also found for viscosity measurements by Joly [91]. Above this critical value, which for BSA was 4 mN/m, the shear viscosity became non-Newtonian.

2.5. Provisional model for shear properties of adsorbed protein layers.

The thin layer of interacting adsorbed protein molecules can be regarded as a very thin three-dimensional protein gel. For the shear properties of such a layer two elements are supposed to play a role, (i) the internal rheological properties of the protein molecule and (ii) the intermolecular interaction. The internal rheological properties of the protein molecule are determined by the strengths and number of intramolecular bonds (hydrophobic interactions, H-bridges, steric interaction and covalent bonds, e.g. S-S bridges). The intermolecular interactions will be determined by the number and type of the adsorbed protein molecules and are probably much weaker.

According to the above picture two models to obtain an adsorbed layer with a high resistance against shear can be imagined (see Fig. 15); (1) a layer of basically flexible molecules with a high number of strong crosslinks between the molecules (rubberlike structure), (2) a layer of molecules, each having a strong internal structure (rigid molecules). With the latter model the required intermolecular interaction depends on the surface concentration. At low surface concentration the molecules must interact strongly to obtain high resistance against shear (model 2a). However, at high adsorbed amount interaction becomes less important. With a close-packed monolayer the individual rigid molecules must deform due to shearing (model 2b).

An adsorbed layer according to model (1) will be mainly elastic because mostly deformation of the chains between the crosslinks takes place. An adsorbed layer according to model (2a) will also be mainly elastic when the interaction between the molecules is strong, because upon shearing, the molecules have to deform. An adsorbed layer according to model (2b) allows the molecules to move along each other if the individual molecules deform to a certain extent. This moving of the molecules will be accompanied by breaking and restoring of intermolecular bonds. Consequently, a layer according to this model will be visco-elastic. The balance between viscous and elastic part will be determined by the nature of intra-molecular bonds. A high viscous part will be caused by weak intermolecular interactions which break and re-form during deformation.

The experimental results discussed above point in the direction of model 2b as the plausible model for rigid globular protein molecules. Summarizing, the arguments are the following:

1   The molecules are rigid
2   A high surface concentration is required for a measurable modulus
3   The modulus is viscoelastic
4   No strong indication for covalent crosslinks.

Additional indication for the latter argument can de deduced from (i) Wierenga et al [15] who found that using ovalbumin with extra sulfhydrylgroups to increase the chance for intermolecular S-S bridges did not result in a higher shear modulus and (ii) Hellebust at al [92] who analyzed the protein layers (skins) around so-called Ultrasound Microbubbles. This analysis revealed few covalent S-S bridges and a considerable amount of hydrophobic interaction.

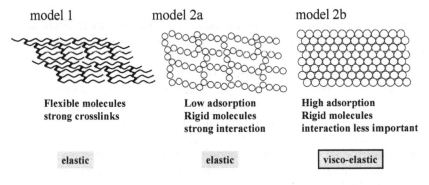

model 1   model 2a   model 2b

Flexible molecules   Low adsorption   High adsorption
strong crosslinks   Rigid molecules   Rigid molecules
                           strong interaction   interaction less important

elastic   elastic   visco-elastic

Fig. 15. Possible structures of an adsorbed protein layer with high shear modulus.

However, by inducing intermolecular cross-links by transglutaminase, covalent bonds that are formed significantly increase the shear resistance [93, 94]. A similar increase of the shear modulus with time, due to intermolecular bond formation, was observed for a photo-polymerisation reaction in an interface [95].

A similar model based on packing density has been described by Cicuta et al [23] and by Edwards and Wasan [96]. In these models, the resistance to shear deformation is the result of a decreased mobility of the particles due to the close packing.

2.6. Relation between dilational and shear modulus

According to the provisional model described in 4.2.5 it is likely that the shear modulus is determined by a thin 3-D network of protein molecules. In three-dimensional rheology [97], the shear modulus measured in extension of an incompressible system is three times higher than the modulus measured in *simple* shear. This means that dilational deformation of a surface carrying such a 3-D network should result in a dilational modulus, E, that is at least equal to three times the surface shear modulus, G:

$$E \geq 3G \tag{10}$$

if we assume the volume of the adsorbed layer to remain unchanged during the deformation. Table 4 compares the values of G and E for BSA, ovalbumin, PMApe and PVA at various concentrations and surface ages. A ratio of three is consistent with an adsorbed surface layer that behaves as a homogeneous gel layer, where G and E are determined by the same interactions between the adsorbed molecules. At high surface coverages (BSA$\geq$0.3g/l and ovalbumin at 0.3g/l) the limiting value of three for E/G is reached and in most cases even surpassed. Values of E/G<3 were only found under conditions when multilayers are formed (BSA, ovalbumin at high concentrations) or with systems that form extremely thick layers (PMApe).

These results suggest that only the first layer, which is in direct contact with the surface, affects the dilational modulus. In the adsorption region where multilayer adsorption takes place, the dilational modulus has reached a maximum value or even decreases somewhat. The effect of a second layer on the shear modulus is obvious, the deformed gel layer has become thicker.

Table 4 also shows that a low E combined with a low G is linked to adsorbed layers of flexible (random coil) molecules viz. Na caseinate and PVA. Adsorbed layers of rigid globular molecules viz. BSA and ovalbumin show a high E combined with a high G. This finding supports the idea that E and G are governed mainly by the same molecular parameters.

The idea that the rigidity of the molecular structure determines to a large extent the magnitude of both moduli is supported by the finding that in the presence of a denaturating agent (6M urea) the moduli of BSA and ovalbumin decrease to the value of Na caseinate.

Table.4 Comparison between shear- and dilational moduli for proteins and polymers. For some proteins the effect of denaturation agents is included; $\omega = 0.42$ s^{-1}

Protein	Conc. (g/l)	Age (h)	Layer Thick- ness (nm)	G (mN/m)	E (mN/m)	E/G
BSA	0.005	0.75		1.4	41	29
		2.75		5.2	56	11
		23		10.6	59	6
BSA	0.03	23	2.5	15	59	3.9
BSA	0.1	23		14	70	5
BSA + 6Murea	0.1	23		0.1	29	290
BSA	0.3	23	11	17	50	2.9
BSA	0.5	1.5		15.8	54	3.4
	0.5	23		24.8	58	2.3
Ovalbumin	0.1	2		11	80	3
	0.1	23	2.5	19	82	4.3
Ovalbumin + 6M urea	0.1	23		0.02	36	180 0
Ovalbumin	0.3	23	16	49	80	1.6
PMApe	0.1	23	25	56	29	0.5
PVA	0.4	23	10	<0.1	11	>11 0
Whole Casein	0.3	23	6	0.2	21	110

PMApe = copolymer of methacrylic acid and its methylester (molar ratio 2:1). and degree of esterification of 0.5.
PVA ex Kurashiki. (molecular weight determined from viscosity=42000).

From ellipsometry the ellipsometric thickness and concentration of the adsorbed layer can be inferred. This offers the possibility to compare the rheology of the adsorbed layer with the rheological behaviour of protein solutions at equal concentration [98]. The results discussed in [98] indicate that, even if we take into account the rather large experimental uncertainties, the shear modulus of the adsorbed layer is some orders of magnitude higher than that of the bulk solution. A possible explanation is that in the adsorbed layer the adsorbed molecules do not only interact in the same way as in a concentrated solution, but that due to the adsorbed segments forming anchor points to the surface which act as extra cross-links, the structure becomes more rigid. Another explanation is that due to conformational changes upon adsorption the interaction between the molecules and the rigidity of the molecules have changed.

2.7. Comparison between dilational and shear properties. Similarities and differences.

The following similarities and differences become apparent.
(i) For both moduli a minimum surface coverage has to be surpassed to obtain a non-zero value. This critical surface coverage is higher for the shear modulus, as shown in Fig. 16. Both moduli increase with further increase of the surface concentration. At a certain surface concentration, depending on protein type, the dilational modulus shows a clear plateau or maximum value, attributed to reconformation or collapse type phenomena. The shear modulus shows a clear plateau value as well. However, if the layer thickness increases further as a result of multilayer formation, the shear modulus also shows a further increase.
(ii) Both moduli become deformation-dependent at about the same, relatively high surface concentration (BSA; 1.2 mg/m^2). At these surface concentrations the dilational modulus has already nearly reached its maximum value, whereas the shear modulus is still very low.
(iii) For all proteins at surface concentrations up to about half saturation the dilational modulus at the given frequency was almost purely elastic. The modulus acquired a minor viscous contribution only at higher surface

concentrations (depending on protein type). In the same surface concentration range the viscous part of the shear modulus is considerable [18].

(iv) The increase of both moduli with time is primarily a surface concentration effect; see Fig. 14. Minor effects of surface age on the dilational modulus (decrease of viscous part) are only found with protein layers adsorbed from higher bulk concentrations (0.1-1g/l). These effects were ascribed to reconformation which is slow at high surface coverage. This slow reconformation may also have a small increasing effect on G.

E. SUMMARY

The rheology of proteinaceous interfacial layers requires special attention to the measuring methodology. For dilational properties this is because these layers can have a considerable resistance against shear deformation. Therefore, measurement of reliable dilational properties requires a method that ensures isotropic area deformations. The square-band barrier (Fig. 2) and the ring trough were designed for this purpose. A more convenient instrument is the drop tensiometer which, in addition, facilitates experiments at oil/water interfaces. As this method allows the simultaneous measurement of Π, E and the viscous phase angle ϕ during the course of the adsorption process, maximum information can be obtained on the development of these layers.

For shear properties, poor reproducibility is the major problem. Insufficient control of layer formation and adherence of layer to measuring body are most likely responsible.

Effects of protein concentration, adsorption time and pH on the rheological properties can all be traced back to the surface concentration (Γ). Another dominating parameter that determines the magnitude of both dilational and shear moduli is the rigidity of the protein molecule.

For all proteins examined E(Π) initially increases linearly. Initial slope and especially the extent of the linear range increases with increasing rigidity of the molecules (PVA<β-casein< globular proteins) and decreasing polarity of the interface (TAG<TD<Air).

Zero phase angles in the linear range, measured at oscillation frequencies of 0.1 Hz and higher, imply that the measured moduli were pure elasticities directly reflecting the surface equation of state. This indicates that the surface pressure adjusts "instantaneously" to the changing adsorption during the compression expansion cycle.

A simple 2-D solution model with a single molecular area of the protein described the data only at fairly low pressures. Much better agreement over the entire elastic range was found with a recently extended version of the model, which accounts for adsorbed proteins adopting smaller molecular areas with increasing surface pressure.

For each protein at different interfaces, the elasticity increased with the interaction parameter H^σ of the equation of state. Elasticity and H^σ both increased with the clean-interface tension, γ^0. At each interface, the elasticity of the different proteins increased with increasing rigidity of the protein molecules, i.e., with decreasing molecular compressibility expressed as variability of the molecular area in the equation of state. The sequence of increasing elasticity was PVA < β-casein < BSA < BLG ≈ ovalbumin

At the three interfaces investigated, within the linear range, equilibration within the surface is much faster than between surface and bulk solution. The characteristic time for reconformations of protein molecules upon adsorption is less than one minute.

Visco-elastic behaviour was found only at high pressures, i.e., in densely packed surfaces. The measured viscous phase angles strongly decreased at still higher pressures, indicating that the rate of the active relaxation mechanism decreased with increasing molecular packing density.

In view of the high Γ needed for a measurable shear modulus G, its high viscous part, the strong effect of molecular rigidity on G and the strong intermolecular interaction as deduced from the dilational properties lead to a layer model in which a high G is only possible if (i) the layer is close-packed, (ii) the molecules are rigid and (iii) the intermolecular bonds reversible. In such a layer, the increase with time of G is primarily a surface coverage effect.

Such a layer can be modelled as a thin homogeneous gel layer (E ≥ 3 G). Exceptions with lower ratios E/G are related to multilayer formation. Higher ratios, ranging from 3 to roughly 1000, are predominantly found for the globular proteins BSA and ovalbumin at low adsorptions and for the flexible molecules PVA and casein.

The interactions between adsorbed protein molecules are electrostatic, hydrogen bridge and steric effects. There is little experimental evidence for a significant contribution of covalent intermolecular S-S bonds.

F. REFERENCES

1. M. van den Tempel, Proceedings of the 3rd International Congress on Surface Active Agents, Vol. II, Cologne, Verlag der Universitätsdruckerei Mainz, Germany, (1960) 573.

2. J. Lucassen and M. van den Tempel, Chem. Eng. Sci., 27 (1972) 1283.

3. J.A. De Feijter and J. Benjamins, J. Colloid Interface Sci., 70 (1979) 375.

4. E. Dickinson, B.S. Murray and G. Stainsby, J. Chem. Soc. Faraday Trans. 1, 84 (1988) 871.

5. R. Miller, R. Wüstneck, J. Krägel and G. Kretzschmar, Colloids Surfaces A, 111 (1996) 75.

6. E.H. Lucassen-Reynders, Dynamic Interfacial Properties in Emulsification, Vol. 4, Marcel Dekker Inc, New York, 1996.

7. F.C. Goodrich, L.H. Allen and A. Poskanzer, J. Colloid Interface Sci., 52 (1975) 201.

8. V. Mohan, B.K. Malvija and D.T. Wasan, Can. J. Chem Eng 54 (1976) 515.

9. D.E. Graham and M.C. Phillips, J. Colloid Interface Sci., 76 (1980) 240.

10. A. Martinez-Mendoza and P. Sherman, J. Dispersion Sci. Techn., 11 (1990) 34

11. V. Kiosseoglou, J. Dispersion Sci. Techn., 13 (1992) 135.

12. J. Benjamins and F.van Voorst-Vader, Colloids Surfaces A, 65(1992)161.

13. B.S. Murray and E. Dickinson, Food Science Techn. Int., 2(3) (1996) 131.

14. S.A. Roberts, I.W. Kellaway, K.M.G. Taylor, B. Warburton and K. Peters, Langmuir, 21 (2005) 7342.

15. P.A. Wierenga, H. Kosters, M.R. Egmond, A.G.J. Voragen and H.H.J. de Jongh, Adv. Colloid Interface Sci., 119 (2006) 131.

16. A.H. Martin, M.A.C. Stuart, M.A. Bos and T. van Vliet, Langmuir 21 (2005) 4083.

17. J. Benjamins and E.H. Lucassen-Reynders. in Studies in Interface Science, D. Möbius and R. Miller (Eds.), Vol. 7, pp. 341-384, Elsevier Science B.V., Amsterdam 1998.

18. J. Benjamins, Static and Dynamic Properties of Proteins Adsorbed at Liquid Interfaces, Ph. D. Thesis, Wageningen University, Wageningen, 2000.

19. M.A. Bos and T. van Vliet, Adv. Colloid Interface Sci., 91 (2001) 43

20. G.A. van Aken and M.T.E. Merks, Progr. Colloid Polymer Sci., 97 (1994) 281.

21. D.E. Graham and M.C. Phillips, J. Colloid Interface Sci., 76 (1980) 227.

22. J. Benjamins, A. Cagna and E.H. Lucassen-Reynders, Colloids Surfaces A, 114 (1996) 245.

23. P. Cicuta and E.M. Terentjev, The Eur. Phys. J. E, 16 (2005) 14

24. J. Benjamins and E.H. Lucassen-Reynders, Special Publication - Royal Society of Chemistry, 284 (2003) 216.

25. J. Benjamins, J. Lyklema and E.H. Lucassen-Reynders, Langmuir 22 (2006) 6181.

26. E.M. Freer, K.S. Yim, G.G. Fuller and C.J. Radke, Langmuir 20 (2004) 10159.

27. J. Maldonado-Valderrama, V.B. Fainerman, M.J. Galvez-Ruiz, A. Martin-Rodriguez, M.A. Cabrerizo-Vilchez and R. Miller, J. Phys. Chem. B, 109 (2005) 17608.

28. J.M.R. Patino, J.M.N. Garcia and M.R.R. Nino, Colloids Surfaces B, 21 (2001) 20

29. F.v. Voorst-Vader and F. Groeneweg. in Food Colloids, R.D. Bee, P.R., J. Mingins (Eds.) Royal Society of Chemistry, London 1989, pp. 218.

30. J.W. Gibbs, Trans. Connecticut Acad., III: (1878) 343-524. Reprinted in: The scientific papers of J.W. Gibbs, Volume 1, Dover Publications Inc., New York. (1961).
31. J. Lucassen and M.van den Tempel, J. Colloid Interface Sci., 41 (1972) 491.
32. J. Lucassen and D. Giles, Journal of Chemical Society, Faraday Trans. 1, 71 (1975) 21
33. L. Ting and K.M. D.T. Wasan, J. Colloid Interface Sci., 107 (1985) 345.
34. K.D. Wantke, K. Lunkenheimer and C. Hempt, J. Colloid Interface Sci., 159 (1993) 28.
35. M.A. Cohen-Stuart, J.T.F. Keurentjes, B.C. Bonekamp and J.G.E.M. Fraaye, Colloids Surfaces A, 17 (1986) 91.
36. B.R. Malcolm, J. Colloid Interface Sci., 104 (1985) 520.
37. J.C. van der Pas and A. Prins., unpublished results (1973).
38. J. Benjamins, J.A. De Feijter, M.T.A. Evans, D.E. Graham and M.C. Phillips, Faraday Discussions Chem. Soc., 59 (1976) 218-29.
39. J.J. Kokelaar, A. Prins and M.d. Gee, J. Colloid Interface Sci., 146 (1991) 50
40. A. Cagna, G. Esposito, C. Rivière, S. Housset and R. Verger. in 33rd Intern. Conf. on Biochemistry of Lipids, Lyon, France 1992.
41. Y. Rotenberg, L. Boruvka and A.W. Neumann, J. Colloid Interface Sci., 93 (1983) 169.
42. C.A. McLeod and C.J. Radke, J. Colloid Interface Sci., 160 (1993) 435.
43. R. Nagarajan, K. Koczo, E. Erdos and D.T. Wasan, AIChE Journal 41 (1995) 915.
44. R. Miller, P. Joos and V.B. Fainerman, Adv. Colloid Interface Sci., 49 (1994) 249.
45. S. Labourdenne, N. Gaundry-Rolland, S. Letellier, M. Lin, A. Cagna, G. Esposito, R. Verger and C. Rivière, Chemistry and Physics of Lipids 71 (1994) 163.
46. M.E. Leser, S. Acquistapace, A. Cagna, A.V. Makievski and R. Miller, Colloids Surfaces A, 216 (2005) 25.

47. E.M. Freer, H. Wong and C.J. Radke, J. Colloid Interface Sci., 282 (2005) 128.

48. E. Dickinson, B.S. Murray and G. Stainsby, J. Colloid Interface Sci., 106 (1985) 259.

49. J.A. De Feijter, J. Colloid Interface Sci., 69 (1979) 375.

50. J. Krägel, S. Siegel, R. Miller, M. Born and K.-H. Schano, Colloids Surfaces A, 91 (1994) 169.

51. F.S. Ariola, A. Krishnan and E.A. Vogler, Biomaterials 27 (2006) 3404.

52. W. Norde, Adv. Colloid Interface Sci., 25 (1986) 26

53. E.H. Lucassen-Reynders, J. Lucassen, P.R. Garrett, D. Giles and F. Hollway, Adv. Chem. Ser., E.D. Goddard (Ed.), 144 (1975) 275.

54. J.A. De Feijter and J. Benjamins. in Special Publication - Royal Society of Chemistry, Food Emulsions Foams, Vol. 58, pp. 72-85 198

55. M. van den Tempel and E.H. Lucassen-Reynders, Adv. Colloid Interface Sci., 18 (1983) 281.

56. E.H. Lucassen-Reynders and J. Benjamins, Special Publication - Royal Society of Chemistry 227 (1999) 195.

57. E.H. Lucassen-Reynders, Colloids Surfaces A, 91 (1994) 79.

58. E.H. Lucassen-Reynders and J. Benjamins, to be submitted.

59. E.H. Lucassen-Reynders, Progress in Surface Membrane Science 10 (1976) 253.

60. V.B. Fainerman and E.H. Lucassen-Reynders, Adv. Colloid Interface Sci., 96 (2002) 295.

61. L.T. Minassian-Saraga and I. Prigogine, Mem. Service Chimie Etat, 38 (1953) 109.

62. H.L. Frisch and R. Simha, J. Chem. Phys., 24 (1956) 652.

63. J.M.H.M. Scheutjens and G.J. Fleer, J. Phys. Chem., 83 (1979) 1619.

64. G.J. Fleer, J.M.H.M. Scheutjens and M.A.C. Stuart, Polymers at Interfaces, Chapman & Hall, London 1994.

65. P. Cicuta and I. Hopkinson, J. Chem. Phys., 114 (2001) 8659.

66. A. Hambardzumyan, V. Aguie-Beghin, I. Panaiotov and R. Douillard, Langmuir, 19 (2003) 72.

67. R. Douillard, M. Daoud and V. Aguie-Beghin, Current Opinion Colloid Interface Sci., 8 (2003) 380.

68. V.B. Fainerman, E.H. Lucassen-Reynders and R. Miller, Adv. Colloid Interface Sci., 106 (2003) 23

69. J.A. De Feijter and J. Benjamins, J. Colloid Interface Sci., 90 (1982) 289.

70. V.B. Fainerman, S.V. Lylyk, A.V. Makievski and R. Miller, J. Colloid Interface Sci., 275 (2004) 305.

71. J.W. Benjamins, B. Jonsson, K. Thuresson and T. Nylander, Langmuir, 18 (2002) 6437-6444.

72. E.H. Lucassen-Reynders, V.B. Fainerman and R. Miller, J. Phys. Chem. B, 108 (2004) 9173.

73. R. Maksymiw and W. Nitsch, J. Colloid Interface Sci., 147 (1991) 67

74. H.B. Bull, J. Colloid Interface Sci., 41 (1972) 305.

75. F. MacRitchie, Adv. Colloid Interface Sci., 25 (1986) 341.

76. G. Serrien, G. Geeraerts, L. Ghosh and P. Joos, Colloids Surfaces A, 68 (1992) 219.

77. J.R. Hunter, P.K. Kilpatrick and R.G. Carbonell, J. Colloid Interface Sci., 142 (1991) 429.

78. R.A. Ganzevles, K. Zinoviadou, T. van Vliet, M.A.C. Stuart and H.H.J. de Jongh, Langmuir, 22 (2006) 10089.

79. E. Dickinson, S.R. Euston and C.M. Woskett, Progress in Colloid Polymer Sci., 82 (1990) 65.

80. J. Chen and E. Dickinson, Food Hydrocolloids, 9 (1995) 35.

81. D.C. Clark, P.J. Wilde, D.J.M. Bergink-Martens, A.J.J. Kokelaar and A. Prins., Food Colloids and Polymers: Stability and Mechanical Properties. E. Dickinson (Ed.), (1993) 354.

82. T. van Vliet, private communication.

83. D.J. Burgess and N.O. Sahin, J. Colloid Interface Sci., 189 (1997) 74.

84. D.J. Burgess and N.O. Sahin, Structure and Flow of Surfactant Solutions. ACS Symposium. Series, 578 (1994) 380.

85. J.V. Boyd, J.R. Mitchell, L. Irons and P.S. P.R. Musselwhite, J. Colloid Interface Sci., 45 (1973) 478.

86. A.H. Martin, M.A. Bos, M.A. Cohen-Stuart and T. van Vliet, Langmuir, 18 (2002) 1238.

87. J. Boyd and P. Sherman, J. Colloid Interface Sci., 34 (1970) 76.

88. H.J. Holterman. Twente University, the Netherlands 1989.

89. A.A. Trapeznikov and N.N. Loznetsova, Kolloidnyi Zh., 47 (1985) 553.

90. A.K. Kenzhebekov and V.N. Izmailova, Vestn. Moscow Univ. Ser.2 (Khim.) 24 (1983) 27

91. M. Joly, in Surface and Colloid Science, E. Matievic (Ed.), Wiley-Interscience, New York 1972, pp. 79.

92. H. Hellebust, C. Christiansen and T. Skotland, Biotechnology And Applied Biochemistry 18 (1993) 227-23

93. M. Færgemand and B.S. Murray, Journal of Agricultural and Food Chemistry 46 (1998) 884.

94. M. Færgemand, B.S. Murray, E. Dickinson and K.B. Qvist, Intern. Dairy J., 9 (1999) 343.

95. H. Rehage and M. Veyssié, Angew. Chemie, 29 (1990) 439.

96. D.A. Edwards and D.T. Wasan, Chem. Eng. Sci., 46 (1991) 124

97. M. Reiner, Deformation and Flow, Lewis, H.K., London 1949, pp. 170.

98. F.J.G. Boerboom, A.E.A. de Groot-Mostert, A. Prins and T. van Vliet, Netherlands Milk & Dairy Journal 50 (1996) 183.

VISCO-ELASTICITY OF MIXED SURFACTANT-POLYMER MONOLAYERS

Dominique Langevin

Laboratoire de Physique des solides, Université Paris sud,
Orsay, FRANCE

Contents

A. INTRODUCTION

When monolayers of amphiphilic molecules are present at a liquid-gas or a liquid-liquid interface, surface tension alone cannot fully describe their response to external perturbations. In addition, several surface visco-elastic parameters need to be introduced, shear and dilational, as introduced in former chapters. Monolayer visco-elasticity, also called monolayer rheology, plays important roles in many practical applications : spray, coating and more generally wetting and dewetting, foaming and emulsification, foam and emulsion stability, Langmuir-Blodgett deposition, liquid-liquid extraction, two-phase flow, etc. In these applications, the liquids are frequently aqueous solutions of surfactants, for the control of surface properties, and polymers, for the control of bulk rheology. Most polymers soluble in water being polyelectrolytes, the majority of existing studies were made on aqueous solutions of surfactants and polyelectrolytes as model systems.

The preceding chapters showed that it is very difficult to investigate the visco-elastic behaviour of monolayers in a fully controlled manner. The mixed polymer-surfactant layers have in addition an unusual behaviour: long equilibration times are found, the adsorption is irreversible and the response to large compressions is highly non-linear. It is therefore especially difficult to obtain reliable determinations of the visco-elastic moduli. This is why extremely few data on the rheology of polymer-surfactant mixed layers can be found in literature.

In the following, we will describe these few experimental data. We will end by discussing the relation between surface visco-elasticity and practical applications.

B. SHEAR VISCO-ELASTICITY

Surface shear properties can be qualitatively probed by very simple tests : looking at the motion of talc particles moving on the surface under the action of air currents. Regismond et al. used this method with several mixed solutions : cationic cellulose ether polymer (JR400) with various alkylsulphate and alkyl sulfonate surfactants, various anionic polymers with alkyl trimethyl ammonium halides surfactants and neutral polymers with either anionic or cationic surfactants. This allowed them to qualitatively distinguishing between fluid, viscous and visco-elastic monolayers [1].

Popular devices for the determination of the surface shear moduli are oscillating disks devices, commercial instruments being now available. The resolution depends on the size of the disk and is typically of the order of 10^{-3} mN.s/m. A new generation of instruments appeared recently, in which magnetic disks or needles floating at the surface are submitted to magnetic torques, instead of being attached to torsion wires. Smaller deformations can thus be detected (strains down to 10^{-7}) and very small shear moduli can be measured.

Experiments using magnetic needles were performed with mixed solutions of dodecyl trimethyl ammonium bromide (DTAB) and polystyrene sulfonate (PSS). In these experiments, a concentration of PSS monomers equal to 2.4 mM was used (i.e., 0.5 g/l of polymer)[2]. In Fig. 1, the values of G' and G'', storage and loss shear surface moduli, measured at a frequency of 0.15 Hz are shown as a function of surfactant concentration C_s. The dotted line in Fig. 1 represents ellipsometry data obtained for the same system and discussed in [3]. G' and G'' had very close values over the entire range of surfactant concentrations investigated, and G' was higher than G'', which is characteristic of an elastic gel-like layer. Both were independent of strain, which in these experiments remained smaller than 3%.

G' and G'' were nearly zero up to a surfactant concentration of 1.45 mM. At this point there was a sharp rise to a maximum of the order of 1 mN/m at 2.4 mM DTAB. The maxima of both elasticity and ellipticity occurred close to the concentration C_0 where the amount of positive surfactant charges compensates the negative charges of the polymer in bulk. The surfactant-polymer bulk aggregates are therefore about neutral and partly hydrophobic because of the polymer styrene groups and the surfactant chains: their bulk solubility is therefore limited, and surface adsorption is favoured. When the surfactant concentration increases further, the solubility of the aggregates should increase and surface adsorption decrease. However, the maximum of adsorption was slightly shifted to higher surfactant concentration in this system. This is probably because the precipitation boundary is close (4mM).

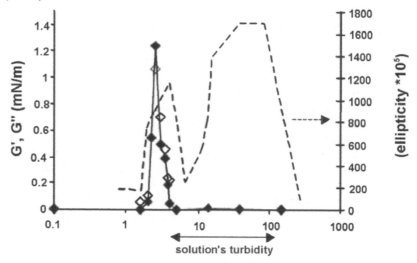

Fig. 1. Storage and loss surface shear moduli, *G'* and *G"* (closed and open symbols, respectively) and ellipticity as a function of DTAB concentration for solutions containing 2.4 mM PSS. The frequency is 0.15 Hz. Reprinted from ref 2 with permission.

In polyelectrolyte-surfactant systems of opposite charge, precipitation does not occur at C_0, because it is not solely controlled by electrostatic effects, hydrophobic effects playing also an important role. Above C_0, more surfactant binds to the polymer, the aggregates become more hydrophobic and adsorb more at the surface, until they become insoluble and precipitate out of the solutions. A sharp drop in adsorption is seen when precipitation starts due to the loss of active material in the precipitated phase. For PSS-DTAB, the peak in the shear moduli was narrower than the first peak in ellipticity, and the value of *G'* already decreased to 0.2 mN/m for a 4 mM DTAB-PSS solution. The maximum of the visco-elastic parameters did not coincide by the maximum in adsorption, maybe because away from neutrality, the surface gel is less entangled .

When C_s was further increased, the precipitate partially resolubilized above 11.5 mM. The bulk aggregates being more charged, they are soluble again; being surface active, they adsorb at the air-water interface, producing a second rise of ellipticity, which peaked near 50 mM DTAB. For much larger C_s, about 300 mM DTAB, the PSS chains were completely solubilized with DTAB in water, and the surface was covered with a monolayer of surfactant only. In the precipitation region, the surface moduli were small, as expected since adsorbed amounts were low. More unexpectedly, the shear surface moduli remained equal to zero between 10 and 100 mM surfactant, where ellipsometry measurements indicated that a thick surfactant-polymer layer is adsorbed at the surface. In neutron reflectivity studies of the same system, Taylor et al. have attributed the second peak in ellipticity to the formation of a structured layer of PSS and DTAB, where a DTAB monolayer is coupled to a PSS subsurface layer, itself bound to DTAB bilayers or micelles [4]. The low values of the surface shear moduli indicate that the subsurface layer is rather mobile and not highly entangled with the surface thus offering little resistance to interfacial flows.

Close to their maxima, G' and G'' varied as a $G' \sim \omega^{n'}$ and $G'' \sim \omega^{n''}$, n' and n'' ranging between 0.3 and 0.6. In general, the lower the exponents n, the more solid-like the material is. Here, Monteux et al. concluded that the adsorbed layers were visco-elastic semi-solids, whose rigidity results from chain entanglements and hydrophobic interactions. Again close to their maxima, G' and G'' increased logarithmically with time. This behaviour is similar to what is observed during aging of jammed systems, such as colloidal gels or polymer glasses, and indicates that slow relaxation mechanisms take place in the surface layer. To probe the slower relaxation mechanisms, creep experiments were performed: a sudden and constant shear stress was applied to the surface, and the displacement of the needle was followed with time. From these experiments, relaxation times less than 100s where found, much smaller than those in other two-dimensional polymer networks, held by covalent links [5] or hydrogen bonds [6]. This comparison suggests that the

DTAB-PSS layers are probably soft physical gels, in which the connections between polymer chains are simple entanglements.

C. DILATIONAL VISCO-ELASTICITY

Dilational properties have been investigated with devices in which capillary waves are excited. Other methods such as oscillations of bubbles or drops and oscillations of barriers in Langmuir troughs were also used, in which the amplitude of the surface deformation is in general much larger than in surface waves methods. The visco-elastic coefficients show large frequency variations in the case of soluble monolayers, because of the important coupling with the bulk. Here, once adsorbed, the monolayers behave as insoluble (they can be compressed reversibly), and the frequency variations are less important. It should be noted that the intrinsic elasticity can be frequency dependent as a result of relaxation processes in the surface (such as surfactant or polymer chain reorientation) and be equal to the Gibbs elasticity E_G only at low frequencies. This frequency dependence is associated to a non-zero intrinsic surface viscosity. For soluble monolayers, the frequency dependence of the intrinsic moduli is smaller than the frequency dependence associated to exchanges between surface and bulk, and the elasticity at infinite frequency E_∞ is usually identified to the Gibbs elasticity E_G. In the case of nonionic surfactants, the diffusion-controlled model [7] for exchanges between surface and bulk is generally in good agreement with experiments [8]. In the case of ionic surfactants, the diffusion process is affected by the surface charges, and the theory is more complicated, but again in good agreement with experiments [9]. Let us also mention that the calculation of the Gibbs elasticity E_G is very delicate, and that the use of an equation of state (empirical relation between surface tension γ and surface concentration Γ) can lead to large errors, even in the simple case of nonionic surfactants [10]. The diffusion-controlled models lead to a maximum of elasticity at a particular surfactant concentration. Maxima of elasticity are also found with polymer-surfactant solutions. However, the concentration of the maximum is much smaller than that predicted by the models. Because the polymer-surfactant layers, once formed, are insoluble, the diffusion-controlled models cannot in fact be used. E_G can in principle be calculated from the derivative of surface pressure with respect to surface area. However, the mixed layers frequently show a large surface pressure hysteresis during compression-expansion cycles, and appreciable time dependences, even after 24 hours, rendering difficult the determination of E_G [11].

1. Excited capillary waves

In most excited capillary wave devices, the waves are excited by electrocapillarity. Typical frequencies ω are between 100 and 1000Hz. In these experiments, the strain is very small, of the order of 10^{-3}: the linearity of the response can be checked by varying the excitation voltage and by analysing the signal of the harmonics $\omega/2$, 2ω, etc.

Series of experiments were made with aqueous solutions containing anionic polymers and DTAB [12] or CTAB (hexadecyl trimethyl ammonium bromide) [11]. Polymers were either xanthan, a polysaccharide with a rigid backbone due to a double helix configuration, and synthetic flexible polymers, copolymers of poly-acrylamido propane sulfonate (AMPS) and poly-acrylamide (AM), with two different ratio of AMPS monomers, 10% or 25%; in the following, these polymers will be denoted PAMPS10 and PAMPS25. The equilibrium surface tensions γ_{eq} of the solutions exhibit plateaus after a concentration which marks the onset of polymer-surfactant aggregation in bulk. This concentration is called critical aggregation concentration (CAC). Provided the polymer concentration is moderate, the surface tension and the layer thickness are independent of the polymer concentration [13]. This behaviour can be explained by a thermodynamic adsorption model based on the exchange of the surfactant counterions by the polymer ions at the surface. The behaviour of DTAB-PSS solutions is very different [14]. First, for given surfactant and polymer concentrations, the decrease of surface tension is much smaller for PSS. Second, there is no evidence of surface tension saturation above a certain polymer concentration, the apparent CAC depends on PSS concentration and is much larger than that determined with specific electrodes (0.02 mM instead of 2.4mM for 0.5g/l PSS)[3]. For a given surfactant concentration, the surface tension decreases upon addition of PSS and increases again after a minimum close to C_0. Obviously, the simple adsorption model valid with PAMPS fails for PSS. PSS is known to take only extended configurations when all the monomers are charged. If the fraction of ionized monomers is less than 1, the chains adopt a pearl-necklace configuration, where the neutral monomers associate into hydrophobic regions similar to surfactant micelle interiors [15]. In the presence of surfactant that neutralizes the PSS charges, such a structure could also form, incorporating the surfactant chains as well. The resulting complexes could be more hydrophilic than those of PAMPS and DTAB, in which the surfactant chains could be dangling. This could perhaps explain the absence of surface tension decrease in the CAC region for PSS.

The results of visco-elasticity measurements for PAMPS and xanthan systems are shown in Fig. 2. The results for solutions with only DTAB are also shown in Fig.3 for comparison. At low surfactant concentrations, the values of the real and imaginary parts of the visco-elastic moduli E' and E'' were small, and sometimes E'' was negative. Negative values of E'' were already reported before [16]. They are not physically absurd, because the surface quantities are excess quantities and a negative E'' only means that the damping of the surface waves is smaller than on the pure substrate. However, up to now, no satisfactory explanation for this surprising behaviour has been found. As C_s is increased, the values of the two coefficients increased, reached a maximum and then decrease again. This is as predicted by the diffusion model, which takes into account the dissolution in the bulk [7]. But the peak in the elasticity was at a much lower value of C_s for the surfactant-polymer solutions, as compared to the pure surfactant solutions (see Fig. 3), and has a different origin. Indeed it occurred close to C_0, as for the shear moduli of the PSS system, probably for similar reasons. Note that in DTAB-xanthan and DTAB-PAMPS systems, C_0 was smaller than the CAC.

The shape of the curves for E' and E'' were similar for PAMPS10 and PAMPS25, but the values of E' are larger for PAMPS25. PAMPS10 is less well coupled to the surface due to the presence of fewer charged groups and forms larger loops beneath [12].

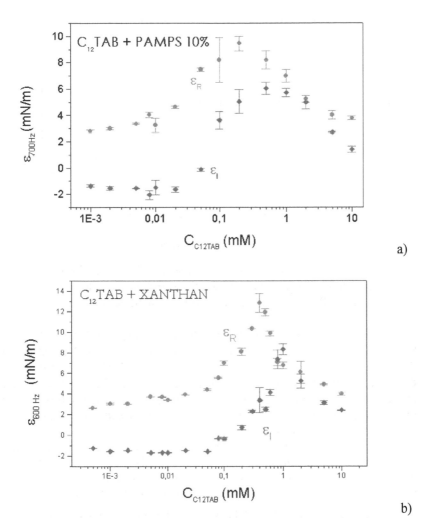

a)

b)

Fig. 2 Surface dilational visco-elasticity as a function of surfactant concentration for DTAB solutions with (a) xanthan and (b) PAMPS10. Reprinted from ref 12a with permission.

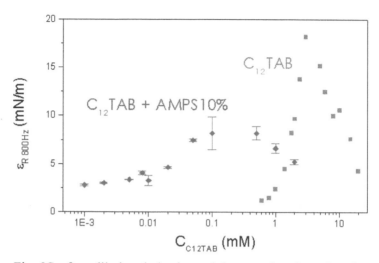

Fig. 3Surface dilational elastic modulus as a function of surfactant
concentration for pure DTAB solutions and mixed solutions with
PAMPS10. Reprinted from ref 12a with permission.

This may be the reason for the lower elasticity of the solutions.
Similar results were obtained for the mixed solutions CTAB-PAMPS. In
this case, two maxima were observed, one below *CAC*, and one close to
C_0. Here, C_0 was larger than the *CAC*, and the competition between bulk
aggregates and surface layers is more complex, but again possible to
rationalize, taking into account the balance between electrostatic and
hydrophobic interactions.

Devices where standing capillary waves (at the difference of the
above, where propagating waves were studied) were developed more
recently. In these experiments, the waves are excited by vibrating the
whole liquid through at frequencies between 10 and 100 Hz. The strain
was also very small, of the order of 10^{-3}. Experiments were performed
with a surface covered by a cationic lipid (insoluble) monolayer on pure
water or on a subphase containing $10^{-8}M$ DNA. The surface
concentration of the lipid was half the maximum concentration, and a
large elastic modulus was found on pure water, of the order of 50mN/m.

In the presence of DNA in the subphase, a mixed monolayer was formed and the elasticity drops by a factor 5, attributed to the monolayer expansion induced by DNA [17].

2. Methods involving large surface deformations

The oscillating bubble or drop method is becoming increasingly popular, commercial instruments being now available. It requires very small amounts of fluid, much smaller than with the waves methods. The frequency is generally small, below 1Hz, but recent instruments produce data up to 500Hz. The typical strain is large, of the order of 10^{-2} to 10^{-1}. In this method, the geometry is radial and pure dilational visco-elastic moduli E_d are determined, at the differences of the wave techniques described above, or of oscillating barriers in Langmuir troughs, where one measures $E = E_d + G$. The use of large amplitudes allows investigating the non-linear behaviour of monolayers, a topic little explored up to now. The method was tested in detail with the DTAB-PAMPS system [18].

2.1. Dynamic surface tension measurements

A typical result of a surface tension measurement performed keeping the bubble volume constant is shown in Fig. 4. The horizontal dotted line corresponds to the value of γ_{eq} measured with a Wilhelmy plate. The time required to reach equilibrium was extremely long, much longer than with the Wilhelmy plate (of the order of 100s) [12]. Second, the time variation of the surface tension depended on the bubble volume and history, unlike what is observed for pure surfactant solutions. Finally, the equilibrium values appeared lower than γ_{eq}. Because volume control implies regular injection (or aspiration) of air, frequent changes in volume occurs, meaning rapid dilatation (or compression) of the surface; this produces new adsorption (or compression of the already adsorbed layer), and continuous uncontrolled changes of the local surface concentration. When the volume (or area) control was suppressed, although the time necessary to reach equilibrium was still longer than with the Wilhelmy plate, the time variation of the surface tension was independent of the bubble volume and the equilibrium value was compatible with the Wilhelmy determination.

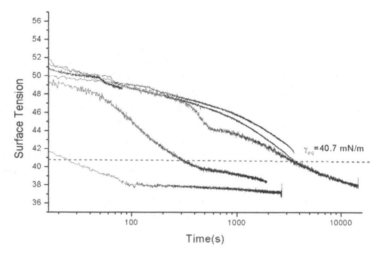

Fig. 4 Dynamic surface tension measurements in which the bubble volume is kept constant. The different curves correspond to different bubble volumes (between 5 and 6μl). Reprinted from ref 17b with permission.

It was already observed that the adsorption kinetics could be faster with the Wilhelmy plate than with the bubble method, for instance in cases of dilute ionic surfactant solutions. Adsorption is slowed down by the formation of a charged surfactant layer at the surface that prevents new surfactant molecules to adsorb: surfactant accumulates below the surface at a distance of the order of the Debye length, and it is only when the concentration in this region reaches a threshold that further adsorption takes place [19]. It is possible that the measurement with the Wilhelmy plate disturbs this region (the plate is moved upwards in the solution) and that the resulting convection produces mixing in the surface region.

Differences between measurements using drops or bubbles have also been reported by Noskov et al. with mixed systems containing DTAB and copolymers AMPS-NIPAM. [20]

2.2. Surface visco-elasticity measurements

When volume oscillations were produced, the equilibrium value that was reached afterwards was also frequently changed, especially if the oscillations were started a long time after the formation of the bubble. When they were started just after bubble formation, the equilibrium surface tension value was reached faster and was closer to the Wilhelmy plate value. The visco-elastic parameters did not seem to be affected by the way in which the surface tension approached equilibrium. Provided oscillations were started when the surface tension has decreased by a few mN/m, the visco-elastic parameters did not vary appreciably with time. For instance in the case of Fig. 5, the elastic modulus determined in region 'a' was (26 ± 1) mN/m and in region 'h', (28 ± 1) mN/m.

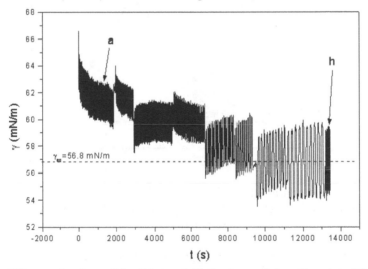

Fig. 5 Determination of the "dynamic" elastic modulus. Reprinted from ref 17b with permission.

The measured values were close even if the surface tension values at which the experiment was started were different and either far or close to the equilibrium value. This behaviour obviously facilitates the visco-elasticity determinations.

The elastic moduli measured at the frequency 0.3 Hz for the system PAMPS10-DTAB are shown on Fig. 6. The elastic moduli were larger for PAMPS25, and the viscous moduli were about ten times smaller for both polymers.

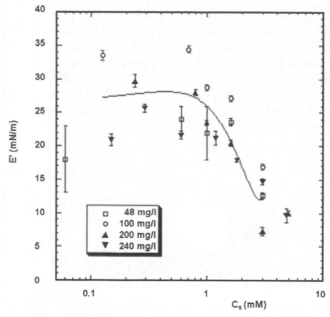

Fig. 6 Surface dilational elastic modulus for mixed aqueous solutions of PAMPS10 and DTAB. The frequency is 0.3Hz. Reprinted from ref 17b with permission.

These E' values were much larger than those measured with wave techniques, although these techniques probe higher frequencies and lead to the sum of dilational and shear moduli. The frequency variation of the elastic moduli was investigated: in some cases, E' decreased when the oscillation frequency increased, in others, it increased. When the frequency was decreased progressively, and at the end of the measurements a determination of the modulus was made again at a large frequency, the E' was found generally higher, by factors up to 2.

Fig. 7 shows Brewster angle microscopy images of mixed polymer-surfactant layers before and after compression: the layer buckled, and it took many hours before the compressed layer became again homogeneous in thickness. This strongly non-linear visco-elastic behaviour might be responsible for the anomalies seen with bubble measurements. The same non-linear behaviour could also explain why the visco-elastic moduli were larger at low frequencies. If the response to a compression of a system is linear, the elastic modulus should increase with increasing frequency. The moduli measured with wave techniques indeed increased with increasing frequency as the deformation amplitudes were small and the response linear. At larger amplitudes such as in the oscillating bubble technique, when the response is non-linear, there is no theoretical argument preventing the moduli to increase with decreasing frequency. The history dependence has probably a similar origin. These mixed layers behave as soft glassy two-dimensional systems, for which elastic moduli increase with the amplitude of the mechanical constraint, until they eventually yield.

The system DTAB-PSS was also studied with the oscillating drop method by Monteux et al. [2]. Fig. 8 presents the values of the dilational storage and loss surface moduli (full and open triangles, respectively) as a function of the DTAB concentration. The frequency of oscillations was 0.05 Hz and the relative amplitude of the deformation was 3%. A maximum in the surface dilational moduli of about 35mN/m was measured again close to C_0 and occurred over a very narrow range of surfactant concentrations. Noskov et al. also studied this system using both the oscillating barrier and drop methods [21]. In the oscillating barrier experiments, the elastic modulus was already large before C_0, of the order of 90mN/m at 0.1mM DTAB and 2.4mM PSS. The difference cannot be due to the fact that in the experiment, the sum $E'+G''$ was measured, because G' is non-zero only near C_0. The oscillating drop measurements were performed at a lower polymer concentration and led to smaller elasticity values. Monteux et al. mention that although they have performed a large number of experiments, the absolute values of the moduli were scattered. This fact and the large difference with the results of Noskov et al. [21] are probably associated to non-linearities and/or hysteresis effects as discussed earlier.

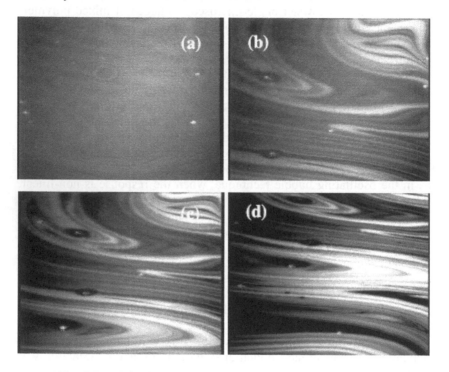

Fig. 7 BAM images under compression. Aqueous solutions of PAMPS25 (200 mg/l) and DTAB at the CAC. The initial surface area was reduced by (a) 0%; (b)25%; (c) 50% and (d) 85%. Reprinted from ref 17b with permission.

Noskov et al. found different types of behaviour with the PAMPS/NIPAM-DTAB system : non monotonous time variation of E' [20] and polyvinylpyridinium chloride- sodium dodecylsulphate (SDS) system : large E", sometimes larger than E' , i.e. more viscoelastic layers

[22]. This variety of behaviour illustrates the sensitivity of surface rheology to the chemical nature of the polyelectrolyte chain.

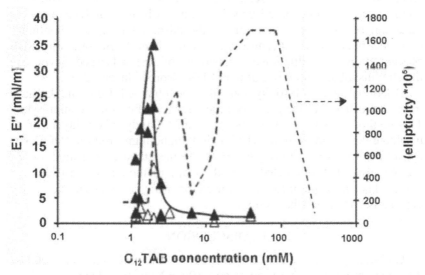

Fig. 8 Storage and loss surface dilational moduli E' and E'' (closed and open symbols, respectively) and ellipticity as a function of DTAB concentration for solutions containing 2.4 mM PSS. The frequency is 0.05 Hz. Reprinted from ref 2 with permission.

Dilational visco-elasticity has been measured by the oscillating bubble method with various polysaccharide systems. In a study of mixed solutions of DTAB with a polysaccharide, λ-carrageenan, and a maximum of dilational modulus was also found around C_0 [23]. In a study of mixed solutions of chitosan, a cationic polysaccharide, with SDS [24], the surface activity was found much higher than that of chitosan modified by an equal number of alkyl groups (polysoap). This observation correlated with the higher rate of increase in the dilational storage modulus E' of the adsorption layers of SDS and chitosan than for the polysoaps. For these systems, the modulus E' was much higher than the loss modulus E'' as for DTAB-PSS or PAMPS adsorption layers at low frequencies. Mixed solutions of gelatin, an amphiphilic polysaccharide, and SDS have also been investigated: here, and as for surfactant-protein mixed layers, binding of SDS molecules disrupts the cross-linked network of adsorbed gelatin molecules and results in a reduction of the elastic modulus of the adsorbed layer that continues until the bulk SDS concentration reaches 1 mM. Beyond this SDS concentration, the dilational properties of the adsorbed layer are indistinguishable from those of pure SDS adsorbed layers [25].

Dilational visco-elasticity can also be obtained from the Laplace transform of the surface pressure time variation after a fast compression. This method was used by Klebanau et al. to study mixed layers of an insoluble surfactant, didodecyl dimethyl ammonium bromide (DODAB) spread onto PSS solutions [26]. Here the measured elastic modulus E' was equal to that calculated from the slope of the surface tension-area curve. The loss modulus E'' varied as $\omega^{-0.5}$, a unusual frequency dependence for a gel-like system.

D. DISCUSSION

Globular proteins such as lysozyme or BSA are characterized by dilational moduli E' between 0 and 100 mN/m and shear moduli G' between 0 and 20 mN/m. Flexible proteins such as β-casein are characterized by E' between 0 and 30 mN/m and G' between 0 and 2 mN/m. For solutions of pure surfactants, G' is zero and lower values of the surface dilational moduli are reported, E' ranging between 0 and 10 mN/m. The maximum values of G' and E' reported for polymer-surfactant mixed surface layers are thus higher than those obtained for pure surfactant solutions and are comparable to those measured for flexible proteins. These results show that the dilational and shear surface moduli are closely related to the kind of intermolecular and intramolecular interactions involved in the adsorbed layers. For example, globular proteins such as lysozyme or BSA are known to form strong intermolecular interactions through hydrogen bonds and disulfide covalent bridges and exhibit the highest shear and dilational moduli. On the contrary, β-casein, which is a much more flexible molecule because of weaker intermolecular interactions, has a configuration close to a random coil and exhibits much lower surface moduli. The polyelectrolyte-surfactant complexes combine chain entanglements as well as hydrophobic interactions, but the absence of stronger bonds such as disulfide bridges or hydrogen bonds explains why their behaviour is closer to that of flexible proteins. We should add that in the case of proteins (BSA), it has been reported that the highest surface shear moduli was measured at a pH close to their isoelectric point, when the net charge of the protein is close to zero and therefore its hydrophobicity is maximal. In polyelectrolyte-surfactant systems, a maximum of visco-elastic moduli is observed for a molar ratio of unity, when polymer chains are nearly saturated with surfactant molecules (i.e., the hydrophobicity of bulk aggregates is maximum).

E. APPLICATIONS

Surface visco-elasticity plays a role in many circumstances, but this role is not frequently recognized. For instance it is known that the velocity of sedimentation of liquid drops or ascension of gas bubbles can vary by 50% when the surface rheological parameters are varied [7]. Among the different visco-elastic parameters, dilational elasticity is likely to play a dominant role for soluble surfactant monolayers, for instance in the spreading coefficient of surfactant solutions on a solid substrate [27], the critical shear rate for breakage of emulsion droplets[28] or the pressure drop in a two-phase flow [29]. Although these different phenomena are not all completely rigorously modelled, the control parameter is clearly the dilational elasticity. So far and to our knowledge, no experiment of this type was done yet with mixed polymer-surfactant solutions.

1. Thin film drainage and stability

The drainage of freely standing horizontal liquid films (diameters less than about 1 mm) made from surfactant solutions has been shown to depend on the Gibbs elasticity E_G [30] : this is because surfactant is taken away along the surface to the menisci and there is not enough surfactant in the foam film to replenish the surface. In the case of mixed polymer-surfactant solutions, the situation is less clear. It has been reported that films made from solutions of surfactants and polyethylene oxide (neutral polymer) drain faster than films made from solutions without polymer [31]. Similar observations were done with foams, and this was attributed to the elongational viscosity of the solutions [32]. To date, no rigourous model is available to account for these observations.

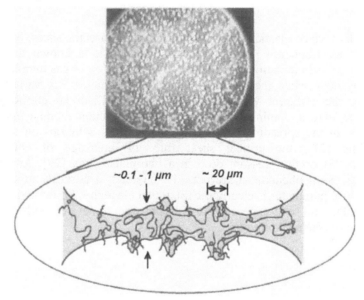

Fig. 9 Photos taken from a single foam film containing 1 mM DTAB and 750mg/l PAMPS25. Schematic sketch of the cross section of the film. Reprinted from ref 32 with permission.

The drainage of horizontal liquid films made with polyelectrolytes is very different according to the respective sign of polymer and surfactant charges : if the charges are opposite, mixed surface layers form at the film surfaces and the drainage is much slower than when the charges are the same or when the surfactant is nonionic [33]. These differences are also not explained up to date. In the case of opposite charges, close to the precipitation concentration, the thickness is very inhomogeneous: the film visibly contains small microgel domains, and the film drainage becomes very slow [34] (Fig. 9). This phenomenon has also be observed with DTAB-PSS solutions between C_0 (2.4 mM) and the precipitation boundary (4 mM), i.e. in a region where the surface visco-elastic moduli are close to zero [2]. The presence of the aggregates bridging the film surfaces apparently provides substantial resistance to flow out of the films.

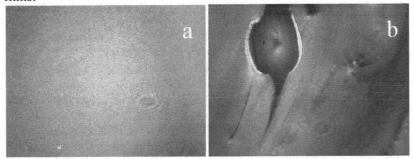

Fig. 10 Brewster angle microscopy pictures of DTAB-DNA system for 1.6mM DTAB and 1mM DNA, a) after equilibration for 24 hours; b) broken after equilibration. Reprinted from ref 34 with permission.

More surprisingly, in the case of species of opposite sign, the films were unstable when the polymer backbone was rigid [35]. They broke after a few seconds in the presence of xanthan or DNA, as pure water films, whereas the films made from pure DTAB solutions lasted a few minutes. In the presence of PAMPS, the films were much more stable and lasted for hours. Close to the precipitation concentration, PAMPS or PSS films are gel-like, they become impossible to break. We have seen in §3.1 that the surface dilational visco-elasticity is very similar for

DTAB with PAMPS or xanthan, meaning that the film stability is not correlated at all with surface visco-elasticity [35]. In the case of DNA, the mixed surface layers are brittle (see Fig. 10) [36] and the non-linear behaviour is very different from that of DTAB-PAMPS layers, more visco-elastic in nature (see Fig. 7).

2. Foam or emulsion drainage (creaming) and stability

It was recognized early that foaming, emulsification, emulsion and foam stability are also related to the surface dilational rheological behaviour: qualitatively, E is a measure of the ability of a film to respond to an increase in surface area. Foam and emulsion properties are also generally related to the stability of the films separating the air bubbles or the liquid drops. Good foamability (emulsification) is achieved when the bubbles (drops) surfaces are rapidly covered by surfactant layers and when these bubbles (drops) do not break too quickly. Foamability (emulsification) and foam (emulsion) stability are therefore closely related. Foam stability is governed by various processes:

a) Drainage: removal of the liquid from the space between the bubbles, and thinning of the films separating the bubbles. Surface dilational visco-elasticity slows down (slightly) the drainage rate [37]. In the emulsion case, creaming can occur instead of drainage, if the continuous phase is denser, but the effect of visco-elasticity is the same.

b) Ostwald ripening: diffusion of the gas (liquid) from the small bubbles (drops) towards the larger ones, leading to a bubble (drop) size growth with time. Gas (liquid) diffusion rate decreases when the surface coverage increases [38] (G and E are closely related to Γ). This idea has been used recently to improve the stability of emulsions against coarsening, by adding polymers that form thick mixed layers with the surfactant.

c) Coalescence: breakage of the film between two adjacent bubbles (drops) to form a single larger bubble (drop). Finite elasticity values protect the films against rupture [39].

All the three phenomena depend on the elasticity and viscosity of the surface, and high elasticities and viscosities are thus expected to stabilize foams and emulsions. Let us however point out that all the above processes occur at different timescales τ and that the surface rheological parameters to be considered are those at a frequency $\omega=1/\tau$.

Results from foaming studies for the DTAB-PAMPS10 and DTAB-xanthan systems are shown in Fig. 11 [12]. For a given gas flow rate, the foam height increased with C_S, the increase becoming very fast close to the critical micellar concentration (*CMC*), foam height being almost identical for 20mM surfactant for all solutions. An intermediate small maximum was seen close to C_0.

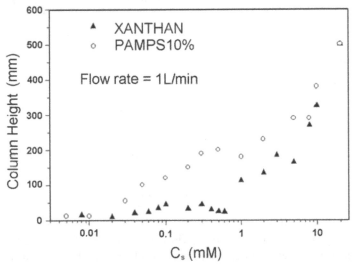

Fig. 11 Foam height as a function of surfactant concentration at a constant gas flow rate of 1l/min: open squares; xanthan, open triangles, PAMPS. Reprinted from ref 12b with permission.

The results of foam stability are shown in Fig. 12. Again the difference in behaviour between the different solutions almost disappeared at 20mM. Although the surface rheological behaviour of the different polymers was similar, the foaming capacity was much larger with PAMPS than with xanthan. The foam lifetime for PAMPS10 was larger than that of PAMPS25, although the surface dilational elasticity was larger for PAMPS25. Around 20mM, the surfactant monolayer is compact and there is no more penetration of the polymer in the surface layer, explaining the common behaviour of the different systems. Surprisingly, there are no special differences on the foam behaviour below and above the precipitation concentration, especially in the region where the foam films are gel-like.

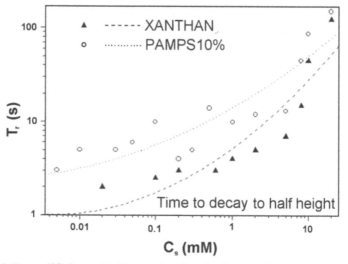

Fig. 12 Foam lifetimes T_r (time taken by the foam column to drop to half its original height after the gas flow is stopped) versus surfactant concentration; open squares, xanthan; open triangles, PAMPS. Reprinted from ref 12b with permission.

There is thus little correlation between the foam stability and the film stability in this type of systems. The rapid increase in foam height and stability upon increasing C_S seems to indicate that the foam behaviour is largely controlled by the amount of surfactant present, whereas the foam film behaviour is related more closely to the presence of the polymer.

In the case of mixed solutions of DTAB and PSS, a maximum in the foam volume was measured close to C_0, (Fig. 13) similar to what has been found with PAMPS and xanthan and with other systems of oppositely charged polymer-surfactants [40]. Between 8 and 20 mM surfactant, when DTAB-PSS precipitates out of solution, the foam volume returned to zero, whereas the foam volume for the pure surfactant solutions attained its maximum. Most of the surfactant is bound to the polymer in the aggregates that have precipitated out of solution, leaving insufficient amounts for high foam stability. Monteux et al. propose that the polymer-surfactant precipitates could act as hydrophobic antifoam particles, which can provoke foam collapse, but in this case, the foam films would also be unstable. After re-solubilisation of DTAB-PSS aggregates, the foam volume increases up to values comparable to those of the corresponding pure surfactant solutions. The first peak in the foam volume occurred close to C_0 as the peaks in G', E' showing that foam stability is here strongly correlated to surface moduli.

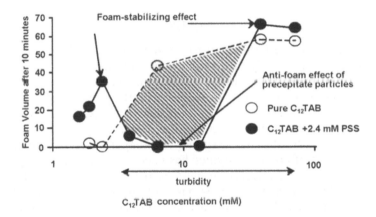

Fig. 13 Foam volume measured with the normalized Ross-Miles test as a function of DTAB concentration, for pure DTAB solutions and for DTAB-PSS solutions. Reprinted from ref 2 with permission.

Several studies have shown that dilational surface moduli can be more important for foam or emulsion stability than shear surface moduli; however, these studies mainly concern pure surfactant solutions, for which shear surface moduli are zero because of the lack of connectivity between molecules in comparison to polymeric systems. With these simpler systems, foam or emulsion stabilization is more strongly dependent on surface tension gradient mechanisms. In the literature, direct comparisons between foam stability and surface rheology (shear and dilational) for systems having high shear surface viscosities are scarce. For protein solutions, it has been shown that foaming cannot be related to these properties, because of the very long times needed to form the visco-elastic layers. In this case, the foaming capacity was related to the formation of gel-like foam films [41]. Surprisingly in the DTAB-PAMPS or PSS systems, the existence of very stable foam films does not seen to influence the foam capacity, perhaps because these gel-like films need long times to be formed, longer than protein films, or because they are less well attached to the surface and are removed during foam drainage.

F. CONCLUSIONS

In conclusion, the field of surface visco-elasticity of monolayers is extremely rich but largely unexplored, especially for mixed solutions of polymers and surfactants. This is at first sight surprising in view of the large number of applications. No studies of wetting, drop breakage, two-phase flow, emulsification and emulsion stability have been yet compared to the surface rheological behaviour. A few attempts to relate foam and foam film stability to surface rheology have been reported. No major influence of polymer addition has been seen for neutral polymers, which role is a mere increase of bulk viscosity. If polyelectrolytes and surfactants have opposite charges, visco-elastic layers are formed and foaming is appreciable even at very low surfactant concentration. Gel-like foam films are observed close to precipitation, but although their stability is remarkable, they do not seem to influence the foam stability. The foam and the foam film stability decreases when polymers with rigid backbones are used, the reason for this being perhaps the brittle nature of the mixed surface layers. Polymer-surfactant mixed layers are metastable systems : once adsorbed, they behave as insoluble layers, and their structure depends on the adsorption history. This explains why it is difficult to obtain reliable visco-elastic data. The elastic moduli appear to increase with the applied strain, until the layers buckle or fracture, depending on the type of polymer used. In summary these mixed layers have unusual properties, with marked hysteresis and non-linear behaviour, making them difficult to study. In view of the numerous applications of the solutions, it is clearly desirable to devote further research efforts in the area.

G. REFERENCES

1. S.T.A.Regismond, F.M.Winnick, E.D.Goddard, Colloids Surfaces A, 119 (1996) 221; S.T.A.Regismond, F.M.Winnick, E.D.Goddard, Langmuir, 13 (1997) 5558; S.T.A.Regismond, F.M.Winnick, E.D.Goddard, Colloids Surfaces A, 141 (1998) 165
2. C.Monteux, G.G.Fuller, V.Bergeron, J. Phys. Chem. B, 108 (2004) 16473
3. C.Monteux, C.E.Williams, J. Meunier, O.Anthony, V. Bergeron, Langmuir 20 (2004) 57; C.Monteux, M.F.Llauro, D.Baigl, C.E.Williams, O.Anthony, V. Bergeron, Langmuir, 20 (2004) 5358; C.Monteux, C.E.Williams, V. Bergeron, Langmuir, 20 (2004) 5367.
4. D.J.F.Taylor, R.K.Thomas, J.Penfold, Langmuir, 18 (2002) 4748.
5. C.F.Brooks, J.Thiele, C.W. Frank, D.F.O'Brien, W.Knoll, G.G.Fuller, C.R.Robertson,. Langmuir, 18 (2002) 2166.
6. C.A.Naumann, C.F.Brooks, G.G.Fuller, J.Lehmann, J.Ruhe, W.Knoll, P.Kuhn, O. Nuyken, C.W. Frank, Langmuir, 17 (2001) 2801; C.A.Naumann, C.F.Brooks, W. Wiyatno, W.Knoll, G.G.Fuller, C.W. Frank, Macromolecules, 34 (2001) 3024.
7. V.G.Levich, "Physico-chemical hydrodynamics", Prentice Hall, 1962, p 413 and 620.
8. J.Lucassen, D.Giles, J.Chem.Soc.Faraday I, 71 (1975) 217-232
9. A.Bonfillon, D.Langevin, Langmuir, 10 (1994) 2965-2971
10. Y.Jayalakshmi, L.Ozanne, D.Langevin, J.Colloid Interface Sci., 170 (1995) 358-366
11. N.Jain, P.A.Albouy, D.Langevin, Langmuir, 19 (2003) 5680; N.J.Jain, P.A.Albouy, D.Langevin, Langmuir, 19 (2003) 8371
12. A.Bhattacharya, F.Monroy, D.Langevin, J.F.Argillier, Langmuir, 16 (2000) 872. H.Ritacco, P.A.Albouy, A.Bhattacharyya, D.Langevin, Phys. Chem. Chem. Phys., 2 (2000) 5243

13. A.Asnacios, D.Langevin, J.F.Argillier, Macromolecules, 29 (1996) 7412; A.Asnacios, D.Langevin, J.F.Argillier, Eur. Phys. J. B, 5 (1998) 905.
14. A.Asnacios, R.V.Klitzing, D.Langevin, Colloids Surfaces A, 167 (2000) 189
15. W.Essafi, F.Lafuma, D.Baigl, C.E.Williams, Eur. Phys. J. B 9 (1999) 261; W.Essafi, F.Lafuma, D.Baigl, C.E.Williams, Eur.Phys.Lett., 71 (2005) 938
16. F.Monroy, J.Kahn, D.Langevin, Colloids Surfaces, 143(1998) 251;. J.Kahn, F.Monroy, D.Langevin, Phys. Rev. E, 60 (1999) 7163.
17. C.Picard, L.Davoust, Rheol. Acta, 45 (2006) 497
18. H.Ritacco, D.Kurlat, D. Langevin, J. Phys. Chem. B, 107 (2003) 9146; H.Ritacco, A.Cagna, D. Langevin, Colloids Surfaces A, 282–283 (2006) 203
19. A.Bonfillon, F.Sicoli, D.Langevin, J. Colloid Interface Sci., 168 (1994) 497
20. B.A.Noskov, G.Loglio, S.-Y.Lin, R.Miller, J. Colloid Interface Sci., 301 (2006) 386
21. B.A.Noskov, G.Loglio, R.Miller, J. Phys. Chem., 108 (2004) 18615
22. B.A.Noskov,A.G.Bykov, D.O.Grigoriev, S.-Y.Lin, G.Loglio, R.Miller, Colloids Surfaces A, 322 (2008) 71
23. T. Morimoto Nobre, K. Wong, M. E. Darbello Zaniquelli, J. Colloid Interface Sci., 305 (2007) 142
24. V.G.Babak, J.Desbrieres, Colloid Polym. Sci., 284 (2006) 745
25. A. Rao, J. Kim, R. R. Thomas, Langmuir, 21 (2005) 617
26. A.Klebanau, N.Kliabanova, F.Ortega, F.Monroy, R.G.Rubio, V.Starov, J. Phys. Chem. B, 109 (2005) 18316.
27. M. Cachile, A.M. Cazabat, Langmuir, 15 (1999) 1515
28. J.J.M.Janssen, A.Boon, W.G.M.Agterof, AICHE J., 40 (1994)1929
29. K.J.Stebe, C.Maldarelli, J.Colloid Interface Sci., 163 (1994) 177
30. A.Sonin, A.Bonfillon, D.Langevin, J.Colloid Interface Sci.,162 (1994) 323
31. R.Bruinsma, J.M.di Meglio, D.Quéré, S.Cohen-Addad, Langmuir, 8 (1992) 3161; S.Cohen-Addad, J.M.di Meglio, Langmuir, 10 (1994) 773

32. M. Safouane, A. Saint-Jalmes, V. Bergeron, D. Langevin, Eur. Phys. J .E, 19 (2006) 195
33. C. Márquez Beltrán, D. Langevin, Phys. Rev. Lett., 94 (2005) 217803
34. V.Bergeron, D.Langevin, A.Asnacios, Langmuir, 12 (1996) 1550.
35. C.Stubenrauch, P.A.Albouy, R.v.Klitzing, D.Langevin, Langmuir, 16 (2000) 3206
36. D.McLoughlin, D.Langevin, Colloids Surfaces A, 250 (2004) 79
37. M.Durand, G.Martinoty, D.Langevin, Phys.Rev.E, 60 (1999) R6307; A.Saint-Jalmes, Y. Zhang, D. Langevin, Eur.Phys.E, 15 (2004) 53
38. S.Mun, D.J.McClements, Langmuir 22 (2006) 1551
39. D.Langevin, Current Opinion Colloid Interface Sci. 3 (1998) 600; D.Langevin in"Encyclopedia of Surface and Colloid Science, A.Hubbard, ed; Marcel Dekker, 2002.
40. M.M.Guerrini, R.Lochhead, W.H.Daly,. Colloids Surfaces. A, 147 (1999) 67.
41. A. Saint-Jalmes, M.-L. Peugeot, H. Ferraz, D. Langevin, Colloids Surfaces A, 263 (2005) 219

SURFACE DILATIONAL RHEOLOGY OF MIXED ADSORPTION LAYERS OF PROTEINS AND SURFACTANTS AT LIQUID INTERFACES

V.I. Kovalchuk[1], E.V. Aksenenko[2], R. Miller[3] and V.B. Fainerman[4]

1   Institute of Biocolloid Chemistry, 42 Vernadsky Avenue, 03680 Kyiv (Kiev), Ukraine
2   Institute of Colloid Chemistry and Chemistry of Water, 42 Vernadsky Avenue, 03680 Kyiv (Kiev), Ukraine
3   Max-Planck-Institut für Kolloid- und Grenzflächenforschung, Am Mühlenberg 1, 14424 Potsdam, Germany
4   Medical Physicochemical Centre, Donetsk Medical University, 16 Ilych Avenue, 83003 Donetsk, Ukraine

A. INTRODUCTION

The addition of surfactants can modify adsorbed protein layers at liquid/fluid interfaces, which leads to changes in interface tension, equilibrium and dynamic adsorption values, and rheological characteristics [1-6]. Although the surface dilational rheology of protein/surfactant mixtures is extremely important from a practical point of view, a corresponding general theory is still not available. For much simpler systems, i.e. surfactant mixtures, it appears possible to predict the rheological behaviour of a mixture using data for the individual components. One of the first ever attempts to analyse theoretically the rheology of surfactant mixtures was made by Lucassen-Reynders [7]. The theoretical analysis of the dilational rheology of surfactant mixtures was later performed by Garrett and Joos [8], who generalised the theory by Lucassen and van den Tempel [9,10]. The theory of Garrett and Joos [8] was further developed in [11-14]. In particular, for the mixture of two surfactants analytical expressions for the complex dilational modulus were derived in [11,12], which contain four partial derivatives of the adsorptions with respect to concentrations [11] or vice versa [12], and two partial derivatives of the surface tension with respect to the adsorptions. In [13,14] an irreversible thermodynamic approach had been applied and a general method for the derivation of equations for the surface elasticity of mixtures of an arbitrary number of surfactants and mixed adsorption kinetics was proposed. It was shown by Joos [12] that by introducing partial elasticities, it becomes possible to express the limiting high-frequency elasticity for the mixture of surfactants in terms of elasticities of the individual components.

The results obtained in the theory for surfactant mixtures are applicable to protein/surfactant mixtures. Clearly, such application should account for the peculiarities of the adsorption and rheological behaviour of the proteins. The theoretical model proposed by Joos [12] for the estimation of the limiting elasticity of mixed surfactant solutions, and also the theoretical model proposed by Garrett and Joos [8] for the description of the dilational elasticity of such mixtures, were used for mixtures of the non-ionic surfactant $C_{10}DMPO$ and the β-lactoglobulin [15]. The theoretical model [8] demonstrates quite satisfactory agreement with experimental results obtained from the oscillating bubble shape method. Also it is shown that the calculated limiting elasticities of the mixtures comply with the experimental data.

In a recent review the expressions for the complex elasticity modulus was derived and a numerical procedure proposed for the calculation of real and imaginary parts of the elasticity modulus [16]. This procedure was illustrated for the protein/non-ionic surfactant mixtures in the framework of the model [17-19] which assumes multiple states of adsorbed protein molecules at the surface and an intrinsic compressibility of the surface layer [20-22].

In this chapter the theory developed in [16] is described in details and generalised to mixtures of protein with ionic surfactant and mixture of surfactants. Theoretical calculations and experimental data are presented for mixtures of a proteins with surfactants and for mixtures of surfactants. The experimental data for mixed adsorption layers containing β-lactoglobulin/C_{10}DMPO, lysozyme/C_{10}DMPO and lysozyme/SDS, and also for mixture of SDS with dodecanol are shown to be in qualitative agreement with the values predicted by the model calculations.

B. THEORY OF SURFACE DILATIONAL ELASTICITY

1. Surface elasticity of mixed surface layers

The surface dilational modulus is defined by an expression originally proposed by Gibbs as the increase in surface tension for a small increase of surface area:

$$E = \frac{d\gamma}{d \ln A},$$ (1)

where γ is the surface tension, and A is the surface area.

The corresponding expressions for the complex elasticity modulus for mixed monolayers were derived in [8, 11, 12] by using similar basic assumptions. In particular, the most general expressions were adopted to mixtures of two surfactants for the surface tension

$$\gamma = \gamma(\Gamma_1, \Gamma_2, T),$$ (2)

and for the adsorption of each component

$$\Gamma_j = \Gamma_j(c_{s1}, c_{s2}).$$ (3)

In these equations Γ_j is the adsorption of component j of the mixture, c_{sj} is the corresponding subsurface concentration, and T is the temperature. A diffusion mechanism of the adsorption process (i.e., the equilibrium between the subsurface concentration of a surfactant and the adsorption) was assumed resulting in an equation of mass balance in the surface layer onto which the harmonic oscillations are imposed with a frequency f and small magnitude ΔA:

$$i\omega\Delta\Gamma_j + i\omega\Gamma_j\Delta A / A + \left(i\omega D_j\right)^{1/2}\Delta c_{sj} = 0 . \tag{4}$$

Here $\omega = 2\pi f$ is the angular frequency, and D_j is the diffusion coefficient of component j. From the equations above, expressions for the complex visco-elasticity modulus were derived in [11,12], which are rather similar to each other. Unfortunately, the final expressions – Eq. (21) in [11] and Eq. (8.105) in [12] were both erroneous. The correct expressions for the case of two components (i,j = 1,2) are considered below in more details.

Equation (21) derived by Jiang et al. in [11] should correctly read:

$$E = \frac{1}{B}\left(\frac{\partial\Pi}{\partial\ln\Gamma_1}\right)_{\Gamma_2}\left[\sqrt{\frac{i\omega}{D_1}}a_{11} + \sqrt{\frac{i\omega}{D_2}}a_{12}\frac{\Gamma_2}{\Gamma_1} + \frac{i\omega}{\sqrt{D_1D_2}}(a_{11}a_{22} - a_{12}a_{21})\right] +$$
$$\frac{1}{B}\left(\frac{\partial\Pi}{\partial\ln\Gamma_2}\right)_{\Gamma_1}\left[\sqrt{\frac{i\omega}{D_1}}a_{21}\frac{\Gamma_1}{\Gamma_2} + \sqrt{\frac{i\omega}{D_2}}a_{22} + \frac{i\omega}{\sqrt{D_1D_2}}(a_{11}a_{22} - a_{12}a_{21})\right] \tag{5}$$

where $B = 1 + \sqrt{i\omega/D_1}\,a_{11} + \sqrt{i\omega/D_2}\,a_{22} + (i\omega/\sqrt{D_1D_2})\cdot(a_{11}a_{22} - a_{12}a_{21})$, $\Pi = \gamma_0 - \gamma$ is the surface pressure, γ_0 the surface tension of solvent, c_1 and c_2 are the bulk concentrations of the components, and $a_{ij} = (\partial\Gamma_i/\partial c_j)\big|_{c_{k\neq j}}$ are the partial derivatives to be determined from the adsorption isotherm.

In turn, the expression (8.105) derived by Joos [12] should read:

$$E = \frac{1}{B_0}\left(\frac{\partial\Pi}{\partial\ln\Gamma_1}\right)_{\Gamma_2}\left[1 + \sqrt{\frac{D_2}{i\omega}}d_{22} - \sqrt{\frac{D_1}{i\omega}}\frac{\Gamma_2}{\Gamma_1}d_{12}\right] +$$
$$\frac{1}{B_0}\left(\frac{\partial\Pi}{\partial\ln\Gamma_2}\right)_{\Gamma_1}\left[1 + \sqrt{\frac{D_1}{i\omega}}d_{11} - \sqrt{\frac{D_2}{i\omega}}\frac{\Gamma_1}{\Gamma_2}d_{21}\right] \tag{6}$$

where $B_0 = 1 + d_{11}\sqrt{D_1/i\omega} + d_{22}\sqrt{D_2/i\omega} + (1/i\omega)\sqrt{D_1D_2}(d_{11}d_{22} - d_{12}d_{21})$, and $d_{ij} = (\partial c_i/\partial\Gamma_j)_{\Gamma_{k\neq j}}$ again are the partial derivatives which should be determined from the adsorption isotherm equation.

Eq. (6) can be transformed into Eq. (5) [16]. From the relations between the small variations of adsorptions and subsurface concentrations which follow from Eqs. (3):

$$\Delta\Gamma_i = a_{ii}\Delta c_{si} + a_{ij}\Delta c_{sj},\tag{7}$$

one obtains expressions for Δc_{si}:

$$\Delta c_{si} = \frac{a_{jj}\Delta\Gamma_i - a_{ij}\Delta\Gamma_j}{a_{11}a_{22} - a_{12}a_{21}},\tag{8}$$

from which the derivatives d_{ij} are obtained:

$$d_{11} = \frac{a_{22}}{a_{11}a_{22} - a_{12}a_{21}}, \quad d_{12} = -\frac{a_{12}}{a_{11}a_{22} - a_{12}a_{21}},$$

$$d_{21} = -\frac{a_{21}}{a_{11}a_{22} - a_{12}a_{21}}, \quad d_{22} = \frac{a_{11}}{a_{11}a_{22} - a_{12}a_{21}}.\tag{9}$$

Substitution of these expressions into Eq. (6) leads to Eq. (5). In the following we will operate only with Eq. (5).

The visco-elastic modulus can be expressed as a complex number: $E = E_r + iE_i$, where the real part represents the storage modulus equal to the dilational elasticity and the imaginary part stands for the loss modulus (see Chapters 1 and 2). Noting that $\sqrt{i} = (1+i)/\sqrt{2}$, one finally obtains expressions for the real and imaginary parts of the visco-elasticity in Eq. (5) as:

$$E_r = (PR + QS)/(P^2 + Q^2), \quad E_i = [PS - QR]/(P^2 + Q^2).\tag{10}$$

and the expressions for the visco-elasticity modulus $|E|$ and phase angle ϕ between stress $(d\gamma)$ and strain (dA):

$$|E| = \sqrt{(R^2 + S^2)/(P^2 + Q^2)}, \quad \phi = arctg(E_i/E_r),\tag{11}$$

with

$$P = 1 + \left(\sqrt{\omega/D_1}\, a_{11} + \sqrt{\omega/D_2}\, a_{22}\right)/\sqrt{2},$$

$$Q = \left(\sqrt{\omega/D_1}\, a_{11} + \sqrt{\omega/D_2}\, a_{22}\right)/\sqrt{2} + \left(\omega/\sqrt{D_1 D_2}\right)\cdot\left(a_{11}a_{22} - a_{12}a_{21}\right),$$

$$P_i = \left[\sqrt{\omega/D_i}\, a_{ii} + \sqrt{\omega/D_j}\, a_{ij}\left(\Gamma_j/\Gamma_i\right)\right]/\sqrt{2},$$

$$Q_i = P_i + \left(\omega / \sqrt{D_1 D_2}\right) \cdot \left(a_{11} a_{22} - a_{12} a_{21}\right),$$

$$R = P_1 \cdot \left(\partial \Pi / \partial \ln \Gamma_1\right)_{\Gamma_2} + P_2 \cdot \left(\partial \Pi / \partial \ln \Gamma_2\right)_{\Gamma_1},$$

$$S = Q_1 \cdot \left(\partial \Pi / \partial \ln \Gamma_1\right)_{\Gamma_2} + Q_2 \cdot \left(\partial \Pi / \partial \ln \Gamma_2\right)_{\Gamma_1}.$$

2. Surface elasticity of individual surfactant layers

Assuming a diffusion controlled exchange of matter mechanism, the rheological dilational characteristics of the surface layer for single surfactants were derived by Lucassen and van den Tempel [9,10]:

$$E = E_0 \left(\frac{d\Gamma}{dc}\sqrt{\frac{i\omega}{D}}\right) \Bigg/ \left(1 + \frac{d\Gamma}{dc}\sqrt{\frac{i\omega}{D}}\right) = E_0 \frac{1 + \zeta + i\zeta}{1 + 2\zeta + 2\zeta^2}, \tag{12}$$

or

$$|E| = E_0 (1 + 2\zeta + 2\zeta^2)^{-1/2}, \quad \phi = arctg[\zeta / (1 + \zeta)], \tag{13}$$

where

$$E_0 = d\Pi / d\ln\Gamma, \quad \zeta = \left(\frac{\omega_D}{\omega}\right)^{1/2}, \quad \omega_D = \frac{D}{2}\left(\frac{dc}{d\Gamma}\right)^2. \tag{14}$$

It can be shown that, if the properties of the two surfactants in a mixture are the same, then Eq. (5) can be transformed to Eq. (12). In this case the equation of state (expressed as functional dependence of Π on Γ, with $\Gamma = \Gamma_1 + \Gamma_2$) and the adsorption isotherm (expressed as functional dependence of c on Γ, with $c = c_1 + c_2$) for mixtures of two surfactants 1 and 2 read:

$$\Pi = \Pi(\Gamma),$$

$$c = c(\Gamma),$$

and the functional dependencies of Π on Γ and c on Γ are exactly the same as those for single surfactants. The differentiation of Eqs. (15) and (16) yields:

$$E_{i0} = \Gamma_i \left(\frac{\partial \Pi}{\partial \Gamma_i}\right)_{\Gamma_j} = \Gamma_i \frac{d\Pi}{d(\Gamma_1 + \Gamma_2)} = \Gamma_i \frac{d\Pi}{d\Gamma}\bigg|_{\Gamma = \Gamma_1 + \Gamma_2}, \tag{17}$$

$$\left[\frac{\partial(c_1+c_2)}{\partial c_i}\right]_{c_j} = 1 = \frac{dc}{d(\Gamma_1+\Gamma_2)}\left[\left(\frac{\partial\Gamma_1}{\partial c_i}\right)_{c_j} + \left(\frac{\partial\Gamma_2}{\partial c_i}\right)_{c_j}\right] = \left(a_{ii}+a_{ji}\right)\frac{dc}{d\Gamma}\bigg|_{\Gamma=\Gamma_1+\Gamma_2}.$$

$$(18)$$

From Eqs. (18) with i,j = 1,2 we get

$$a_{11} + a_{21} = a_{12} + a_{22} = \frac{d\Gamma}{dc}\bigg|_{\Gamma=\Gamma_1+\Gamma_2} = \frac{d\Gamma}{dc}\bigg|_{c=c_1+c_2}. \qquad (19)$$

We can express a_{22} via Eq. (19) as $a_{22} = a_{11} + a_{21} - a_{12}$, and therefore obtain:

$$a_{11}a_{22} - a_{21}a_{12} = (a_{11} + a_{21})(a_{11} - a_{12}). \qquad (20)$$

Introducing this equality into Eq. (5) we finally get:

$$E = (\Gamma_1 + \Gamma_2)\frac{d\Pi}{d\Gamma}\bigg|_{\Gamma=\Gamma_1+\Gamma_2} \frac{\sqrt{\frac{i\omega}{D}}(a_{11}+a_{21})}{1+\sqrt{\frac{i\omega}{D}}(a_{11}+a_{21})} = E_0(\Gamma)\frac{\sqrt{\frac{i\omega}{D}}\frac{d\Gamma}{dc}\bigg|_{\Gamma=\Gamma_1+\Gamma_2}}{1+\sqrt{\frac{i\omega}{D}}\frac{d\Gamma}{dc}\bigg|_{\Gamma=\Gamma_1+\Gamma_2}}. \qquad (21)$$

with $E_0(\Gamma) = \Gamma\dfrac{d\Pi}{d\Gamma}\bigg|_{\Gamma=\Gamma_1+\Gamma_2}$.   Therefore,   for $\Gamma = \Gamma_1 + \Gamma_2$ (and   also

$c_1 + c_2 = c$) Eq. (21) becomes equal to Eq. (12) which expresses the visco-elasticity for a single surfactant solution. The same result also follows directly from Eqs. (2)-(4).

C. THEORY OF ADSORPTION AND VISCO-ELASTIC PROPERTIES OF SURFACE LAYERS

The expressions for the surface visco-elastic properties, Eq. (12) for the single surfactant monolayer and Eqs. (5) and (10) for mixtures, involve two and six derivatives, respectively. To determine these values, one should assume certain models of the surface layer which correspond to certain equations of state and adsorption isotherms. Here we will consider three types of systems: (i) adsorption of an individual protein; (ii) adsorption of an individual surfactant; and (iii) competitive adsorption of a protein and a surfactant. As first, however, it is necessary to numerically calculate the quantities involved in the expressions above, using a suitable procedure for the calculation of the respective surface pressure and adsorption isotherms.

1. Numerical calculation of partial derivatives

If the system considered involves a single surfactant only, the numerical calculation of the derivatives involved in Eqs. (12) and (14) is straightforward. Provided that a suitable procedure exists which enables one to calculate the surface pressure and adsorption at any value of the surfactant concentration, the relevant derivatives can be calculated from finite differences.

For a system which involves two (or more) surfactants the procedure is more involved. Let us assume that for each pair of concentrations $\{c_1, c_2\}$ it is possible to calculate the surface pressure $\Pi = \Pi(c_1, c_2)$ and adsorption of the components $\Gamma_1 = \Gamma_1(c_1, c_2)$ and $\Gamma_2 = \Gamma_2(c_1, c_2)$. For small deviations of the concentrations δc_1 and δc_2 in the vicinity of a point $\{c_1, c_2\}$, i.e., in the points $\{0\} = \{c_1, c_2\}$, $\{1\} = \{c_1 + \delta c_1, c_2\}$ and $\{2\} = \{c_1, c_2 + \delta c_2\}$ one can calculate the values:

$$\Pi^{\{0\}} = \Pi(c_1, c_2), \quad \Pi^{\{1\}} = \Pi(c_1 + \delta c_1, c_2), \quad \Pi^{\{2\}} = \Pi(c_1, c_2 + \delta c_2), \qquad (22)$$

$$\Gamma_i^{\{0\}} = \Gamma_i(c_1, c_2), \Gamma_i^{\{1\}} = \Gamma_i(c_1 + \delta c_1, c_2), \Gamma_i^{\{2\}} = \Gamma_i(c_1, c_2 + \delta c_2), (i = 1,2) \ (23)$$

Then the partial derivatives of adsorptions with respect to concentrations $a_{ij} = (\partial \Gamma_i / \partial c_j)\big|_{c_{k\neq j}}$ in Eq. (5) can be approximately calculated from finite differences:

$$a_{11} = (\partial \Gamma_1 / \partial c_1)_{c_2} = (\Gamma_1^{\{1\}} - \Gamma_1^{\{0\}})/\delta c_1, \quad a_{12} = (\partial \Gamma_1 / \partial c_2)_{c_1} = (\Gamma_1^{\{2\}} - \Gamma_1^{\{0\}})/\delta c_2,$$
$$a_{21} = (\partial \Gamma_2 / \partial c_1)_{c_2} = (\Gamma_2^{\{1\}} - \Gamma_2^{\{0\}})/\delta c_1, \quad a_{22} = (\partial \Gamma_2 / \partial c_2)_{c_1} = (\Gamma_2^{\{2\}} - \Gamma_2^{\{0\}})/\delta c_2, \qquad (24)$$

To calculate the partial derivatives $(\partial \Pi / \partial \ln \Gamma_i)_{\Gamma_{j\neq i}} \equiv (\partial \Pi / \partial \ln \Gamma_i)_{\ln \Gamma_{j\neq i}}$ we perform a change of variables, noting that the surface pressure $\Pi = \Pi(c_1, c_2)$ can be considered as a function of two other variables, namely of $\ln(\Gamma_1)$ and $\ln(\Gamma_2)$, which are taken as independent, instead of the concentrations. Therefore, the small deviations of Π which correspond to small deviations of $\ln(\Gamma_1)$ and $\ln(\Gamma_2)$ (in turn caused by small deviations of c_1 and c_2) can be expressed as $(k = 1,2)$:

$$(\delta \Pi)^{\{k\}} = (\partial \Pi / \partial \ln \Gamma_1)_{\ln \Gamma_2} \cdot (\delta \ln \Gamma_1)^{\{k\}} + (\partial \Pi / \partial \ln \Gamma_2)_{\ln \Gamma_1} \cdot (\delta \ln \Gamma_2)^{\{k\}}, \quad (25)$$

where the superscript $\{k\}$ denotes the points $\{1\}$ and $\{2\}$ defined above, and $(\delta \Pi)^{\{k\}} = \Pi^{\{k\}} - \Pi^{\{0\}}$, $(\delta \ln \Gamma_i)^{\{k\}} = \ln \Gamma_i^{\{k\}} - \ln \Gamma_i^{\{0\}}$. Finally, one can calculate the partial derivatives $(\partial \Pi / \partial \ln \Gamma_i)_{\Gamma_{j\neq i}} \equiv (\partial \Pi / \partial \ln \Gamma_i)_{\ln \Gamma_{j\neq i}}$ from the equations (25) with the known coefficients determined from Eqs. (22) and (23) provided that a suitable procedure is available for the numerical calculation of the surface pressure and the adsorption values.

2. Individual protein solutions

Protein molecules can adsorb in a number of states with different molar area, varying from a maximum (ω_{max}) to a minimum value (ω_{min}), which is described by the following equation of state for the surface layer [17]:

$$-\frac{\Pi \omega_0}{RT} = \ln(1-\theta_P) + \theta_P(1-\omega_0/\omega_P) + \alpha_P \theta_P^2, \tag{26}$$

where R is the gas law constant, α_P is the intermolecular interaction parameter, ω_0 is the molar area of the solvent, or the area occupied by one segment of the protein molecule,

$$\Gamma_P = \sum_{i=1}^{n} \Gamma_{Pi}, \tag{27}$$

is the total adsorption of proteins in all n states, and

$$\theta_P = \omega_P \Gamma_P = \sum_{i=1}^{n} \omega_i \Gamma_{Pi}, \tag{28}$$

is the total surface coverage by protein molecules. Here ω_P is the average molar area of the adsorbed protein, $\omega_i = \omega_1 + (i-1)\omega_0$ ($1 \le i \le n$) is the molar area in state i, assuming $\omega_1 = \omega_{min}$, $\omega_{max} = \omega_1 + (n-1)\omega_0$. The equations for the adsorption isotherm for each state (j) of the adsorbed protein are:

$$b_{Pj}c_P = \frac{\omega_P \Gamma_{Pj}}{(1-\theta_P)^{\omega_j/\omega_P}} \exp\left[-2\alpha_P(\omega_j/\omega_P)\theta_P\right]. \tag{29}$$

Here c_P is the protein bulk concentration and b_{Pj} is the equilibrium adsorption constant for the protein in the jth state. When we assume that the b_{Pi} are equal to each other: $b_{Pj} = b_P$ for any j (and therefore the adsorption constant for the protein molecule as a whole is $\Sigma b_P = nb_P$), then from Eq. (29) one can calculate the distribution function of various adsorption states of the protein molecules [17]:

$$\Gamma_{Pj} = \Gamma_P \frac{(1-\theta_P)^{\frac{\omega_j-\omega_1}{\omega_P}} \exp\left[2\alpha_P \theta_P \frac{\omega_j-\omega_1}{\omega_P}\right]}{\sum_{i=1}^{n}(1-\theta_P)^{\frac{\omega_i-\omega_1}{\omega_P}} \exp\left[2\alpha_P \theta_P \frac{\omega_i-\omega_1}{\omega_P}\right]}. \tag{30}$$

The model given by Eqs. (26)-(30) describes the evolution of states of protein molecules with increasing adsorption, which agrees in many details with known experimental results [17], including also the rheological behaviour of protein adsorption layers [18].

The procedure used for the numerical calculation of the thermodynamic quantities is as follows. Introducing the dimensionless variable $\Omega = \omega_P/\omega_0$ and dimensionless parameter $\Omega_1 = \omega_1/\omega_0$, from Eq. (30) (written for $j = 1$), Eq. (27) and (28) we obtain

$$\Omega = [\Xi(\theta_P, \Omega)]^{-1} \cdot \sum_{i=1}^{n} (\Omega_1 + i - 1)(1 - \theta_P)^{(i-1)/\Omega} \exp[2\alpha_P(i-1)\theta_P/\Omega], \qquad (31)$$

$$b_P c_P = \frac{\theta_P \exp(-2\alpha_P \theta_P \Omega_1/\Omega)}{(1-\theta_P)^{\Omega_1/\Omega} \Xi(\theta_P, \Omega)}, \qquad (32)$$

where we introduced the auxiliary function

$$\Xi(\theta_P, \Omega) = \sum_{i=1}^{n} (1 - \theta_P)^{(i-1)/\Omega} \exp[2\alpha_P(i-1)\theta_P/\Omega] \qquad (33)$$

Then, for any set of model parameters ω_{min}, ω_{max}, ω_0, α_P, b_P at any protein concentration c_P one can eliminate θ_P and Ω from the equations (31), (32) to calculate $\Gamma_P = \theta_P/(\Omega\omega_0)$ and Π.

3. Individual surfactant solutions

As was shown in a number of experimental studies with different surfactant systems, the limiting elasticity E_0 obtained from rheological studies levels off or even passes through a maximum with increasing concentration, whereas that predicted by the Frumkin isotherm increases continuously and can be larger even by some orders of magnitude. A new interpretation was proposed recently [20-22], which explains this effect on the basis of a finite intrinsic compressibility of the surfactant adsorption layer. The rigorous theoretical model given in [21] can be simplified by neglecting the contribution of non-ideality of entropy. In this case the equations of state and adsorption isotherm turn into the ordinary Frumkin model:

$$b_S c_S = \frac{\theta_S}{(1-\theta_S)} \exp(-2\alpha_S \theta_S), \qquad (34)$$

$$\Pi = -\frac{RT}{\omega_{S0}} [\ln(1-\theta_S) + \alpha_S \theta_S^2], \qquad (35)$$

where $\theta_S = \omega_S \cdot \Gamma_S$ is the surface coverage by surfactant molecules, where Γ_S is the adsorption of the surfactant, b_S is the adsorption equilibrium constant, α_S is the interaction constant. The molar area of a surfactant ω_S can be approximately represented by a linear dependence on surface pressure Π [20, 21]:

$$\omega_S = \omega_{S0}(1 - \varepsilon \Pi \theta_S), \tag{36}$$

where ω_{0S} is the molar area at zero surface pressure and ε is the two-dimensional relative surface layer compressibility coefficient, which characterises the intrinsic compressibility of the molecules in the surface layer. This intrinsic compressibility, for example, reflects the change of the tilt angle of the molecules upon surface layer compression, accompanied by an increase in the thickness of the surface layer [22]. In contrast to the expression for ω_S used in the theory described in [20-22], Eq. (36) involves the additional factor θ_S, which roughly accounts for the fact that the composition of the adsorption layer (of an individual surfactant and for the mixture of two or more surfactants or for surfactant/protein mixtures) depends on the composition of the saturated monomolecular layer of an individual surfactant. For the saturated monolayer of an individual surfactant or for a surfactant/protein mixture the value θ_S (or the total coverage) is equal to unity, and Eq. (36) becomes identical to the expression used in [20] which was derived on the basis of Grazing Incidence X-ray Diffraction experiments for condensed insoluble monolayers. Accordingly, one can express θ_S in Eqs. (34), (35) as:

$$\theta_S = \Gamma_S \omega_{S0}(1 - \varepsilon \Pi \theta_S). \tag{37}$$

Then, for any set of model parameters α_S, ω_{S0} and b_S, one can solve Eq. (34) at any value c_S. From this solution $\theta_S = \theta_S(c_S)$ the values of Π and Γ_S are calculated via Eqs. (35) and (37).

4. Mixtures of protein with surfactant

With the approximation $\omega_0 \cong \omega_S$, the following equation of state for a protein/non-ionic surfactant mixture was derived in [19]:

$$-\frac{\Pi \omega_0^*}{RT} = \ln(1 - \theta_P - \theta_S) + \theta_P(1 - \omega_0 / \omega_P) + \alpha_P \theta_P^2 + \alpha_S \theta_S^2 + 2\alpha_{PS}\theta_P\theta_S, \tag{38}$$

where α_{PS} is a parameter which describes the interaction between the protein and surfactant molecules. A small difference between ω_0 and ω_S can be accounted for by introducing

$$\omega_0^* = \frac{\omega_0 \theta_P + \omega_{S0} \theta_S}{\theta_P + \theta_S}. \tag{39}$$

It should be noted that, in contrast to the model developed earlier in [16], Eq. (39) involves the parameter ω_{S0} rather than ω_S. Equation (39) in its present formulation provides better agreement with the theory described in [21], because it disregards the compressibility of the solvent molecules. The numerical calculations have shown that both models give very close results. For the protein adsorbed in state $j = 1$ and the surfactant, the adsorption isotherms read [19]:

$$b_{P1} c_P = \frac{\omega_P \Gamma_{P1}}{(1 - \theta_P - \theta_S)^{\omega_1/\omega_P}} \exp\left[-2\alpha_P (\omega_1/\omega_P)\theta_P - 2\alpha_{PS}\theta_S\right], \tag{40}$$

$$b_S c_S = \frac{\theta_S}{(1 - \theta_P - \theta_S)} \exp\left[-2\alpha_S \theta_S - 2\alpha_{PS}\theta_P\right], \tag{41}$$

where the subscripts S and P refer to parameters characteristic for the individual surfactant and protein. The distribution of protein adsorptions over the states j is given by the expression [19]:

$$\Gamma_{Pj} = \Gamma_P \frac{(1 - \theta_P - \theta_S)^{\frac{\omega_j - \omega_1}{\omega_P}} \exp\left[2\alpha_P \theta_P (\omega_j - \omega_1)/\omega_P\right]}{\sum_{i=1}^{n} (1 - \theta_P - \theta_S)^{\frac{\omega_i - \omega_1}{\omega_P}} \exp\left[2\alpha_P \theta_P (\omega_i - \omega_1)/\omega_P\right]}, \tag{42}$$

where the total adsorption of protein molecules and the total surface coverage by protein molecules obey Eqs. (27) and (28). Also, the surface pressure dependence of a surfactant with molar area ω_S and adsorption Γ_S is assumed to obey Eqs. (36) and (37). Therefore, the problem of the theoretical description for a mixture can be formulated as follows: given the known values of T, ω_0, ω_{min}, ω_{max}, α_P, b_P, ε, ω_{S0}, α_S and b_S for the individual components and α_{PS} for the mixture, the dependencies of the parameters ω_P, Γ_P, Γ_S, θ_P, θ_S and Π as a function of the concentrations c_S and c_P should be calculated.

The procedure used for the numerical calculations is rather similar, although much more complicated, to that described in Section 3.2. In terms of the dimensionless quantities $\Omega = \omega_P/\omega_0$ and $\Omega_1 = \omega_1/\omega_0$, Eqs. (27), (28) and (40) can be combined to give a relation between θ_0 and Ω:

$$\Omega = \left[\Xi_M(\theta_P, \Omega)\right]^{-1} \cdot \sum_{i=1}^{n} (\Omega_1 + i - 1)(1 - \theta_P - \theta_S)^{(i-1)/\Omega} \exp\left[2\alpha_P(i-1)\theta_P/\Omega\right], \tag{43}$$

where

$$\Xi_M(\theta_P,\Omega) = \sum_{i=1}^{n}(1-\theta_P-\theta_S)^{(i-1)/\Omega}\exp[2\alpha_P(i-1)\theta_P/\Omega], \qquad (44)$$

and the adsorption isotherms for the protein in any state Eq. (40) becomes:

$$b_P c_P = \frac{\theta_P \exp(-2\alpha_P\theta_P\Omega_1/\Omega - 2\alpha_{PS}\theta_S)}{(1-\theta_P-\theta_S)^{\Omega_1/\Omega}\Xi_M(\theta_P,\Omega)} \qquad (45)$$

For any set of model parameters the set of transcendental equations (41), (43) and (45) with the function $\Xi_M(\theta_P,\Omega)$ defined by Eq. (44) can be solved numerically to yield the dependencies of Ω, θ_P and θ_S on c_P and c_S. With these values of Ω, θ_P and θ_S, one can solve the equation that follows from Eq. (38) into which the expressions for ω_0^* from Eq. (39) and for ω_S from Eq. (37) are introduced:

$$\frac{\Pi}{RT}\frac{\omega_0\theta_P + \omega_{S0}\theta_S}{\theta_P + \theta_S} +$$
$$\ln(1-\theta_P-\theta_S) + \theta_P\left(1-\frac{1}{\Omega}\right) + \alpha_P\theta_P^2 + 2\alpha_{PS}\theta_P\theta_S + \alpha_S\theta_S^2 = 0 \qquad (46)$$

with respect to Π. Finally, the adsorption values are determined as $\Gamma_P = \theta_P/(\Omega\omega_0)$ and $\Gamma_S = \theta_S/[\omega_{S0}(1-\varepsilon\Pi(\theta_S+\theta_P))]$.

The behaviour of mixed ionic surfactant/protein solutions is essentially different from that of non-ionic surfactant/protein mixtures. When a protein molecule with m ionized groups at a concentration of c_P interacts with ionic surfactant molecules of concentration c_S, Coulomb forces cause the formation of complexes. These complexes are determined by the average activity of ions $(c_P^m c_S)^{1/(1+m)}$ participating in the reaction. The respective equation of state of the surface layer is similar to mixed non-ionic surfactant/protein solutions (38) [23]:

$$-\frac{\Pi\omega_0^*}{RT} = \ln(1-\theta_{PS}-\theta_S) + \theta_{PS}(1-\omega_0/\omega) + a_{PS}\theta_{PS}^2 + a_S\theta_S^2 + 2a_{SPS}\theta_{PS}\theta_S \quad (47)$$

The corresponding adsorption isotherms for protein/surfactant complexes in state j=1 (similar isotherms can be obtained for any of the possible i states) and for the free surfactant not bound to the protein read [23]

$$b_{PS}\left(c_P^m c_S\right)^{1/(1+m)} = b_{PS} c_P^{m/(1+m)} c_S^{1/(1+m)} = \frac{\omega \Gamma_1}{\left(1 - \theta_{PS} - \theta_S\right)^{\omega_1/\omega}} \exp\left[-2a_{PS}\left(\omega_1/\omega\right)\theta_{PS} - 2a_{SPS}\theta_S\right]$$

$$(48)$$

$$b_S\left(c_S c_C\right)^{1/2} = \frac{\theta_S}{\left(1 - \theta_{PS} - \theta_S\right)} \exp\left[-2a_S\theta_S - 2a_{SPS}\theta_{PS}\right] \qquad (49)$$

Here $\theta_{PS} = \omega\Gamma$ is the coverage of the interface by adsorbed protein/surfactant complexes, c_C is the surfactant counter-ion concentration, and a_{SPS} is the parameter which describes the interaction of the non-associated surfactant with the protein/surfactant complexes. The subscript PS refers to the protein/surfactant complex, and the subscript S to the free surfactant.

5. Mixture of two surfactants

If for surfactants 1 and 2 the approximation $\omega_{10} \cong \omega_{20}$ is applicable, the following generalized Frumkin equation of state for a non-ionic surfactants mixture (neglecting the contribution of non-ideality of entropy) results [24]:

$$-\frac{\Pi \omega_0^*}{RT} = \ln(1 - \theta_1 - \theta_2) + \alpha_1\theta_1^2 + \alpha_2\theta_2^2 + 2\alpha_{12}\theta_1\theta_2, \qquad (50)$$

with

$$\omega_0^* = \frac{\omega_{10}\theta_1 + \omega_{20}\theta_2}{\theta_1 + \theta_2}, \qquad (51)$$

where $\theta_i = \omega_i \cdot \Gamma_i$ are the surface coverage by surfactants molecules, Γ_i are the adsorptions of the surfactants, b_i are the adsorption equilibrium constants, α_i are the interaction constants. The molar area of a surfactant ω_i can be approximately represented by a linear dependence on surface pressure Π and the total surface coverage θ, and therefore we get

$$\theta_1 = \Gamma_1\omega_1 = \Gamma_1\omega_{10}\left[1 - \varepsilon_1\Pi\theta\right]; \quad \theta_2 = \Gamma_2\omega_2 = \Gamma_2\omega_{20}\left[1 - \varepsilon_2\Pi\theta\right] \qquad (52)$$

$$\omega_1 = \omega_{10}(1 - \varepsilon_1\Pi\theta); \quad \omega_2 = \omega_{20}(1 - \varepsilon_2\Pi\theta) \qquad (53)$$

where $\theta = \theta_1 + \theta_2$. The adsorption isotherms turn into the generalized Frumkin equation [24]:

for surfactant 1:

$$b_1 c_1 = \frac{\theta_1}{(1-\theta_1-\theta_2)} \exp\left[-2\alpha_1\theta_1 - 2\alpha_{12}\theta_2\right] \tag{54}$$

for surfactant 2:

$$b_2 c_2 = \frac{\theta_2}{(1-\theta_1-\theta_2)} \exp\left[-2\alpha_2\theta_2 - 2\alpha_{12}\theta_1\right] \tag{55}$$

where c_i are the bulk concentrations of the surfactants.

Equations (50)-(55) can also be used to describe the mixture of two ionic surfactants or the mixture of ionic and non-ionic surfactant at an excess of inorganic electrolyte. These equations are also applicable to mixtures of an ionic surfactant (1) with a non-ionic surfactant (2) in the absence of inorganic electrolyte, however, we have to assume that the molar area ω_1 is equal to the area of a surface active ion, while the actual area per molecule is two times higher [24].

The procedure used for the numerical calculations is quite straightforward: given the known values of T, ω_{10}, ω_{20}, α_1, α_2, α_{12}, ε_1, ε_2, b_1 and b_2, a computation procedure was developed which for any given values of surfactant concentrations c_1 and c_2 eliminates θ_1 and θ_2 from Eqs. (54) and (55). The values θ_1 and θ_2 are then used to calculate the values of individual adsorption via Eqs. (52) and, finally, the surface pressure via Eqs. (50) and (51).

D. COMPARISON OF EXPERIMENTAL RESULTS WITH THEORETICAL MODEL

Let us employ the theoretical model presented above for the interpretation of experimental rheological results obtained for mixtures of non-ionic surfactant $C_{10}DMPO$ / β-lactoglobulin (β-LG), and for model calculations of the rheological characteristics of mixtures of β-LG with $C_{12}DMPO$ and $C_{14}DMPO$. Using experimental results for the systems lysozyme/SDS and lysozyme/$C_{10}DMPO$, we then demonstrate that a satisfactory agreement between the theoretical model and experimental findings can be obtained. At the first stage, the rheological behaviour of the individual substances solutions should be considered.

1. Adsorption and rheology of individual β-LG and C10DMPO solutions

Figure 1 presents the surface tension isotherm plotted for a β-LG solution in the presence of phosphate buffer (0.01 M of Na_2HPO_4 and NaH_2PO_4, pH 7.0) [5, 25]. These measurements were carried out by the drop shape method (PAT1 tensiometer from SINTERFACE Technologies, Berlin). The solution drops were formed at the tip of a PTFE capillary immersed into a cuvette filled with a water-saturated atmosphere. These experimental dependencies are well described by Eqs. (26)-(30), leading to the following parameters: $\omega_0 = 3.5 \cdot 10^5$ m^2/mol, $\omega_{min} = 5.3 \cdot 10^6$ m^2/mol, $\omega_{max} = 1.2 \cdot 10^7$ m^2/mol, $\alpha_P = 0.5$, $b_P = 1.9 \cdot 10^6$ l/mol (or $\Sigma b_P = 1.9 \cdot 10^6 \cdot 20 = 3.8 \cdot 10^7$ l/mol for the β-LG molecule as a whole).

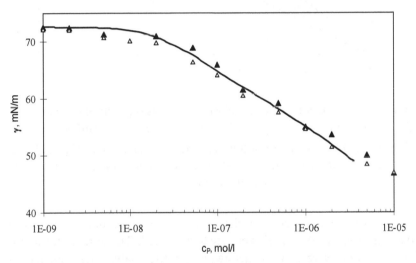

Fig. 1.   Surface tension isotherm of β-LG in phosphate buffer (from [5,24, theoretical isotherms were calculated from Eqs. (26)-(30) using the parameters given in the text.

The theoretical model given by Eqs. (13) and (26)-(30) with the same set of parameters, reproduces satisfactory the dilational rheological characteristics of the β-LG solutions. Figure 2 illustrates the theoretical dependence of limiting elasticity $E_0 = d\Pi/d\ln\Gamma$ on surface pressure for β-LG. This dependence agrees rather well with the experimental data obtained by the Wilhelmy method in a circular trough at an oscillation frequency of 0.13 Hz [26,27] (shown in Fig. 2 as dotted line and squares). Similar experimental data were obtained also by the dynamic drop tensiometer at a frequency of 0.1 Hz [28] also presented in Fig. 2. It follows from the calculations of phase angle that at a frequency 0.13 Hz the visco-elasticity modulus approaches the limiting elasticity modulus.

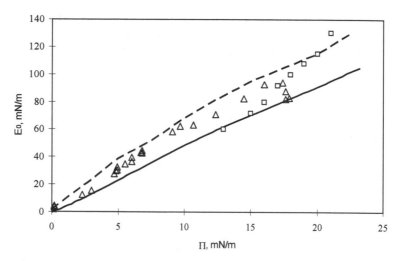

Fig. 2.   Dependence of visco-elastic modulus on surface pressure for β-LG. Solid line –limiting elasticity modulus $E_0 = d\Pi/d\ln\Gamma$ calculated from Eqs. (14), (26)-(30); dotted line and □ – data from [25] and [26], respectively, for a frequency of 0.13 Hz, △ refer to data from [27] at a frequency of 0.1 Hz

   The rheological characteristics of β-LG solution at a fixed concentration of 10^{-6} mol/l were studied in [15]. The surface tension of β-LG solution were again measured by profile analysis tensiometry. To study the dilational elasticity, after having reached the adsorption equilibrium (about 20 000 s) the solution drop was subjected to harmonic oscillations with a magnitude of ΔA = ± (7-8)% and frequencies in the range between 0.01 and 0.2 Hz. The results of the experiments with harmonic oscillations of the surface area were analysed using the Fourier transformation [29, 30]:

$$\varepsilon(i\omega) = A_0 \frac{F[\Delta\gamma]}{F[\Delta A]},\qquad(56)$$

where A_0 is the initial area of the drop surface. The initial experimental data were first filtered to exclude scattering errors. Also a constant drift of the surface tension with time was eliminated, caused by some small deviation of the system from equilibrium when starting the harmonic oscillations. Figure 3 illustrates the experimental dependencies of the visco-elasticity modulus and phase angle on frequency f. The experimental results are in a good agreement with the calculations made with Eqs. (10) for $c_S = 0$ (individual protein solution) and $D_P = 10^{-12}$ m²/s, or with Eqs. (13), (14). It is seen from this data that at a frequency of 0.13 Hz (similarly to [26,27]) the visco-elasticity modulus approaches the limiting elasticity, and the phase angle decreases to 5°, i.e. almost to zero. The diffusion coefficient D_P obtained here is much lower than expected for β-LG (about 10^{-10} m²/s). Calculations using such large D_P value would not agree with the experimental data in Fig. 3. This shows that in addition to the diffusional exchange with the bulk phase other relaxation effects take place in the adsorption layer, such as molecular reconformation, aggregation, etc. These processes are not analysed in detail here, instead we have used an effective diffusion coefficient.

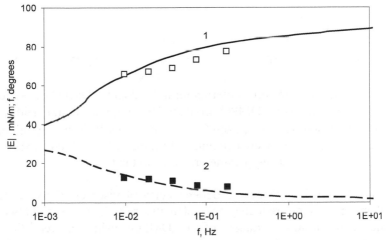

Fig. 3. Dependencies of visco-elastic modulus (Curve 1) and phase angle (Curve 2) on frequency f for β-LG at a concentration of 10^{-6} mol/l calculated from Eqs. (10), (11) at $c_S = 0$; experimental points are taken from [15].

Figure 4 illustrates the experimental surface tension isotherm for C_{10}DMPO (taken from [19]). To calculate the theoretical curves the following parameters of Eqs. (34)-(37) were used: $\omega_{S0} = 2.66 \cdot 10^5$ m²/mol, $\alpha_S = 0$, $b_S = 2.12 \cdot 10^4$ l/mol, and $\varepsilon = 0.012$ m/mN, and a quite satisfactory agreement with the experimental results is found. Moreover, the same set of model parameters for the equations of state and adsorption isotherm satisfactorily describes the C_{10}DMPO dilational rheological characteristics.

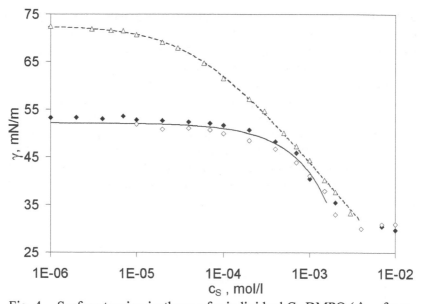

Fig. 4. Surface tension isotherms for individual $C_{10}DMPO$ ($\triangle$ – from
[19]) and β-LG/$C_{10}DMPO$ mixtures ($\blacklozenge$ – without sodium azide;
$\diamond$ – with sodium azide, from [15]) *vs* the $C_{10}DMPO$ concentration;
theoretical isotherms were calculated from Eqs. (34)-(37) and (38)-(42)
using parameters given in the text.

Figures 5 and 6 illustrate the dependencies of visco-elasticity modulus
and phase angle on frequency f as calculated from Eqs. (13) and (14) in a
wide concentrations range of $C_{10}DMPO$ using a realistic for
$D_S = 3 \cdot 10^{-10}$ m^2/s. It is seen from Fig. 5 that the frequency increase leads
to a monotonic increase of the visco-elasticity modulus |E|, however, its
concentration dependence is non-monotonous with a maximum. At the
same time, the dependence of ϕ on f decreases monotonously, and at
higher $C_{10}DMPO$ concentrations larger phase angles ϕ are observed.
Such behaviour of the visco-elasticity modulus and phase angle ϕ were
studied first by Lucassen and van den Tempel in [9].
The |E| values for $C_{10}DMPO$ calculated from the theory agree
satisfactorily with experimental values taken from [31] and added to
Fig. 5. These experimental values of $|E|$ for the $C_{10}DMPO$ were
measured for 0.5, 1.0 and 2.0 mmol/l using the oscillating bubble
method, and are quite close to the corresponding theoretical curves.
Moreover, they reproduce the anomalous behaviour characteristic to the
increasing $C_{10}DMPO$ concentration.

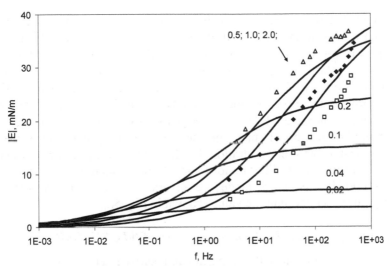

Fig. 5. Dependence of |E| on frequency f for $C_{10}DMPO$ solutions at various concentrations (labels at curves are concentrations in mmol/l) calculated from Eqs. (13) and (14); experimental points for concentrations of $\triangle$ – 0.5 mmol/l; $\blacklozenge$ – 1 mmol/l; $\square$ – 2 mmol/l are taken from [30]

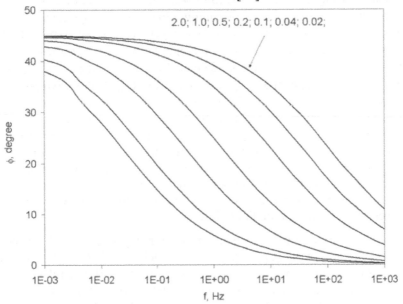

Fig. 6. Dependence of ϕ on frequency f for $C_{10}DMPO$ solutions at various concentrations (labels at curves are concentrations in mmol/l) as calculated from Eqs. (13) and (14).

This behaviour can be attributed to the fact that a concentration increase leads to an increase in the characteristic frequency given by the diffusion relaxation frequency ω_D, cf. Eq. (14). Therefore the high frequency elasticity limit is attained at higher frequencies. The intrinsic compressibility of $C_{10}DMPO$ molecules in the monolayer also results in the slower increase of the elasticity at high concentrations with frequency increase. For $\varepsilon = 0$ Eqs. (34)-(37) become identical to Frumkin's equation of state, and the high frequency elasticity increases much faster with concentration [20-22]. In this case, for $C_{10}DMPO$ concentrations of 0.5-2.0 mmol/l the elasticity modulus values in the 100-1000 Hz frequency range are several times (about one order of magnitude) higher than the values shown in Fig. 5.

2. Adsorption and rheological behaviour of β-LG/C10DMPO mixtures

Figure 4 illustrates the experimental surface tension isotherms for β-LG/$C_{10}DMPO$ mixtures within a range of $C_{10}DMPO$ concentration at a fixed β-LG concentration of 10^{-6} mol/l (reproduced from [15]). The two experimental series refer to solutions with and without added sodium azide. The theoretical curves for the mixtures shown in Fig. 4, were calculated using the model Eqs. (38)-(42) with parameters for β-LG and $C_{10}DMPO$ as given above, assuming either $\alpha_{PS} = 0$ (because strong inhomogeneities in the mixed surface layer have to be expected), or $\alpha_{PS} = (\alpha_S + \alpha_P)/2 = 0.25$, i.e., the average between the corresponding values for the protein and the surfactant. Both indicated α_{PS} values give rather similar dependences γ vs c_S for the β-LG/$C_{10}DMPO$ mixtures. The protein and surfactant molecules practically do not mix at the surface but form domains which contain essentially one of the components [32,33]. Additional phenomena are possible for protein/surfactant mixtures as well, such as complex formation, two- and three dimensional aggregation at the interface, etc. [17, 19, 34-37]. The given models assume mainly homogeneous mixed layers, in accordance with the models discussed in [1], due to which protein/surfactant aggregates compete with free surfactant molecules at the liquid interface.

Figure 7 shows the dependence of the monolayer coverage by protein and surfactant, calculated from the model Eqs. (38)-(42). At low $C_{10}DMPO$ concentrations (and also for single β-LG solution) the surface coverage by β-LG can be as high as 98%, while for a $C_{10}DMPO$ concentration of about 2 mmol/l the parts of the surface occupied by β-LG and $C_{10}DMPO$ are approximately the same. In this case the overall monolayer coverage becomes higher: 99%.

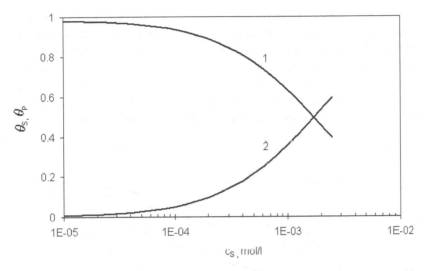

Fig. 7. Dependence of surface layer coverage by β-LG (curve 1) and by C$_{10}$DMPO (curve 2) in a β-LG/C$_{10}$DMPO mixture *vs* C$_{10}$DMPO concentration.

The dependencies of the dilational elasticity modulus $|E|$ on the oscillation frequency at various C$_{10}$DMPO concentrations in the β-LG/C$_{10}$DMPO mixtures are shown in Fig. 8, as reproduced from [15]. To study the dilational elasticity the solution drop was subjected to harmonic oscillations with a magnitude of $\Delta A = \pm(7-8)\%$ and frequencies in the range between 0.01 and 0.2 Hz. It follows from Fig. 8 that, with increasing C$_{10}$DMPO concentration, the elasticity modulus of the β-LG/C$_{10}$DMPO mixture decreases significantly. For example, at f = 0.1 Hz the modulus for β-LG mixed with 0.7 mmol/l C$_{10}$DMPO is 20 times lower than that for pure β-LG. The theoretical dependencies, also shown in Fig. 8, were calculated from Eq. (11) with model parameters of individual β-LG and C$_{10}$DMPO solutions listed above, $\alpha_{PS} = 0\div0.25$ and a surfactant diffusion coefficient of $D_S = 10^{-10}$ m^2/s (which is somewhat lower than the value $D_S = 3\cdot10^{-10}$ m^2/s used for the calculation of rheological characteristic of pure C$_{10}$DMPO solution) for the lowest C$_{10}$DMPO concentration in the mixture, and $D_S = 3\cdot10^{-10}$ m^2/s for $c_S > 0.2$ mmol/l. For this lower diffusion coefficient the agreement between theory and experiment becomes better. Fig. 8 shows that the agreement becomes quite satisfactory, especially in the range of low C$_{10}$DMPO concentrations.

At C$_{10}$DMPO concentrations above 0.1 mmol/l the agreement between theory and experiment becomes worse. To obtain a better correspondence, one has to use (as one of the possibilities) a higher diffusion coefficients for the protein, e.g., $D_P = (10^{-10}\text{--}10^{-11})$ m^2/s instead of 10^{-12} m^2/s. Probably, in presence of a surfactant the processes of protein reconformation and aggregation in the surface layer are accelerated, which increase the effective diffusion coefficient.

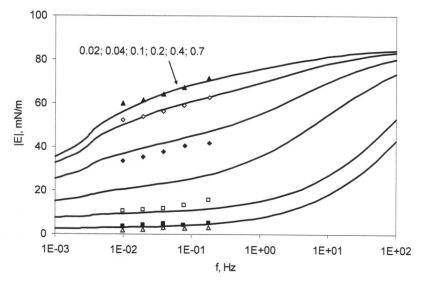

Fig. 8.   Dependencies of |E| on frequency f at various C$_{10}$DMPO concentrations (labels at curves are concentrations in mmol/l) in β-LG/C$_{10}$DMPO mixtures; theoretical curves were calculated from Eq. (11) with parameters indicated in the text; experimental points are data taken from [15] for ▲ – 0.02; ◇ – 0.04; ◆ – 0.1; □ – 0.2; ■ – 0.4; △ – 0.7 mmol/l C$_{10}$DMPO.

It should be noted that the desorption of β-LG from the adsorption layer is extremely slow, and surface oscillations lead to an increase in the adsorbed amount of protein in the surface layer, while desorption of this protein during the surface compression stage is very weak [34]. It was noted in [32,33,35,36] that three-dimensional domains of β-LG are formed upon the competitive adsorption of a surfactant at the surface. If a surfactant is desorbed from the surface layer (e.g., due to a washing off process by the solvent), then three-dimensional β-LG domains can transform again into a two-dimensional structure [34]. It is quite probable therefore that the equilibrium between subsurface and adsorption layer for protein molecules can be attained faster during the expansion/compression process in presence of a surfactant. It should be noted that the average size of β-LG domains in the presence of a surfactant is of the order of 1 μm or lower [37], which is significantly smaller than the diffusion layer thickness (for a diffusion coefficient of 10^{-10} m^2/s and frequencies below 1 Hz the diffusion layer thickness exceeds 10 μm). Therefore, one can neglect the non-homogeneity of the adsorption layer and accept the model given by Eqs. (5)-(11) which is valid for homogeneous layer.

The dependencies of the phase angle ϕ on frequency f for β-LG/C$_{10}$DMPO mixtures as calculated from Eq. (11) are shown in Fig. 9. For low surfactant concentrations a monotonous decrease of ϕ can be observed. However, with increasing C$_{10}$DMPO concentration a maximum in the ϕ vs f curve is obtained. Note′, the higher is the C$_{10}$DMPO concentration, the higher is the maximum ϕ value. This last effect is attributable to the increase of the fraction of the area covered by C$_{10}$DMPO (see Fig. 7), because for pure C$_{10}$DMPO solutions in the frequency range studied a viscous behaviour is observed (see Fig. 6), i.e. the corresponding elasticities are very small.

From Fig. 9 we see that the experimental phase angles agree rather well with the theoretical values calculated from Eq. (11). For the highest C$_{10}$DMPO concentrations studied (0.4 and 0.7 mmol/l) the experimental data scattering was very high (in the range of 0 to 50°) because the visco-elasticity modulus was extremely small (2-4 mN/m, see Fig. 8); therefore these data are not shown in Fig. 9.

One feature, characteristic to the theoretical calculations of the visco-elasticity modulus and phase angle from Eqs. (10), (11) should be noted here. For the indicated parameters of the β-LG/C$_{10}$DMPO mixtures, and at C$_{10}$DMPO concentrations above 1 mmol/l (i.e., for surface layer coverage > 0.985 and surface tension γ < 40 mN/m, cf. Figs. 4 and 7) these equations lead to negative values of E$_r$, while the phase angle exceeds 90°. From a detailed analysis of Eqs. (10) and (11) it follows that this result is due to the fact that one of the parameters P$_i$ becomes negative.

Note that partial derivatives a$_{11}$ and a$_{22}$ are always positive, while the derivatives a$_{12}$ and a$_{21}$ are always negative, however the term (a$_{11}$a$_{22}$ − a$_{12}$a$_{21}$) is positive even for negative E$_r$ values obtained in the calculations. We believe that negative E$_r$ values obtained in this range of C$_{10}$DMPO concentrations are simply artefact caused by some features of the theoretical model used, which affect the values of these derivatives.

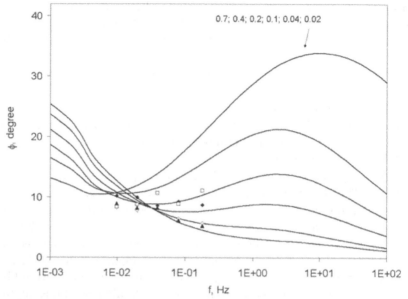

Fig. 9. Dependencies of phase angle ϕ on frequency f at various $C_{10}DMPO$ concentrations (labels at curves are concentrations in mmol/l) in β-LG/$C_{10}DMPO$ mixtures; theoretical curves were calculated from Eq. (11) with parameters indicated in the text; experimental points are data taken from [15] with addition of: ▲ – 0.02; ◇ – 0.04; ◆ – 0.1; □ – 0.2 mmol/l $C_{10}DMPO$.

For example, for β-LG/$C_{10}DMPO$ mixtures with $C_{10}DMPO$ concentrations above 2 mmol/l (i.e. almost at the CMC) the E_r values remain positive in the γ range up to 30 mN/m if one assumes that the molar areas in Eq. (38) are exactly equal to each other ($\omega_0 = \omega_S$ instead of the approximate relation $\omega_0 \cong \omega_S$, when Eq. (39) based on an additive model is valid).

3. Adsorption and Dilational Rheology of β-LG/C_nDMPO mixtures

Let us consider now the influence of the hydrocarbon chain length in the C_nDMPO molecule on the dilational rheological behaviour of mixtures with β-LG. The adsorption equilibrium and dilational rheological characteristics of $C_{12}DMPO$ and $C_{14}DMPO$ were studied in [20-22,24,31]. The adsorption and rheological characteristics of these surfactants are well described by the model of Eqs. (34)-(37) with the following parameters:

C_{12}DMPO: $\omega_{S0} = 2.48 \cdot 10^5$ m^2/mol, $\alpha_S = 0.38$, $b_S = 1.49 \cdot 10^5$ l/mol, and $\varepsilon = 0.01$ m/mN;

C_{14}DMPO: $\omega_{S0} = 2.52 \cdot 10^5$ m^2/mol, $\alpha_S = 0.6$, $b_S = 1.59 \cdot 10^6$ l/mol, and $\varepsilon = 0.01$ m/mN.

In the calculations of the rheological behaviour of β-LG/C_{12}DMPO and β-LG/C_{14}DMPO mixtures the average (for the protein and surfactant) α_{PS} values were used; the diffusion coefficient was assumed to be $D_S = 10^{-10}$ m^2/s.

The results obtained for mixtures of 10^{-6} mol/l β-LG with C_{12}DMPO at various concentrations are presented in Figs. 10 and 11.

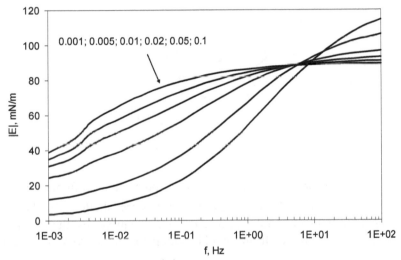

Fig. 10. Dependencies of $|E|$ on frequency f at various C_{12}DMPO concentrations (labels at curves are concentrations in mmol/l) in the β-LG/C_{12}DMPO mixtures; theoretical curves were calculated from Eq. (11) with parameters given in the text.

Throughout the C_{12}DMPO concentrations range chosen for the calculations, the surface tensions of the mixture decrease down to 40 mN/m, and for the maximum C_{12}DMPO concentration (0.1 mmol/l) the fraction of the surface covered by the surfactant exceeds 40%. Comparing the results shown in Fig. 10 with those presented in Fig. 8 one can see that a maximum visco-elastic modulus is attained at a lower oscillation frequency than for the β-LG/C_{10}DMPO mixtures. Also, the shape of the maximum in the dependence of phase angle on frequency in Fig. 11 is less pronounced, and occurs at a lower oscillation frequency as compared to Fig. This shift of the curves obtained for β-LG/C_{12}DMPO mixtures is attributable to the fact that the individual surfactant exhibits more elastic behaviour as compared to $C_{10}DMPO$ [31].

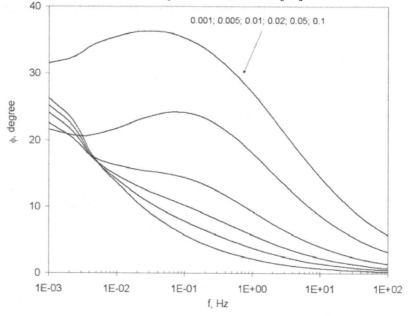

Fig. 11. Dependencies of phase angle ϕ on oscillation frequency at various $C_{12}DMPO$ concentrations (curves are labelled with the concentrations in mmol/l) in β-LG/$C_{12}DMPO$ mixtures; theoretical curves were calculated from Eq. (11) with parameters given in the text.

For β-LG/$C_{14}DMPO$ mixtures the changes in the rheological behaviour are yet more pronounced. For all surfactant concentrations in the range between 0.0001 and 0.015 mmol/l (with a decrease in surface tension down to 35 mN/m at the maximum concentration studied) the limiting visco-elastic modulus is observed at frequencies above 1 Hz. This behaviour is accompanied by a monotonous decrease of the phase angle; at this frequency the phase angle value does not exceed 5° at all $C_{14}DMPO$ concentrations.

4. Rheological behaviour of lysozyme mixed with $C_{10}DMPO$ or SDS

Results of lysozyme and mixed solutions of lysozyme ($7 \cdot 10^{-7}$ mol/l) with SDS and $C_{10}DMPO$ were recently published in terms of dynamic and equilibrium surface tension, adsorption and dilational rheology and discussed on the basis of the theoretical models described above [38].

The surface pressure isotherm for lysozyme/C_{10}DMPO and lysozyme/SDS mixtures are presented in Figs. 12 and 13, respectively, including the isotherms of the pure surfactants. The theoretical dependencies of the surface pressure for lysozyme/C_{10}DMPO mixtures shown in Fig. 12 were calculated from Eqs. (38)-(42). The parameters used in these calculations were reported in [38] as: $\omega_0=5\cdot10^5$ m^2/mol, $\omega_1=7.5\cdot10^6$ m^2/mol, $\omega_{max}=2.5\cdot10^7$ m^2/mol, $\alpha_P=0$ and $b_P=5\cdot10^3$ m^3/mol. It is seen from Fig. 12 that in the surfactant concentration range $c_S < 5\cdot10^{-4}$ mol/l the theoretical curve (bold solid line) agrees very well with the experimental data, while at concentrations above $5\cdot10^{-4}$ mol/l the theoretical values are slightly higher than the experimental data. To improve the agreement the adsorption equilibrium constant for the protein should be decreased by one order of magnitude, as shown by the bold dotted line in Fig. 12. This fact indicates that less hydrophobic protein/surfactant complexes are formed at higher C_{10}DMPO concentrations.

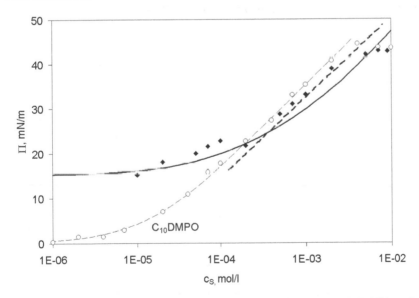

Fig. 12: Equilibrium surface pressure isotherm for C_{10}DMPO ($\bigcirc$) and $7\cdot10^{-7}$ mol/l lysozyme/C_{10}DMPO ($\blacklozenge$) at the air/water interface, lines are theoretical values (see text).

The interaction between a protein and an ionic surfactant obeys different mechanisms. At pH 7 each lysozyme molecule has 8 positive net charges [39]. Due to electrostatic interactions the protein and surfactant molecules at low surfactant concentration form complexes which are more hydrophobic and hence have a higher surface activity than the original protein molecules. In Figure 13 the experimental surface pressure for mixtures of lysozyme with SDS and pure SDS solutions are shown. The curved correspond to theoretical calculations from the model Eqs. (47)-(49), using the parameters for lysozyme as given in Fig. 12 (additional parameter is α_{SPS}), and for SDS in buffer those reported in [38]: $\omega_{S0} = 2.85 \cdot 10^5$ m^2/mol, $b_S = 19$ m^3/mol, $\alpha_S = 0$ and $\varepsilon = 0.012$ m/mN. The thin solid line corresponds to a model without any complex formation, i.e. assuming that the surface activity of the protein/surfactant complex is the same as in individual protein solutions. The bold solid line corresponds to complexes formed by one protein molecule and eight SDS molecules (m=8) with a tenfold increased surface activity for the complex. The bold dotted line was calculated using a ten times lower activity of the protein/surfactant complex (as compared with the individual protein solution) due to the hydrophilisation of the complex, assuming 30 or more SDS molecules bound to the protein via hydrophobic interaction.

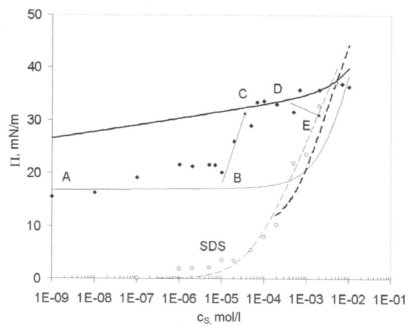

Fig. 13: Equilibrium surface pressure isotherm for SDS (○) and $7 \cdot 10^{-7}$ mol/l lysozyme/SDS (◆) at the air/water interface, lines are theoretical values (see text).

These theoretical curves in Fig. 13 agree quite well with the experimental findings. In the concentration range between the points A and B, electrostatic binding of SDS to the protein is negligible. At SDS concentrations $c_S > 10^{-6}$ mol/l (point B) the inter-ionic binding of SDS becomes significant, the hydrophobicity of the complex increases, resulting in an increased adsorption activity (range B-C-D in Fig. 13). Further increase of the SDS concentration leads to a hydrophilisation of the complexes caused by the hydrophobic interactions (range D-E), however, this does not result in any appreciable surface pressure changes.

The surface dilational modulus for a fixed lysozyme concentration of $7 \cdot 10^{-7}$ mol/l is shown as a function of the surfactant concentration for lysozyme/C_{10}DMPO mixtures in Figure 14 and lysozyme/SDS mixtures in Figure 15, respectively, measured at a fixed oscillation frequency of 0.08 Hz (similar data are presented in [38] for 0.01, 0.04, 0.16 and 0.4 Hz). For both systems the elasticity slightly increases with increasing oscillation frequency, while the phase angle decreases, as expected. At all frequencies the phase angles remains below 12°. For lysozyme/C_{10}DMPO mixtures the elasticity modulus decreases with increasing surfactant concentration. The dependencies $E(c_S)$ are shown in Figs. 14 and 15, calculated with the theoretical model (11). The respective curves have been obtained in the same way as those in Figs. 12 and 13. For the theoretical calculations model parameters identical to those for the equilibrium surface pressure calculation were used [38] with the following diffusion coefficients: $D = 10^{-12}$ m²/s for lysozyme, and $D = 4 \cdot 10^{-10}$ m²/s for the surfactants. The theoretical curve in Fig. 14 describes the experimental results very well. From the surface tension isotherm in Fig. 1 we see that at C_{10}DMPO concentrations above 10^{-4} mol/l a partial hydrophilisation of the protein/surfactant complexes can take place, which decreases the protein adsorption and the dilational elasticity considerably decreases, as shown by the bold dotted line in Fig. 14.

However, for lysozyme/SDS mixtures there is a surfactant concentration interval, in which the elasticity of the mixed layer is significantly increased (see Fig. 15). In agreement with the situation shown in Fig. 13, the arrows indicate the transitions from non-associated lysozyme (A-B) to the more hydrophobic complex (B-C-D) and then to the hydrophilised complex (D-E). Note that the transition points B and D in both Figs. 13 and 15 correspond to approximately the same SDS concentrations.

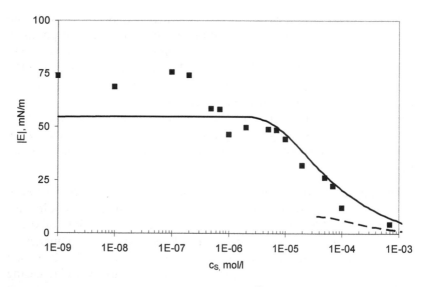

Fig. 14: Surface dilational modulus of $7 \cdot 10^{-7}$ mol/l lysozyme/C_{10}DMPO versus C_{10}DMPO concentration at an oscillation frequency of 0.08 Hz (■), lines are theoretical values (see text).

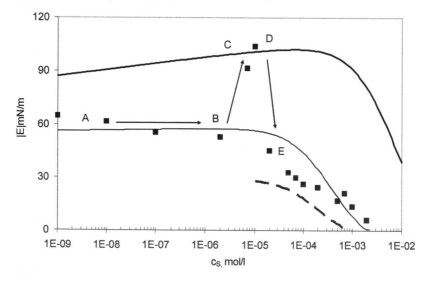

Fig. 15: Surface dilational modulus of $7 \cdot 10^{-7}$ mol/l lysozyme/SDS versus SDS concentration at an oscillation frequency of 0.08 Hz (■), lines are theoretical values (see text).

It is essential that the model calculations allow us to explain the maximum observed in the dependence of the elasticity $E(c_S)$ for lysozyme/SDS mixed adsorption layers as shown in Fig. 15. This maximum can be attributed essentially to the adsorption of hydrophobised lysozyme/SDS complexes. The step by step transition to complexes in point E which are significantly hydrophilised due to the hydrophobic interaction (with an activity of the complex comparable with the pure protein or even by a factor of 10 lower) is accompanied by a decreased visco-elasticity modulus.

5. Adsorption and rheological behaviour of SDS/dodecanol mixture

Let us consider first the characteristics of individual SDS and dodecanol solutions. The fitting of the corresponding surface tension isotherms was performed by using Eqs. (34)-(37). For the considered surfactants these equations are only approximations, because they disregard both the average activity of SDS ions and the cluster formation in dodecanol monolayers. The experimental surface tensions of SDS solutions were taken from [24,40] while those for dodecanol from [41]. In [40] also data for the limiting (high-frequency) elasticity for of SDS and dodecanol solutions were presented. These data were also used to determine the parameters of the theoretical model:

SDS: $\omega_0 = 1.7 \cdot 10^5$ m^2/mol (per one ion), $\alpha = 1.4$, $b = 3.0 \cdot 10^2$ l/mol, and $\varepsilon = 0.013$ m/mN;

Dodecanol: $\omega_0 = 1.2 \cdot 10^5$ m^2/mol, $\alpha = 1.55$, $b = 1.12 \cdot 10^5$ l/mol, and $\varepsilon = 0.004$ m/mN;

It is seen from Fig. 16 that with these values and the Eqs. (34)-(37) both the surface tension isotherms and limiting elasticities E_0 on the surfactant concentration can be well reproduced. Figures 17 and 18 illustrate the dependencies of the visco-elasticity modulus and phase angle on frequency for dodecanol solutions, as calculated from Eqs. (13)-(14) and (34)-(37) with $D = 5 \cdot 10^{-10}$ m^2/s. At a frequency of 10 Hz the phase angle becomes equal to zero. The calculated dependencies shown in Fig. 17 for concentrations 0.005, 0.006 and 0.008 mmol/l are almost identical to the experimental data presented in [40]. Also the calculated phase angles shown in Fig. 18 are similar to the experimental data, i.e. for f > 10 Hz the phase angles are zero [40].

The dependencies of visco-elasticity modulus on frequency for SDS solutions calculated from Eqs. (13)-(14) for $D = 4 \cdot 10^{-10}$ m^2/s agree quite well with experimental data [40]. In the frequency range 0.001–1000 Hz the theory predicts a monotonous increase of the modulus, and a monotonous decrease of the phase angle, however, even at 1000 Hz the phase angle remains nonzero. The theoretical predictions for the frequency dependence of the phase angle disagree with the experimental data. It was observed in [40] (see Fig. 4 therein) that the phase angle increases with frequency between 1 and 500 Hz. It will be shown below that this experimental behaviour is possibly be provoked by the presence of trace quantities of dodecanol (at a level of 0.01 % mol) in the SDS solution.

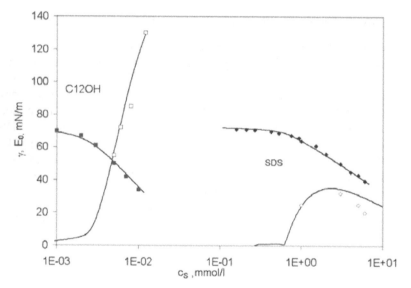

Fig. 16. Dependencies of equilibrium surface tension (filled points) and limiting elasticity (open points) of SDS and dodecanol solutions on concentration c_S (data taken from [24,40,41]); theoretical curves were calculated from Eq. (34)-(37) with parameters given in the text.

Figures 19 and 20 illustrate the theoretical dependencies of the visco-elasticity modulus and phase angle on frequency for SDS/dodecanol mixtures calculated from Eqs. (11) and (50)-(55) with parameters of the individual surfactants as listed above. The α_{12} value was taken to be 1.5, i.e., the average between those for dodecanol and SDS. With increasing frequency the modulus also increases (see Fig. 19), while the phase angle attains first a minimum, and then a maximum value (see Fig. 20). All theoretical curves shown in Figs. 19 and 20 for SDS/dodecanol mixtures exhibit a similar behaviour, and are in a qualitative agreement with experimental data presented in [40] for the same concentrations of SDS and dodecanol in the mixtures.

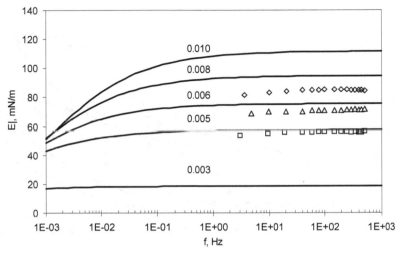

Fig. 17. Dependencies of dilational elasticity modulus |E| on oscillation frequency at various dodecanol concentrations (labelled give concentration in mmol/l); experimental points (for the concentrations 0.005, 0.006 and 0.008 mmol/l) are taken from [40]; theoretical curves calculated from Eq. (13) with parameters given in the text.

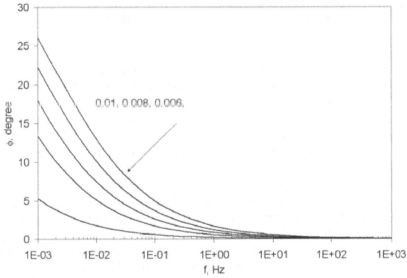

Fig. 18. Dependencies of phase angle ϕ on oscillation frequency at various dodecanol concentrations (curves labelled with the concentrations in mmol/l) calculated from Eq. (13) with parameters given in the text.

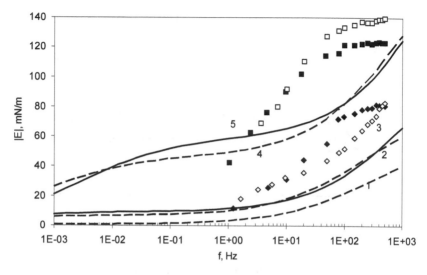

Fig. 19 Dependencies of dilational elasticity modulus |E| on oscillation frequency for SDS/dodecanol mixtures at various concentrations; theoretical curves calculated from Eq. (11) with parameters given in the text; experimental points taken from [40]; Curves labelled as mmol/l SDS+ mmol/l dodecanol: (1) 3 + 0.0003 ; (2, ◇) 3 + 0.0015; (3, ◆) 5 + 0.003; (4, □) 3 + 0.012; (5, ■) 5 + 0.02.

For slightly lower SDS and dodecanol diffusion coefficients the agreement between theoretical values and experimental data becomes even better. However, having in mind that a quite approximate model for the surfactants was employed here, one could hardly expect a perfect agreement between theory and experiment.

The theoretical curves shown in Figs. 19 and 20 were calculated from Eq. (10) for SDS solutions in the presence of dodecanol. 0.0003 mol/l dodecanol corresponds 0.01 mol % in the solution. In Fig. 20 it is shown that even a small admixture of dodecanol leads to a phase angle increase in a certain frequency range. These results are in qualitative agreement with the data reported in [42] on pure SDS solutions. It can be shown by model calculations that only for dodecanol concentrations lower than 0.001 mol % (one alcohol molecule per 100.000 SDS molecules), the presence of dodecanol does not affect the rheological properties of SDS. Therefore, it is reasonable to assume that the anomalous character of the phase angle dependence on frequency for SDS (see Fig. 4 in [40]) is caused by the presence of dodecanol, which cannot be eliminated from the solution due to a permanent hydrolysis process.

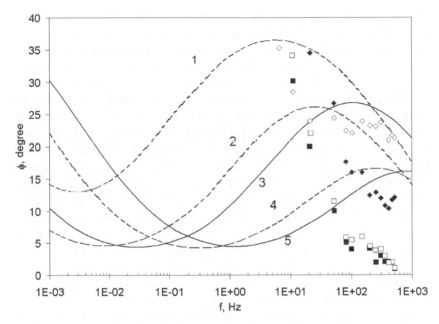

Fig. 20. Dependencies of phase angle ϕ on oscillation frequency for SDS/dodecanol mixtures at various concentrations; theoretical curves calculated from Eq. (11) with parameters given in the text; experimental points taken from [40]; Curves labelled as mmol/l SDS + mmol/l dodecanol: (1) 3 + 0.0003 ; (2, ◇) 3 + 0.0015; (3, ◆) 5 + 0.003; (4, □) 3 + 0.012; (5, ■) 5 + 0.02.

F. CONCLUSIONS

Expressions for the complex elasticity modulus of the mixture of two surfactants as presented by Jiang et al. [11] and Joos [12] are derived more accurately and it is also shown that these equations are equivalent to each other, and can be transformed into equations derived by Lucassen and van den Tempel [9,10] for single surfactant solutions. A numerical procedure is proposed for calculating the real and imaginary parts of the elasticity modulus, which can be applied to any model of the interfacial layer provided the model enables the determination of the surface tension and adsorption dependence on concentration of the components. This procedure is illustrated for protein/non-ionic and protein/ionic surfactant mixtures in the framework of a model proposed earlier [17-19,23]. This model assumes multiple states of the protein molecule in the surface layer and an intrinsic compressibility of the surface layer [20-22]. A procedure for numerical solution of the set of equations which govern the adsorption behaviour of the mixed protein/surfactant solution is developed. The rheological characteristics of mixed adsorption layers containing a protein and a non-ionic surfactant (β-lactoglobulin and alkyl dimethyl phosphine oxide C_nDMPO) were calculated in the frequency range 0.001 to 100 Hz, while the model involves only parameters of the individual components determined from results obtained in separate studies of surface tension, adsorption and dilational rheology of individual solutions. The experimental data obtained from slow oscillating drop studies for mixed β-lactoglobulin/C_{10}DMPO adsorption layers are in good agreement with values predicted by model calculations. It is shown that the competitive adsorption depends also on the surfactant's nature – non-ionic or ionic. For the non-ionic surfactant C_{10}DMPO the concentration increase results in a monotonous decrease of the equilibrium surface tension, dilational elasticity and adsorption of β-lactoglobulin and lysozyme. For the ionic surfactant SDS (in mixture with lysozyme) the surface tension exhibits a sharp decrease in the SDS concentration range between 10^{-5} and 10^{-4} mol/l, accompanied by maxima in dilational elasticity and adsorption. This phenomenon can be attributed to the formation (due to electrostatic interaction) of highly surface active hydrophobic complexes of lysozyme/SDS. The experimental dependencies agree well with the theoretical model.

The theoretical analysis of the rheological behaviour of a surfactant mixture is presented using the well-known system SDS/dodecanol. The dependencies of the elasticity modulus and phase angle on frequency calculated using the model parameters of SDS and dodecanol and assuming an intrinsic compressibility agree qualitatively with experimental data from literature [40] performed using the oscillating bubble method in the frequency range between 1 and 500 Hz.

Additional demonstrations for the good agreement of the proposed theoretical models for proteins, protein/surfactant and surfactant/surfactant mixtures with experimental data are presented for example in the publications [42-44].

G. DISTRIBUTION OF SOFTWARE USED FOR NUMERICAL CALCULATIONS

To perform the calculations according to the theoretical approach described in Sections 3.1-3.5, a set of software packages was developed which implement the corresponding models. In particular, using these programs one can calculate the thermodynamic and rheological characteristics of individual proteins and surfactants, and for mixtures of proteins with non-ionic and ionic surfactants; some of the programs provide for a visual comparison between the experimental data and calculated curves.

These program packages (and also those described earlier [45]) include software, manuals and examples of initial data files, and are available free of charge as downloads via http://www.sinterface.de/ or http://www.mpikg.mpg.de/gf/miller/.

ACKNOWLEDGEMENTS

The work was financially supported by projects of the European Space Agency (FASES MAP AO-99-052) and the DFG (Mi418/16-1). The authors express their thanks to Emmie Lucassen-Reynders and Jaap Lucassen for many instructive discussions.

H. REFERENCES

1. R. Miller, V.B. Fainerman, A.V. Makievski, J. Krägel, D.O. Grigoriev, V.N. Kazakov and O.V. Sinyachenko, Adv. Colloid Interface Sci., 86 (2000) 39.

2. A. Dussaud, G.B. Han, L. Ter Minassian-Saraga and M. Vignes-Adler, J. Colloid Interface Sci., 167 (1994) 247.

3. N.J. Turro, X.-G Lei, K.P. Ananthapadmanabhan and M. Aronson, Langmuir, 11 (1995) 2525.

4. R. Wüstneck, J. Krägel, R. Miller, P.J. Wilde and D.C. Clark, Colloids Surfaces A, 114 (1996) 255.

5. J. Krägel, M. O'Neill, A.V. Makievski, M. Michel, M.E. Leser and R. Miller, Colloids Surfaces B, 31 (2003) 107.

6. E. Dickinson, Colloids Surfaces B, 15 (1999) 161.

7. E.H. Lucassen-Reynders, J. Colloid Interface Sci., 42 (1973) 573.

8. P.R. Garrett and P. Joos, J. Chem. Soc. Faraday Trans. 1, 72 (1976) 2161.

9. J. Lucassen and M. van den Tempel, Chem. Eng. Sci. 27 (1972) 1283.

10. J. Lucassen and M. van den Tempel, J. Colloid Interface Sci. 41 (1972) 491.

11. Q. Jiang, J.E. Valentini and Y.C. Chiew, J. Colloid Interface Sci. 174 (1995) 268.

12. P. Joos, *Dynamic Surface Phenomena*, VSP, Dordrecht, The Netherlands, 1999.

13. B.A. Noskov and G. Loglio, Colloids Surfaces A, 141 (1998) 167.

14. I.B. Ivanov, K.D. Danov, K.P. Ananthapadmanabhan and A. Lips, Adv. Colloid Interface Sci. 114–115 (2005) 61.

15. R. Miller, M.E. Leser, M. Michel and V.B. Fainerman, J. Phys. Chem., 109 (2005) 13327.

16. E.V. Aksenenko, V.I. Kovalchuk, V.B. Fainerman and R. Miller, Adv. Colloid Interface Sci., 122 (2006) 57.

17. V.B. Fainerman, E.H. Lucassen-Reynders and R. Miller, Adv. Colloid Interface Sci., 106 (2003) 237.

18. E.H. Lucassen-Reynders, V.B. Fainerman and R. Miller, J. Phys. Chem., 108 (2004) 9173.

19. V.B. Fainerman, S.A. Zholob, M. Leser, M. Michel and R. Miller, J. Colloid Interface Sci., 274 (2004) 496.

20. V.B. Fainerman, R. Miller and V.I. Kovalchuk, J. Phys. Chem. B., 107 (2003) 6119.

21. V.B. Fainerman, V.I. Kovalchuk, E.V. Aksenenko, M. Michel, M.E. Leser and R. Miller, J. Phys. Chem. B., 108 (2004) 13700.

22. V.B. Fainerman, R. Miller and V.I. Kovalchuk, Langmuir, 18 (2002) 7748.

23. V.B. Fainerman, S.A. Zholob, M.E. Leser, M. Michel and R. Miller, J. Phys. Chem. B, 108 (2004) 16780.

24. V.B. Fainerman, R. Miller, E.V. Aksenenko and A.V. Makievski, In "Surfactants – Chemistry, Interfacial Properties and Application", Studies in Interface Science, V.B. Fainerman, D. Möbius and R. Miller (Eds.), Vol. 13, Elsevier, 2001, p. 189-286.

25. J. Krägel, R. Wüstneck, F. Husband, P.J. Wilde, A.V. Makievski, D.O. Grigoriev and J.B. Li, Colloids Surfaces B, 12 (1999) 399.

26. M.J. Ridout, A.R. Mackie and P.J. Wilde, J. Agric. Food. Chem., 52 (2004) 3930.

27. B. Rippner Blomqvist, M.J. Ridout, A.R. Mackie, T. Wärnheim, P.M. Claesson and P.J. Wilde, Langmuir, 20 (2004) 10150.

28. J. Benjamins and E.H. Lucassen-Reynders. In: *Food Colloids, Biopolymers and Materials*. E. Dickinson and T. van Vliet (Editors), Royal Society of Chemistry, Cambridge, 2003.

29. G. Loglio, P. Pandolfini, R. Miller, A.V. Makievski, P. Ravera, M. Ferrari and L. Liggieri. In: *Novel Methods to Study Interfacial Layers*. D. Möbius and R. Miller (Editors), Elsevier Science, Amsterdam, 2001.

30. G.M. Jenkins and D.G. Watts. *Spectral Analysis and its Applications.* Holden Day, San Fransisco, 1969.

31. K.-D. Wantke and H. Fruhner, J. Colloid Interface Sci., 237 (2001) 185.

32. A.R. Mackie, A.P. Gunning, M.J. Ridout, P.J. Wilde and V.I. Morris, Langmuir, 17 (2001) 6593.

33. A.R. Mackie, A.P. Gunning, M.J. Ridout, P.J. Wilde and J.R. Patino, Biomacromolecules, 2 (2001) 1001.

34. V.B. Fainerman, M.E. Leser, M. Michel, E.H. Lucassen-Reynders and R. Miller, J. Phys. Chem. B., 109 (2005) 9672.

35. A.R. Mackie, A.P. Gunning, P.J. Wilde and V.J. Morris, Langmuir, 16 (2000) 2242.

36. J.M.R. Patino, M.R.R. Nino, C.C. Sanchez and M.C. Fernandez, Langmuir, 17 (2001) 7545.

37. A.R. Mackie, A.P. Gunning, L.A. Pugnaloni, E. Dickinson, P.J. Wilde, and V.J. Morris, Langmuir, 19 (2003) 6032.

38. V.S. Alahverdjieva, D.O. Grigoriev, V.B. Fainerman, E.V. Aksenenko, R. Miller and H. Möhwald, *J. Phys. Chem.*, 112 (2008) 2136.

39. M.D. Lad, V.M. Ledger, B. Briggs, R.J. Green and R.A. Frazier, Langmuir, 19 (2003) 5098.

40. K.-D. Wantke, H. Fruhner and J. Ortegren, Colloid and Surfaces A, 221 (2003) 185-195.

41. R.-Y. Tsay, T.-F. Wu and S.-Y. Lin, J. Phys. Chem. B, 108 (2004) 18623.

42. E.V. Aksenenko, V.I. Kovalchuk, V.B. Fainerman and R. Miller, J. Phys. Chem. C, 111 (2007) 14713

43. R. Miller, V.S. Alahverdjieva and V.B. Fainerman, Soft Matter, 4 (2008) 1141.

44. Cs. Kotsmar, J. Krägel, V. I. Kovalchuk, E. V. Aksenenko, V. B. Fainerman| and R. Miller, J. Phys. Chem. C, 113 (2009) 103.

45. E.V. Aksenenko, In "Surfactants – Chemistry, Interfacial Properties and Application", Studies in Interface Science, V.B. Fainerman, D. Möbius and R. Miller (Eds.), Vol. 13, Elsevier, 2001, p. 619-648.

INTERFACIAL SHEAR RHEOLOGY – AN OVERVIEW OF MEASURING TECHNIQUES AND THEIR APPLICATIONS

J. Krägel[1] and S.R. Derkatch[2]

[1]Max Planck Institute of Colloids and Interfaces, Potsdam-Golm,
Germany
[2]Murmansk State Technical University, Murmansk, Russian Federation

Contents

A.   INTRODUCTION

In this chapter at first the basic concepts of shear interfacial rheology are discussed. A general introduction into two-dimensional rheology was already given in Chapters 1 and 2. The further aim is to give an overview of different measuring techniques to characterise the flow behaviour of interfaces under shear deformations, however, any detailed theoretical background of the different techniques is not given here. The description of different measuring techniques is mainly focused on torsion pendulum methods. Also some selected examples of experimental results are presented to demonstrate the application of different techniques to various chemical systems. The micro-rheological techniques which use optical or magnetic tweezers are given in Chapter 16.

Many colloidal systems comprise large interfaces, such as foams and emulsions, and hence there is a remarkable impact of interfacial rheology to the bulk rheology of such fluids. In general interfacial rheology describes the functional relationship between the deformation of an interface, the stresses exerted in and on it, and the resulting flows in the adjacent fluid phases.

The shear rheology of interfacial active molecules adsorbed at air/liquid and liquid/liquid phase boundaries is relevant in a wide range of technical applications such as mass transfer, monolayers, foaming, emulsification oil recovery, or high speed coating. Interfacial rheological data have been used for model calculation e.g. for foam drainage and Ostwald ripening in emulsions [1, 2]. A variety of measuring techniques have been proposed in the literature to measure interfacial shear rheological properties. The different techniques can be classified in indirect and direct methods. Indirect techniques base on the displacement of tracer particles placed at the interface is recorded and analyzed by image analysis. While with direct techniques the displacement or torque of a probe located within the interface is directly measured. One point of importance in the design of such instruments is to provide adequate sensitivity to detect stresses in the interfacial layer in presence of stresses in the adjacent subphases. Therefore the analysis of hydrodynamic flow fields in different measuring techniques has been a subject of many experimental and theoretical studies and will be discussed under the relevant subsections. Many of the existing developed measuring methods are based on wave behaviour, classified as capillary wave damping, surface shear and longitudinal wave experiments. These techniques and also the methods which utilize spectroscopic, AFM or particle tracking techniques to measure rheological properties at interfaces are not discussed in this chapter.

B. INDIRECT METHODS

Indirect methods measure velocity profiles using visible inert particles from which the interfacial shear viscosities can be determined. They are more or less restricted to the gas/liquid interface, although some modifications of the deep channel technique were also used at the liquid/liquid interface. In this section we discuss mainly four types of viscometers, from which only the first two techniques have been widely applied.

1. Channel surface viscometer

For insoluble spread monolayers the interfacial shear viscosity η_S can be measured in a way entirely analogous to the Hagen-Poiseuille law for bulk viscosity of liquids. It is based on the determination of the flow rate of a film through a narrow channel or slit under an applied two-dimensional pressure difference $\Delta\Pi$. This technique can only be used at the air/liquid interface. The classical channel surface viscometers, first proposed by Myers and Harkins [3] and Dervichian and Joly [4] are based on this idea, schematically shown in Fig. 1.

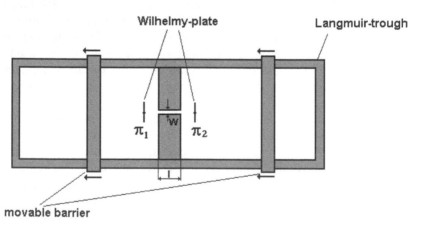

Fig. 1 Schematic principle of channel viscometer

η_S is calculated from the rate of film flow Q through the channel of width w and length l according to Eq. (1).

$$\eta_s = \frac{\Delta\Pi w^3}{12lQ} \tag{1}$$

An improved form which includes a correction term for the drag of the underlying liquid having a bulk viscosity η was given by Harkins and Kirkwood [5].

$$\eta_s = \frac{\Delta\Pi w^3}{12lQ} - \frac{w\eta}{\pi} \tag{2}$$

Dervichian and Joly [6] emphasized that the flow in a narrow channel is much more complicated. Eq. (2) is valid only if the channel is narrow (1 » w) and the walls are smooth and parallel. A more detailed analysis of the flow of films and the subphase has been given by Hansen [7] and later by Goodrich [8] and Stone [9].Other assumptions are that there is no slip of the film along the walls, and that the flow is Newtonian with constant surface viscosity within the channel. It is not easy to achieve all these theoretical conditions experimentally. Hühnerfuss [10] suggested a redefinition of the parameter "surface viscosity", which includes the coupled system surface film/film-induced water layer. Based on this definition it is possible to measure time-dependent interfacial viscosities of both relaxing and relaxed surface films by this method.

A first systematic experimental confirmation of the hydrodynamic coupling as described by Eq. (2) was provided by Sacchetti et al. [11] for a complete incompressible surface film. Therefore further experiments with fatty acids, lipids and polymer monolayers are performed in the liquid-condensed phase in order to comply with the incompressibility requirement. Turati et al. [12] used the channel technique to study the influence of subphase viscosity and channel width at different surface pressure on the flow of bolaform lipids and oleic acid monolayers through a slit.

Over the long period of time the classical channel viscometer has been used for a lot of different systematic studies of insoluble monolayers. As an example the studies of long-chain aliphatic amides, amines, alcohols and carboxylic acids or polydimethylsiloxane monolayers [13] should be mentioned here. Hühnerfuss [10] measured the time-dependent surface viscosity of both relaxing and relaxed surface films by means. Monolayer film viscosities for fatty acids have been investigated by many experimentalists, who clarified the effect of alkyl chain length and metal ions added to a aqueous subphase. It has been found out that the monolayer properties are essentially affected by their hydrophobic and hydrophilic interactions. In other studies Hühnerfuss [14] outlined the optimum hydrophobic interaction between the alkyl chains resulting in an increase in surface viscosity. The surface viscosity values measured for a series of homologous compounds were compared and an optimum hydrophobic interaction was found. In more recent studies fatty acid monolayer were investigated. Fatty acids display a variety of phases as a function of surface pressure and temperature. The two-dimensional phase behaviour is well understood and one can relate the rheological response to the known molecular organization. These phases are characterized by a local lattice structure of the molecular headgroups as well as by long-range order of the molecular tilt angle giving rise to a domain structure. Schwartz et al. [15] used a fluorescence microscope to observe directly the velocity profile in the channel, which was invariant with increasing area fraction of the condensed phase. A semi elliptical velocity profile was obtained and thereby demonstrated that the flow is controlled by a coupling to the subphase. These observations are used for an analysis of the fluid motion of monomolecular films in a channel by Stone [9], and the monolayer velocity profiles were given as a function of the viscosity contrast between the monolayer and the subphase. Kurnaz and Schwartz [16] observed the surface-pressure driven flow of eicosanoic acid Langmuir monolayer trough a narrow channel using Brewster angle microscopy (BAM). At low surface pressure the velocity profile is parabolic for low flow rates. This implies that the surface viscosity dominates the coupling to the aqueous subphase as a source of dissipation and that the monolayer behaves Newtonian. At extremely high shear rates, a flattened velocity profile is observed, similar to a plug flow, which is the evidence of a non-Newtonian flow behaviour. The flow-induced orientational alignment of 2D crystallites has been studied by the same authors [17]. Fatty acid crystallites are formed during slow monolayer compression in the coexistence region. Observations of the channel flow show, that the crystallites were generally aligned in shear direction and a slight dependence of the crystallite alignment upon shear rate has been proposed. Ivanova et al. [18] extended these to different mesophases as a function of temperature, flow rate and alkyl chain length. At a steady-state flow unusual behaviour of the velocity profile was typical for phases with different tilted angle. It is concluded that the flow behaviour can be related to details of the molecular packing in the particular mesophases and not to the interaction of the monolayer with the subphase. The role of domain boundaries of fatty acid monolayers during shear flow was examined also by Ivanova et al. [19]. These results suggest a fundamental correspondence between the macroscopic monolayer rheology and domain-level processes. The coupling between the structure of the tilted condensed phases of docosanoic acid monolayers and an external shear flow has been studied in [20] using BAM. In general, the coupling is discussed as a result of a reorientation of the alkyl chains, although different kinetics are observed depending on the thermodynamic phase, shear rate, and surface pressure. The coupling can result in continuous orientational changes or in abrupt reorientation of the molecular tilt angle. These molecular-level effects can be connected to the macroscopic texture as well, and shear can result in either domain fragmentation or annealing.

Changes in domain geometry of a flowing monolayer trough were visualized by Olson and Fuller [21]. It has been demonstrated that the flow in a channel is a combination of shear and dilation. Streamline snapshots and flow velocity profiles of a monolayer with Newtonian and Non-Newtonian behaviour were also presented and discussed. As Newtonian system arachidyl alcohol and as Non-Newtonian poly(octadecyl methacrylate) monolayers were used and their interfacial shear rheology characterized by the oscillating needle technique (cf. paragraph C.4.).

The dynamic interfacial behaviour of emulsifier films (e.g. surfactants, polymers, proteins, lipids and their mixtures) has been recognized as important for the formation and stability of food colloids and can be characterised by interfacial rheology [22]. The structure, topography and surface shear rheology of monoglycerides and milk proteins (for example β-casein) at the air/water interface were studied by Patino et al. [23, 24] using surface film balance technique, channel viscosimetry and BAM. In general, η_S increases with the surface pressure Π. At higher surface pressures, collapsed β-casein residues may be displaced from the interface by monoglyceride molecules with important effects on the shear characteristics of the mixed films. The displacement of β-casein by monoglycerides is facilitated under shear conditions, especially by monoolein. Shear-induced changes in the topography of monoglyceride and β-casein domains and segregation between domains of the film forming components are discussed by the authors.

2. Deep channel surface viscometer

The classical channel technique is limited essentially to studies of insoluble monolayers at the gas/liquid interface and less of adsorbed films because a surface pressure gradient is required to maintain the flow. A different type of channel viscometer has been described by Davies and Mayers [25] that utilizes viscous-traction forces to produce flow without surface pressure gradients. The basic principle is to use a circular surface viscometer with continuous rotary movement of the surface, so that there is no change in area of the adsorbed interfacial layer. The proposed apparatus consisted of a channel formed by two concentric sharp-edged cylinders or two rings of thin wires. The sharp edges or rings just touch the liquid surface and are held stationary and concentrically in the interface. The dish is rotated around its axis with a turning table. The bulk motion of the fluid tends to provide traction on the liquid interface between the two stationary walls. Differences between the surface velocity of talc particles on a clean and on a film-covered substrate were used to provide an indication of the surface viscosity. By using different rotation speeds, the shear rate can be varied. One of the problems encountered in these measurements is that, due to the viscosity of the liquid, not only the surface is set in motion but also some liquid close to the surface. The hydrodynamic equations describing the flow in the channel are very complicated. Such measurements cannot be easily used to deduce directly the relevant rheological coefficients, and therefore the apparatus was calibrated with monolayers of known surface viscosity. The technique can be used at the water/oil interface as well.

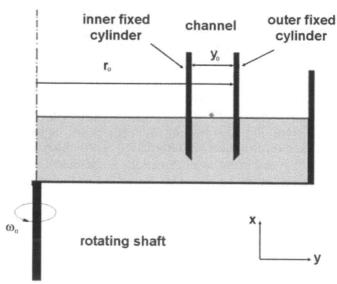

Fig. 2 Schematic principle of deep channel viscometer

An improved design was proposed by Burton and Mannheimer [26], consisting of two concentric, stationary vertical cylinders and a rotating flat-bottomed dish, forming a channel with a moving floor. The floor moves with a constant angular velocity. The fluid in the channel tends to rotate owing to the rotation of the dish and is sheared by the channel walls. This enhances the effect of the interface and makes it possible to use such an instrument to determine the interfacial shear rheology. The centreline surface motion within the channel is monitored by a very small inert particle placed within the fluid surface. A schematic sketch is shown in Fig. 2. The peculiarity is that the cylinders are not placed in the interface – they almost touch the bottom of the rotating dish. This small change allows the hydrodynamic equations to be solved.

The axisymmetric geometry allows the interfacial shear viscosity to be isolated since it is the only interfacial property that appears in the azimuthal component of the tangential stress balance. Of course this condition is only fulfilled for slow motions of the floor. At this inertialess limit any flow in other directions is negligible. In order to determine the interfacial shear viscosity, one has to analyse the velocity profiles in the channel. The interfacial viscosity can be related to the velocity at the centreline of the interface, the channel width, the height of liquid in the channel and the floor velocity. By means of a hydrodynamic analysis, the shear stress exerted by the moving bottom via the viscous liquid on the surface can be calculated. Hence the surface velocity in the centre of the channel is a measure of the interfacial shear viscosity η_s.

For laminar, Newtonian, time-independent flow in the channel, the centreline velocity V_c was given by Gupta and Wasan [27]

$$V_c = \frac{8e^{-\pi x_0}(1 + \pi \frac{\eta_s}{\eta})}{\pi(1 + \pi \frac{\eta_s}{\eta})} \tag{3}$$

The interfacial shear viscosity η_s is related to the experimentally available velocities V_c^* and V_c for the pure and loaded interface as follows:

$$\eta_s = \frac{\eta_0 y_0}{\pi}(\frac{V_c^*}{V_c} - 1) \tag{4}$$

Both equations illustrate the principle of the measuring technique, however, the complete mathematical descriptions are more complicated. Eqs. (3) and (4) are based on a first step analysis of Pintar et al. [28] and were deduced from a basic model for the deep-channel surface viscometer using assumptions that do not very closely approach reality. First the assumption of a flat shape of the fluid interface cannot hold if the separation width of the concentric cylinders is small. Further problems are the contact angle between the interface and the walls and the small gap between the rotating dish and the channel walls, which will be discussed later.

The deep-channel surface viscometer is considered as a technique with a high sensitivity down to 10^{-4} mNs/m. Due to the slow motion of the floor, measurements of larger values of η_s are not possible for low viscosity liquids such as water.

The analysis of Burton and Mannheimer's [26] was limited to a steady Newtonian flow in an infinitely long channel having a flat gas/liquid interface. Measurements of the flow characteristics of different pure liquids show that the agreement between theory and experiment depends on the width of the channel. This limitation is caused by the curved meniscus near the wall and impreciseness in the measurement of the liquid channel depth/width ratio. An improvement for the hydrodynamic analysis was proposed by Osborne [29] and Mannheimer and Schechter [30]. Their comparison of a theoretical analysis for linear and annular channel geometry has clearly sown that the deviation from a completely flat liquid surface is the main reason for discrepancies between theory and experiment. The theory was either limited to liquid films that exhibited the property of Newtonian surface viscosity or to solid films that exhibited the property of surface rigidity. To bridge the gap between these two idealized models, Mannheimer and Schechter [31] performed an analysis for a Bingham plastic surface flow in a channel viscometer, suitable to interpret shear-dependent observations in terms of a yield value and Newtonian viscosity. The measurement of surface yield value is complicated by the fact that the film yields first at the inner wall. Mannheimer and Schechter[32] described two methods to measure the visco-elastic interfacial properties with the viscous-traction channel viscometer. The first method is based on a comparison of the calculated and observed frequency response of the fluid interface to an impressed sinusoidal motion at the channel floor. An alternative technique by which the visco-elastic properties are obtained from relaxation measurements are also analyzed and discussed for several linear visco-elastic surface models.

Due to the potential of the viscous traction type technique to study interfacial shear rheology Mannheimer and Schechter [32] improved also the design of the deep channel method. To maintain a flat liquid surface a small step was machined into the walls of the channel and the surface of the top step has non-wetting properties. The precise control of the depth/width channel ratio was achieved within two micrometers. The centreline velocity was measured by observing small PTFE particles of about 0.04 mm in diameter which are deposited at the surface.

For studies of non-linear interfacial behaviour with the deep channel technique Hedge and Slattery [33] suggested, instead of a Bingham plastic model, to use a method in which the surface velocity distribution can be used to derive the dependence of the surface shear stress upon the surface shear rate provided that the bulk liquid exhibits Newtonian behaviour.

Pintar et al. [28] studied the interfacial viscosity of non-Newtonian polymer solutions with the deep channel technique. It was found that non-Newtonian surface viscosities and the corresponding interfacial shear rate can be determined from experimentally measurable quantities and fit to the Powell-Eyring model. A first approach to apply the technique also to the liquid/liquid interface is presented in this work. To validate the theory, measurements with different water soluble polymers were performed at the water/air and water/benzene interface. The results show, that the model is limited to small deformational flow. The constitutive equation presented by Gardner et al. [34] relates surface stress to the deformational history of an interface. The equation has the capability of representing complex interfacial visco-elastic properties. To verify the applicability of the equation to real interfacial films, experimental studies involving spread and adsorbed monolayers have been performed with oscillatory and constant floor motion [35].

For measurement at liquid/liquid interfaces modifications of the deep channel viscometer were proposed by Wasan et al. [36] and Deemer et al. [37]. The hydrodynamic equations of motion were solved to yield the steady state velocity distribution at liquid/liquid interfaces under the assumption of laminar flow resulting from slow rotation of the dish containing the liquids [36]. This analysis takes into account the finite depth of the two liquids and permits interpretation of the experimental results in terms of interfacial shear viscosities at the liquid/gas interface in addition to the liquid/liquid interface. The results clearly show a coupling of interfacial and bulk fluid flows. Measurements with xylene-water and hexadecan-water interfaces are in excellent agreement with the theory. Deemer et al. [37] extended the analysis to the liquid-liquid-gas case with two interfaces by taking into account the viscous effects in all three phases. By measuring the gas-liquid surface velocity distribution, the liquid-liquid interfacial behaviour can be determined, as long as the shear stress distribution at the gas-liquid interface has already been found.

The early instruments were mainly applied for measurements of the interfacial shear viscosity under constant shear conditions. The deep channel technique was further developed for oscillatory shear [32, 35]. The principles of the complex interfacial rheology of liquid/liquid interfaces have been formulated by Wasan et al. [38, 39]. The deep channel rheometer fitted with an oscillatory floor was used to examine the interfacial rheology in the presence of surfactants. The hydrodynamic analysis assumed that the floor oscillates sinusoidally and with small velocities. The study shows that an adsorbed layer of an oil-soluble macromolecular surfactants at the decane/water interface is visco-elastic even though the adjacent phases are Newtonian liquids.

Another discrepancy for the viscous traction method utilized in the deep channel viscometer is the generation of surface-bounded vortices by the rotating floor. Therefore at higher floor velocities the inertialess conditions are not fulfilled and a sum of interfacial shear and dilational viscosity is measured. New non-invasive measuring techniques have been applied by Hirsa et al. [40] to obtain the surfactant concentration and the velocity field at a planar surface even in presence of vortices. The optical technique of second-harmonic generation (SHG) has been used to determine the time-dependent surfactant concentration at the interface. The digital particle image velocimetry (DPIV) technique has been used for the measurement of the velocity field. Lopez and Hirsa [41] utilized the deep channel technique to study the coupling between bulk flow and a Newtonian gas/liquid interface in presence of an insoluble surfactant using a numerical Navier-Stokes model for the flow. The floor of the annular channel is rotated fast enough so that the flow is non-linear and drives the initially uniformly distributed surfactant towards to the inner cylinder. The boundary conditions at the interface are functions of the interfacial tension, interfacial shear and dilation viscosity. A fully coupled visco-elastic interfacial flow with a swirling bulk flow has been modelled numerically for flow and material properties corresponding to a physical air/water system with a hemicyanine monolayer [42]. The stress balance in the radial direction was found to be dominated by the Marangoni stress. In the azimuthal direction, the tangential stress balance is solely determined by the interfacial shear viscosity. For a sinusoidal motion of the floor the Navier-Stokes computations with the Boussinesq-Scriven surface model have been presented by Lopez and Hirsa [43] utilizing the equation-of-state measured for a vitamin K_1 monolayer.

The coupling between a bulk vertical flow and a surfactant-loaded air/water interface has been examined with an optical annular channel rheometer by Hirsa et al. [44]. The geometry of this device is similar to the conventional deep-channel viscometer, however, the optical part is not restricted to small annular gap-to-depth ratios and allows experiments of non-Newtonian interfacial hydrodynamics with non-linear flow. The flow at the air/water interface is measured using digital particle image velocimetry (DPIV). For channel flow with inertia a technique is proposed to determine the interfacial shear viscosity from azimuthal velocity measurements at the interface which extends the range of interfacial shear viscosity to be measured using a deep channel viscometer in the usual Stokes flow regime by exploiting flow inertia. A Navier-Stokes-based model of bulk flow coupled to a Newtonian interface that has interfacial shear viscosity as the only interfacial property was developed. This is achieved by restricting the flow to regimes where the interfacial radial velocity vanishes. Measurements on vitamin K_1 and stearic acid monolayer were described.

Gupta and Wasan [27] presented the results of extensive measurements of the surface shear viscosity of a number of different soluble surfactant systems. None of the soluble single-component surfactants investigated demonstrated a significant surface shear viscosity. However, dodecanoic acid and dodecyl alcohol, are practically insoluble in water but solubilised by sodium dodecyl sulphate, yielded significant surface shear viscosities at certain concentrations and proportions. The surface rheological behaviour is often dependent on the rapid aging of surfactant solutions. A paper by Mohan et al. [45] deals with the interpretation of the interfacial shear flow data in the deep channel surface viscometer in terms of aging effects on surface viscosity. It was found that the surface shear viscosity increases considerably with surface aging and at the same age, the surface shear viscosity increases with surfactant concentration.

A technique for measuring the interfacial visco-elasticity at liquid/liquid interfaces, proposed by Mohan and Wasan [38], consists in imparting sinusoidal or on-off motion to the floor of the deep channel viscometer and observing the centreline displacement either at the liquid/air or the liquid/liquid interface. The authors claimed that the interfacial shear visco-elasticity at a liquid/liquid interface can be determined from displacement data at liquid/air interface alone, so that this method is particularly useful for opaque systems such as crude oil/aqueous solutions. The feasibility of this technique was tested with monolayers of bovine serum albumin spread at the water/benzene interface. It was detected that interfacial shear viscosity and relaxation time, compared to the interfacial pressure, are much more sensitive to changes in molecular packing.

The shear rheological of various interfacial layers was investigated by the deep channel viscometer, such as spread dipalmitoyl lecithin monolayers as model lung surfactants [46] and real lung extracts [47], long chain fatty acids [34] and other spread and adsorbed monolayers [35], diethanolamine derivatives of n-alkylsuccinic anhydrides [48], SDS in presence and absence of organic salts [49], oil-soluble surfactants at the decane/water interface [39], and many more.

3. Rotating knife-edge in wall surface viscometer

The rotating wall knife-edge surface viscometer introduced by Goodrich et al. [50] measures the ratio of displacement between the surface velocity and the angular velocity of the rotating outer ring, from which the interfacial shear viscosity can be calculated. The method is capable of achieving a high sensitivity $(10^{-5} mNsm^{-1} < \eta_s < 0.1 mNsm^{-1})$ and does not suffer from the difficulties of placing a knife "in" the interface, yet without breaking the plane of the interface. As with the deep-channel viscometer, a small particle is placed within the surface to determine the rotational speed of the fluid interface.

The basic experimental design of the rotating wall knife-edge surface viscometer is depicted in Fig. 3. A cylindrical dish is filled with a liquid to a level at least as high as the radius of the dish. The upper surface of the liquid contacts the wall of the container along a knife-edge ring that has been inserted into the wall. Rotation of the ring creates a viscometric flow in the vicinity of the interface. A small talc or Teflon particle is placed to the centre part of the interface, and the period of rotation is monitored. From the ratio of the displacement velocity to the applied angular velocity of the rotating ring, the interfacial shear viscosity is determined.

The simple geometry of Fig. 3 is used in the theoretical analysis by Goodrich et al. [50] which yields a simple analytical expression. For very small interfacial shear viscosities and under the conditions of a small gap the surface viscosity near the cylindrical axis is asymptotically given by

$$\eta_s = 0.5631(\Omega_p / \Omega_r)\eta R_r \qquad (5)$$

where R_r is the ring radius and (Ω_p / Ω_r) is the ratio of particle to ring angular rotation. For practical purposes, a least-square polynomial fit of the general integral expression was proposed

$$\eta_s = \eta R_r \left[0.5631 \left(\frac{\Omega_p}{\Omega_r} \right) + 1.1189 \left(\frac{\Omega_p}{\Omega_r} \right)^3 - 0.6254 \left(\frac{\Omega_p}{\Omega_r} \right)^5 + 3.4489 \left(\frac{\Omega_p}{\Omega_r} \right)^7 \right]$$

$$(6)$$

which is valid to within an accuracy of 0.3%.

Some experiments with this device on stearic acid monolayers spread on water have been reported in [51] and it was found that the surface shear viscosity is very low up to high surface concentrations. Over all ranges of surface concentrations studied the rheology was Newtonian. Experiments on SDS solutions with and without added dodecanol were performed as well [52], and the technique was even applied to the oil/water interface [53].

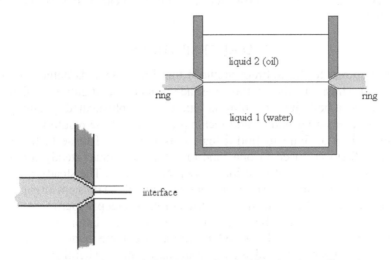

Fig. 3 Schematic principle of a rotating knife-edge in wall surface viscometer

4. Transient rotating cylinder apparatus

The method proposed by Krieg et al. [54] consists in monitoring the decay of surface motion after a sudden cessation of a rigid body motion of a rotating cylindrical container filled with liquid, extended later to liquid-liquid interfaces by Hassager and Westborg [55]. Both fluid phases are assumed to be Newtonian and the interface too. The two key parameters for the design and sensitivity of this type of viscometer are the height and radius of the cylinder. The angular displacements of interfacial elements were obtained from direct visual observation and from video recordings of small polystyrene spheres as tracer particles. By applying oscillatory shear deformations the interfacial visco-elastic response is obtained, the theoretical basis of which was presented by Prieditis et al. [56]. The method consists of measuring the amplitude and phase shift of oscillations imposed on a cup of liquid and it is most sensitive for geometries where the cup radius-to-liquid height ratios are about 0.5.

With another approach the liquid/gas viscometer proposed by Krieg et al. [54] was extended to liquid/liquid interfaces by Cambridge et al. [57] where the two fluids are contained in a slowly turning cylindrical cell. The motion of the cell is abruptly stopped and the deformation of the interface is measured via tracer particles. Assuming Newtonian flow behaviour and laminar flow the interfacial shear viscosity can be calculated.

C. DIRECT METHODS

In contrast to the indirect methods the direct methods determine the torsional stress of the liquid interface under investigation which is stressed by a body just touching it. Due to the sophisticated geometry of different measuring bodies and techniques to detect the resulting motion or stress it is difficult to find a logic classification of these techniques. Most of them are used in one kind of rotational mode: steady rotation, transient regimes of deformation or oscillating mode. The steady rotation can be applied after the Searl or Couette principle. Transient regimes of deformation can be performed as relaxation or creep experiments. The oscillatory mode can be applied as forced oscillation or damped oscillation. Most of the described techniques are home made instruments, which have been designed for a certain application. Therefore not with all instruments the complete spectrum of rheological experiments can be performed.

The classical instruments for the direct determination of interfacial shear properties consist of a circular measuring body suspended from a thin torsion wire such that it just touches or is positioned in the interface of the liquid contained in a circular measuring vessel. In this section the torsion pendulum techniques are divided mainly into sharp-edge, biconical disk, oscillating disk, oscillating needle and oscillating ring geometries.

1. Torsion pendulum techniques

The determination of surface shear viscosity by measuring the damping of an oscillating torque pendulum touching the interface is one of the oldest methods in interfacial rheology [58]. The classical knife-edge surface viscometer of Brown et al. [59] consists of a knife-edge bob suspended from a torsion wire such that the circular knife just touches the surface of a solution contained in a cylindrical vessel. The measuring vessel is forced to rotate, and the torsional stress on the bob is measured in order to determine the surface shear viscosity. Mannheimer and Burton [60] improved the theoretical analysis of the typical "knife edge" torsional viscometer. Based on this principle, two more versions are known, the double knife-edge [61] and the blunt knife-edge surface viscometer [62]. The sensitivity of these viscometers is of the order of $\eta_s \geq 0.01 mNs / m$.

Many surface shear viscometers described in literature utilise torsional stress measurements applied to a rotating disk [62,63,64,65] suspended from a torsion wire into the interface. The measuring vessel is rotated with an angular velocity and a torque is exerted on the disk by both the surfactant film and the viscous liquid. Such instruments allow also for creep, stress relaxation or yield stress measurements. If a ring or disk is forced to oscillate in the interface, the rheological parameters are better defined, particularly when the interface is bounded by a concentric outer ring [66]. The equation of motion for either free or forced oscillations has been solved by Tschoegl [67]. For free oscillation, the additivity of resistance due to the interface and the adjoining phase is assumed and the surface shear viscosity can be determined by measuring the decrements of damping and periods of oscillation of a free and surfactant loaded interface. In forced oscillation experiments the amplitude ratio and phase angle are the experimentally determined variables. These methods of measuring the surface viscosity are in principle invalid as long as the viscous interactions between the interface and the adjoining phases are not taken into account. The improved theoretical analysis by Mannheimer and Burton [60] estimated an additional torque due to the viscous interaction between the flowing monolayer and the substrate. Calculations based on this analysis indicate that surface viscosities reported from current torsional formulae - all of which assume that the film slips over the substrate - may be in serious error. Examples are also presented that illustrate how comparative torque measurements can lead to anomalous results when the effects of temperature on surface viscosity are considered.

Such Couette type interfacial rheometers are the most widely used equipment for investigating interfacial shear rheological properties. The measuring interval can easily be changed across a wide range, which is very important because of large differences in the interfacial rheological behaviour of different surface active materials. Very different techniques have been developed to characterise these properties and to determine both viscous and elastic parameters over a broad time interval. Proteins are the most frequently studied polymers at liquid interfaces [68,69]. The importance of structure formation and interfacial rheological properties of mixed protein-surfactant interfacial adsorption layers have been recognized and reviewed by Izmailova [70]. The interfacial rheological behaviour of protein adsorption layers has been extensively studied by many groups. Instructive summaries of such studies have been given recently by Izmailova and Yampolskaya [71,72], Dickinson [73,74],

Murray [75], Bos and van Vliet [76], Derkatch et al. [77] or Krägel et al. [78].

Torques which are involved in interfacial rheometry are usually very small. Therefore torsion pendulum techniques have been often used to characterise the visco-elasticity of comparatively un-rigid interfacial layers. A disadvantage of such setups is the limitation of the angular frequency due to the torsion constant of the suspension wire. To cover a wider range of frequencies, different wires must be used. In addition, to get smooth torsional oscillation of the pendulum the elasticity constant of the wire and the moment of inertia of the rheological probe must be in a certain ratio. Therefore for a change in frequency the complete pendulum must be exchanged. Consequently, frequency sweep experiments are not possible.

1.1. Sharp edge geometry for measurements at the air/liquid interface

The knife-edge surface viscometer of Brown et al. [59] is a rotational viscometer with a knife-edge bob (Fig. 4). The cup is rotated and the torsional stress on the bob measured to determine the interfacial shear viscosity from the steady state situation. A constant rate of torque produces a constant rate of shear in the film. The viscometer operated in this way is the two-dimensional analogue of the well known bulk rotational viscometer (Couette type apparatus). The film can be sheared at a constant rate only if it is either purely viscous or after the elastic forces have been ruptured. The limiting stress necessary to overcome the elastic forces is called surface yield stress and the shearing element will show a constant angular displacement in steady-state. Measurements at constant rate of torque in steady-state cannot provide information on the elastic properties of the surface film.

Mannheimer and Burton [60] proposed the following expression for the torque on the torsional wire in the single knife-edge experiment

$$
knife\ torque = \frac{\pi a^2 R_r \eta \Omega_r}{\left[\sum_{i=1}^{\infty} \frac{\left[J_1(a\xi_i / R_r) \right]^2}{\left[J_0(\xi_i) \right]^2 \left[\frac{\eta_s}{\eta R_r}(\xi_i)^2 + \xi_i \coth(h\xi_i / R_r) \right]} \right]} \tag{7}
$$

where $J_0(\xi_i)$ and $J_1(a\xi_i / R_r)$ are the 0th and 1st order Bessel-functions of the first kind, respectively, and ξ_i are the roots of $J_1(\xi_i) = 0$.

In literature also more sophisticated geometries were described, for example the double knife-edge and blunt knife-edge, for which the theoretical description of the respective viscometric flow was provided by Lifschutz et al. [79] and Briley et al. [62].

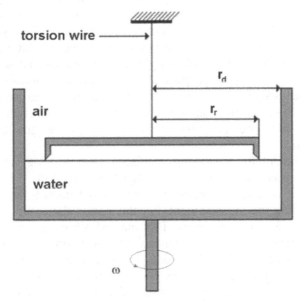

Fig. 4 Schematic principle of single knife-edge torsion pendulum technique

For each of these knife-edge viscometers it is crucial that the knife is in contact with, but not break through the fluid interface. Hence it has to be non-wetting, though the problem of knife placement is still cumbersome [60].

First surface rheological measurements with flat rings and blunt edges suspended on a torsion wire have been carried out by several authors and are described in literature e.g. by Karam [80]. An axisymmetric technique using a small blunt knife PTFE edge was introduced by Abraham et al. [81]. This apparatus enables one to measure simultaneously the surface shear modulus, the surface shear viscosity and the surface pressure of a monomolecular film, particularly suitable for insoluble monolayers. To measure the quasi-static shear properties, the cup is rotated with a small angle and the residual deflection of the blunt knife edge is determined. The dynamic shear response is probed by studying transient torsional oscillations generated by a sudden angular displacement. The apparatus has a resolution of 5×10^{-4} mNs/m of shear viscosity and 10^{-4} mN/m of static shear modulus. First measurements of classic monolayer systems (stearyl alcohols and nonadecanoic acid) spread on water have been studied as a function of density by Abraham et al. [82,83]. An exact solution to the visco-elastic damping of a torsion pendulum by a surface film was given by Feng et al. [84]. The surface shear viscosity of different systems have been studied with this type of viscometer, such as valinomycin [85], dimyristoyl phosphatidylserin and bovine brain phosphatidylserin [86], poly(methyl methacrylate), poly(vinyl stearate), octadecanol and behenic acid [87]. The visco-elastic parameters of polymer monolayers have much higher values than those of single amphiphiles and a clear correlation was found between the shear visco-elasticity of the monolayers and the presence of surface pressure gradients upon compression.

An instrument for measuring interfacial shear rheology at very low shear-rates was described by Krägel et al. [88]. Rheological investigations require equipment which does not disturb the interfacial structure during the measurement. For this reason the torsion pendulum technique is very advantageous. This method allows experiments with very small mechanical deformations of the adsorption layer. Its principle is based on a ring with a sharp edge hanging at a torsion wire. When applying an impulsive torque by an instantaneous movement of the torsion head the pendulum performs damped oscillations with the damping factor α and the circular frequency ω. This kind of experiment provides simultaneously information on the surface shear coefficient of viscosity and the surface shear modulus of rigidity from a single experiment. Off course, experiments involving free damped oscillations are a simpler alternative for measuring the complex modulus, because only the displacement of the rheological probe is measured. In general these experiments are easier to perform than those in the forced oscillation mode.

For the torsional oscillation of a pendulum the stimulation of oscillation, the elasticity of the torsion wire, the moment of inertia of the measuring system, and the elastic and friction forces in the interface are important. In the present instrument a thin tungsten wire transfers the deformation, produced by the torsion head, via the edge onto the interface. By deflecting the torsion head a torque is transferred to the interface which leads to a shearing of the interface in the slit between the edge and the wall of the measuring vessel. The acting shear stress is given by

$$S = \frac{M}{4\pi}\left(\frac{1}{r_r^2} + \frac{1}{r_d^2}\right) \tag{8}$$

where S is the shear stress, M is the transferred torque, r_r is the outer radius of the edge, and r_d is the inner radius of the measuring vessel. Fig. 5 shows schematically the shear field between the wall of sample container and the moving measuring body. The mathematical relationships for an oscillating torsion pendulum in surface films were derived by Tschoegl [67] who used the concept of a linear visco-elastic theory for two-dimensional systems. Under the assumption of a simple Volgt model (spring and dash-pot in parallel) to describe the visco-elastic behaviour of the film, it results in the circuit diagram for the torsion pendulum as shown in Fig. 6.

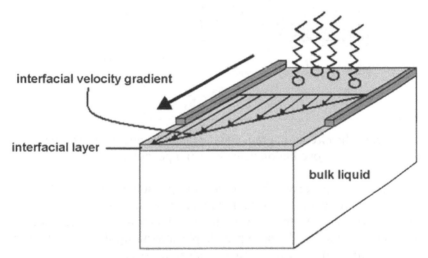

interfacial velocity gradient

interfacial layer

bulk liquid

Fig. 5 Schematic shear field in the gap between the wall of the sample container and the moving measuring body

The general equation of oscillation of a pendulum is given by

$$I_r \cdot \ddot{y} + (F_r + \eta_s / H_s) \cdot \dot{y} + (E_r + G_s / H_s) \cdot y = \Theta(t) \tag{9}$$

where the function y(t) describes the oscillational movement of the measuring body. To solve this differential equation it is important to discriminate between different forms of stimulus $\Theta(t)$. Only for sinusoidal angular displacements of the driver (the pendulum performs forced periodic oscillations) and for the application of an impulsive torque to deflect the pendulum (the pendulum performs free damped oscillations) is it possible to determine the surface shear viscosity and the surface shear elasticity from a single experiment.

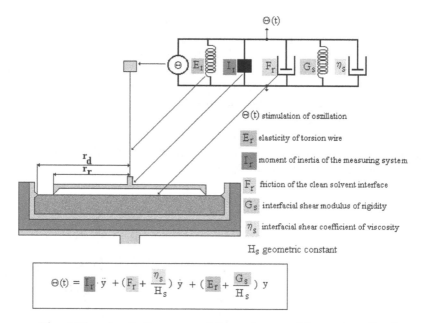

$$\Theta(t) = I_r \cdot \ddot{y} + (F_r + \frac{\eta_s}{H_s}) \dot{y} + (E_r + \frac{G_s}{H_s}) y$$

Fig. 6 The circuit diagram and the main quantities to describe the oscillation behaviour of a pendulum

Other types of drive for the torsion pendulum are the "top" or "bottom" drive with a constant circular velocity. A surface torsion pendulum operated in this way is the two-dimensional analogue of the rotational viscometer. To set up a simple measuring device with short measuring time an impulsive torque is advantageous.

Provided that the moment of inertia of the measurement system, I_r, is not too small, a free damped oscillation of the pendulum can be observed after the deflection of the torsion head. At very small values of I_r, a creep curve results. In all practical applications of the torsion pendulum, clean solvent surfaces show oscillatory behaviour. Also for interfaces covered with an adsorption layer or spread film, free damped oscillations are obtained under suitable measurement conditions.

The free damped oscillation as the solution of Eq. (9) is given by

$$y(t) = y_0 \cdot \exp(-\alpha \cdot t) \cdot \sin(\beta \cdot t - \varphi) + c \qquad (10)$$

where t is the time, y_0 is the amplitude, and α, β, φ, c are parameters. It can be rewritten to

$$y(t) = y_0 \cdot \exp(-\frac{t}{a}) \cdot \sin\left(\omega \cdot (t - \delta)\right) + c \qquad (11)$$

where a is the decay time, ω is the oscillation frequency, δ is the phase lag and c is an offset, with

$$a = \frac{1}{\alpha}, \quad \omega = \beta, \quad \delta = \frac{\varphi}{\beta} \text{ and } T = \frac{2\pi}{\omega} = \frac{2\pi}{\beta} \tag{12}$$

T is the period of oscillation, φ the phase shift in angle units. From Eq. (9) we get

$$\alpha = \frac{F_r + \dfrac{\eta_s}{H_s}}{2 \cdot I_r}, \quad \beta = \sqrt{\frac{E_r + \dfrac{G_s}{H_s}}{I_r} - \alpha^2} \tag{13}$$

The constants H_s, I_r, E_r and F_r, are geometric values (H_s) or can be determined by calibration measurements (I_r, E_r and F_r). These problems will be discussed later. Knowing the values of the constants, the rheological parameters η_s and G_s, of the surface layer can be calculated from α and β, which can be determined by fitting the experimental oscillation curve. The geometric constant H_s results from the measuring geometry, I_r and E_r can be measured by two independent measurements in air, and F_r from a measurement with the pure solvent.

For the measurement in water $\eta_s = 0$ and $G_s = 0$ – with the equations (13) and with known values for E_r and I_r the friction of the pure water surface and an additional quantity can be calculated. From eq. (13) follows

$$\alpha_0 = \frac{F_r}{2} \cdot \frac{1}{I_r}, \quad \beta_0 = \sqrt{\frac{E_r}{I_r} - \alpha_0^2} \tag{14}$$

and

$$F_r = 2\alpha_0 \qquad I_r, E_r = I_r(\alpha_0^2 + \beta_0^2), \tag{15}$$

With the known parameters H_s, I_r, E_r and F_r from eq. (13) follows

$$\eta_s = 2 \cdot H_s \cdot I_r \cdot (\alpha - \alpha_0) \tag{16}$$

$$G_s = H_s \cdot I_r \cdot (\alpha^2 - \alpha_0^2 + \beta^2 - \beta_0^2) \tag{17}$$

The schematic set-up of the rheometer is shown in Fig. 7. The active part (stepper motor, gearing, motor controller) comprises a drive for the deflection, at which the torsion wire with a circular measuring body is fixed.

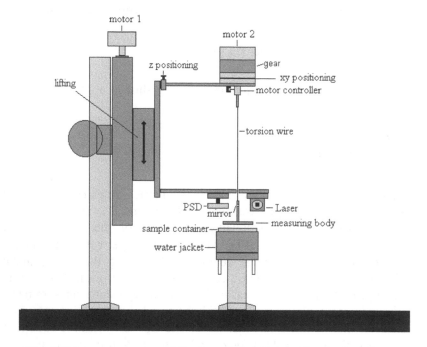

Fig. 7 Schematic set-up of the Interfacial Shear Rheometer ISR-1

The body touches the liquid surface or is situated at the liquid/liquid interface. The angular position of the body is registered by means of a mini laser and a position-sensitive photosensor with an accuracy to ±0.01° at deflection angles between 0.1° and 2°. The instrument is completely computer-driven and the software controls all calibrations, measurements including data acquisition and analyses (SINTERFACE, Berlin, Germany).

The sensitivity and accuracy of this torsion pendulum apparatus has been demonstrated by different model experiments [88,89,90,91,92]. The instrument is applicable also to liquid/liquid interfaces and due to its high sensitivity also investigations of insoluble monolayers is possible [93,94]. Fig. 8 shows the measuring principle in combination with a Langmuir trough.

To perform forced oscillation experiments with the torsion pendulum method, Ghaskadvi et al. [95] proposed to attach the sharp knife edge ring via a rigid rod to a rectangular coil. This rotor is placed in contact with the monolayer and suspended by a thin torsion wire. The implemented coil carries a direct current. The voltage induced in this coil by a high-frequency external magnetic field is used to measure the angle of rotation. A low-frequency external magnetic field is used to drive the rotor into forced oscillations at the desired frequency (1mHz – 1Hz), or the rotor may oscillate freely with the natural frequency variable over the range 0.01 - 3Hz using a static external magnetic field. The method has been combined with a circular trough equipped with an elastic barrier for radial compressions [96,97]. Due to the construction of apparatus the effective torsion constant can also be changed electromagnetically by using a static external magnetic field. An external torque is applied which is proportional to the angle and allows a frequency sweep.

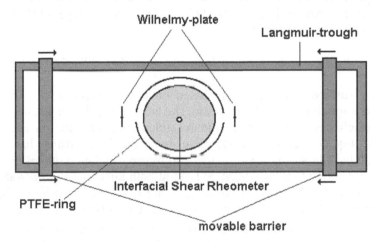

Fig. 8 Schematic principle of interfacial shear rheology of insoluble monolayers on liquid subphases

With this type of technique the non-Newtonian behaviour of red blood cells [98] and the lipid-protein interaction in mixed monolayers from phospholipid and proteins have been studied [99]. Krägel et al. [93] used the version combined with a Langmuir trough to measure the surface shear viscosity of DPPC and DMPE spear monolayers. The determined shear elasticity was independent of the surface pressure and close to zero. With increasing surface pressure Π the viscosity increases from zero first almost linearly and then abruptly to reach a plateau value at very high surface pressures. For the DMPE an abrupt further increase to very high viscosity values is observed after the transition point from a liquid condensed (LC) to a solid-like state at about $\Pi = 21\,mN/m$. For DPPC a similar behaviour is observed. Although a transition from the LC-phase to a solid-like film at about $\Pi = 32\,mN/m$ was not reported so far in literature, the shear viscosity rises up sharply in this point.

Comparisons of the results for DPPC monolayers with the dilational rheology shows that both shear and dilational rheology correlate with the morphology of the monolayer as observed by optical methods, such as Brewster angle microscopy.

Experiments on two different model lung surfactants spread on a buffered subphase have been discussed by Grigoriev et al. [100]. Further studies on various interfacial layers were reported in literature, such as layer of gelatine in absence and presence of surfactants [101, 102, 103], human serum albumin [88], different food proteins [89,104, 105, 106, 107, 108, 109, 110], polymeric humic substances [92], plant proteins [111], mixed protein-lipid layers [112], spread protein layers [113] or phytoplankton systems [114].

1.2. Biconical geometry for measurements at the liquid/liquid interface

Many surface viscometers are based on measurements of torsional stress on a rotating or oscillating disk. The edge of the disk is located in the interfacial region between two immiscible liquids or at the surface of a single liquid. Several studies have been published using flat or biconical disks in steady rotation, free and forced oscillation modes or in creep recovery and stress relaxation tests at both air/water and oil/water interfaces [69,115,116]. Such viscometers are moderately sensitive with an accuracy of $\eta_s \geq 0.01\text{mNs/m}$ but avoid the difficult tracking of particles to determine the interfacial velocity - the primary drawback of the deep-channel surface viscometer. Torsional devices are, therefore, useful for measuring interfacial shear viscosities of highly viscous surfactant monolayers [117].

In the classical thin disk surface viscometer a thin flat circular cylindrical disk rotates within the plane of the interface with the angular velocity Ω_D. A torque M is exerted upon the disk by both the surfactant film and the viscous liquid. According to the theoretical model, the torque exerted may be decomposed into a lineal surface traction torque along the rim of the disk (trough which the interfacial shear viscosity η_s imparts a direct influence), and a tractional torque owing to the bulk liquid along the base of the disk. Standard analyses do not account for contact angle effects or for the effects of the finite disk thickness.

In the limit of zero Boussinesq numbers

$$Bo \equiv \frac{\eta_s}{\eta \cdot R_C} \to 0, \tag{18}$$

(R_C is the characteristic distance of flow geometry) with

$$\frac{r_r}{h} \to 0 \ and \ \frac{r_r}{R_C} \to 0, \tag{19}$$

the torque M exerted upon the rotating disk is provided to leading order by the expression [118]

$$M = (16/3)r_r^3\eta\Omega_D. \tag{20}$$

For large Boussinesq numbers we have

$$Bo \to \infty, \ at \ \frac{r_r}{h} \to 0 \ and \ \frac{r_r}{R_C} \to 0 \tag{21}$$

and the following relation holds:

$$M = (8/3)r_r^3\eta\Omega_D + 4\pi r_r^2\eta_s\Omega_D. \tag{22}$$

Ω_D is the angular velocity and r_r the disk radius. General torque expressions were provided by Briley et al. [62], Shail [63], and Oh and Slattery [64].

A second type of disk surface viscometers [119] employs a thin disk rotating below the fluid interface. This technique may exhibit an advantage relative to the preceding designs because it avoids the delicate placement of the disk "in" the interface. It also eliminates possible contact angle complications. The sensitivity of this configuration to interfacial shear viscosity is reduced however because the disc is placed in the bulk fluid at a finite distance from the interface. A similar analysis has been carried out by Davis et al. [65] for a rotating sphere beneath a surfactant loaded fluid interface.

A variation of the flat disk viscometer is the biconical bob interfacial viscometer shown schematically in Fig. 9. This geometry has some advantages for the practical use and is therefore more common for measurements at the liquid/liquid interface, applied also at air/liquid interfaces. The theoretical analysis [64,117] for steady state rotations is quite similar to that of the disc viscometer, and the technique has similar advantages and limitations. A theoretical analysis for an oscillatory disk torsional interfacial viscometer was given by Ray et al. [120], Lee et al. [121] and Nagarajan et al. [116]. A sinusoidal angular oscillation is imposed on dish. The bob is induced to oscillate with the same frequency as the dish. From measurements of the amplitude ratio and phase angle between these two oscillations, the surface shear viscosity can be determined.

Using the biconical disk method various experimental studies were published, such as mesquite gum films adsorbed at the liquid paraffin/water interface as a function of gum concentration, pH, and electrolyte concentration [122], sodium caseinate mixed with mono- and diglyceride at the oil/water interface [123], adsorbed films of casein and gelatine at the n-hexadecane-water interface [115], myosin [124] or proteins like α-lactalbumin, β-lactoglobulin and β-casein [125] at the tetradecane/water interface, aqueous gelatine-surfactant solutions at the air/water interface [121], ovalbumin, β-lactoglobulin, glycinin [126] or chemically modified ovalbumin [127] at the air/water interface, proteins at the oil/water interface in presence of glycerol trioctadecanoate crystals [128], two-dimensional polymerisation process of diacrylate diesters at the oil/water interface [129]. Burger and Rehage also reported on studies at the oil/water interface of rubber-elastic, glass-like and transient membranes, which have been stabilized or crosslinked by physical or chemical contacts [130]. Also investigations of the crude oil-water interface were published by different authors [131, 132].

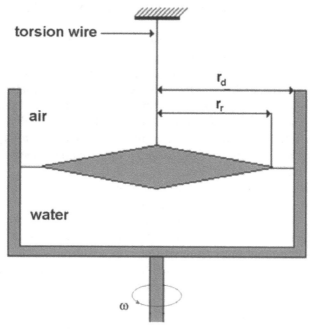

Fig. 9 Schematic principle of biconical disk geometry

As two more application of the biconical disk viscometer for measurements of the interfacial shear rheology we want to mention here studies of recombinant fragments of human dystrophin, corresponding to a single structural repeating unit of the rod domain with membrane lipids [133], and of three-phase partitioning systems [134].

2. Biconical disk coupled with a conventional rheometer

In the last decade significant progress was achieved in the development of research rheometer and a variety of viscometers and rheometers with rapid response times, broad ranges of measuring geometries, and accurate velocity control have been developed. Mainly due to the use of better and faster electronic components, improved measuring principles and new solution to approach frictionless bearing, the flexibility and sensitivity of such rheometer have been improved. Since the torques involved in interfacial shear rheology are usually very small, most of the used devices are based on disk suspended by torsion wires. The improved sensitivity of conventional bulk rheometer offers the opportunity of a rigid coupling of the disk with a top-driven motor and torque transducer. In all experiments the biconical geometry has been used for measurements at the air/water or oil/water interface. The direct coupling offers some advantages, such as the option of rotational, relaxation or oscillation mode without exchanging the sample or any component (e.g. torsion wire). Fig. 10 shows the schematic principle of the rigid coupling of a biconical disk.

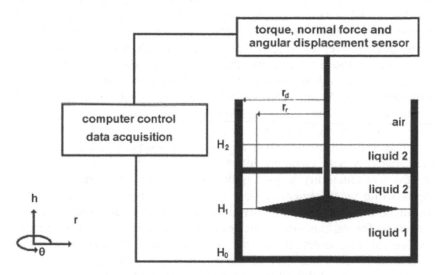

Fig. 10 Schematic of a biconical disk coupling with a commercial bulk rheometer, according to Erni et al. [135]

In rheometers for research two principles are realised, the controlled rate or strain CR, and the controlled stress approach CS. In CR devices the displacement or speed (strain or strain rate) is applied to the sample and the resulting torque is measured separately by a transducer, whereas in CS devices a force is applied and the resulting displacement or speed is measured. Certain measuring principles enable measurements at very small shear stresses and strains. Torques of about 0.02 μNm and deflection angles of 0.1 μrad can be attained in a range needed for interfacial rheometry, for example with rigid layers. The direct strain oscillatory mode has been applied for the development of the sensitive rheometer MCR 300 (Anton Paar Germany GmbH). An extended report about the development of an interfacial rheometer for both CR and CS measurements at liquid/liquid and gas/liquid interfaces using the MCR 300 was given by Erni et al. [135]. The experimental setup consisting of a biconical disk and an axis rigidly coupled with a low friction electronically commutated motor system. The disk is rotating or oscillating at a controlled torque or rotational speed while the cup remains stationary. For measurements at the liquid/liquid interface the measuring position of the disk at the interface can be detected with the normal force sensor of the rheometer. Therefore, the system can be easily aligned in order to ensure reproducible measuring conditions.

An improved numerical method for the calculation of the velocity distribution in the measuring cell was also proposed by Erni et al. [135]. The interfacial shear stress, viscosity, and dynamic moduli are obtained by solving the Stokes equations at low Reynolds number along with the Boussinesq-Scriven interfacial stress tensor, which is used as boundary conditions at the interface. Results from steady shear and oscillatory experiments as well as creep recovery and stress relaxation tests at the air/water and oil/water interface with adsorbed ovalbumin layers and spread sorbitan tristearate monolayers have been presented to illustrate the functionality and flexibility of the instrument. There are, however, limits of the instrument, such as the fixed geometry does not allow measurements with changing interfacial area of spread monolayers of, and the reproducibility is good only with comparatively rigid interfacial layers while measurements at less rigid systems become more delicate due to the extremely small involved torque. Erni et al. used the MCR 300 rheometer for various systematic investigations, such as for CS and CR rheological measurements of the water-insoluble surfactant sorbitan tristearate at air/water and oil/water interface [136, 137], lysozyme adsorption layers at the oil/water interface 138, and some other systems [139, 140, 141].

Besides the Paar-Physica rheometer MCR 300, also other commercial rheometers were modified for interfacial studies. For example Lakatos-Szabo and Lakatos [142] studied the interfacial rheological properties of model crude oil/water systems in presence of sodium hydroxide with a

Contraves Low Shear 30. Rehage et al. applied a Rheometrics Scientific RFS II fluid spectrometer for shear rheological measurements of flat membranes at the oil/water interface [143]. Wilde et al. [144] and Gunning et al. [145] made their investigations with a Bohlin CS 10 controlled stress rheometer equipped with a IFG10 interfacial geometry bob (Malvern-Instruments) and studied the shear rheology of beer alone and in presence of C6 and C16 fatty acids, and also the displacement of β-lactoglobulin by the nonionic surfactant Tween 20 and the zwitterinonic surfactant lysophosphatidylcholine-lauroyl. Kilpatrick and Spiecker [146] applied a biconical disk rheometry to highly elastic films formed at asphaltene-containing, model oil/water interfaces, utilizing a TA-Instruments AR-1000 stress rheometer.

3. Oscillating disk techniques

A constant stress, parallel plate viscometer without bearing for very low shear stresses has been proposed by van Vliet et al. [147]. The rotating part consists of a ferromagnetic disc which floats on the liquid surface and is driven and centred by a rotating magnetic field. The shear stress can be varied continuously by changing the distance between the disc and the rotating magnet.

An oscillating disk rheometer for measurements of spread monolayers at low frequencies (0.4 mHz – 1 Hz) has been proposed by Gaub and McConnell [148]. It uses a small hydrophobized glass disk equipped with a small magnet in the centre. The round compartment of a trough in which the interfacial viscosity is measured is covered by a Plexiglas ring. The liquid surface in this round compartment has a slight concave profile that keeps the hydrophobic disk positioned in the centre. In the rectangular compartment of the trough the lateral pressure in the monolayer is adjusted by a movable barrier. The disk is made to rotate by an external magnetic field generated by two pairs of coils. The surface viscosity is measured by determining the viscous drag on the rotating disk that floats on top of a spread monolayer. The surface viscosity is calculated from the torque at steady-state condition. The authors described a relative procedure to discuss the change in surface viscosity of different monolayers systems in arbitrary units. For each system the minimum external magnetic field at which the disk stopped was determined by measuring the current trough the coils. An improved version of this apparatus was reported by Müller et al.149.

An interesting and unique device for characterising the rheological behaviour of foam lamellas was designed by Zotova and Trapeznikov [150]. A foam lamella was positioned in the gap between a ring hanging on a torsion wire and a second ring on the wall of the cell. The device has later been used to characterise both surfactant and protein lamellas. Another rheological study of freely suspended foam lamellas was reported by Bouchama and di Meglio [151], who formed the foam lamella in a porous ring made of sintered glass. In the centre of this lamella a small circular magnetic piece was deposited and an oscillating magnetic field applied on the magnetic disk. The amplitude of the rotation of the disk has been studied by following the displacement of a laser beam reflected from a tiny mirror glued onto the disk. The whole system is kept in a closed chamber to avoid evaporation and is shielded from the natural magnetic field. To demonstrate the functionality the visco-elastic properties of sodium dodecyl sulphate (SDS) and a fluorinated betaine surfactant were studied. The incorporation of polymer chains into the foam lamella does not change significantly its visco-elastic properties.

For monolayer studies in small circular troughs Zakri et al. [152] and Venien-Bryan et al. [153] used a small disk of paraffin-coated aluminium on which a small magnet and mirror was fixed (see Fig. 11).

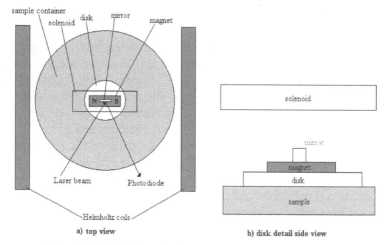

Fig. 11 Schematic principle of small oscillating disk technique according to Zakri et al. [152]

The disk is floating in the centre of the circular trough on the subphase and is surrounded by the monolayer under study. To avoid the drift of the disk away from the trough centre, a small solenoid is placed above the disk, applying a permanent magnetic field which keeps it centred. The contribution of the small solenoid is equivalent to a torsion wire giving an equilibrium position and orientation to the disk. Due to this construction any physical link with the disk is avoided. A current in large peripherical Helmholtz coils applies a torque to the small disk, which tends to force it to rotations. The amplitude of rotation of the disk, measured with a laser beam reflected on the small mirror, is directly related to the rigidity of the monolayer. A sinusoidal torque excitation can be applied to the disk in the frequency range between 0.01 and 100 Hz.

The oscillating disk technique was applied to many system, such as for understanding the phase transitions in monolayers of L-α-dimyristoyl phosphatidyl choline (DMPC), L-α-dipalmitoyl phosphatidyl choline (DPPC), and palmitic acid [148], the chemically cross-linking of actin networks in dependence on frequency [149], the growth of 2D protein crystals in a lipid monolayer [153], thus surface induced polymerization of actin [154], the formation of beta-sheet network at the air/water interface of adsorbed native ovalbumin and S-ovalbumin [155]. Croguennec et al. [156] studied the differences in the interfacial shear elasticity of native and previously heat-denatured ovalbumin.

4. Oscillating needle

Another interfacial viscometer has been proposed by Shahin [157], in which a thin rod is placed within the planar surface of a solution. A magnetic field forces the rod to move within the interface. Small particles are placed within the wake of the moving rod to track the interfacial shear field. By establishing a relation between the force exerted upon the rod and the shear field created, interfacial shear viscosity can be detected. Shahin [157] cites a sensitivity of 10^{-5} mNs/m.

The oscillating needle technique has been modified later by Brooks et al. [158] (Fig. 12). In the centre of a conventional Langmuir trough a flow cell is mounted consisting of half glass tube. The rheological probe for this type of rheometer is a magnetized needle or rod residing between the two glass slides of the flow cell. A concave meniscus is formed between the edges of the half tube which keeps the needle self-centred. The needle is set in motion by applying a magnetic field gradient throughout the space between two large Helmholtz coils surrounding the Langmuir trough. The needle is oscillated longitudinally back and forth in a sinusoidally. As the needle glides parallel to its axis between the edges of the half tube, an interfacial shear flow will be created. It is assumed that the drag experienced by the needle predominantly arises from shear stresses. The rheological measurements are performed by determining the displacement of the moving needle as a function of time, accomplished by imaging the tip onto a position sensitive sensor.

Alternatively a steady force can be applied resulting in a creep experiment.

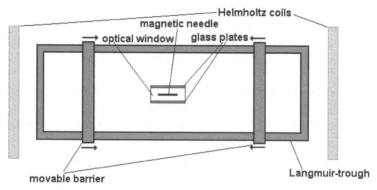

Fig. 12 Schematic principle of oscillating needle technique according Brooks et al. [158]

The detection of the needle movement allows the determination of the interfacial strain e_{xy} defined as the amplitude of the needle displacement divided by the distance between the needle and the vertical glass wall. For small periodic displacements, the complex interfacial shear modulus G_s^* is obtained as the linear proportionality constant between the applied stress and response strain. The interfacial stress can be measured in real time, allowing the study of time-dependent flows and subsequently, the determination of the elasticity and viscosity of monolayers. The measuring technique is available as the commercialized instrument ISR 400 (KSV-Instruments). Recently an analysis of this magnetic rod interfacial stress rheometer was published [159].

A simplified magnetic needle viscometer capable of quick measurements of the steady interfacial shear viscosity at constant surface pressure has been proposed by Ding et al. [160, 161]. Fig. 13 shows the schematic principle of this technique. Instead of an oscillating force, a constant force is applied to the magnetic needle by two parallel magnetic coils. The interfacial shear viscosity is extracted from the terminal velocity of the needle. In a Langmuir trough along the axis of the magnetic field, there is an open-ended channel made from two hydrophilic glass plates in which the magnetic needle floats. The meniscus between the glass plates helps direct the magnetic needle along the axis of magnetic field. The movement of the Teflon coated needle is recorded by a video camera. Thin films of silicone oils of known viscosity have been used to calibrate the viscometer.

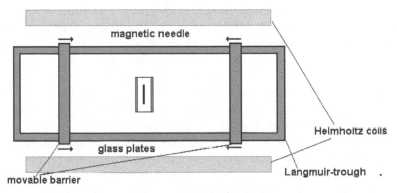

Fig. 13 Schematic principle of a needle viscometer, according to
[160]

Another modification of an interfacial stress rheometer based on a magnetic needle has been proposed by Cristofolini and Fontana [162], where the needle is displaced by a magnetic dipole which is placed underneath the Langmuir trough. The needle orientation is monitored by a charge-coupled device camera that also collects the light emitted by a light-emitting diode, which indicates the status of the driving magnetic dipole moment. The digitized images are then analyzed to determine the oscillation frequencies. The needle oscillation amplitude and its phase lag with respect to the driving excitations are recorded.

The theoretical analysis of the mobility of extended bodies [163] and the drag on moving needles [164,165] in a viscous film shows that in the high-interfacial-viscosity regime the sensitivity of a disk and a needle is the same. The drag coefficient tensor for such objects is isotropic and independent of their size. For larger objects the analysis of Levine et al. [163] shows that the drag tensor becomes size dependent and anisotropic for rodlike objects. The analysis by Fischer [164] underlines that needle viscometer are more sensitive to interfacial shear viscosities than disk viscometers if the ratio between the interfacial and bulk viscosity lies between width and length of the needle. For a viscosity ratio smaller than the needle width, the sensitivity is lost.

Oscillating needle based shear rheology studies were published in literature for various interfacial layers and a complete review cannot be given here. Thus, we can mention here only a subjective selection, such as the work by Brooks et al. [158] who showed that the presence of rigid-rod polymer changes the Newtonian behaviour of an eicosanol monolayer to a Non-Newtonian behaviour. The gelation of lung surfactant induced by epithelial cells exposed to air pollution or oxidation were reported by Anseth et al. [166]. The effect of shear flow on the structure and dynamics of particles suspended at the interface between decane and water have been examined by Stancik et al. [167]. It was shown that in an undisturbed surface layer the monodisperse spherical polystyrene particles arrange themselves on a hexagonal lattice due to strong dipole-dipole repulsion while in a shear flow two distinct regimes are observed: at low particle concentrations or high shear rates, nearest neighbours in the lattice align in flow direction and at high concentrations or low shear rates the particles stay in their lattice positions. The formation of a two-dimensional physical gel of β-casein layers at the air/water interface was studied by Bantchev and Schwartz [168] in a narrow concentration range. Finally, Freer et al. [169,170] present an extended report about the interfacial rheology under both shear and dilation deformation of lysozyme and β-casein at the hexadecane/water interface has been published by using the interfacial stress rheometer for characterisation under shear deformation.

5. Oscillating ring

Sherriff and Warburton [171] developed an oscillating ring interfacial rheometer which can exploit mechanical resonance. Due to an electronic feed-back system it is possible to shift the resonant frequency and to restore the system to resonance at any arbitrary frequency within the range over which the resonance can be shifted. This procedure results in simpler expressions for the visco-elastic parameters [172,173,174,175]. The technique measures the oscillatory properties of a small Pt-Ir ring (Du Noüy ring) located in the surface of solution under study, utilizing the principle of normalized resonance. The measuring head consisted of the Pt-Ir ring is suspended from the moving coil a galvanometer. Measurements were carried out by placing the plane of the ring at the surface of liquid and oscillating at resonance frequency through a small angle. Oscillations can be set up by passing an input sinusoidal voltage through the galvanometer head, and the resulting movement is recorded as an output voltage using a proximity probe and a displacement sensor unit. The interfacial elastic modulus can be calculated from the change in the resonant frequency of the system monitored by electronic feed-back circuit used to achieve normalized resonance. The instrument was called Mark II surface rheometer (Surface Science Enterprises). The equation of motion of the ring is

$$I\frac{d^2\Theta}{dt^2} + \phi\frac{d\Theta}{dt} k\Theta = M \tag{23}$$

where θ is the angle turned trough by the movement, t is the time, ϕ is the total friction, k is the total restoring torque, M is the driving torque and I is the moment of inertia. The amplitude of motion of the ring is measured by a proximity transducer and automatic analysis of the signal generated gives the frequency dependent dynamic surface elastic modulus G_s'. G_s' is defined as

$$G_s' = g_f \, I \, 4\pi\left(\omega^2 - \omega_0^2\right) \tag{24}$$

where I is the moment of inertia, ω is the sample interfacial resonance frequency, ω_0 is the reference interfacial resonance frequency and g_f is the geometric factor. g_f is defined as

$$g_f = \frac{1}{4\pi}\left[\frac{\left(r_d + r_r\right)\left(r_d - r_r\right)}{\left(r_r^2 \cdot r_d^2\right)}\right] \tag{25}$$

where r_d is inner radius of the sample cell and r_r is the outer radius of the ring.

Later the measuring technique was improved commercialised as computer controlled instrument CIR 100 (Camtel). Fig. 14 shows the schematic principle of the oscillating ring technique.

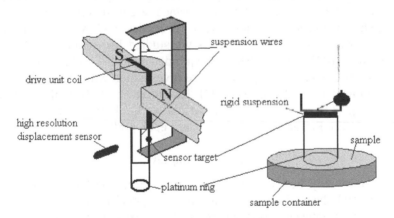

Fig. 14 Schematic principle of oscillating ring technique, according to [176]

The rheometer operates in a force oscillation mode between 10^{-3} and 20 Hz. A sinusoidal torque is applied, and the corresponding angular displacement recorded. Below 2 Hz, the natural frequency of the instrument, the surface storage $G_s'(\omega)$ and loss $G_s''(\omega)$ moduli are obtained from the amplitude ratio and phase lag between torque and displacement. Above 2 Hz, $G_s'(\omega)$ and loss $G_s''(\omega)$ are determined according to the method of normalized resonance which uses a feedback network to restore the resonance condition at a given frequency and amplitude of angular displacement. Before measuring the properties of the film-covered surface, the response of the pure water surface is recorded as a reference [176]. Moduli of the monolayers are extracted by subtracting the response recorded for the pure water surface from the response of the film-covered surface. This procedure assumes that the torques transmitted to the film and to the subphase is additive and decoupled.

Very recently TA-Instruments offered an interfacial accessory kid for there very sensitive bulk rheometer AR-G2 which is based on a magnetic thrust bearing technology for ultra-low, nano-torque control. The measurement is done with a Du Noüy ring positioned at the interface between two liquids, or the surface of a liquid in a circular glass dish [177]. With this technique Cox et al. [178] studied the adsorption behaviour of class II hydrophobine HFBII.

The oscillating ring surface rheometer was applied in many studies of various systems, such as for the adsorption of acacia gum at the aqueous solution/air interface [179], for blood proteins, bovine serum albumin (BSA) and human immunoglobulin G (HIG) [180], or for surfactant adsorption layers at the oil/water interface water in order to find a relationship between interfacial film properties and W/O emulsion stability [181]. Langmuir monolayers we studied for example by Luap and Goedel [182] with a CIR-100 equipped with a special Langmuir trough. Further interesting systems were investigated with the oscillating ring rheometer by Wilde et al. [144], Ridout et al. [183], Golding and Sein [184], Roberts et al. [185], Ariola et al. [186], Wooster and Augustin [187], Piazza et al. [188], or Caro et al. [189].

6. Other techniques

A conceptually new and highly sensitive technique for determining the interfacial shear viscosity of adsorbed or spread interfacial layers at the gas/liquid interface was proposed by Petkov et al. [190]. The interfacial viscosity is calculated from the drag coefficient of a submillimeter-sized particle floating at the surface under the action of an external capillary force of known magnitude. In this technique the particle, partially submersed in the subphase, is not used as a tracer. The motion of the particle is observed by a horizontal and vertical long-focus microscope. The force in this technique arises from the lateral capillary interaction between the particle and a vertical plate. In the vicinity of the plate the air/water interface is not perfectly flat. Its curvature can be determined solving the Laplace equation of capillarity with the respective boundary conditions. Under certain conditions an attractive capillary force appears whose magnitude can be calculated [191]. From the determined drag coefficient the interfacial shear viscosity can be calculated by a respective theory [192,193]. The sensitivity of the method is of the order 10^{-8} Ns/m. For the determination of the yield stress of gel-like protein films with the attached particle technique Petkov et al. [194] proposed a modified Bingham model to describe this particular interfacial rheological behaviour.

Barentin et al. [195] proposed a measuring technique which measures the drag force acting on a macroscopic translating disk. A circular thin disk is forced to translate with a constant velocity at the interface. The measurements have been performed under the condition that the depth of the subphase is small as compared to the disk radius and the velocity field in the subphase is analyzed. Hydrodynamic calculations are discussed to obtain the interfacial shear viscosity from the measurements. Different hydrodynamic models have been proposed for soluble and spread interfacial layers to describe the experimental systems. Experimental results obtained with this technique were published for an adsorbed polymer (polyethylenoxide) and an insoluble monolayer (stearic acid) [196] and a two-dimensional foam [197].

D. INTERFACIAL SHEAR RHEOLOGY OF PROTEIN LAYERS

During the description of interfacial shear rheology instrumentations few applications to investigate protein systems have already been discussed very briefly. Here some selected systems containing proteins will be discussed here in more detail.

It has been shown that interfacial shear rheology studies of high accuracy can provide very valuable information on intermolecular interaction processes and structure changes in interfacial layers. Due to the missing standardization of the different techniques it is very difficult to compare the obtained data quantitatively. The inconsistency of results reported in literature on interfacial shear rheology has been recognised and discussed by several authors [168,198, 199]. This problem is caused not only due to the variety of different systems (soluble or insoluble compounds at air/water or oil/water interfaces). Even for a certain measuring technique, e.g. the torsion pendulum technique, the used measuring bodies (biconical disk, sharp edge, ring) have different shapes and sizes and are often made from different materials (glass, PTFE, stainless steel, titanium, gold, or platinum). Some approaches to ascribe a two-dimensional shear viscosity to the interface are "experiment-dependent" and can represent measuring artefacts. Often the geometry of the apparatus varies and the data analysis takes not properly into account the influence of the adjacent subphase. Therefore the experimental techniques have been often used only for relative measurements. Another source for inconsistent results is the generation of a radial flow due to a fast deformation of the interfacial layer and by that not only an interfacial tension gradient is generated, but in addition the shear deformation is inhomogeneous. There are interfacial systems which react very sensitive on deformation. If the deformation is very strong the interfacial structure can be destroyed. For such systems a continuous shear can not be applied. The maximum shear stress which can be applied to an interfacial layer must be checked for example by an amplitude sweep in the oscillatory mode to guarantee that the experiment is performed in the linear visco-elastic regime. Often instruments are unable to perform such tests. On the other hand the reproducibility of measurements is sometimes very poor due to the different wetting conditions to the rheological probe or due to the fact that soluble macromolecular protein layers are more three-dimensional layers at the interface. Although for insoluble systems the interfacial shear viscosity of monolayers is clearly defined the reproducibility comparability is often also insufficient. For such systems not only the geometry and the position in of rheological probe but also the wetting and the adhesion conditions to the rheological probe play an important role for a clear rheological flow field.

Nevertheless the different measuring techniques have most often been applied to protein adsorption layers at the air/water or oil/water interface to study the adsorption layer formation of pure protein or mixed protein/surfactant systems. The reason for this is that proteins play an important role as macromolecular surfactants to stabilize foams or emulsions. Especially in food, cosmetic and pharmaceutical products they are common in use due to the lack of other efficient foaming agents or emulsifiers. The activity of proteins in these applications is determined by their structure and properties in the adsorption layers: Typical food proteins are mixtures of several protein components. Therefore the interaction between these components in the adsorption layer also influences their ability as macromolecular surfactants to stabilize dispersed systems. The formation and stabilisation of emulsions requires the presence of an emulsifier that can effectively reduce the interfacial tension between the oil and aqueous phase. Proteins are generally less surface active than small molecular weight surfactants. The less interfacial activity of proteins is related to their complex structural properties. Proteins are macromolecular organic compounds made by different amino acids. Therefore proteins contain hydrophilic and hydrophobic groups which are randomly distributed in their primary structure. In the secondary and tertiary folded conformation some of them exist as segregated patches on the surface of the protein molecule. There are not clearly defined hydrophobic tail and hydrophilic head as known for typical surfactants. When a protein adsorbs at the oil/water interface, only the hydrophobic residues of the molecule with are positioned to the oil/water interface. Most of the protein molecule is oriented into the aqueous bulk phase. Due to conformational constrains to properly orient the hydrophilic and hydrophobic groups at the interface and improper packing at the interface proteins are not able to significantly reduce the interfacial tension. The differences in structural and other physicochemical properties of proteins greatly affect their interfacial activity and mechanical properties of interfacial layer. Due to the complexity of protein unfolding during the adsorption process, this process is so far understood only at a phenomenological level. To act as an efficient surfactant, the protein should be able to adsorb readily to interfaces, it must be able to unfold easily at the interface and it should be able to form a cohesive film at the interface via intermolecular interactions. Therefore in practical systems often mixtures of protein with small surfactants are used. In such systems competitive adsorption processes and interactions between the components happen in the bulk and at the interface [200, 201, 202]. To predict or model the behaviour of such complex relationships between interfacial properties and foam/emulsion behaviour, various approaches exist, most of which being purely empirical. A bottom-up approach, i.e. the description of the interfacial phenomena on a molecular and supra-molecular level, suggests, that knowledge on the dynamic and equilibrium properties of adsorption layers at the water/air and water/oil interface are needed for understanding the interaction between individual droplets and the stability of liquid films between them as elementary steps in the process of formation and stabilization or destabilization of an emulsion.

The proteins most frequently studied by interfacial shear rheology are β-casein and β-lactoglobulin. β-casein has 209 amino acid residues and a molecular weight is 24 kDa. Due to the absence of sulphuric amino acids, e.g. cysteine, and is therefore unable to form intramolecular covalent disulfide bonds, in contrast to β-lactoglobulin. Most of the hydrophobic amino acids are located within the first 50 residues and the structure is often described as a random coil copolymer.

Bovine β-lactoglobulin has 162 residues and a molecular weight is 18.4 kDa. Several genetic variants have been identified, the main ones are labelled A and B. Mackie et al. [203] demonstrated that the different molecular structure of these genetic variants has a strong influence on the interfacial activity. Bovine β-lactoglobulin contains sulphuric residues and has therefore a complex secondary and ternary structure. The charged residues are distributed over the whole molecule. It is predominately dimeric and in native state it belongs to the globular proteins.

The globular protein β-lactoglobulin adsorbs comparatively fast from aqueous solutions (pH 7, in phosphate buffer with an ionic strength of 10mM) at the liquid/air interface and forms visco-elastic adsorption layers. With increasing β-lactoglobulin concentration first an increase of interfacial shear viscosity is observed. At a concentration of 10^{-6} mol/l the shear viscosity increases very fast before levelling off. At lower concentration the viscosity increases slower and the plateau values are significantly lower. For higher concentration the viscosity starts at comparatively large values and decreases with time. The decrease may be explained by the structure of the β-lactoglobulin in the adsorption layer. At low concentration the adsorbed protein molecules have time to unfold and hence can form a layer of increasing viscosity. The higher the protein concentration the faster the adsorption, therefore adsorbing molecules get less and less time to unfold. Consequently the molecules arrange a more and more compact intermolecular conformation in the adsorption layer which leads to a decrease in interfacial shear viscosity. The time dependencies of the surface shear viscosity are similar but the absolute values are lower than those for β-lactoglobulin. While the β-lactoglobulin layer is visco-elastic the β-casein layers exhibit a purely viscous behaviour. The adsorption data, given as dynamic surface tensions, confirm this picture drawn from the shear rheology. Increasing viscosity is connected with unfolding, which on the other hand leads to a rather thin adsorption layer and hence small surface tension decrease.

This picture of structure changes at higher β-lactoglobulin concentration has been confirmed by recently performed experiments in forced oscillating mode using the biconical disk rheometer MCR-301-IRS (Anton Paar Germany GmbH). The relative deformation is for the MCR-301-IRS with 0.2% higher than 0.09% for the torsion pendulum apparatus.

In Fig. 15 the time dependencies of the interfacial shear storage modulus $G_i{'}$ and lost modulus $G_i{''}$ for different β-lactoglobulin concentrations ($2 \cdot 10^{-6}$ mol/l, $5 \cdot 10^{-6}$ mol/l and 10^{-5} mol/l) are shown. With increasing concentration higher values of interfacial modules are reached in shorter adsorption times. The measured curves show that at short adsorption times for the lower β-lactoglobulin concentrations $2 \cdot 10^{-6}$ mol/l and $5 \cdot 10^{-6}$ mol/l the interface behaves more viscous. After a certain time however a cross over of G' and G'' can be observe and the interface starts to behave more elastic. This cross over time is shorter for the higher concentration of 5.10^{-6} mol/l. For the 10^{-5} mol/l β-lactoglobulin solution both interfacial modules increase very fast at short adsorption times and after approximately 80 min a steep decrease of both modules is measured.

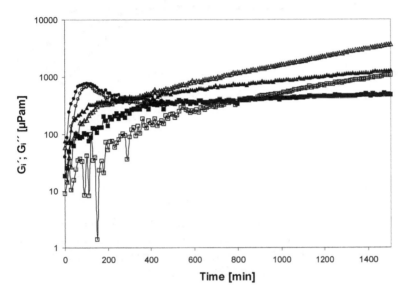

Fig. 15 Time dependence of interfacial shear storage modulus G_i' and interfacial shear loss modulus G_i'' for three different β-lactoglobulin concentrations at pH 7, I=10mM, γ=0.2%, ω=0.7rad/s. 2.10^{-6} mol/l (■) G_i'', (□) G_i'; 5.10^{-6} mol/l (▲) G_i'', (△) G_i'; 1.10^{-5} mol/l (●) G_i'', (○) G_i'

As mentioned earlier interfacial layers are very sensitive against deformation. Hence, to be sure that interfacial structures are not destroyed the maximum shear stress which can be applied to the interface must be checked out for example by an amplitude sweep. For the torsion pendulum technique this has been done by experiments with increasing deflection angle. In a certain range the rheological results should be independent of the deflection angle. For the torsion pendulum technique, which transfers a lower torque to the interface all experiments have been performed in the linear visco-elastic regime. This was checked for the biconical disk rheometer MCR-301-IRS by an amplitude sweep experiment. Fig. 16 shows the results for two differently pre-aged β-lactoglobulin solutions with the same concentration of 5.10^{-6} mol/l. Both amplitude sweeps show that the adsorption layers have elastic properties.

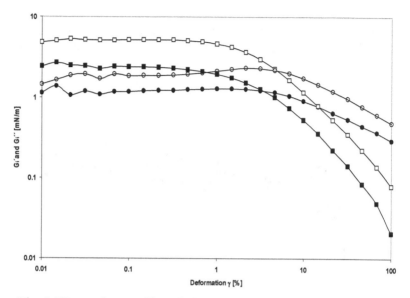

Fig. 16 Dependence of interfacial shear storage G_i' and loss modulus G_i'' on deformation for two different pre-aged β-lactoglobulin solutions of $5 \cdot 10^{-6}$ mol/l. After 5h pre-aging (●) G_i'', (■) G_i' and after 20h pre-aging (○) G_i'', (□) G_i'

The values for the 20h pre-aged solution are higher than for the 5h pre-aged solution, however, the maximum deformation which can be applied to the interface without breaking any structure is very similar. The data for the interfacial shear storage modules decrease already at a relative deformation of 0.8%. Hence, the time dependent oscillation experiments presented in Fig. 15 were performed in the linear visco-elastic regime.

A comparison between literature data from different origin is quite difficult due to various reasons, including the peculiarities of the experimental set-ups and measured samples. In addition, often the solution conditions were very different, such as pH, ionic strength. Despite these problems, we try here to make such comparison for the interfacial shear rheology of adsorption layers formed from β-casein solutions.

Bantchev and Schwartz [168] studied the aging effect of β-casein adsorption layers the oscillating needle technique and compared their data with those published by Murray [204] and Krägel et al. [107], who used the torsion pendulum technique. The resulting interfacial shear viscosities in dependence of adsorption time showed a similar qualitative trend in the aging behaviour of β-casein layers regardless of the differences in the measuring conditions. Murray's data were obtained at the water/tetradecane interface under steady state conditions. Krägel's data refer to a β-casein concentration of 10^{-6} mol/l while Bantchev and Schwartz used a solution with different pH and ionic strength. If one would compare the different data more carefully, an even better agreement can be achieved. In Fig. 17 the interfacial shear viscosities as a function of the adsorption time of Bantchev and Schwartz for a $3.7 \cdot 10^{-7}$ mol/l β-casein solution are shown together with the data for β-casein solutions of 10^{-7} mol/l and $3 \cdot 10^{-7}$ mol/l published in [104]. In addition, the results of Ridout et al. [183] for a $2 \cdot 10^{-7}$ mol/l β-casein solution are included in Fig. 17. These data have been obtained using the oscillating ring technique. To be able to compare interfacial shear viscosities the data of Ridout et al. had to be recalculated from the interfacial shear modules using the expression $G''/\omega = \eta_s$.

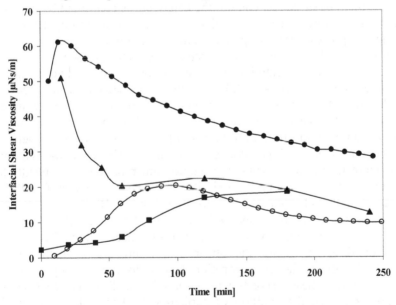

Fig. 17 Comparison of time dependent interfacial shear viscosity data of β-casein solution from literature (▲) data from Bantchev and Schwartz [168] $3.7 \cdot 10^{-7}$ mol/l, pH 7.4, I=100mM, ω=1.57 rad/s; (■) data from Ridout et al. [183] $2 \cdot 10^{-7}$ mol/l, aqueous solution, ω=18.85 rad/s; (O) $1 \cdot 10^{-7}$ mol/l data from Wüstneck et al. [104] and (●) $3 \cdot 10^{-7}$ mol/l unpublished data, for pH 7, I=10mM, ω=0.69 rad/s.

The data obtained for β-casein with three different oscillatory measuring techniques are in a more than qualitative agreement, although the measuring conditions were very different. First the β-casein samples were from different sources and the solution conditions were different as well. While Ridout et al. just solved the β-casein in pure water the other groups used phosphate buffered solutions. Bantchev and Schwartz worked at pH 7.4 and ionic strength of 100 mM, Krägel et al. at pH 7 and a lower ionic strength of 10mM. Regarding the oscillatory frequency, the lowest frequency of 0.69 rad/s has been used by Krägel et al. while Bantchev and Schwartz applied a frequency of 1.57 rad/s and Ridout et al. used the resonance frequency of the instrument of 18.85 rad/s. In all cases the applied maximum strain was kept very low to remain in the linear visco-elastic regime.

The dynamic adsorption behaviour of mixed protein/surfactant systems is very complex and depends most of all on the composition. At high protein or surfactant content the behaviour is controlled by the main component while in the intermediate range the adsorption dynamics is a complex process involving adsorption, desorption, interaction and conformational rearrangement in dependence on the surface coverage, as discussed from a thermodynamic, adsorption dynamics and dilational rheology point of view [202]. Results of respective shear rheology studies have been discussed for example in [106, 107, 110], for example in terms of dependencies of the interfacial shear viscosity on the adsorption time for different mixtures of β-lactoglobulin with the anionic surfactant SDS (sodium dodecyl sulphate) and with the cationic surfactant CTAB (cetyl trimethyl ammonium bromide). The results for mixtures of β-lactoglobulin with the non-ionic surfactant $C_{10}DMPO$ (decyl dimethyl phosphine oxide) are shown in Fig. 18.

These three different kinds of surfactants have a similar CMC and were used at the same fixed β-lactoglobulin concentration of 10^{-6} mol/l. For all three surfactant the shear elastic properties of the mixed β-lactoglobulin/surfactant adsorption layers disappear completely and showed only a shear viscosity was observed. The different dependencies demonstrate that the addition of surfactant lead to a decreases in the interfacial shear viscosity. For the ionic surfactants (SDS and CTAB) already at very low surfactant concentration low viscosities are obtained. For the non-ionic surfactant ($C_{10}DMPO$), due to the peculiarities of the protein/surfactant interactions, the viscosity first increases reaches a maximum and than decreases. With increasing surfactant concentration the decrease sets in earlier and becomes faster. The presence of surfactant leads to a competitive adsorption of low-molecular weight compounds with the protein. The surfactant molecules replace the protein in the adsorption layers.

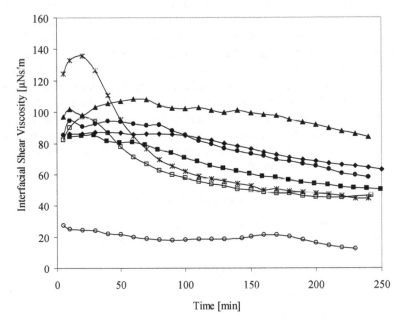

Fig. 18 Dependence of interfacial shear viscosity on the C_{10}DMPO concentration in β-lactoglobulin / C_{10}DMPO mixtures (fixed β-lactoglobulin concentration 10^{-6} mol/l); ■- pure β-lactoglobulin; ♦- 10^{-6} mol/l C_{10}DMPO; ●- $2\,10^{-6}$ mol/l C_{10}DMPO; ▲- $5\,10^{-6}$ mol/l C_{10}DMPO; *- 10^{-5} mol/l C_{10}DMPO; □- $2\,10^{-5}$ mol/l C_{10}DMPO; ○- $3\,10^{-5}$ mol/l C_{10}DMPO) as a function of adsorption time

In different other studies of protein/surfactant mixtures it was observed that even very low concentrations of surfactants cause often a deep decrease in the values of interfacial viscosity and elasticity. With increasing surfactant/protein ratio the protein is displaced from the interface for which different mechanisms are discussed in literature. The displacement of protein by surfactant depends on the strength of the protein/protein interaction and on the surfactant's nature. There are evidences that the random coiled protein β-casein will be easier displaced from the interface than the globular protein β-lactoglobulin. The displacement of proteins by ionic surfactant is much more pronounced than by non-ionic surfactants. The direct binding of ionic surfactant to the protein changes the conformation of the protein molecules to a much more unfolded state. Depending on the surfactant concentration less protein molecules will remain in their native structure in the bulk solution while the conformation of protein-surfactant complexes can be very different from the native protein. At higher ratios of surfactant/protein mixtures the surfactant dominates at the interface and the interfacial shear rheological values are very low. Note, the shear rheological parameters of pure surfactant adsorptions layers are smaller that the sensitivity of typical instruments and hence values are close to zero. The interaction of non-ionic surfactants with proteins is a much more specific effect as demonstrated in Fig. 18. Due to the hydrophobic interaction at low surfactant concentrations a certain complex formation at the interface leads to an enhancement of interfacial shear viscosity. At higher non-ionic surfactant concentration the surfactant molecules become dominant at the interface.

ACKNOWLEDGEMENTS

This work was partially supported by the German Science Foundation DFG SPP 1273 Mi 418/16-1 and the ESA MAP-FASES project.

E. LIST OF SYMBOLS

A	surface area
A_c	contact area
a	decay time
Bo	Boussinesq number
c	offset
$\mathbf{D}^{xy}$	interfacial rate of deformation tensor
$\mathbf{div}_{xy}$	interfacial divergence
E_r	elasticity of torsion wire
e_{xy}	interfacial shear strain
$\dot{e}_{xy}$	rate of interfacial shear strain
F_r	friction of clean solvent interface
G_s	interfacial shear modulus
$G^{\bullet}$	complex interfacial modulus
$G_s^{\bullet}$	complex interfacial shear modulus
G'	interfacial storage modulus
G''	interfacial lost modulus
g_f	geometric factor (oscillating ring geometry)
H_s	geometric factor (sharp edge or disk geometry)
h	thickness of a fluid layer
I	moment of inertia
I_r	moment of inertia of measuring system
k	total restoring torque

L	length of a surface element
L_c	characteristic length
l	length of a channel
M	torque
$\mathbf{P}$	interfacial projection tensor
P_c	contact perimeter
R_C	characteristic distance of flow geometry
R_f	ring radius (wall edge viscometer)
r_d	inner radius of sample cell
r_r	outer radius of ring, sharp edge or disk
S	shear stress
T	period of oscillation
$\mathbf{T}^{(xy)}$	interfacial stress tensor
t	time
Q	film flow rate
V	characteristic velocity
V_C	centreline velocity
V^{xy}	interfacial velocity
w	width of a channel
$y(t)$	oscillation function
y_0	amplitude
α	damping parameter
β	frequency parameter
γ	surface or interfacial tension
γ_0	surface or interfacial tension of a pure solvent system
δ	phase lag
η	bulk viscosity
η_s	surface or interfacial shear viscosity
η_d	surface or interfacial dilational viscosity
Θ	deflection angle
$\Theta(t)$	stimulation of oscillation
ϕ	total friction
ξ_i	displacement of an interfacial element
Π	surface pressure
π	3.1416.....
σ_{xy}	interfacial shear stress
φ	phase shift parameter
Ω_D	angular velocity of disk
Ω_r	angular rotation of ring
Ω_P	angular rotation of particle
ω	frequency

F. REFERENCES

1. D. Langevin, Adv. Colloid Interface Sci. 88 (2000) 209.
2. J. Maldonado-Valderrama, A. Martin-Molina, A. Martin-Rodriguez, M.A. Cabrerizo-Vilchez, M.J. Galvez-Ruiz and D. Langevin, J. Phys. Chem. C 111 (2007) 2715.
3. R.J. Myers and W.D. Harkins, J. Chem. Phys., 5 (1937) 601.
4. D.G. Dervichian and M. Joly, Comptes rendus, 2004 (1937) 1318.
5. W.D. Harkins and J.G. Kirkwood, J. Chem. Phys. 6 (1938) 53.
6. D.G. Dervichian and M. Joly, J. Chem. Phys., 6 (1938) 226.
7. R.S. Hansen, J. Phys. Chem., 63 (1959) 637.
8. F.C. Goodrich, In: Solution Chemistry of Surfactants, K.L. Mittal (ed.), Vol. 2, Plenum Press, New York, 1979, pp. 733.
9. H.A. Stone, Phys. Fluids, 7 (1995) 2931.
10. H. Hühnerfuss, J. Colloid Interface Sci., 107 (1985) 84.
11. M. Saccetti, H. Yu and G. Zografi, J. Chem: Phys., 99 (1993) 563.
12. V. Turati, C. Ferrari, A. Relini and R. Rolandi, Langmuir, 14 (1998) 1963.
13. N.L. Jarvis, J. Phys. Chem., 70 (1966) 3027.
14. H. Hühnerfuss, J. Colloid Interface Sci., 126 (1988) 384.
15. D.K. Schwartz, C.M. Knobler and R. Bruinsma, Phys. Rev. Lett, 73 (1994)2841.
16. M.L. Kurnatz and D.K. Schwartz, Phys. Rev. E, 56 (1997) 3378.
17. M.L. Kurnatz and D.K. Schwartz, J. Rheol., 41 (1997) 1173.
18. A.T. Ivanova, M.L. Kurnatz and D.K. Schwartz, Langmuir, 15 (1999) 4622.
19. A.T. Ivanova, J.Ignes-Mullol and D.K. Schwartz, Langmuir, 17 (2001) 3406.
20. J. Ignes-Mullol and D.K. Schwartz, Langmuir, 17 (2001) 3017.
21. D.J. Olsen and G.G. Fuller, J. Non-Newtonian Fluid Mech., 89 (2000) 187.
22. J. Benjamins and E.H. Lucassen-Reynders, In: Studies in Interface Science Series, Vol. 7, Proteins at Interface, D. Möbius and R. Miller (eds.), Elsevier, Amsterdam, 1998, pp. 341.
23. J.M. Rodriguez Patino and C. Carrera Sanchez, Biomacromolecules, 5 (2004) 2065.
24. J.M. Rodriguez Patino, M. Cejudo Fernandez, C. Carrera Sanchez and M.R. Rodriguez Nino, Ind. Eng. Chem. Res., 45 (2006) 1886.
25. J.T. Davies and G.R.A. Mayers, Trans Faraday Soc., 56 (1960) 691.
26. R.A. Burton and R.J. Mannheimer, "Ordered Fluids and Liquid Crystals", Advances in Chemistry Series No. 63, American Chemical Society, Washington (DC), 1967, pp. 315.

27. L. Gupta and D.T. Wasan, Ind. Eng. Chem. Fundam., 13 (1974) 26
28. A.J. Pintar, A.B. Israel and D.T. Wasan, J. Colloid Interface Sci., 37 (1971) 52.
29. M.F.M. Osborne, Kolloid Z. Z. Polym., 224 (1968) 150.
30. R.J. Mannheimer and R.S. Schechter, J. Colloid Interface Sci., 27 (1968) 324.
31. R.J. Mannheimer and R.S. Schechter, J. Colloid Interface Sci., 25 (1967) 434.
32. R.J. Mannheimer and R.S. Schechter, J. Colloid Interface Sci., 32 (1970) 195.
33. M.G. Hedge and J.C. Slattery, J. Colloid Interface Sci., 35 (1971) 593.
34. J.W. Gardner, J.V. Addison and R.S. Schechter, AIChE J., 24 (1978) 400.
35. J.V. Addison and R.S. Schechter, AIChE J., 25 (1979) 32.
36. D.T. Wasan, L.Gupta and M.K. Vora, AIChE J., 17 (1971) 1287.
37. A.R. Deemer, J.D. Chen, M.G. Hedge and J.C. Slattery, J. Colloid Interface Sci., 78 (1980) 87.
38. V. Mohan and D.T. Wasan, In: "Colloid and Interface Science", Vol. 4, M. Kerker (ed.), Academic Press, New York, 1976, pp. 439.
39. R. Nagarajan and D.T. Wasan, Rev. Sci. Instrum., 65 (1994) 2675.
40. A.Hirsa, G.M. Korenowski, L.M. Logory and C.D. Judd, Langmuir, 13 (1997) 3813.
41. J.M. Lopez and A.H. Hirsa, J. Colloid Interface Sci., 229 (2000) 575.
42. J.M. Lopez, R. Miraghaie and A.H. Hirsa, J. Colloid Interface Sci., 248 (2002) 103.
43. J.M. Lopez and A.H. Hirsa, J. Colloid Interface Sci., 242 (2001) 1.
44. A.H. Hirsa, J.M. Lopez and R. Miraghaie, J. Fluid Mech., 470 (2002) 135.
45. V. Mohan, L. Gupta and D.T. Wasan, J. Colloid Interface Sci., 57 (1976) 496.
46. A.T. Kott, J.W. Gardener and R.S. Schechter, J. Colloid Interface Sci., 47 (1974) 265.
47. C. Meban, Respiration Physiology, 33 (1978) 219.
48. A.K. Chattopadhyay, L. Ghaicha, S.G. Oh and D.O. Shah, J. Phys. Chem., 96 (1992) 6509.
49. I.Blute, M. Jansson, S.G. Oh and D.O. Shah, J. Amer. Oil Chemist Soc., 71 (1994) 41.
50. F.C. Goodrich, L.H. Allen and A. Poskanzer, J. Colloid Interface Sci., 52 (1975) 201.

51. A. Poskanzer and F.C. Goodrich, J. Colloid Interface Sci., 52 (1975) 213.
52. A. Poskanzer and F.C. Goodrich, J. Phys. Chem., 79 (1975) 2122.
53. F.C. Goodrich and D.W. Goupil, J. Colloid Interface Sci., 75 (1980) 590.
54. R.D. Krieg, J.E. Son and R.W. Flumerfeld, J. Colloid Interface Sci., 79 (1981) 14.
55. O. Hassager and H. Westborg, J. Colloid Interface Sci., 119 (1987) 524.
56. J. Perieditis, N.R. Amundson and R.W. Flumerfelt, J. Colloid Interface Sci., 119 (1987) 303.
57. V.J. Cambridge, W.D. Constant and J.M. Wolcott, Chem. Eng. Comm., 70 (1988) 137.
58. I. Langmuir, Science, 84 (1936) 378.
59. A.G. Brown, W.C. Thuman and J.W. McBain, J. Colloid Sci., 8 (1953) 491.
60. R.J. Mannheimer and R.A. Burton, J. Colloid Interface Sci., 32 (1970) 73.
61. Lifschutz, N., M.G. Hedge and J.C. Slattery, J. Colloid Interface Sci., 37 (1971) 73.
62. P.B. Briley, A.R. Deemer and J.C. Slattery, J. Colloid Interface Sci., 56 (1976) 1.
63. R. Shail, J. Engng. Math., 12 (1978) 59.
64. S.G. Oh and J.C. Slattery, J. Colloid Interface Sci., 67 (1978) 516.
65. A.M.J. Davis and M.E. O'Neill, Int. J. Multiphase Flow, 5 (1979) 413.
66. J. Ross, J. Phys. Chem., 62 (1958) 531.
67. N.W. Tschoegl, Kolloid-Zeitschr., 181 (1962) 19.
68. W.N. Cumper and A.E. Alexander, Trans. Faraday Soc., 46 (1950) 235.
69. D.E. Graham and M.C. Phillips, J. Colloid Interface Sci., 76 (1980) 240.
70. V.N. Izmailova, In: Progr. in Surface and Membrane Sci., D.A. Cadenhead and J.F. Danielli (eds.), Vol. 13, Academic Press, London, 1979, pp. 141.
71. V.N. Izmailova and G.P. Yampolskaya, In: Studies in Interface Science Series, Vol. 7, Proteins at liquid interfaces, D. Möbius and R. Miller (eds.), Elsevier, Amsterdam, 1998, pp. 103.
72. V.N. Izmailova and G.P. Yampolskaya, Adv. Colloid Interface Sci., 88 (2000) 99.
73. E. Dickinson, Food Hydrocolloids, 17 (2003) 25.
74. E. Dickinson, Soft Matter, 2 (2006) 642.

75. B.S. Murray, Curr. Opin. Colloid Interface Sci., 7 (2002) 426
76. M.A. Bos and T. van Vliet, Adv. Colloid Interface Sci., 91 (2001) 437
77. S.R. Derkatch, J. Krägel and R. Miller, Colloid Journal (Russian), 71 (2009) 1
78. J. Krägel, S.R. Derkatch and R. Miller, Adv. Colloid Interface Sci., 144 (2008) 38
79. Lifschutz, N., M.G. Hedge and J.C. Slattery, J. Colloid Interface Sci., 37 (1971) 73.
80. H.J. Karam, J. Appl. Polym. Sci., 18 (1974) 1693.
81. B.M. Abraham, K. Miyano, S.Q. Xu and J.B. Ketterson, Rev. Sci. Instrum., 54 (1983) 213.
82. B.M. Abraham, K. Miyano, S.Q. Xu and J.B. Ketterson, Phys. Rev. Lett., 49 (1982) 1643.
83. B.M. Abraham, K. Miyano and J.B. Ketterson, Ind. Eng. Chem. Prod. Res. Dev., 23 (1984) 245.
84. S.S. Feng, R.C. MacDonald and B.M. Abraham, Langmuir, 7 (1991) 572.
85. B.M. Abraham and J.B. Ketterson, Langmuir, 1 (1985) 461.
86. B.M. Abraham and J.B. Ketterson, Langmuir, 2 (1986) 801.
87. J.B. Peng, G.T. Barnes and B.M. Abraham, Langmuir, 9 (1993) 3574.
88. J. Krägel, S. Siegel, R. Miller, M. Born and K.-H. Schano, Colloids & Surfaces A, 91 (1994) 169.
89. J. Krägel, D.C. Clark, P.J. Wilde and R. Miller, Prog. Coll. Polym. Sci., 98 (1995) 239.
90. D.C. Clark, F. Husband, P.J. Wilde, M. Cornec, R. Miller, J. Krägel and R. Wüstneck, J. Chem. Soc. Faraday Trans., 91 (1995) 1991.
91. J. Krägel, R. Wüstneck, D.C. Clark, P.J. Wilde and R. Miller, Colloids & Surfaces A, 98 (1995) 127.
92. J. Krägel, A.M. Stortini, N. Degli-Innocenti, G. Loglio and R. Miller, Colloids & Surfaces A, 101(1995)129.
93. J. Krägel, G. Kretzschmar, J.B. Li, G. Loglio, R. Miller and H. Möhwald, Thin Solid Films, 284/285 (1996) 361.
94. J. Krägel, J.B. Li, R. Miller, M. Bree, G. Kretzschmar and H. Möhwald, Colloid Polym. Sci., 274 (1996) 1183.
95. R.S. Ghaskadvi, J.B. Ketterson, R.C. MacDonald and P. Dutta, Rev. Sci. Instrum., 68 (1997) 1792.
96. T.M. Bohanon, J.M. Mikrut, B.M. Abraham, J.B. Ketterson, L.S. Flosenzier, J.M. Torkelson, and P. Dutta, Rev.Sci.Instrum., 63 (1992) 1822.

97. R.S. Ghaskadvi, T.M. Bohanon, P. Dutta, and J.B. Ketterson, Physical Review E, 54 (1996) 1770.

98. P.W. Chun, S.Y. Shiao, E.E. Saffen, D.O. Shah, W.J. Taylor and R.J. DiTore, Anal. Biochem., 76 (1976) 648.

99. T. Mita and Y. Sairyo, Biosci. Biotech. Biochem., 56 (1992) 1971.

100. D.O. Grigoriev, J. Krägel, A.V. Akentiev, B.A. Noskov, U. Pison and R. Miller, Biophysical Chem., 104 (2003) 633.

101. R. Wüstneck, Colloid Polymer Sci., 262 (1984) 821.

102. R. Wüstneck, H. Hermel and G. Kretzschmar, Colloid Polymer Sci., 262 (1984) 827.

103. R. Wüstneck and L. Zastrow, Colloid Polymer Sci., 263 (1985) 778.

104. R. Wüstneck, J. Krägel, R. Miller, P.J. Wilde and D.C. Clark, Colloids and Surfaces A, 114 (1996) 255.

105. R. Wüstneck, J. Krägel, R. Miller, V.B. Fainerman, P.J. Wilde, D.K. Sarker and D.C. Clark, Food Hydrocolloids, 10 (1996) 395.

106. J. Krägel, M. Bree, R. Wüstneck, A.V. Makievski, D.O. Grigoriev, O. Senkel, R. Miller and V.B. Fainerman, Nahrung-Food, 42 (1998) 229.

107. J. Krägel, R. Wüstneck, F. Husband, P.J. Wilde, A.V. Makievski, D.O. Grigoriev and J.B. Li, Colloids and Surfaces B, 12 (1999) 399.

108. J. Krägel, M. O'Neill, A.V. Makievski, M. Michel, M.E. Leser and R. Miller, Colloids and Surfaces B, 31 (2003) 107.

109. D.O. Grigoriev, S.R. Derkatch, J. Krägel and R. Miller, Food Hydrocolloids, 21 (2007) 823.

110. R. Miller, J. Krägel, R. Wüstneck, P.J. Wilde, J.B. Li, V.B. Fainerman, G. Loglio and A.W. Neumann, Nahrung, 42 (1998) 224.

111. J.-P. Krause, J. Krägel and K.D. Schwenke, Colloids and Surfaces B, 8 (1997) 279.

112. J.B. Li, J. Krägel, A.V. Makievski, V.B. Fainerman, R. Miller and H. Möhwald, Colloids and Surfaces A, 142 (1998) 355.

113. J. Krägel, D.O. Grigoriev, A.V. Makievski, R. Miller, V.B. Fainerman, P.J. Wilde and R. Wüstneck, Colloids and Surfaces B, 12 (1999) 391.

114. V. Kuhnhenn, J. Krägel, U. Horstmann and R. Miller, Colloids and Surfaces B, 47 (2006) 29.

115. E. Dickinson, B.S. Murray and G. Stainsby, J. Colloid Interface Sci., 106 (1985) 259.

116. R. Nagarajan, S.I. Chung and D.T. Wasan, J. Colloid Interface Sci., 204 (1998) 53.

117. T.S. Jiang, J.D. Chen and J.C. Slattery, J. Colloid Interface Sci., 96 (1983) 7.

118. F.C. Goodrich and A.K. Chatterjee, J. Colloid Interface Sci., 34 (1970) 36.
119. R. Shail and D.K. Gooden, Int. J. Multiphase Flow, 7 (1981) 245.
120. Y.C. Ray, H.O. Lee, T.L. Jiang and T.S. Jiang, J. Colloid Interface Sci., 119 (1987) 81.
121. H.O. Lee, T.L. Jiang and K.A. Avramidis, J. Colloid Interface Sci., 146 (1991) 90.
122. E.J. Vernon-Carter and P. Sherman, J. Dispersion Sci. Techn., 2 (1981) 399-413 and 415.
123. G. Doxastakis and P. Sherman, Colloid & Polymer Sci., 264 (1986) 254.
124. E. Dickinson, B.S. Murray, G. Stainsby and C.J. Brock, Int. J. Biol. Macromol., 9 (1987) 302.
125. E. Dickinson, S.E. Rolfe and D.G. Dalgleish, Int. J. Biol. Macromol., 12 (1990) 189.
126. A.Martin, M.A. Bos, M., Cohen Stuart and T. van Vliet, Langmuir, 18 (2002) 1238.
127. P.A. Wierenga, H. Kosters, M.R. Egmond, A.G.J. Voragen, H.H.J. de Jongh, Adv. Colloid Interface Sci., 119 (2006) 131.
128. L.G. Ogden and A.J. Rosenthal, J. Colloid Interface Sci., 191 (1997) 38.
129. H. Rehage and M. Veyssie, Angew. Chem., 102 (1990) 497.
130. A. Burger, H. Leonhard, H. Rehage, R. Wagner and M. Schwoerer, Macromol. Chem. Phys., 196 (1995) 1.
131. R.A. Mohammed, A.I. Bailey, P.F. Luckham and S.E. Taylor, Colloids and Surfaces A, 91 (1994) 129.
132. M. Li, M. Xu, M. Lin and Z. Wu, J. Dispersion Sci. Techn., 28 (2007) 189.
133. C. DeWolf, P. McCauley, A.F. Sikorski, C.P. Winlove, A.I. Bailey, E. Kahana, J.C. Pinder and W.B. Gratzer, Biophysical Journal, 72 (1997) 2599.
134. R. Borbas, B.S. Murray and E. Kiss, Colloids And Surfaces A, 213 (2003) 93.
135. P. Erni, P. Fischer, E.J. Windhab, V. Kusnezov, H. Stettin and J. Läuger, Rev. Sci. Instr., 74 (2003) 4916.
136. P. Erni, P. Fischer, P. Heyer, E.J. Windhab, V. Kusnezov and J. Läuger, Progr. Colloid Polym. Sci., 129 (2004) 16.
137. P. Erni, P. Fischer and E.J. Windhab, Langmuir, 21 (2005)10555.
138. P. Erni, P. Fischer and E.J. Windhab, Appl. Phys. Lett., 87 (2005) 244104-1.
139. P. Erni, E.J. Windhab, R. Gunde, M. Graber, B. Pfister, A. Parker, and P. Fischer, Biomacromolecules 8 (2007) 3458-3466.

140. R. Krishnaswamy, S. Majumdar, R. Ganapathy, V.V. Agarwal, A.K. Sood, and C.N.R. Rao, Langmuir 23 (2007) 3084-3087.
141. R. Krishnaswamy, V. Rathee, and A.K. Sood, Langmuir 24 (2008) 11770-11777.
142. J. Lakatos-Szabo and I. Lakatos, Colloids and Surfaces A, 149 (1999) 507.
143. H. Rehage, M. Husmann and A. Walter, Rheol. Acta, 41 (2002) 292.
144. P.J. Wilde, F.A. Husband, D. Cooper, M.J. Ridout, R.E. Muller and E.N.C. Mills, J. Am. Soc. Brew. Chem., 61 (2003) 196.
145. P.A. Gunning, A.R. Mackie, A.P. Gunning, N.C. Woodward, P.J. Wilde, and V.J. Morris., Biomacromolecules, 5 (2004) 984.
146. P.M. Spiecker and P.K. Kilpatrick, Langmuir, 20 (2004) 4022.
147. T. van Vliet, A.E.A. de Groot-Mostert and A. Prins, J. Phys. E: Sci. Instrum., 14 (1981) 745.
148. H.E. Gaub and H.M. McConnel, J. Phys. Chem., 90 (1986) 6830.
149. O. Müller, H.E. Gaub, M. Bärmann and E. Sackmann, Macromolecules, 24 (1991) 3111.
150. K.V. Zotova and A.A. Trapeznikov, Kolloid. Zh., 26 (1964) 190.
151. F. Bouchama and J.-M. di Meglio, Colloid Polym. Sci., 278 (2000) 195.
152. C.Zakri, A. Renault and B. Berge, Physica B, 248 (1998) 208.
153. C.Venien-Bryan, P.-F. Lenne, C. Zakri, A. Renault, A. Brisson, J.-F. Legrand and B. Berge, Biophysical J., 74 (1998) 2649.
154. A.Renault, P.-F. Lenne, C. Zakri, A. Aradian, C. Venien-Bryan and F. Amblard, Biophysical J., 76 (1999) 1580.
155. A.Renault, S. Pezennec, F. Gauthier, V. Vie and B. Desbat, Langmuir, 18 (2002) 6887.
156. T. Croguennec, A. Renault, S. Beaufils, J.J. Dubois and S. Pezennec, J. Colloid Interface Sci., 315 (2007) 627.
157. G.T. Shahin, PhD Thesis, University of Pennsylvania, Philadelphia, 1986.
158. C.F. Brooks, G.G. Fuller, C.W. Frank, and C.R. Robertson, Langmuir 15 (1999) 2450.
159. S. Reynaert, C.F. Brooks, P. Moldenaers, J. Vermant, and G.G. Fuller, J.Rheology 52 (2008) 261-285.
160. J. Ding. H.E. Warriner and J.A. Zasadzinski, Phys. Rev. Lett., 88 (2002) 168102-1.
161. J. Ding. H.E. Warriner, J.A. Zasadzinski and D.K. Schwartz, Langmuir, 18 (2002) 2800.
162. L. Cristofolini and M.P. Fontana, Philosophical Magazine, 84 (2004) 1537.

163. A.J. Levine, T.B. Liverpool and F.D. MacKintosh, Phys. Rev. E, 69 (2004) 021503-1.
164. T.M. Fischer, J. Fluid Mech., 498 (2004) 123.
165. C. Alonso and J.A. Zasadzinski, Phys. Rev. E, 69 (2004) 021602-1.
166. J.W. Anseth, A.J. Goffin, G.G. Fuller, A.J. Ghio, P.N. Kao, and D. Upadhyay, American Journal of Respiratory Cell and Molecular Biology, 33 (2005) 161.
167. E.J. Stancik, A.L. Hawkinson, J. Vermant, and G.G. Fuller, J. Rheology, 48 (2004) 159.
168. G.B. Bantchev and D.K. Schwartz, Langmuir, 19 (2003) 2673.
169. E.M. Freer, K.S. Yim, G.G. Fuller, and C.J. Radke, Langmuir, 20 (2004) 10159.
170. E.M. Freer, K.S. Yim, G.G. Fuller, and C.J. Radke, J. Phys. Chem. B, 108 (2004) 3835.
171. M. Sherriff and B. Warburton, Polymer, 15 (1974) 253.
172. M. Sherriff and B. Warburton, In: Theoretical Rheology, J. Holten, J.R.A. Pearson and K. Walters (eds.), Applied Science, London, 1975, pp 299.
173. B.Warburton, In: Techniques in Rheological Measurements, A.A. Collyer (ed.), Chapman and Hall, London, 1993, pp. 55.
174. B.Warburton, In: Rheological Measurement, A.A. Collyer and D.W. Clegg (eds.), Chapman and Hall, London, 1998, pp. 723.
175. B.Warburton, Curr. Opin. Colloid Interface Sci., 1 (1996) 481.
176. C.Moules, Principles of Interfacial Rheology & Measurement with CIR-100, TechNote1, Camtel Ltd, Royston, UK, 1998.
177. A.Franck, Nachrichten aus der Chemie, 55 (2007) 163.
178. A.R. Cox, D.L. Aldred, and A.B. Russell, Food Hydrocolloids 23 (2009) 366-376.
179. C.A. Moules and B. Warburton, In „Rheology of Food, Pharmaceutical and Biological Materials with General Rheology", R.E. Carter (Ed.), Elsevier Applied Science, 1990, pp 211.
180. D.J. Burgess and N.O. Sahin, J. Colloid Interface Sci., 189 (1997) 74.
181. F.O. Opawale and D.J. Burgess, J. Colloid Interface Sci., 197 (1998) 142.
182. C. Luap and W.A. Goedel, Macromolecules, 34 (2001) 1343.
183. M.J. Ridout, A.R. Mackie, and P.J. Wilde, J.Agric.Food Chem., 52 (2004) 3930.
184. M. Golding and A. Sein, Food Hydrocolloids, 18 (2004) 451.
185. S.A. Roberts, I.W. Kellaway, K.M.G. Taylor, B. Warburton, and K. Peters, Langmuir, 21 (2005) 7342.

186. F.S. Ariola, A. Krishnan and E.A. Vogler, Biomaterials, 27 (2006) 3404.
187. T.J. Wooster and M.A. Augustin, Food Hydrocolloids, 21 (2007) 1072.
188. L. Piazza, J. Gigli and A. Bulbarello, J. Food Eng., 84 (2008) 420.
189. A.L. Caro, M.R. Rodriguez Nino, and J.M. Rodriguez Patino, Colloids and Surfaces A, 327 (2008) 79-89.
190. J.T. Petkov, K.D. Danov, N.D. Denkov, R. Aust and F. Durst, Langmuir, 12 (1996) 2650.
191. P.A. Kralchevsky, V.N. Paunov, N.D. Denkov and K. Nagayama, J. Colloid Interface Sci., 167 (1994) 47.
192. K.D. Danov, R. Aust, F. Durst and U. Lange, Chem. Eng. Sci., 50 (1995) 263-277.
193. K.D. Danov, R. Aust, F. Durst and U. Lange, J. Colloid Interface Sci., 175 (1995) 36.
194. J.T. Petkov, T.D. Gurkov, and B.E. Campbell, Langmuir, 17 (2001) 4556.
195. C. Barentin, C. Ybert, J.-M. di Meglio and J.-F. Joanny, J. Fluid Mech., 397 (1999) 331.
196. C. Barentin, P. Muller, C. Ybert, J.-F. Joanny, and J.-M. di Meglio, Eur. Phys. J. E, 2 (2000) 153.
197. S. Courty, B. Dollet, F. Ellias, P. Heinig and F. Graner, Europhys. Lett., 64 (2003) 709.
198. J. Benjamins, PhD –Thesis, Wageningen University, Wageningen, 2000.
199. P. Stevenson, J. Colloid Interface Sci., 290 (2005) 603.
200. R. Miller, V.B. Fainerman, A.V. Makievski, J. Krägel, D.O. Grigoriev, V.N. Kazakov and O.V. Sinyachenko, Adv. Colloid Interface Sci., 86(2000)39
201. V.B. Fainerman, S.A. Zholob, M. Leser, M. Michel and R. Miller, J. Colloid Interface Sci., 274 (2004) 496
202. R. Miller, V.S. Alahverdjieva and V.B. Fainerman, Soft Matter, 4 (2008) 1141
203. A.R. Mackie, F.A. Husband, C. Holt and P.J. Wilde, Intern. J. Food Sci. Techn., 34 (1999) 509.
204. B.S. Murray, In: Studies in Interface Science Series, Vol. 7, Proteins at Interfaces, D. Möbius and R. Miller (eds.), Elsevier, Amsterdam, 1998, pp. 179.

NEW ASPECTS IN THE CONSIDERATION OF COMPRESSIBILITY FOR THE CHARACTERIZATION OF LANGMUIR MONOLAYERS

D. Vollhardt[1] and V. B. Fainerman[2]

[1]Max Planck Institute of Colloids and Interfaces,
D-14424 Potsdam/Golm, Germany
[2]Donetsk Medical University, Donetsk 83003, Ukraine

Content

Over decades, the two-dimensional compressibility defined on the basis of the surface pressure – molecular area (Π - A) isotherms has been the main source to obtain information about the phase behaviour and its rheological features of Langmuir monolayers. Since the last two decades, fundamental progress was attained in the experimental determination of the morphological and structural characteristics of condensed monolayer phases in microscopic and molecular scale. Numerous experimental studies have shown that already smallest changes in the molecular structure of the amphiphile can result in changes in the molecular arrangement in the monolayer and thus, in changes of the main characteristics of the monolayer such as, the surface pressure –area per molecule (π - A) isotherms, the shape and texture of the condensed phase domains and the two-dimensional lattice structure.

The experimental progress required improved thermodynamic possibilities to describe the Π - A isotherms, as the classical equations of state allowed only the characterisation of the fluid (gaseous, liquid-expanded) state. Thermodynamically based equations of state, which consider also the aggregation of monolayer material to condensed phase, have been developed to filling this gap. A large number of monolayers with one condensed phase has been analysed on this basis.

The present review focuses particularly to amphiphilic monolayers, indicating the existence of two condensed phases in the Π - A isotherms. Corresponding experimental results which demonstrate differences in the structure features and phase properties are presented. For the thermodynamic description, the equation of state for Langmuir monolayers has been generalized for the case that one, two or more phase transitions in the monolayer take place.

The results obtained on the basis of the generalized equation of state are in agreement with the experimental results. The two-dimensional compressibility of the condensed phases undergoes a jump at the phase transition, whereas the compressibility is proportional to the surface pressure within the condensed phases. An example is presented which explains the procedure of the theoretical analysis of Π - A isotherms indicating the existence of two condensed phases. An element of the procedure is the application of the general principle that the behaviour of any thermodynamic system is determined by the stability condition.

New interesting problems are discussed arising by the anisotropy of monolayers. For the case that the lattice structure is a rectangular unit cell, a concept of linear compressibilities has been introduced. The consequences of another anisotropy of the compressibility, revealed by GIXD studies of the S-phase of octadecanol monolayers, are discussed. In this case, the anisotropy is concluded from the dependencies of the positions of the diffraction peaks of GIXD studies on the surface pressure at different temperatures. Similar studies performed close to the LS-S phase transition would result in a thermodynamically impossible negative compressibility so that the compressibility cannot be determined from the positions of the peak maxima. The results are discussed by a disordered state of the monolayer close to the LS-S phase transition characterised by elastic distortions which are caused by fluctuations with the structure of the new phase in the surrounding matrix without destroying the quasi-long-range positional order.

A. INTRODUCTION

In the last two decades, fundamental progress in the experimental determination of the main characteristics of Langmuir monolayers in microscopic and molecular scale allows new insights into their phase and structural features. It has been found that already smallest changes in the molecular structure of the amphiphile can result in changes in the molecular arrangement in the monolayer and thus, in changes of the main characteristics of the monolayer such as, the surface pressure –area per molecule (Π - A) isotherms, the shape and texture of the condensed phase domains and the two-dimensional lattice structure.

The existence of different two-dimensional phases similar to the three-dimensional state could be concluded from the shape of the Π - A isotherms, already in the early periods of research on Langmuir monolayers and over many decades,. Correspondingly, Gibbs pointed out that there is an elasticity associated with a liquid film if the surface tension varies with the area of the surface [1]. For an insoluble monolayer, the equilibrium elasticity is related to the compressibility of the monolayer.

Analogously to the bulk compressibility, the compressibility of monolayers is defined as [2]

$$C = -\frac{1}{A}\left(\frac{dA}{d\Pi}\right)_T \tag{1}$$

where A is the molecular area. In the case of a pure surface, the compressibility C is infinite because the surface tension is not changed with area. To characterize the monolayer properties, the reciprocal quantity of C, denoted as surface compressional modulus K, is often used.

$$K = \frac{1}{C} = -A\left(\frac{d\Pi}{dA}\right)_T \tag{2}$$

Thus the compressibility properties of Langmuir monolayers can be determined in a simple way from the slope of the Π - A isotherms. It is easy to see that the compressional modulus K precisely coincides with the value of the limiting high-frequency dilational modulus of soluble surfactants monolayers $E_0 = \Gamma(d\Pi/d\Gamma)_T$, where adsorption $\Gamma=1/AN$, $d\ln A = -d\ln\Gamma$, and N is Avogadro's number. In this limiting (high-frequency) case, the value of E_0 depends only on the equilibrium equation of state, as at these frequencies there is no significant exchange of the surfactant between the surface and the bulk (conservation of the adsorbed mass of surfactant).

Over decades, surface tension measurements have been the main source of information about insoluble monolayers. Based on the differences in the compressibility, Harkins [3] and Dervichian [4] introduced four different monolayer phases designated as gaseous (G), liquid expanded (LE), liquid condensed (LC),and solid (S) phases, whose compressibility follows the sequence G > LE > LC > S. Measurements of the Π - A isotherms under equilibrium conditions have been the basis not only for the determination of the monolayer elasticity but also for the development of equations of state. However, the occurrence of several phases and corresponding transition regions complicated the situation so that an effective description of the experimental isotherms has been limited only on the fluid (gaseous, liquid-expanded (LE)) state of monolayers. New progress in the extension of the equations of state on the transition to the condensed state has been made for the condition which describes the equilibrium between the monomers and aggregates considering the free surface existing in the monolayer. A summarised description is given in the theoretical part.

Generally, the time-dependent resistance to deformation of insoluble monolayers has opened up the possibility for the rheological characterisation. For example, relaxation processes manifested as a compression rate dependent hysteresis in a compression-expansion cycle of the Π - A measurement can be due to a dilatational surface viscosity [2]. Furthermore, the flow and shear behaviour in the plane of the surface is dependent on the phase properties of monolayers and show often viscoelastic features [5 - 7]. In this case, the systems have been treated as a combination of viscous and elastic components including, for example, the surface compressibility modulus K and the surface shear modulus.

In recent work, the compressibility of monolayers has been of renewed interest for the characterisation of lattice structures and their changes at the corresponding phase transitions. Thus, the consideration of compressibility can provide new information about thermodynamics and structural characteristics. An important problem, discussed on the basis of grazing incidence X-ray diffraction results, is the anisotropy of the monolayer which is concluded from linear compressibilities along each diffraction vector.

This review is organized as follows. At the beginning a few examples are given to demonstrate the large effect of small structural changes on the main characteristics of monolayers. Then, a survey is given of the thermodynamic possibilities to characterize the monolayers at the air/water interface. In the next section, experimental results of monolayers indicating only one condensed phase are thermodynamically characterised. In the subsequent section, monolayers having more than one condensed phases are discussed. At first, structure features and phase properties at the existence of two condensed phases are demonstrated. After that, the effect of the anisotropy of monolayers on the compressibility and the lattice is discussed. Next, possibilities of the theoretical treatment of Π-A isotherms in the presence of two condensed phases is demonstrated. Finally some conclusions are drawn.

B. MAIN CHARACTERISTICS OF MONOLAYERS SENSITIVELY AFFECTED BY SMALL STRUCTURAL CHANGE OF THE AMPHIPHILE

The molecular arrangement of the monolayer and thus, the main characteristics of the monolayer such as, the Π - A isotherms, the shape and texture of the condensed phase domains and the two-dimensional lattice structure can be largely affected already by smallest changes in the molecular structure of the amphiphile. This has been demonstrated, for example, by a systematic alteration of the head group structure in four monoglycerol amphiphiles by an amide, ether, ester and amine group [8], by four phospholipids of the same chain length which are different only with respect to the number of methyl groups at the nitrogen of the head group [9] or even by exchange of the two substituents at the acid amide group studied by two very similar amphiphiles   N-tridecyl-β-hydroxypropionic   acid   amide $(C_{13}H_{27}-NH-CO-C_2H_4OH)$ and   N-(β-hydroxyethyl)tridecanoic   acid amide $(C_{13}H_{27}-CO-NH-C_2H_4OH$ [10]).

Chiral discrimination effects observed in monolayer studies of numerous chiral amphiphile substantiate the high sensitivity of the monolayer characteristics to the specific stereochemistry. The chiral nature of the enantiomeric and racemic amphiphiles is apparent not only in characteristic differences in the domain shape and the sense of curvatures in the textures but also in differences of the lattice structures. A survey of the effect of molecular chirality on the morphology and two-dimensional lattice structure of Langmuir monolayers is reviewed in ref. [11].

Also the characteristic features of n-hydroxyoctadecanoic acids are essentially affected if the OH-substitution in mid-position of the alkyl chain is only slightly changed [12, 13]. These small changes in the chemical structure give rise to large differences in the monolayer features, as in the following demonstrated by comparison of phase behavior, morphological texture, and lattice structure of 9-, 11-, and 12-hydroxyoctadecanoic acid monolayers at different temperatures.

The OH-substitution in mid position of the alkyl chain is obviously responsible for the common special characteristics of the Π - A isotherms deviating from those of typical amphiphilic monolayers. This conclusion is corroborated by a comparative study of 1-(12-hydroxy)stearoyl-rac-glycerol monolayers which show a similar flat and extended plateau region over a large area range although the unsubstituted 1-stearoyl-rac-glycerol monolayers behaves as usual amphiphilic monolayers [14]. An extended flat plateau region, the extension of which only slightly decreases with the increase of temperature, is typical for the Π- A isotherms of all three hydroxyoctadecanoic acids. Despite the general similarities there are clear differences in the π - A isotherms. In this review, only the isotherm set of 12--hydroxyoctadecanoic acid (Fig. 1) and 9-hydroxyoctadecanoic acid (Fig. 2) is compared. The comparison shows that, at the same temperature, the extension of plateau region increases from 9- to 12-hydroxyoctadecanoic acid whereas the plateau pressure decreases. The kink point at $A = A_c$ is characteristic for the onset of the main phase transition of first order. Correspondingly, at $A > A_c$ the monolayers exist in the fluid (gaseous, LE) state. Theoretical calculations of the Π - A isotherms were performed to explain the strong effect of the position of the OH-substitution on the thermodynamic properties of the monolayers. The calculations provide reasonable agreement between the theoretical predictions and the experimental Π - A isotherms [15].

The temperature dependence of the main phase transition pressure, π_t of the octadecanoic acids OH-substituted in the mid-position provides access to the enthalpy (ΔH) and entropy (ΔS) changes of the phase transition [12, 16, 17]. Linear π_t (T) relations exist for all three n-hydroxyoctadecanoic monolayers (Fig. 3).

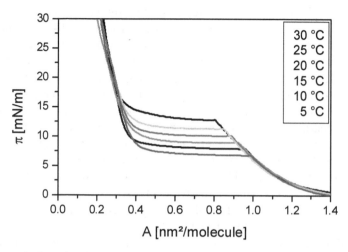

Figure 1. Π-A isotherms of 9-hydroxyoctadecanoic acid monolayers spread on pH 3 water and measured at different temperatures.

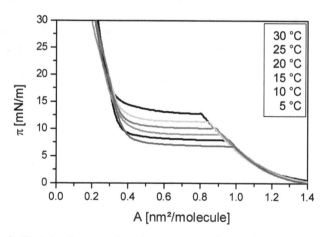

Figure 2. Π-A isotherms of 12-hydroxyoctadecanoic acid monolayers spread on pH 3 water and measured at different temperatures.

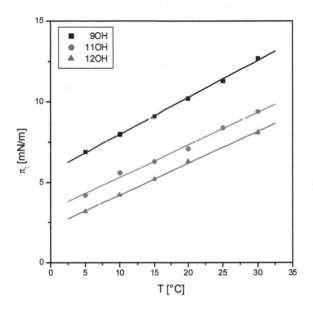

Figure 3. Temperature dependence of the main phase transition pressure (Π_t) of 9-, 11- and 12-hydroxyoctadecanoic acid monolayers.

The $d\pi_t/dT$ value are very similar: 0.23 mN/m per K for 9-hydroxyoctadecanoic acid, 0.20 mN/m per K for 11-hydroxyoctadecanoic acid and 0.20 mN/m per K for 12-hydroxyoctadecanoic acid. In comparison to usual amphiphiles with one alkyl chain (with $d\pi_t/dT \sim 1$ mN/m per 1 K) the $d\pi_t/dT$ slopes of the three 9-, 11- and 12-hydroxyoctadecanoic acid monolayers are very small and only very slightly dependent on the position of the OH-substitution.

It is interesting to refer to a special feature of the π - A isotherms of 9-hydroxyoctadecanoic acid monolayer occurring at low temperatures, as presented for 5°C in Fig. 4. In addition to the main phase transition (arrow I in Fig. 4), a kink at ~18 mN/m indicate a phase transition between two condensed phases (arrow II in Fig. 4). This interesting second phase transition at $\Pi = 18$ mN/m between two condensed phase has been satisfactorily described by the theoretical model under consideration of the two-dimensional compressibility of the condensed monolayer discussed in Section 6.2. The GIXD results presented later in this Section provide detailed information on this phase transition.

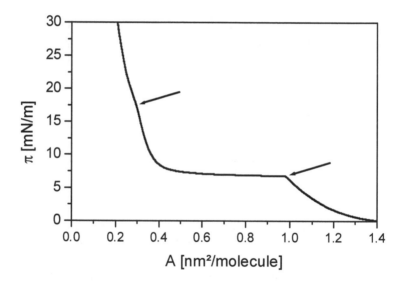

Figure 4. Π-A isotherm of 9-hydroxyoctadecanoic acid monolayer spread on pH 3 water and measured at T = 5 °C. The arrows indicate two phase transition points.

Considerable differences between the monolayers of n-hydroxyoctadecanoic acids OH-substituted in mid-position of the alkyl chain exist also in the domain morphology, as visualized by Brewster angle microscopy (BAM) studies. Representative examples for the domains of 9-hydroxyoctadecanoic acid grown in the two-phase coexistence region at 5, 20 and 25 °C are presented in Figure 5. Despite essential differences in the domain shape, at all temperatures a center exists and the homogeneous reflectivity of the domains indicate the absence of an inner texture. Whereas at low temperatures (5 °C) irregular compact domains are formed, at higher temperatures (20 and 25 °C) four-arm structures with two-fold symmetry are developed. The domains grow rather irregular with more or less developed side arms. At 20 °C two small acute angles and two large obtuse angles between the main arms are formed in opposite direction. With increasing temperature (25 °C) these angles approach each other and become approximately 90 degree.

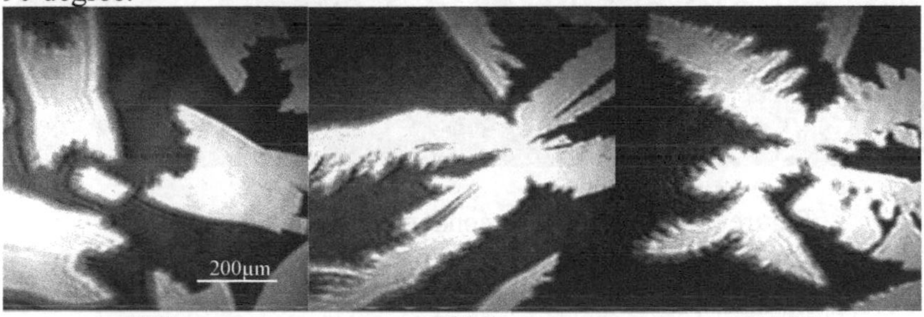

Figure 5. Representative condensed phase domains of 9-hydroxyoctadecanoic acid monolayers at different temperatures. Image size: 750 × 750 μm

Finally it is seen from the BAM images of the 12-hydroxyoctadecanoic acid domains their main characteristics. Typical domain textures of 12-hydroxystearic acid monolayers measured at 10, 20 and 30 °C (Fig. 6) make obvious that small changes in the position of the OH-substitution change considerably the domain morphology. Again, the domains of 12-hydroxyoctadecanoic acid are homogeneously reflecting but, in the case of 12OH-substitution, they develop several arms with the tendency to form curvatures, especially in the medium temperature region, and grow rather irregularly with differences in the growth direction. The comparison with 1-(12-hydroxy)stearoyl-*rac*-

glycerol monolayers reveals similar phase and structural features [14]. The similarity of the π - A isotherms and the domain morphology of both amphiphiles demonstrates the dominating effect of the alkyl chain substitution by the hydroxyl group in mid-position on the monolayer properties whereas the head group effect is of minor significance.

Figure 6. Representative condensed phase domains of
12-hydroxyoctadecanoic acid monolayers at different temperatures.
Image size: 750 × 750 μm.

The GIXD data confirm the conclusions on the effect of small changes in the OH-position in the alkyl chain on the lattice structure of the condensed monolayer phase. The contour plots of the corrected diffraction intensities as a function of the in-plane (Q_{xy}) and out-of-plane (Q_z) components of the scattering vectors reveal considerable differences between 9-, 11- and 12-hydroxyoctadecanoic acid monolayers (9-OH, 11-OH and 12-OH, respectively) (Fig. 7). The structure data calculated for different surface pressures of the three OH-substituted are listed in Table 1, wherein a, b and γ are the unit cell parameters, A_{xy} is the in-plane molecule area, t is the polar tilt angle, td is the tilt direction, ψ_a is the angle between azimuthal tilt direction and a-axis, and A_0 is the cross-section area of alkyl chain.

The contour plots of 9-hydroxyoctadecanoic acid are of special interest because the kink in the π - A isotherm at 5 °C indicates a phase transition of two condensed phases at ~18 mN/m (see Fig. 4). The two reflexes of the contour plots indicate a centered rectangular lattice but according to their position in the lower surface pressure region at π=10 mN/m (both reflexes with Q_z > 0) the molecular tilt is in NNN (next nearest neighbor) direction, whereas at higher surface pressures (π=20 mN/m) the molecules are tilted towards NN (nearest neighbor) direction (Q_z = 0 and Q_z > 0). This change of the lattice structure provides evidence for the phase transition between the two condensed phases which was indicated by the kink in the π - A isotherm at 5 °C at ~18 mN/m.

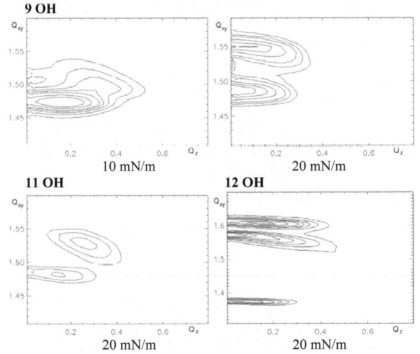

Figure 7. Contour plots of the corrected diffraction intensities as a function of the in-plane (Q_{xy}) and out-of-plane (Q_z) components of the scattering vectors at 5 °C;

Table 1. Lattice structure data of 9-, 11- and 12- hydroxyoctadecanoic acid monolayers.

9-hydroxyoctadecanoic acid

Conditions	Π mN/m	a Å	b Å	γ °	Axy Å^2	t °	td	A$_0$ Å^2
pH 3, 5 °C	10	4.79	4.89	120.7	20.7	12.6	NNN	20.2
	20	4.63	4.82	122.6	19.6	4.8	NN	19.5

11-hydroxyoctadecanoic acid

Conditions	Π mN/m	a Å	b Å	γ °	Axy Å^2	t °	td	A$_0$ Å^2
pH 3, 5 °C	10	4.95	4.99	120.5	21.4	21.5	NNN	19.9
	20	4.69	4.85	122.1	19.9	9.3	NNN	19.7

12-hydroxyoctadecanoic acid

Conditions	Π mN/m	a Å	b Å	γ	Axy Å^2	t °	Ψ_a	A$_0$ Å^2
pH 3, 5 °C	6	4.61	4.99	112.3	21.3	20.3	28	20.0
	10	4.60	4.99	112.3	21.2	19.5	25	20.0
	20	4.40	5.01	114.0	20.2	5.5	30	20.0

Π - surface pressure, a, b, γ - lattice constants, A$_{xy}$ - molecular area, t - polar tilt angle, A$_0$ - cross section area of alkyl chain, td – tilt direction, ψ_a - angle between azimuthal tilt direction and a-axis.

In agreement with the π - A isotherms, a phase transition between condensed monolayer phases of the 11- and 12OH substituted octadecanoic acids was not found using GIXD results (see Table 1). In that case, the lattice type of the condensed monolayer phase is unchanged over the accessible surface pressure range. The contour plots of 11-hydroxyoctadecanoic acid show two reflexes with $Q_z > 0$ (demonstrated for 20 mN/m in Fig. 7) characteristic of a centered rectangular lattice with tilt of the alkyl chains towards NNN direction. The polar tilt of the alkyl chains is somewhat larger than in the case of the 9-OH substitution at the same surface pressure. The three reflexes ($Q_z > 0$) in the contour plots of 12-hydroxyoctadecanoic acid indicate an oblique lattice of the alkyl chains over the entire surface pressure range.

C. THEORY

Under corresponding conditions, the behaviour of Langmuir monolayers formed by insoluble amphiphiles is characterised by the existence of a sharp break in the surface pressure (Π) *vs* area per one mole or molecule (A) isotherm in the region. The surface pressure *vs* area per molecule ((Π-A) isotherms of Langmuir monolayers show a sharp break under appropriate conditions, where the main phase transition from a gaseous (G) or liquid-expanded (LE) to the liquid-condensed (LC) states takes place. Possibly, the first theoretical description of the Π-A isotherm in the 2D phase transition region was given by De Boer [19]. It was shown that the van der Waals equation for the non-ideal 2D gas predicts the existence of metastable states for high values of the intermolecular constant. These states should be characterised by the stepwise decrease of the area A during the 2D condensation, while the surface pressure value remains unchanged or decreases only weekly.

Several attempts were made to give an explanation for the non-horizontal shape of Π-A isotherms in 2D transition region [20-28]. In these studies, LE-LC transitions were treated in the framework of a quasi-chemical approach, where the mass action law was employed for the description of the monomer/aggregate equilibrium. For calculating the free energy of the 2D transition some constituents of standard free energy were taken into account in ref. [24], in particular, the energy associated with the transition of hydrocarbon chains from the gaseous phase to the liquid phase, the conformational energy of these chains in the monolayer and their interaction with the aqueous phase, steric and electrostatic interactions between the polar groups. A still more detailed analysis of the free energy constituents was presented by Ruckenstein and Li in ref. [25] and they obtained good agreement with e experimental data. At the same time, the implementation of the theory for the description of actual monolayers is hampered by the fact that the resulting equations involve 5 to 7 unknown parameters (domain radius, dipole-dipole, electrostatic and van der Waals interaction constants, parameters which define the conformational free energy and the entropy of mixing etc.). The combination of a theoretical approach on the basis of a quasi-chemical approach with model considerations was proposed in refs. [18, 22, 26, 27], resulting in rather simple equations which contain a few parameters only.

An equation of state for the fluid (gaseous, G, or liquid-expanded, LE) states of Langmuir monolayers was derived in ref. [18] by the simultaneous solution of the differential equation for the chemical potential of components within surface layer μ_i^s (Butler's differential equation):

$$d\mu_i^s = RTd\left(ln\, f_i^s + ln\, x_i^s\right) - \Omega_i d\gamma , \qquad (3)$$

and Gibbs' adsorption equation:

$$d\Pi = \sum_{i=1}^{n} \Gamma_i d\mu_i \qquad (4)$$

where T is temperature, R is the gas law constant, f_i is the activity coefficient, Ω_i is the partial molar area, x_i are the molar fraction, γ is the surface tension, Π is the surface pressure, Γ_i is the adsorbed amount, and index I concerns component i. The general equation derived in this way in ref. [18], was then used to obtain Frumkin's, van der Waals' and Volmer's [29] equations of state. The equation of state for monolayers in the fluid (G or LE) state is represented by the Volmer-type equation [16, 18, 28, 30]:

$$\Pi = \frac{nkT}{A - \omega} - \Pi_{coh} \qquad (5)$$

where k is the Boltzmann constant, ω is the partial molecular area for monomers (or the limiting area of molecule in the gaseous state), A is the area per molecule, Π_{coh} is the cohesion pressure, which accounts for the intermolecular interaction under the condition of a constant value of the enthalpy activity coefficient in the monolayer $f_i^s = const$ [18], and the n value accounts for the association (n<1, and 1/n is the aggregation number of small aggregates [28]) or dissociation (n>1, and n is the number of kinetically independent units: fragments or ions in a molecule [30, 31]) degrees of amphiphilic molecules in the monolayer. For example, for some insoluble proteins in a liquid-expanded monolayer state, a value of n = 20–100 was obtained in ref. [31].

In ref. [18], the generalised Volmer's equation which describes the multicomponent insoluble monolayer of monomer or aggregates was derived:

$$\Pi = RT \frac{\sum_i \Gamma_i}{1 - \sum_i \Gamma_i \omega_i} - B \qquad (6)$$

where R is the gas law constant, Γ_i is the adsorption (surface concentration) value of the ith component or state, and B is the integral Volmer's constant. The equation of state for the case of monomers and large aggregates (area A<A_c) using a combination of the generalised Volmer's equation (6) and the quasi-chemical monomers/aggregates equilibrium model was obtained in ref. [18]. It was assumed that the area per one molecule in the aggregate can differ from the area per free monomer molecule. Also, the condition which describes the equilibrium between the monomers and aggregates was formulated considering the free surface existing in the monolayer. The following equation of state valid for the bimodal distribution (large clusters and monomers) was derived [16, 18, 28, 30]:

$$\Pi = \frac{nkT\alpha\beta}{A - \omega[1 + \varepsilon(\alpha\beta - 1)]} - \Pi_{coh} \qquad (7)$$

In this equation the parameter α expresses the dependence of the aggregation constant on the surface pressure:

$$\alpha = \frac{A}{A_c} exp\left[-\varepsilon\frac{\Pi - \Pi_c}{kT}\omega\right], \qquad (8)$$

and the parameter β is the fraction of the monolayer free from aggregates:

$$\beta = 1 + \omega(1 - \varepsilon)(\alpha - 1)/A, \qquad (9)$$

where $\varepsilon = 1 - \omega_{cl}/\omega$, ω_{cl} is the area per monomer in a cluster, and A_c the molecular area which corresponds to the onset of the phase transition (i.e., at $\Pi = \Pi_c$). If $A_c < 2\omega$ than holds $\beta = \alpha$ [18]. The area per molecule in the cluster can be different from the limiting area per molecule in the gaseous state. This fact is accounted for by the parameter ε. However, to reproduce the experimental behaviour of ω_{cl} vs Π for a number of amphiphiles [32], the ε value should consist of two terms, $\varepsilon = \varepsilon_0 + \eta\Pi$:

$$\omega_{cl} = \omega(1 - \varepsilon) = \omega(1 - \varepsilon_0 - \eta\Pi), \qquad (10)$$

where ε_0 is the relative jump of the area per molecule during phase transition, and η is a two-dimensional compressibility (or intrinsic compressibility) coefficient of the condensed monolayer. For example the intrinsic compressibility can be physically understood by the change of the tilt angle of molecules upon surface layer compression, accompanied by a corresponding increase in the surface layer thickness [32]. In subsequent publications these equations were further generalised to take into account the further phase transitions between condensed phases [15, 32], and they were successfully used to describe the experimental surface pressure isotherms for various amphiphilic monolayers [31, 33-36].

Recently the equation of state for the monolayer comprising molecules of different sizes (water and biopolymers) (see ref. [37]) has been modified to describe the liquid-expanded state of the insoluble molecules monolayer. In contrast to the equation of state derived earlier in ref. [18], this equation does not involve either the Gibbs' adsorption equation or the differential equation for the chemical potential of the insoluble component, but it is based only on equations for the chemical potential of the solvent in the bulk and in the surface layer. This is possible because in the case of the insoluble monolayer only the solvent is the substance which is present both in the bulk phase and in the surface layer. This equation (slightly corrected to account for the non-ideality of enthalpy) was applied in satisfactory agreement with the experimental Π-A isotherms to insoluble monolayers of amphiphilic molecules [38], and to describe the monolayer of insoluble particles for a wide range of particle diameters between 75 μm and 7.5 nm [38].

Equating the chemical potentials of the solvent (i = 0) in the solution bulk (α) and in the surface layer (s) it follows:

$$\mu_0^{0S} + RT \ln f_0^S x_0^S - \gamma \Omega_0 = \mu_0^{0\alpha} + RT \ln f_0^\alpha x_0^\alpha \tag{11}$$

where $\mu_0^{0\alpha} = \mu_0^{0\alpha}(T, P)$ and $\mu_0^{0s} = \mu_0^{0s}(T, P)$ are standard chemical potentials of the solvent (usually a pure component is assumed). From Eq. (11) one derives the equation of state for surface layer for any number of components with any geometry [37-39]:

$$\Pi = -\frac{kT}{\omega_0} \left(\ln x_0 + \ln f_0 \right), \tag{12}$$

where ω_0 is the molecular area of the solvent molecule ($\omega_0 \leq \omega$), and it depends on the choice of the position of the dividing surface. Approximations for the activity coefficients and the molar fraction were given in refs. [37-39]. So, instead of Eq. (5) we have the next expression for the Π-A isotherm of the fluid (LE) monolayer:

$$\Pi = -\frac{kT}{\omega_0} \left[\ln \left(1 - \frac{\omega}{A} \right) + \left(\frac{\omega}{A} \right) \left(1 - \frac{\omega_0}{\omega} \right) \right] - \Pi_{coh} \tag{13}$$

The main difference between Eqs. (5) and (13) consists in taking into account the contribution of entropy non-ideality in Eq. (13), caused by the difference in the sizes between the solvent and amphiphilic molecules. It is seen that Eq. (5) which is an approximation of Eq. (13), when assuming low monolayer coverage and neglecting the non-ideality of entropy. Eq. (13) yields positive values for Π, and, therefore, it is only applicable at surface coverages above a minimum value of $(\omega/A)_{min}$, determined by the condition $\Pi = 0$. For $(\omega/A) < (\omega/A)_{min}$, i.e. $A > A_{min}$, the contributions of ideal and non-ideal entropy (the first term in the right hand side of Eq. (13)) are compensated by the interaction between the components (Π_{coh}). As one can see from Eq. (13), the surface pressure depends weakly on the particle size but it is mainly determined by the monolayer coverage ω/A.

Eq. (13) can be extended for the case of phase transition from the fluid (gaseous or liquid-expanded) to the condensed states taking into consideration that the area of clusters essentially exceeds the area of monomers $(A_c > A)$. That is, in this model process of clusterization (2D main phase transition) affects the surface pressure only by means of the non-ideality of entropy of mixing. The phase transition takes place at $\theta = \theta_c$. Approximately, believing in Eq. (13) $\omega_0 = \omega$, and taking into account that for $\varepsilon = 0$ the fraction of the monolayer free from aggregates is equal $\beta = A/A_c$ (see Eq. (9)), we can obtain the contribution of entropy non-ideality in Eq. (13), caused by the difference in the sizes between the amphiphilic molecules and the large clusters in the next form: $\theta - \theta_c^*$, where $\theta_c^* = \theta_c (\theta_c / \theta)$. That is, for the 2D phase transition region (area $A < A_c$) we obtain:

$$\Pi = -\frac{kT}{\omega_0}\left[ln(1 - \theta) + (\theta - \theta_c^2 / \theta) \right] - \Pi_{coh},\qquad(14)$$

where $\theta = \omega/A$, and $\theta_c = \omega/A_c$. The Π-A isotherms calculated from the equation of state (14) for various values of θ_c and fixed values of $\omega_0 (=\omega)$ and Π_{coh} are presented in Fig. 8. The theoretical dependences correspond reasonably to numerous experimental data. It is seen that an increase of A_c (decrease of θ_c) results in a decrease of the isotherm slope in the 2D transition region.

For the low θ value approximation $(-ln(1-\theta) \approx \theta)$, we obtain from Eq. (14):

$$\Pi = \frac{kT(\theta_c^2 / \theta)}{\omega_0} - \Pi_{coh} \qquad (15)$$

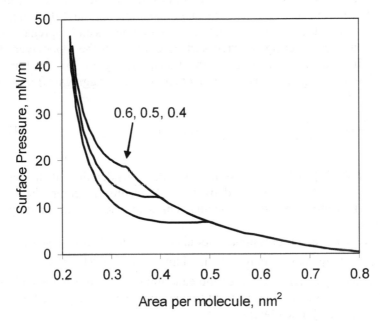

Figure 8. Π-A isotherms calculated from the equation of state Eq. (11) for various values of θ_c and a fixed value of ω_0 (=ω) and Π_{coh}.

It is interesting, that the same result follows from Eq. (7). For the simplest case of $\varepsilon=0$, when $\alpha=\beta=A/A_c$, Eq. (7) is transformed to:

$$\Pi = \frac{nkT(A/A_c)^2}{A - \omega} - \Pi_{coh} = \frac{nkT(\theta_c^2 / \theta)}{\omega(1 - \theta)} - \Pi_{coh} \qquad (16)$$

where $n = \omega/\omega_0$. In the case of the low θ values approximation $((1-\theta) \approx 1)$, we obtain from Eq. (16) exactly Eq. (15). Thus, both models give very close results.

The rigorous equation of state for monolayers in the (G, LE)/LC transition region which takes into account the entropy nonideality of mixing of monomers and clusters was derived recently [40]. Similar to Eq. (7), it was assumed that the area per one molecule in the aggregate can differ from the area per free monomer molecule, and the condition that describes the equilibrium between the monomers and aggregates was formulated considering the free surface existing in the monolayer. The following equation of state in the transition region for the bimodal distribution (large clusters and monomers or small aggregates) takes the form [40]:

$$\Pi = \frac{1}{m} \frac{kT\alpha\beta}{A - \omega[1 + \varepsilon(\alpha\beta - 1)]} - \frac{kT}{A}(1 - \alpha\beta) - \Pi_{coh} \tag{17}$$

In ref. [40], it has been shown that the consideration of the mixing entropy non-ideality provided by Eq. (17) results (in comparison with Eq. (7)) in more realistic molecular characteristics of the amphiphilic compounds.

The surface compressional modulus $K = -A(d\Pi/dA)_T$ (see Eq. (2)) or high-frequency dilational modulus $E_0 = \Gamma(d\Pi/d\Gamma)_T$ can be calculated by differentiation of the equation of state. For the equation of state in the case of monomers and large aggregates (2D phase transition region, Eq. (7)), one obtains:

$$K = \frac{A\alpha\beta - \dfrac{\alpha}{A}[A - \omega(1 - \varepsilon)][A + \omega(1 - \varepsilon)\alpha]}{\dfrac{\{A - \omega[1 + \varepsilon(\alpha\beta - 1)]\}^2}{nkT} - \omega\eta\alpha\beta(\alpha\beta - 1) + \dfrac{\alpha}{A}[A - \omega(1 - \varepsilon)]B} \tag{18}$$

where

$$B = \left\{ \omega\eta(\alpha - 1) + [A + \omega(1 - \varepsilon)(2\alpha - 1)]\left[\frac{\omega\eta(\Pi - \Pi_C)}{kT} + \frac{\omega\varepsilon}{kT}\right]\right\}.$$

Eq. (18) can be essentially simplified for $\beta = \alpha$ and the extremely compressed monolayer when $A \to \omega_{cl} = \omega(1 - \varepsilon)$:

$$K = \frac{(1-\varepsilon)}{\left(\frac{\omega(\varepsilon\alpha)^2}{nkT}\right) - \eta(\alpha^2 - 1)}. \tag{19}$$

As usually $\varepsilon^2 \ll 1$, and in the limit of $A \to \omega_{cl} = \omega(1-\varepsilon)$, we have (for $A_c \geq 2\omega$) $\alpha^2 \ll 1$, so that from Eq. (19) it approximately follows:

$$K - \frac{1}{C} \cong \frac{(1-\varepsilon)}{\eta} \tag{20}$$

For K in the limit of $A \to \omega_{cl}$, a similar expression follows also from the equation of state (17). Thus, in the case of the extremely compressed monolayer, the compressibility C of the whole monolayer is approximately equal to the two-dimensional compressibility of the condensed monolayer η. It is a quite expected result reconfirming the correctness of the equations of state Eqs. (7) and (17).

D. THERMODYNAMIC CHARACTERISATION OF MONOLAYERS WITH ONE CONDENSED PHASE

The characteristic features of Langmuir monolayers depend sensitively on the chemical structure of the amphiphiles and the monolayer state is mainly determined by surface pressure and temperature. As known, the precondition for the development of well-shaped two-dimensional morphology (domain textures) of the condensed phase is the existence of a 'plateau' region in the Π-A isotherm at different temperatures in the accessible range that represents thermodynamically the coexistence between the fluid (G, LE) and the condensed state. In a large number of such Π-A isotherms, the continuous strong increase in the surface pressure with decreasing area values after the phase transition region indicates thermodynamically the existence of a condensed phase. The experimental isotherms of some various amphiphiles have been thermodynamically analysed on the basis of the equation of state for the case of monomers and large aggregates (area $A<A_c$) (Eqs. (7)–(9)) obtained by a combination of the generalised

Volmer's equation (Eq. 6)) and the quasi-chemical monomers/aggregates equilibrium model [15, 16, 18, 28, 30].

$2C_nH_{2n+1}$-melamine monolayers spread on water behave thermodynamically also as usual amphiphilic monolayers [33]. Here, we discuss the characteristics of selected $2C_nH_{2n+1}$-melamine monolayers as example for monolayers with one condensed phase. A 'plateau' region in the Π-A isotherm indicates the transition from the fluid to the condensed state at different temperatures. Such a two-phase transition region exists only in three homologues of the $2C_nH_{2n+1}$-melamine series with the alkyl chain lengths $C_{10}H_{21}$, $C_{11}H_{23}$, $C_{12}H_{25}$ over the whole accessible temperature range. The extent of coexistence region decreases clearly as the temperature increases. This is demonstrated by Fig. 9 which shows eight Π-A isotherms of $2C_{11}H_{23}$-melamine between 10.2 °C and 31.9 °C The symbols represent selected values of the experimental curves (although continuously measured) whereas the solid lines of Fig. 9 are the theoretical curves obtained on the basis of the equations of state (Eqs. (5) - (7)). The corresponding parameters of the isotherms are listed in Table 2.

Table 2: Parameter values of Eqs. (7)-(9) calculated for $2C_{11}H_{23}$-melamine monolayers at different temperatures

Temperature, °C	12.5	15.8	20.0	23.1	25.4	29.2	31.9
ω_1, nm^2	0.36	0.36	0.36	0.36	0.36	0.36	0.35
A_c, nm^2	0.58	0.55	0.52	0.504	0.498	0.484	0.476
Π_{coh}, mN/m	16.2	15.7	16.6	16.6	16.6	15.6	18.4
ε_0	0.19	0.18	0.22	0.24	0.22	0.21	0.18
n	1.00	1.02	0.97	0.93	0.91	0.91	0.72
$-\Delta H^0$, kJ/mol	8.9	8.8	8.5	8.4	8.2	8.1	8.0

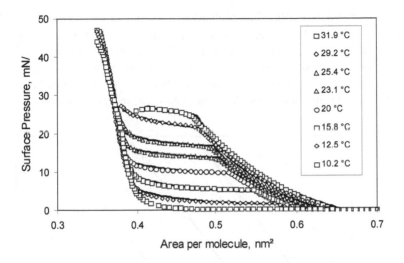

Figure 9. Π-A isotherms of $2C_{11}H_{23}$-melamine monolayers spread on water measured in the temperature range between 10.2 °C und 31.9 °C.

The calculated values show that, in agreement with the experimental results (see Fig. 9) the area per molecule A_c which corresponds to the onset of the phase transition decreases as the temperature increases. These results apply to the monolayers of all three melamine-type homologues [33]. The aggregation number for small aggregates n existing in the fluid state at $A>A_c$, decreases also as the temperature increases. In the case that $n\approx1$ monomers exist whereas the case $n<1$ indicates relative freedom of the two alkyl chains of the melamine-type amphiphiles.

The presentation of the temperature dependence of the phase transition point Π_c (the kink point in the Π-A isotherm at onset of the phase transition) indicates clearly (Fig. 10) that the fluid/condensed phase coexistence region of the homologous melamine amphiphiles with the alkyl chain lengths between $C_{10}H_{21}$ and $C_{12}H_{25}$ is in the measurable temperature range. Here, the symbols represent the experimental Π_c-T dependence for the three homologous melamine amphiphiles

$2C_{10}H_{21}$-melamine, $2C_{11}H_{23}$-melamine and $2C_{12}H_{25}$-melamine and the solid lines show the excellent linear fit of the data. The presented $\Pi_c(T)$ shift with increasing alkyl chain length allows conclusion that already for the monolayers of $2C_{13}H_{27}$-melamine and those with still larger alkyl chain length, the phase coexistence region must be expected in a temperature range inaccessible to the used measuring conditions.

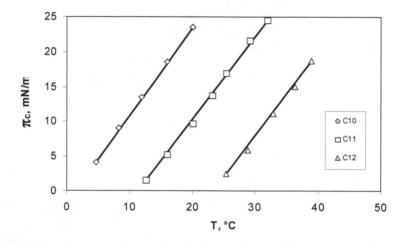

Figure 10. Temperature dependence of the phase transition pressure Π_c of the monolayers of $2C_{10}H_{21}$-melamine-, $2C_{11}H_{23}$-melamine and $2C_{12}H_{25}$-melamine spread on water.

For the calculation of the standard enthalpy of two-dimensional phase transition ΔH^0 (see Table 2) the equation rigorously derived in ref. [33] can be used:

$$\Delta H^0 = -\frac{RT^2}{x}\frac{d\Pi_c/dT}{d\Pi_c/dx} = -\frac{RT^2}{x}\frac{dx}{dT} \qquad (21)$$

where $x = \omega_{cl} / A_c$ and R is the gas constant. The values of $x = \omega_{cl} / A_c$ are calculated using the experimental values of A_c, and $\omega_{cl} = 0.36$ nm^2 for $2C_{11}H_{23}$- melamine and the different temperatures. x depends linearly from T for all three melamine homologues which have similar dx/dT values. Therefore a constant dx/dT value for each melamine homologue can be used. for the calculation of ΔH^0.

E. ANISOTROPY OF MONOLAYERS AND COMPRESSIBILITY

The application of the GIXD technique allows interesting information about the compressibility effects. There are several work which suggest that intermolecular forces and conformational defects can affect the elastic properties of monolayers. So far, two-dimensional compression and shear elastic moduli have been determined by applying the shear to a large polycrystalline sample [41,42], bending a monocrystal [43], or analyzing the shape of Bragg singularities [44]. For example, because of the inhomogeneity of large polycrystalline samples, their elastic moduli were found to be much smaller than those obtained with monocrystalls. Similar conclusions were also drawn for the compression modulus obtained from the Π - A isotherms and GIXD measurements, respectively, on the basis of the molecular areas for corresponding surface pressures [45].

Another important problem not considered by above papers is the anisotropy of monolayers. Therefore, it is generally possible to affect the organization of Langmuir monolayers by the two dimensions.

An example for the anisotropy of the compressibility is demonstrated by the interesting work of Kaganer et al. [46]. They used the compressibility of long-chain alcohol monolayers at several temperatures for the calculation of the exponent η of the correlation function $G(x) \propto x^{-\eta}$ to obtain information about the phase transition of the two condensed phases of octadecanol monolayers. The higher-temperature phase, LS, shows a single first-order diffraction peak due to hexagonal local arrangement of the molecules. The lower temperature phase, S, possesses two first-order peaks of the centred rectangular unit cell because of ordering of the short molecule axes. According to the definition in Eqs. (1) and (2), in the case of the hexagonal phase, the compressibility C and the compressibility modulus K can be directly expressed by the change in the position of the diffraction peak on the applied isotropic surface pressure

$$K = \frac{1}{C} = -\frac{1}{2} Q_0 \left(\frac{d\Pi}{dQ_0} \right) \tag{22}$$

In phases with a rectangular unit cell, the linear compressibility along each diffraction vector Q_{hk} can be determined as [47]

$$C_{hk} = -Q_{hk}^{-1} \left(\frac{dQ_{hk}}{d\Pi} \right) \tag{23}$$

The surface pressure dependencies of the peak positions of the S- and LS-phases of the in octadecanol and hexadecanol monolayers are demonstrated in Fig. 11. In the case of the S-phase (in Fig. 11, left presented for 5.2 °C), the dependencies of the positions of the two diffraction peaks on the surface pressure reveal an interesting anisotropy of the compressibility.

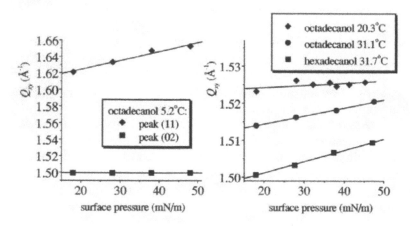

Figure 11. Surface pressure dependence of the peak positions in monolayers of octadecanol and hexadecanol [47].

The two peaks of the S-phase react differently to the applied surface pressure. The compressibility in direction of the degenerate (11) peak is $C_{11} = 0.67$ m/N, whereas the position of the nondegenerate (02) peak remains unchanged within the accuracy of the measurements. The compressibility of area per molecule $C = 0.67$ m/N and the corresponding compressibility modulus $K = 1.18$ N/m were calculated using the relation $C = C_{11}/\cos^2(\gamma^*/2)$, where $\gamma^* \cong 55°$ is the angle between the Q_{11} and $Q_{1\bar{1}}$ reciprocal lattice vectors (cf. [48]). A similar anisotropy with corresponding compressibility values was found for the S-phase of behenic acid [47].

The compressibility of the LS phase was found to be much lower (Fig. 11, right). The compressibility values obtained by linear fits were found to be $C = 0.41$, 0.29, and 0.08 m/N for hexadecanol at 31.7 °C and octadecanol at 31.1 and 20.3 °C, respectively. These compressibility values correspond to values of the compression modulus $K = 1/C$ of $K = 2.4$, 3.45, and 12.5 N/m, respectively.

At the constant temperature of 11.5 °C, the LS-S phase transition of octadecanol monolayers occurs on compression at a transition pressure of ~40 mN/m [46]. The corresponding diffraction peaks, presented in Fig. 12 for $\Pi = 18$, 28, and 38 mN/m, show that the peak position Q_0 of the LS-phase decreases from 18 to 38 mN/m. This would correspond to a thermodynamically impossible negative compressibility. The authors of ref. [47] discuss this behaviour as caused by a disordered state of the monolayer close to the LS-S transition so that the compressibility cannot be determined from the positions of the maxima.. This is experimentally supported by significant peak broadening in this region. The determination of the correlation functions of positional order in octanol monolayers revealed a sharp increase of the exponent η at the LS-S transition that points to backbone ordering as a source of additional positional disorder in the system. The unusual broad peaks were attributed to elastic distortions caused by fluctuations (with the structure of the new phase) in the surrounding matrix which do not destroy the quasi-long-range positional order but cause local strains.

For phases with a rectangular unit cell, Fradin et al. introduced the concept of linear compressibilities (see Eq. (23)) [47]. They determined the unit cell parameters of the different phases of behenic acid and myristinic acid monolayers and substantiated the existence of differences in the linear compressibilities. Without going into detail, it should be mentioned that in the case of the two-dimensional fatty acids monolayers, four linear compressibilities have been determined in dependence on their phases and direction of crystallisation. Two-dimensional rectangular unit-cells were found in all phases, studied in ref.[47], of the phase diagrams of behenic acid [45, 49] and myristic acid [50, 51]. The molecules are tilted towards one of their six nearest neighbours (NN) in the L_2 and L_2'' phases, they are tilted towards one of their six next-nearest neighbours (NNN) in the L_2' phase, and they are not tilted in the more ordered S and CS phases. The most interesting linear linear compressibilities are those along the unit cell axes a and b with $C = C_a + C_b$. Transverse linear compressibilities are mainly interesting in the tilt direction where they differ from the *in-plane* compressibilities. It is interesting to note that the authors of ref. [47] determined negative compressibilities in the transverse plane for the L_2' and L_2'' phases and they attributed this result to subtle reorganisations upon untilting the molecules.

The GIXD results obtained in respect to the anisotropy of Langmuir monolayers suggest further interesting questions to be solved. This concerns the role of the interactions between the head groups for the compressibility of tilted phases, the dependence of the compressibility on the monolayer thickness in the nontilted phases, or the reasons and mechanism of the decrease of compressibility with increasing temperature.

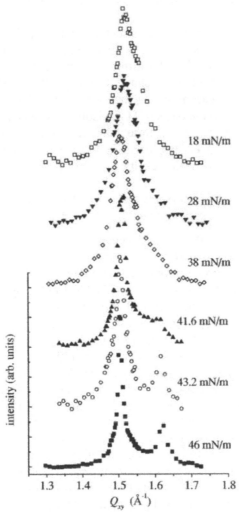

Figure 12. Diffraction peaks along the 11.5 °C isotherm. A phase transition is seen between 38 and 41.6 mN/m [47].

F. EXISTENCE OF MORE THAN ONE CONDENSED PHASES

GIXD studies of the lattice structure of fatty acid monolayers have shown that, despite the same rectangular unit cell, several phases can be distinguished characterized by differences in the molecular ordering and the tilt direction of the alkyl chains (see e.g. [45, 49-51]). In that case, there are several reasons that a thermodynamic characterization of the phase transitions on the basis of Π - A isotherms has not been performed. However, the existence of more than one condensed phase of other amphiphilic monolayers has been manifested by combination of Π - A isotherms, BAM and GIXD results.

1. Structure features and phase properties for the case of two condensed phases indicated by the Π - A isotherms

The existence of two condensed phases can be impressively demonstrated by the studies of the phase and structural features of N-tetradecyl-β-hydroxy-propionic acid amide (TDHPA) monolayers [52]. Fig. 13 shows the Π - A isotherms of TDHPA monolayers between 5 and 20 °C with the inserted chemical formula of the TDHPA amphiphile. The Π - A isotherms reveal two pronounced phase transitions at temperatures of ≤15 °C which are indicated by two plateau regions. At large areas per molecule, the transition from the low-density fluid phase to a condensed phase is revealed by the break in the slope of the isotherm followed by the large plateau region. At small areas per molecule and a surface pressure of ~16 mN/m the isotherms show, after a second break in the slope of the isotherm, a second small plateau with a width of ~0.02 nm². As a break in the slope of the isotherm followed by a plateau region indicates a first order phase transition, it can be suggested that the transition between the two condensed phases is also of first order.

Representative BAM images taken for different surface pressures at 10 °C are shown in Fig. 14. Development and growth of the condensed phase domains within the main phase transition between the fluid and the condensed phases is clearly seen in the BAM images of Fig. 14, above. The bright crystalline domains of condensed phase are surrounded by the continuous fluid phase of lower density and, thus, lower reflectivity (dark regions). The dendritic domains grow in 3 – 6 main growth directions. The proportion of 6 main growth directions increases with the temperature. At 25 °C, almost only 6 main growth direction were observed, while at 5 °C the low-arm domains predominate.

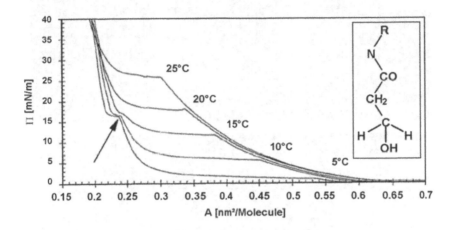

Figure 13. Π-A isotherms of TDHPA monolayers measured in the temperature range between 5 °C and 25 °C. The arrow indicates the transition point between two condensed phases. The formula of the TDHPA molecule (R=$C_{14}H_{29}$) is inserted.

The effect of the second phase transition on the domain textures, observed in the Π - A isotherms of TDHPA monolayers at ≤15 °C, is of particular interest. A conspicuous change in the textures due to the second phase transition could not be observed. The domain morphology before (Fig. 14, above right; 12 mN/m) and the end of the second phase transition (Fig. 14, below left, 19 mN/m) is generally similar. However, it is obvious that, because of the second phase transition, the domains start to grow in lateral direction giving rise to an step-like increase in the intensity of the reflected signal. The fact that the domain texture does not change at the second phase transition allows conclusion that the azimuthal tilt angle of the TDHPA molecules is not essentially changed. However the step-like increase of the reflected intensity indicates a decrease in the polar tilt angle and, consequently a corresponding increase of the packing density of the TDHPA molecules.

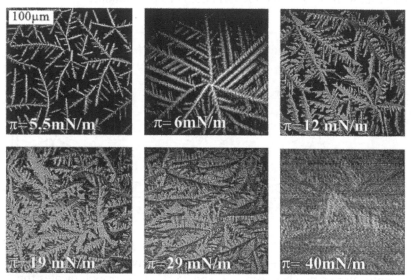

Figure 14. BAM imaging of the TDHPA monolayer at T = 10 °C at different surface pressures. Above: at surface pressure below the second phase transition. Below: at surface pressure above the second phase transition.

At temperatures of ≤15 °C, the GIXD contour plots of the corrected diffraction intensities as a function of the in-plane (Q_{xy}) and out-of-plane (Q_z) components of the scattering vectors show three scattering peaks for all surface pressures investigated (presented in Fig. 15 for 10 °C). Three scattering peaks indicate generally an oblique lattice structure for the two condensed phases at area values A separated by the second phase transition [10]. However, the peak positions change drastically in the region of the second phase transition at $\Pi \sim 16$ mN/m. At smaller surface pressures, one scattering peak at a Q_z value of almost zero and two peaks at lower Q_{xy} but higher Q_z are seen (e.g. Fig. 15, above). At surface pressures above the second phase transition, all peaks are shifted to significantly higher Q_{xy} positions and the two peaks with the higher Q_z value are now located at higher Q_{xy} positions (e.g. Fig. 15, below). Accordingly, at low temperatures of ≤ 15 °C the main characteristics of the lattice data change abruptly at the surface pressure, indicating a second phase transition (see Table 3). It is clearly seen in Table 3 that the lattice constants a, b, c, the polar tilt angle t, and the unit cell area A_{xy} decrease only weekly both below the phase transition pressure of ~16 mN/m and above it, whereas they change jumplike in the phase transition region.

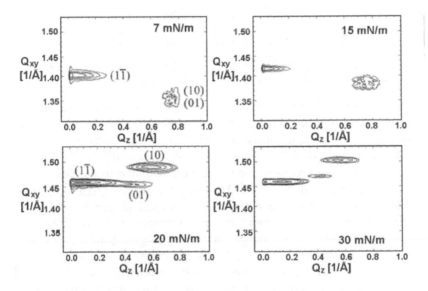

Figure 15. Contour plots of the corrected diffraction intensities as a function of the in-plane (Q_{xy}) and out-of-plane (Q_z) components of the scattering vectors of TDHPA monolayers at different surface pressures.

Table 3. Values of the lattice parameters of TDHPA monolayers at 10 °C: a,b,c – lattice spacings, t – polar tilt angle of the molecules, A_{xy} – molecular area parallel to the interface, A_0 – cross-sectional area per alkyl chain perpendicular to the chain axis

Π [mN/m]	a [Å]	b [Å]	c [Å]	t [°]	A_{xy} [Å²]	A_0 [Å²]
7	5,166	5.237	5.385	36.2	23.96	19.33
12	5,127	5.192	5.293	34.8	23.44	19.25
15	5.095	5.158	5.232	33.2	23.06	19.31
20	4.897	5.019	4.912	24.1	21.15	19.32
30	4.887	4.998	4.852	22.1	20.89	19.35
40	4.884	4.979	4.808	20.9	20.69	19.33

Thus the second phase transition between the two condensed phases is based on an abrupt transition between two different oblique lattice structures. In the oblique lattice formed after reduced from t > 33 ° to T < 24 ° and the TDHPA molecules occupy essentially smaller areas $A_{xy} \leq 21.15$ Å². The cross-sectional area perpendicular to the chain axis A_0 remains constant at $A_0 \approx 19.3$ Å² during the second phase transition so that the transition between the two condensed phases is obviously attributed to a change in head group conformation of the TDHPA molecule, probably to a change in the hydrogen bonding systems between the headgroups of the TDHPA amphiphiles. The head group consists of two hydrophilic groups, namely an amid group (-NH-CO-) bond to the tetradecyl chain, and an hydroxyl group (-OH), separated by a methylene group ($-CH_2-$) from the amide group. The OH- and –NH-CO- groups can form endless hydrogen-bonded chains independent of one another. It is possible that a hydrogen bonding network between the endless hydrogen-bonded chains of amide and the hydroxyl groups can be formed [53]. Therefore it is probable that, for temperatures $T \le 15\,°C$, local minima of the binding energy exist at the two phase transition points which can result in two different conformation states of the head groups. At temperatures $T > 16\,°C$, the second phase transition disappears and the condensed phase changes its density continuously with increasing surface pressure. The continuous change of the lattice parameters indicates that no change in the head group conformation occurs.

2. Theoretical description of Π - A isotherms indicating two condensed phases

The theoretical section has given a survey about the possibilities of thermodynamic description of the Π - A isotherms of Langmuir monolayers. The equation of state for monolayers in the fluid state is represented by the Volmer-type equation Eq. (5). The equations of state valid for bimodal distribution (aggregated molecules/monomers) Eqs. (7) and (17) applies to the fluid/condensed phase coexistence.

In ref. [32] the attempt was made to generalize the model proposed in ref. [18] onto phase transitions for which the area per molecule in the condensed state exhibit either a linear decrease with increasing surface pressure or a sharp jump. To describe two phase transitions, the theory was developed in ref. [32] which assumes two states of the condensed phase with different parameters which characterise the molecular compressibility in the condensed state. The transition to these two states takes place from the LE state of the monolayer. This theory also implies the condition of the thermodynamic stability of the system, $d\Pi/dA \le 0$ [54], which is necessary to make the transition between the two theoretical Π-A isotherms possible.

To describe the second phase transition by another method we can assume that the transition which takes place from one condensed phase into another condensed phase is accompanied by the change of molecular compressibility parameters in the condensed state [55]. The main difference between the model developed in ref. [55] and that described in ref. [32] is the dependence of the parameter ε on surface pressure which accounts for the existence of two phase transitions. For the first phase transition, i.e., at $\Pi \geq \Pi_{c1}$ or $A \leq A_{c1}$ the expression (10) for ε is:

$$\varepsilon_1 = \varepsilon_{01} + \eta_1 (\Pi - \Pi_{c1}) \tag{23}$$

In the surface pressure range above the critical pressure of the second phase transition (i.e., for $\Pi \geq \Pi_{c2}$ or $A \leq A_{c2}$) the ε value, as given by Eq. (10), is expressed as:

$$\varepsilon_2 = \varepsilon_{02} + \varepsilon_{01} + \eta_2 (\Pi - \Pi_{c2}) + \eta_1 (\Pi_{c2} - \Pi_{c1}) \tag{24}$$

In these expressions, ε_{0i} is the relative jump of the area per molecule during i-th phase transition, and η_i is the relative two-dimensional compressibility of the condensed monolayer in i-th state, $\eta_i = -d \ln \omega_{cl}/d\Pi$. Note that the transition from the condensed phase 1 to the condensed phase 2 at $\Pi = \Pi_{c2}$ $(A = A_{c2})$ is thermodynamically possible, because the mechanical stability condition $d\Pi/dA \leq 0$ is valid in this case.

In the preceding section the phase and structural properties of TDHPA monolayers, the Π-A isotherms of which indicate a second phase transition between two condensed phases, have been characterized in detail. Similar features were also observed with THPA monlayers discussed in the section '*Effect of small structural changes on the main characteristics of monolayers*'. In both cases, the two-dimensional compressibility undergoes a jump at the second phase transition, whereas within one of the two condensed phases this compressibility is proportional to the surface pressure.

For the case of two phase transitions, the Π-A isotherm of the TDHPA monolayer measured at 10 °C was selected to demonstrate the procedure of the thermodynamic analysis. In the selected Π-A isotherm, shown in Fig. 16, above, the positions of the two phase transitions are marked by arrows. The dark points correspond to the values of the surface pressure and area per molecule, for which the GIXD experiments were performed.

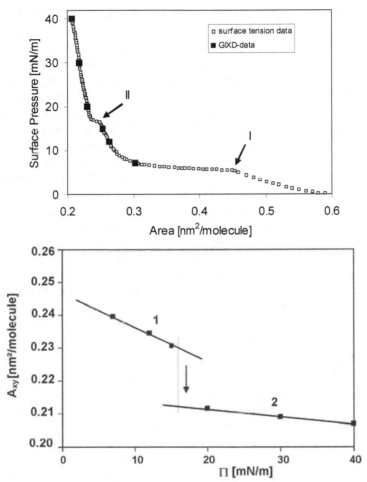

Figure 16. Experimental Π-A isotherm of the TDHPA monolayer at T = 10°C (*cf* Fig. 14), above. Dark points correspond to the values of pressure and area per molecule for which the GIXD experiments were performed. The arrows indicate the two phase transitions. Dependence of the TDHPA molecular area parallel to the surface, A_{xy}, on the surface pressure obtained from the GIXD experiments, bottom.

The dependence of the A_{xy} values (projected area parallel to the surface) of the TDHPA molecule on the surface pressure is shown in Fig. 16, bottom. In the surface pressure range between the two phase transitions shown in Fig. 16, below (5.5 – 16.5 mN/m), the area per molecule in the condensed state (A_{xy}) decreases linearly with the increase in the surface pressure. Above the second phase transition (16.5 mN/m and higher) the molecular area (A_{xy}) decreases also in the higher surface pressure range, but with a slope lower than in the region of the first condensed phase. At the surface pressure value of 16.5 mN/m, which corresponds to the second phase transition point in Fig. 16, above, the area per TDHPA molecule, obtained from the lattice structure, changes drastically shown by the arrow in Fig. 16, bottom. The abrupt change in the lattice structure (intermolecular distances, polar tilt angle), at the second phase transition is discussed in the preceding section. As shown in ref. [53] for the TDHPA and for a number of other amphiphiles (see, for example refs. [10, 56-60]), the polar tilt angle of the molecules varies monotonously within the range, where a definite condensed phase exists. This behavior correlates to the two-dimensional compressibility of the corresponding monolayer phase.

It follows from the GIXD data of the THPAA monolayer at 10 °C (see Fig. 16, bottom) that $\omega = 0.247$ nm^2 (intersection point of the straight line 1 with the ordinate), $\eta_1 = 0.004$ m/mN, $\eta_2 = 0.001$ m/mN (slopes of the straight lines 1 and 2, respectively), $\varepsilon_{02} = 0.13$ (jump between phase transition. The experimental values of the THPAA monolayer compressibility parameters in the LC state (from GIXD data) are almost coincident with the corresponding parameters of Eqs. (5), (8), (9), (17), (23) and (24) used for the calculation of the theoretical curves (see Table 4): $\omega = 0.24$ nm^2, $\eta_1 = 0.004$ m/mN, $\eta_2 = 0.0015$ m/mN and $\varepsilon_{02} = 0.11$. This coincidence indicates the validity of the proposed theoretical model.

It is interesting to look at the compressibilities of the Π - A isotherm shown in Fig 17.

Table 4. Characteristics of the TDHPA monolayer at 10 °C

Phase transition	1	2
ω, nm^2	0.24	0.24
A_c, nm^2	0.446	0.446
η_i, m/mN	0.004	0.0015*
ε_{0i}	0.07	0.11*
n	1.4	1.4
Π_{coh}, mN/m	7.7	7.7

*Distinguished parameters of the second phase transition.

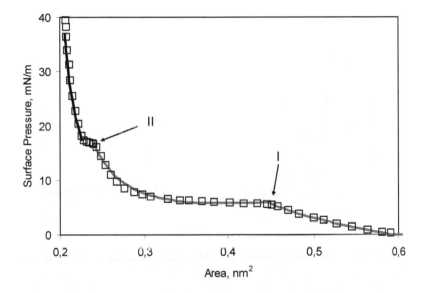

Figure 17. Comparison between the experimental (points) and theoretical Π-A isotherms of the TDHPA monolayer. I – first phase transition point, II – second phase transition point. The theoretical curves (red and blue) were calculated from the model of Eqs. (5) and (17) with the values of parameters listed in Table 4

Fig. 18 shows the experimental and theoretical dependencies of the surface compressional modulus K on the area per molecule, A of TDHPA calculated on the basis of Eq. (2). These dependencies were obtained using the experimental and theoretical curves in Fig. 17. As well as in Fig. 17, both phase transitions are shown by arrows. As can be seen in Fig. 18, the value of surface compressional modulus K reaches a maximum in the two initial phase transition points and then, at further decrease of the area per molecule within the phase transition region it decreases nearly to zero. That is, maximal compressibility of monolayer C=1/K exists in this region after beginning the phase transition.

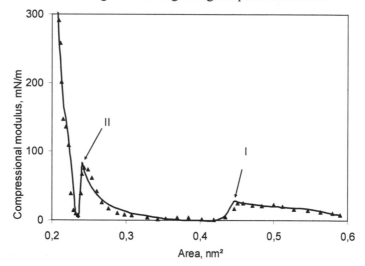

Figure 18. Experimental (symbols) and theoretical (solid line) dependencies of the surface compressional modulus K on the area per TDHPA molecule. I – first phase transition point, II – second phase transition point.

The limiting experimental value of K in the condensed state for the second phase transition point are in satisfactorily agreement with estimations obtained on the basis of Eq. (20). For the conformation state

II it follows a limiting theoretical value of K $\cong$ 750 mN/m. It is necessary to underline that these K values are evaluated from independent GIXD experiments. For the state II the experimental value of K, as follows from Fig. 18, grows quickly at further decrease of area per TDHPA molecule. Justifying for the condensed state II a tendency to the limiting area per TDHPA molecule of 0.205 nm^2 (at Π=40 mN/m), the experimental K value verges on theoretical estimations on the basis of Eq. (20).

G. CONCLUSIONS

Since the early periods of research on Langmuir monolayers over many decades, surface tension measurements have been the main source of information about insoluble monolayers. The existence of different condensed monolayer phases could be concluded from the features of the Π - A isotherms supported essentially on differences in the compressibility

New insights into their phase and structural features of Langmuir monolayers have been gained by fundamental progress in the experimental determination of the main characteristics in microscopic and molecular scale in the last two decades. Already smallest changes in the molecular structure of the amphiphile can result in changes in the molecular arrangement in the monolayer and thus, in changes of the main characteristics of the monolayer such as, the surface pressure –area per molecule (π - A) isotherms, the shape and texture of the condensed phase domains and the two-dimensional lattice structure. Such essential changes of the main characteristics have been demonstrated by smallest alteration of the head group structure of phospholipids, by systematic variation in four monoglycerol amphiphiles by an amide, ether, ester and amine group, even by exchange of the position of the two substituents at the acid amide group of the headgroup, or by chiral discrimination effects. In this chapter, it is reviewed that only slight changes of the OH-substitution in mid-position of the alkyl chain give rise to large differences in the monolayer features, as demonstrated by coupling of π - A isotherms, BAM imaging and GIXD measurements.

The classical equations of state allowed only the characterisation of the fluid (gaseous, liquid-expanded) state. However the experimental progress required improved thermodynamic possibilities for describing the Π - A isotherms. To filling this gap, thermodynamically based equations of state, which consider also the aggregation of the monolayer material to the condensed phase, have been developed. Now those parts of the π - A isotherm which represent the transition region to the condensed phase (decisive for the features of the monolayer) can be thermodynamically characterized. A large number of monolayers with one condensed phase has been analysed on this basis.

The present review focuses particularly to amphiphilic monolayers, indicating the existence of two condensed phases in the Π - A isotherms. For this case, the experimental results of the differences in the structure features and phase properties are discussed. The equation of state for Langmuir monolayers is generalized for the case that one, two or more phase transitions in the monolayer take place. In the case of two phase transitions, the experimental results show that the two-dimensional compressibility of the condensed phases undergoes a jump at the phase transition, whereas the compressibility is linear proportional to the surface pressure within the condensed phases. The procedure of the theoretical analysis of Π - A isotherms, which indicate the existence of two condensed phases, applies the general principle that the behaviour of any thermodynamic system is determined by the stability condition. The experimental and theoretical determination of the surface compressional modulus K in dependence on the area per molecule, A shows that the K value reaches a maximum in the two phase transition points and then, at further decrease of the area per molecule within the phase transition region it decreases nearly to zero.

New interesting problems are based on the anisotropy of monolayers resulting in the introduction of the concept of linear compressibilities for the case of rectangular unit cells. An interesting anisotropy of the compressibility is revealed by GIXD studies of the S-phase of octadecanol monolayers concluded from the dependencies of the positions of the diffraction peaks on the surface pressure at different temperatures. Similar studies performed close to the LS-S phase transition would result in a thermodynamically impossible negative compressibility. The disordered state the monolayer close to the LS-S phase transition (attributed to elastic distortions caused by fluctuations with the structure of the new phase in the surrounding matrix which do not destroy the quasi-long-range positional order) may be the reason that the compressibility cannot be determined from the positions of the maxima because of a disordered state of the monolayer due to elastic distortions close to the LS-S phase transition which are caused by fluctuations with the structure of the new phase in the surrounding matrix without destroying the quasi-long-range positional order.

H. REFERENCES

1. Gibbs JW. Collected Works, Vol. I, p.301.
2. Gaines GL jr. Insoluble Monolayers at Liquid-Gas Interfaces, Wiley Interscience, New York 1966.
3. Harkins WD, Young TF, Boyd E. J Chem Phys 1940; 8: 954.
4. Dervichian LE. J Chem Phys 1939; 7: 931.
5. Tschoegl NW. Australian J Phys 1958; 11: 154.
6. Tschoegl NW. J Colloid Sci 1958; 13: 500.
7. Inokuchi K. Bull Chem Soc Japan 1953; 26: 500; 1955; 28: 453.
8. Thirumoorthy K, Nandi N, Vollhardt D J Phys Chem B 2005; 109: 10820.
9. Weidemann G, Vollhardt D Biophys J 1996; 70: 2758
10. Vollhardt D, Wagner R J Phys Chem B 2006; 110: 14881.
11. Nandi N, Vollhardt D Chem Rev 2003; 103: 4035.
12. Vollhardt D, Siegel S, Cadenhead DA Langmuir 2004; 20: 7670.
13. Vollhardt D, Siegel S, Cadenhead DA J Phys Chem B 2004; 108: 17448.
14. Vollhardt D, Weidemann G, Lang S J Phys Chem B 2004; 108: 3781.
15. Vollhardt D, Fainerman VB J Phys Chem B 2004; 108: 297.
16. Vollhardt D, Fainerman VB J Phys Chem B 2002; 106: 12000.
17. Fainerman VB, Vollhardt D J Phys Chem B 2005; 109: 11706.
18. Fainerman VB, Vollhardt D. J Phys Chem B 1999; 103: 145.

19. De Boer JH. The Dynamical Character of Adsorption; Oxford University Press, London 1945.
20. Smith T. Adv. Colloid Interface Sci 1972; 3: 161.
21. Ruckenstein E, Bhakta A. Langmuir 1994; 10: 2694.
22. Israelachvili JN. Langmuir 1994; 10: 3774.
23. Ruckenstein E, Li B. Langmuir 1995; 11: 3510; 1996; 12: 2309.
24. Ruckenstein E, Li B. , J Phys Chem 1996; 100: 3108.
25. Ruckenstein E, Li B. J Phys Chem 1998; 102: 981.
26. Fainerman VB, Vollhardt D, Melzer V. J Phys Chem 1996; 100: 15478.
27. Fainerman VB, Vollhardt D, Melzer V. J Chem Phys 1997; 107: 242.
28. Fainerman VB, Vollhardt D, Emrich G. J Phys Chem B 2000; 104: 8536.
29. Volmer M. Z Phys Chem (Leipzig) 1925; 115: 253.
30. Vollhardt D, Fainerman VB, Siegel S. J Phys Chem B 2000; 104: 4115.
31. Wüstneck R, Fainerman VB, Wüstneck N, Pison U. J Phys Chem B 2004; 108: 1766.
32. Fainerman VB, Vollhardt D. J Phys Chem B 2003; 107: 3098.
33. Vollhardt D, Fainerman VB, Liu F. J Phys Chem B 2005; 109: 11706.
34. Fainerman VB, Vollhardt D, Aksenenko EV, Liu F. J Phys Chem B 2005; 109: 14137.
35. Fainerman VB, Vollhardt D. in "Organized Monolayers and Assemblies: Structure, Processes and Function ", Studies in Interface Science, Möbius D and Miller R (Eds.), Vol. 15, Elsevier, 2002, p.105–160.
36. Vysotsky YuB, Bryantsev VS, Fainerman VB, Vollhardt D. J Phys Chem B 2002; 106: 11285.
37. Fainerman VB, Lucassen-Reynders EH, Miller R. Adv Colloid Interface Sci 2003; 106: 237.
38. Fainerman VB, Kovalchuk VI, Lucassen-Reynders EH, Grigoriev DO, Ferri JK, Leser ME, Michel M, Miller R, Möhwald H. Langmuir 2006; 22: 1701.
39. Fainerman VB, Vollhardt D. J Phys Chem B 2006; 110: 10436.
40. Fainerman VB, Vollhardt D. J Phys Chem B 2008; 112: 1477.

41. Abraham BM, Miyano K, Ketterson JB, Xu J J Chem Phys 1983; 78: 4776.
42. Abraham BM, Miyano K, Ketterson JB, Xu J Phys Rev Lett 1983, 51: 1975.
43. Bercegol H, Meunier J Nature 1992; 356: 226.
44. Berge B, Konovalov O, Lajzerowicz J, Renault A, Rieu JP, Vallade M, Als-Nielsen J, Grübel G, Legrand JF Phys Rev Lett 1994 ; 73 : 1652.
45. Bommarito GM, Foster WJ, Pershan PS, Schlossman ML J Chem Phys 1996; 105: 5265.
46. Fradin C, Daillant J, Braslau A, Luzet D, Alba M, Goldmann M. Eur Phys J B 1998; 1: 57.
47. Kaganer VM, Brezesinski G, Möhwald H, Howes PB, Kjaer K. Phys Rev E 1999; 59: 2141.
48. Kjaer K, Als-Nielsen J, Helm CA, Tippman-Krayer P, Möhwald H. J Phys Chem 1989; 93: 3200.
49. Kenn RM, Böhm C, Bibo AM, Peterson IR, Möhwald H, Als-Nielsen J, Kjaer K J Phys Chem 1991; 95: 2092.
50. Bibo AM, Peterson IR Adv Mater 1990; 2: 309.
51. Akamatsu S, Rondelez F J Phys II France 1991; 1: 1309.
52. Melzer V, Weidemann G, Vollhardt D, Brezesinski G, Wagner R, Struth B, Möhwald H. Supramolec Sci 1997; 4: 391.
53. Rudert R, André Ch, Wagner R, Vollhardt D. Z Kristallogr 1997; 212: 752.
54. Rusanov AI. Fazovye Ravnovesija i Poverchnostnye Javlenija, Khimija, Leningrad, 1967.
55. Fainerman VB, Vollhardt D. J Phys Chem B; submitted.
56. Gehlert U, Vollhardt D, Brezesinski G, Möhwald H. Langmuir 1996; 12: 4892.
57. Gehlert U, Weidemann G, Vollhardt D, Brezesinski G, Wagner R, Möhwald H. Langmuir 1998; 14: 2112.
58. Weidemann G, Brezesinski G, Vollhardt D, DeWolf C, Möhwald, H. Langmuir 1999; 15: 2901.
59. Krasteva N, Vollhardt D, Brezesinski G, Möhwald H. Langmuir 2001; 17: 1209.
60. Gehlert U, Vollhardt D. Langmuir 2002; 18: 688.

DILATATIONAL RHEOLOGY OF THIN LIQUID FILMS

V.I. Kovalchuk[1], J. Krägel[2], P. Pandolfini[3], G. Loglio[3], L. Liggieri[4],
F. Ravera[4], A.V. Makievski[5] and R. Miller[2]

[1] Institute of Bio-Colloid Chemistry, 03142 Kiev, Ukraine
[2] Max Planck Institut für Kolloid- und Grenzflächenforschung, D-14424
Potsdam, Germany
[3] Department of Organic Chemistry, University of Florence, Sesto, Italy
[4] CNR - Istituto per l'Energetica e le Interfasi, 16149 Genova, Italy
[5] SINTERFACE Technologies, 12489 Berlin, Germany

Contents

A. INTRODUCTION

The dynamics of thin liquid films was, and still is, the subject of extensive studies [1-6]. Various factors which affect the film dynamics, in particular the adsorption/desorption of surfactants, mobility and deformation of film surfaces, interaction forces which exist between the surfaces etc. were analysed. These problems are of paramount technological importance because the stability of foams and emulsions is determined mainly by the dynamics of films which arise between approaching bubbles or drops. The decrease of film thickness finally results either in its rupture or in the formation of black (common or Newtonian) films which are also destroyed after a time. The film lifetime is one of the most important characteristics which determine the stability of foams and emulsions.

It was noted by Mysels *et al.* [2] that the film thinning process can take place in a different way depending on whether the film is mobile or rigid (immobile). In mobile films different film elements can move with respect to each other. In these films the thinning process is much more rapid because the gravitational convection causes the thinner parts of the film to ascend relatively to its thicker parts, and also because of the so called marginal regeneration, the spontaneous convection which exists in the transition zone between the film and the meniscus (Plateau borders).

In rigid ('immobile') films the thinning takes place much slower. The main mechanisms which govern the decrease of the film thickness in this case are the viscous flow of liquid between the film surfaces, the evaporation of liquid and the film stretching caused by external forces. Another process which leads to the film thinning is related to the formation and extension of black films. These mechanisms can also play a certain role in the thinning of mobile films. In what follows, we consider mainly the thinning of rigid films.

The viscous flow between the film surfaces (drainage) arises due to gravitational forces and also due to the capillary suction caused by the pressure difference between the film and the meniscus. This leads to a continuous decrease of the amount of liquid in the film; however, the film surface area A remains almost constant (Fig. 1a). The rate of this process depends on the film thickness h: with decreasing thickness the rate becomes essentially lower. If the evaporation of liquid and the film stretching are negligibly small, the film lifetime is determined by the drainage rate:

$$\tau = \int_{h_{cr}}^{h_{in}} \frac{dh}{V} \tag{1}$$

where h_{in} is the initial film thickness, h_{cr} is the critical thickness of film rupture, V is the velocity at which the film surfaces approach each other due to the outflow of liquid from the film. For a plane-parallel film stabilised by a surface active substance this velocity was calculated in [7] as:

$$V = V_{Re}(1 + b + h_S/ h) \tag{2}$$

where $V_{Re} = 2h3\Delta P/3\mu R2$ is the velocity at which two rigid disks of radius R approach each other due to the force F calculated according to the Reynolds formula; $\Delta P = F/\pi R^2$, $b = 3\mu D/h_a E_0$, $h_S = 6\mu D_S/E_0$, μ is the liquid viscosity, D and D_S are the coefficients of bulk and surface diffusion, respectively, $E_0 = -d\gamma / d \ln \Gamma$ is the limiting surface elasticity, $h_a = d\Gamma/dc$, γ is the surface tension, Γ is the adsorption, c is the bulk concentration of the surfactant solution in the film. It follows from Eq. (2) that the velocity of mutual approach is roughly proportional to the cube of the film thickness. Therefore, if the thinning is governed solely by viscous drainage, the film lifetime is much higher as compared with that when other thinning mechanisms are also relevant. The influence of various factors on the film thinning governed by viscous drainage, in particular the shape and deformation of the film surface, surface viscosity, non-diffusional adsorption kinetics, the presence of micelles etc. was studied and reported in a number of publications (see e.g. [4, 8-12]). In the present consideration we do not discuss these factors in detail; it is just important to note that in all cases the drainage rate becomes essentially lower with the decrease of film thickness.

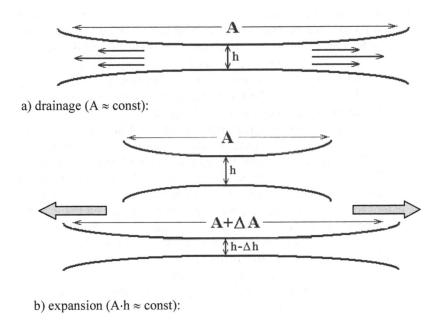

a) drainage (A ≈ const):

b) expansion (A·h ≈ const):

Fig.1. Film thinning mechanisms (immobile films)

The evaporation of liquid leads to the decrease of the solvent amount in each film element; this should lead to the increase of the surfactant concentration, which, in turn, would decrease the surface tension. However, this would deteriorate the mechanical equilibrium between the film and the meniscus, as the tension at the meniscus surface remains approximately constant. Therefore, the film is stretched, which results in the maintaining of constant surface tension, surface and bulk concentration of surfactant (due to additional adsorption at the surfaces) in each film element. Thus, a decrease of film thickness during evaporation should take place not only due to the decrease of the solvent amount, but also due to film stretching [2]. If two surfactants with different activities are present in the solution, the situation is somewhat different: the surface tension is maintained constant due to the adsorption of the substance which is more surface active. This results in a variation in the solution composition: the concentration of the less active substance becomes higher [2]. However, in general terms the film thinning mechanism remains the same. The evaporation of liquid enhances film thinning, thus leading to a decrease in film stability. However, this process could be much slower if the gas phase adjacent to the film becomes saturated by the solvent vapour.

The increasing external force imposed to the film leads to a film stretching, increase of its surface and increase in surface tension (Fig. 1b). This process requires the consumption of some mechanical work. Film thinning caused by stretching can result in its rapid rupture (much more rapid than that caused by viscous drainage or evaporation). The extent to which the film becomes stretched depends on the external force character, and also on the ability of the film to counteract this stretching, i.e., the film elasticity [1-3,5,6]. In most studies the elasticity is considered to be one of the main factors which affect its stability. However, the relation between film elasticity and its stability still remains somewhat obscure, while it is generally understood that this relation should be rather complicated and involved. A good example here is the fact that, while monolayers of insoluble or slightly soluble surfactants exhibit high elasticity modulus values, these monolayers are poor foam stabilisers [5]. This ambiguity in what regards the stabilising role of elasticity as compared with other factors can be attributed to the fact that, on one hand, experiments on the film elasticity are extremely complicated and sensitive to the ambient conditions and, on the other hand, the properties of actual films could be essentially different from those of the films prepared under laboratory conditions.

One of the factors which complicate the experiments on film elasticity is the film non-uniformity, leading to the fact that the local elasticity in different regions of the film could be quite different. Another problem is the fact that, in contrast to black films (which are stabilised by surface forces), thicker films are essentially non-equilibrium entities: their properties vary with time due to drainage and/or evaporation. Moreover, the film properties essentially depend on the method of film preparation. Additional complications are introduced by the fact that the film is always kept by the meniscus which exists in mechanical equilibrium with the film, therefore stretching of the film is accompanied by stretching of the meniscus surfaces. Also the experimental values could be affected even by small admixtures of highly active surfactants. Therefore, the results for the same systems studied by different authors could be essentially different [13-18], and only qualitative agreement between theory and experiment can be expected.

On the other hand, it is to be noted that the problem with the agreement between theory and experiment is also brought about by the application of simplified adsorption and surface tension isotherms for the surfactants, while these simplified theories fail to provide a sufficiently correct description of the behaviour of actual systems. Usually, the Langmuir model (or one of its modifications) is used in the calculations. This isotherm disregards interactions between adsorbed surfactant molecules, and also some other important effects remain unaccounted for. However, the elasticity of films, similarly to the surface elasticity of surfactant solutions [18-22], is much more sensitive to the particular properties of adsorption layers than the surface tension is. Therefore, the application of simplified adsorption isotherms could lead to large discrepancies between the calculated properties and experimental data. It is important also to note that the theoretical calculations of elastic properties should be based on exact values of surfactant concentrations just in the film, while the values which refer to the quantities in the solution used for film preparation would result in large bias. The above considerations are very important in view of the fact that these factors determine the possibility of film rupture during its stretching, which in turn plays a crucial role in the problem of foam and emulsion stability.

B. DILATATIONAL ELASTICITY OF FILMS

The importance of elastic properties of films was first noted by Gibbs more than a century ago [1]. According to Gibbs, the dilatational elasticity of a film is related to the decrease of the bulk and surface concentration of surfactant due to film stretching. Unlike the process of viscous outflow of a liquid from the film, stretching could happen much more rapidly (because the stretching time is determined by the characteristics of external action). Therefore, during the stretching, the area of a film element is increased without any significant change of its volume, and also the total amount of surfactant in this element remains constant. However, as the surface area becomes larger, some surfactant molecules are transferred from the film bulk to its surface to compensate partially the decreased surface concentration. This leads to a decrease of the surfactant bulk concentration, resulting in an increase of surface tension during film stretching.

According to Gibbs the elasticity modulus of a film (or any of its elements) is determined by:

$$E_f = A \frac{d\gamma_f}{dA} \tag{3}$$

where A is the area of the film (or its small element), dA is the area increase, and $d\gamma_f$ is the increase in film tension. If the film is thick enough and symmetric, and the influence of surface forces is negligibly small, this increase is twice as large as the corresponding surface tension increase: $d\gamma_f = 2d\gamma$ (Fig. 1b). During film stretching the total amount of surfactant in each film element remains constant:

$$N_S = cV_L + 2\Gamma A \tag{4}$$

The film volume $V_L = Ah$ also remains approximately constant. Assuming that during the stretching process a local equilibrium between the adsorption layer and the volume bulk remains non-violated, using Eqs. (3) and (4), and taking into account the dependencies of adsorption on concentration $\Gamma(c)$ and surface tension on adsorption $\gamma(\Gamma)$, we obtain the expression (see [3,5]):

$$E_f = \frac{2E_0}{1 + \frac{h}{2}\frac{dc}{d\Gamma}} \tag{5}$$

where E_0 is the limiting surface elasticity. Eq. (5) is the simplified form of an expression for the film elasticity, because it disregards the solution non-ideality and exact position of the dividing surfaces which correspond to zero adsorption of the solvent for the Gibbs model. More general expressions for the film elasticity were presented in [6,14,23]. However, any corrections are essentially insignificant, and therefore, these are usually disregarded when Eq. (5) is used in the analysis of experimental data. Also, Eq. (5) does not account for the final time necessary to establish the equilibrium between the adsorption layer and the solution bulk. Therefore this equation corresponds to the case when deformations are rather slow. Effects related to a slow equilibration will be considered below.

Comparing the expression for the film elasticity, Eq. (5), with the expression for the dilatational surface elasticity $E_d = E_0 \left[1 + \frac{dc}{d\Gamma} \sqrt{\frac{D}{i\omega}} \right]^{-1}$,

where D is the diffusion coefficient and ω is the angular frequency, one can easily see that these expressions have a similar form, and both depend on two parameters: E0 and $d\Gamma/dc$ [3,24]. Therefore, to calculate theoretical values of the elasticity of a film or a surface layer exact values of these two parameters are necessary. It is important to note that the drainage velocity also depends on the parameters E0 and $d\Gamma/dc$ (according to Eq. (2) or even to a stronger dependence, see [11]). The parameters E0 and $d\Gamma/dc$ are determined by the surfactant type and its concentration and could, in principle, be calculated by a differentiation of the adsorption isotherm $\Gamma(c)$ and the equation of state of the surface $\gamma(\Gamma)$. Hence, both film elasticity and dilatational surface elasticity should depend on the parameters involved in the adsorption isotherm and the equation of state of the surface.

However, it was shown in [18-22] that the dependence of rheological characteristics of surface layers on the parameters of the adsorption isotherm and the equation of state of the surface is much stronger than the dependence of surface tension $\gamma(c)$ on these parameters. In other words, the rheological characteristics of surfaces and films are much more sensitive to the state and interaction of molecules in the adsorption layer than the equilibrium surface tension is. Therefore, in practice measurements of equilibrium surface tension are insufficient to calculate exactly the values of E_0 and $d\Gamma/dc$ and, therefore, to determine the surface and film elasticities.

Fig. 2 shows some calculations of the limiting elasticity E_0 and surface pressure $\Pi = \gamma_0 - \gamma$ [19, 25]. It is seen that the shape of the E_0 vs. Π dependence is affected by the non-ideality (Frumkin) parameter a to a far greater extent than the shape of the Π vs. log c dependence.

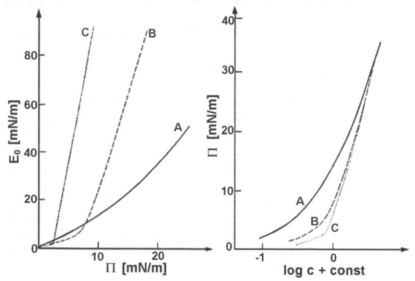

Fig.2. Effect of deviations from Szyszkowski-Langmuir behaviour on the limiting elasticity E_0 and surface pressure $\Pi = \gamma_0 - \gamma$; the curves are calculated according to Frumkin's equation of state with different interaction parameters: a = 0 (A), 1.55 (B), 2.175 (C) (according to [19, 25]).

A detailed analysis of measured E_0 vs. Π curves for a number of surfactants in [19, 25] led to the conclusion that the equation of state is far more complex than given by the widely used Frumkin isotherm. It was demonstrated that a single quadratic non-ideality term in the equation of state is unable to describe the complex interactions within the adsorption layer, leading to the observed behaviour of limiting elasticity, and that terms of higher order are necessary even at very low surface pressure. Such important information about the surface behaviour could not be obtained from equilibrium data $\Pi(c)$.

As well, the equilibrium curves $\Pi(c)$ do not give information about the intrinsic (two-dimensional) surface layer compressibility. It is known, that at high surface pressure the experimentally measured limiting elasticity E0 is usually much smaller than the values predicted by the Langmuir or Frumkin model [26-33]. The same contradiction with the experimental data is observed also for the derivative $d\Gamma/dc$ [29, 30, 32, 33]. An example of such data is presented in Fig.3.

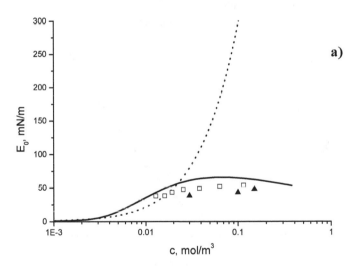

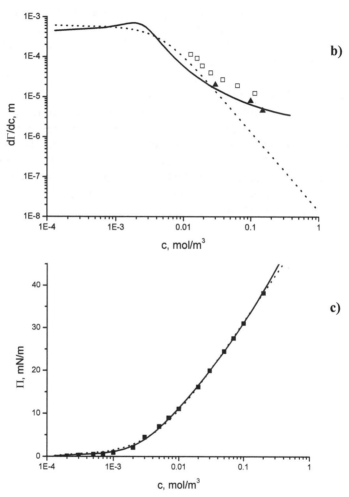

Fig. 3. Limiting elasticity (a), dΓ/dc (b) and surface pressure (c) vs. concentration for dodecyl dimethyl phosphine oxide solutions. Experimental data are from [30, 32] (▲), [33] (□), and [39] (■); theoretical dependencies are calculated for Frumkin isotherm (dotted lines) and for the model accounting for intrinsic monolayer compressibility (solid lines) [32].

It was shown recently [22, 32-38] that this contradiction can be explained by the effect of an intrinsic 2D surface layer compressibility ε, which influences only slightly the form of the surface pressure vs. concentration dependence (Fig.3c) but can lead to dramatic changes in the limiting elasticity E_0 and the derivative $d\Gamma/dc$ (Fig.3a, 3b).

The 2D compressibility reflects the variation of the molar area of adsorbed molecules in a condensed monolayer, $\Omega = 1/\Gamma_\infty$, with the surface pressure Π:

$$\Omega = \Omega_0 \left(1 - \varepsilon \Pi\right) \tag{6}$$

Here Ω_0 is the molar area determined by extrapolation to zero surface pressure Π, and ε is the two-dimensional relative surface layer compressibility coefficient. Such variation of Ω can for example be associated with the variation of the tilt angle of adsorbed molecules (accompanied by a variation of the adsorption layer thickness) with increasing surface pressure and is confirmed by experimental data showing that Γ continues to increase at high surface coverage [38, 40]. The variation of the molecular area due to variable molecular orientation at high surface pressures is assumed also in [41-43].

The examples presented above demonstrate that the behaviour of the limiting surface elasticity E_0 and the derivative $d\Gamma/dc$ is much more complicated than that expected on the basis of the measured values of equilibrium surface tension or pressure. Accordingly, the behaviour of the dilatational elasticity of films E_f (or surfaces E_d) should also be much more complicated. However, this important fact is usually disregarded in studies of film elasticity and its influence on the stability of foams and emulsions.

To demonstrate the effect of 2D compressibility on the film characteristics we compare the results of calculations in the framework of the Frumkin model (Ω = const) with those of a model accounting for the dependency of the molar surface area on the surface pressure (Eq.(6)).

The Frumkin model is described by the set of equations

$$bc = \frac{\Theta}{(1-\Theta)} exp(-2a\Theta)$$
(7)

$$\frac{\Pi\Omega}{RT} = -ln(1-\Theta) - a\Theta^2$$
(8)

where $\Theta = \Gamma\Omega$ is the surface layer coverage, b is the adsorption constant, a is the (Frumkin) interaction constant, R is the gas constant, and T is the temperature. Integration of the Gibbs adsorption equation together with Eqs. (6) and (7) results in the following semi-empirical equation of state instead of Eq. (8) [32, 35]

$$\frac{\Pi\Omega_0(1-\varepsilon\Pi/2)}{RT} = -ln(1-\Theta) - a\Theta^2$$
(9)

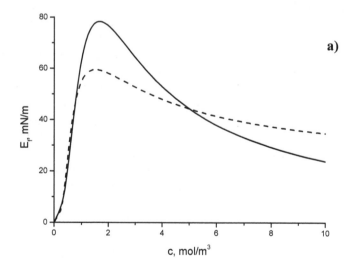

a)

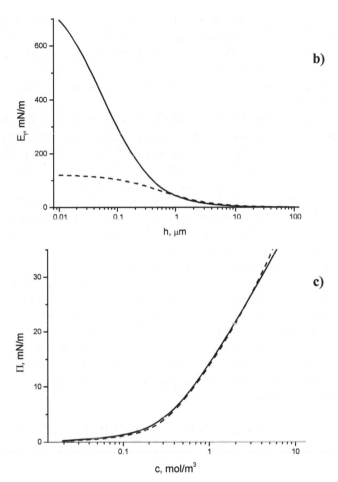

Fig.4. Film elasticity vs. concentration for h = 1 μm (a) and vs. film thickness for c = 5 mol/m^3 (b) and surface pressure vs. concentration (c) calculated for the Frumkin model (full lines, $\Omega = 2 \cdot 10^5$ m^2/mol, b = 1 m^3/mol, and a = 1), and for the 2D compressibility model (dashed lines, $\Omega_0 = 2.5 \cdot 10^5$ m^2/mol, b = 1 m^3/mol, a = 1.2, and $\varepsilon = 0.01$ m/mN) (according to [21]).

A more rigorous approach to account for 2D surface layer compressibility is proposed in [35], which is based on the two-dimensional solution theory. However, the results, which follow from more rigorous models, are qualitatively similar to those represented by Eqs. (6), (7) and (9), and therefore we consider here only this approximation.

The film elasticities shown in Fig. 4a and 4b are calculated either for a fixed film thickness (h = 1 μm) and varying concentration or for a fixed concentration (c = 1 mol/m^3) and varying film thickness h [21]. It is seen that the film elasticities E_f calculated for the two considered adsorption layer models are very different. In spite of that the respective surface pressure isotherms calculated for the chosen sets of parameters coincide quite well (Fig. 4c). These numerical examples demonstrate that the film elasticity E_f, similarly to the surface dilational modulus E_d, is much more sensitive to the adsorption layer model and, in particular, to the intrinsic surface layer compressibility, than the dependence $\Pi(c)$. This is confirmed by direct measurements of film elasticity [17, 40]. Thus, in order to describe correctly the elasticity of films stabilised by surfactants it is necessary to know the adsorption layer properties much more accurately than it is typically available from studies of equilibrium surface tension. In this respect studies of the surface rheology of surfactant solutions are very important as they can provide the necessary accurate information about the parameters E_0 and $d\Gamma/dc$, which can be used for studies on film rheology.

C. SURFACTANT CONSERVATION DURING EXPANSION OF A FILM

The film elasticities, shown in Fig. 4, were calculated under the assumption that the film thickness and concentration are independent of each other. However, in each film element the local bulk concentration varies upon expansion and hence it cannot be independent of the film thickness [3, 18, 44, 45]. The variation of concentration with thickness should affect the film elasticity and, therefore, its stability. The number of surfactant molecules within a film element N_s and the volume of the film element V_L practically do not change during the expansion (if it is sufficiently rapid to neglect evaporation and viscous drainage). Therefore, according to Eq.(4), the nominal (effective) concentration, defined as

$$c_{ef} = \frac{N_S}{V_L} = c + \frac{2\Gamma}{h}, \tag{10}$$

should be constant during the expansion process. If interfacial equilibrium restores instantaneously (as expected in sufficiently thin films), one obtains from Eq.(10) the relationship

$$h = \frac{2\Gamma(c)}{c_{ef} - c} \tag{11}$$

which establishes a dependence of bulk concentration on film thickness c(h). For thick films the major part of surfactant is dissolved in the bulk and $c_{ef} \approx c$. But during the thinning of a film the concentration in its bulk can decrease by more than one order of magnitude (Fig.5a).

Such depletion of solution has a considerable influence on the film elasticity [3, 44]. Substituting the dependence c(h) into Eq. (5) one obtains a dependence of film elasticity on film thickness for the given surface equation of state $\Pi(\Gamma)$ and adsorption isotherm $\Gamma(c)$, which accounts for the depletion effect [18, 21, 44]. In Fig. 5b such dependencies are presented for two particular models, the Frumkin and 2D compressibility model, respectively. These dependencies have a sharp maximum at a film thickness of approximately 1 μm. Left to the maximum the film elasticity decreases fast due to depletion of surfactant from the solution. At small film thicknesses the surfactant is mainly adsorbed at the interfaces and one obtains from Eq.(11): $\Gamma \approx c_{ef} h / 2$.

Accordingly, the asymptotic for the film elasticity gives

$$E_f \approx 2E_0 \approx R_g T c_{ef} h \tag{12}$$

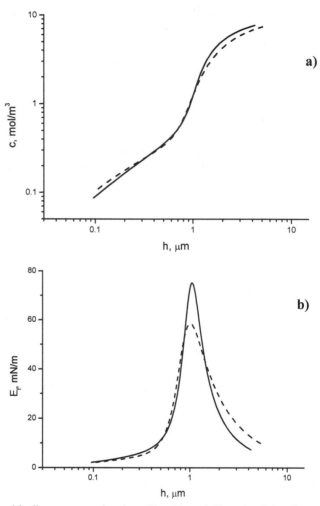

Fig.5. Local bulk concentration in a film (a) and film elasticity (b) calculated according to Eqs.(5) and (11) for the Frumkin model (full lines), and for the 2D compressibility model (dashed lines) with the same sets of parameters as given in Fig.4; c_{ef} = 10 mol/m^3 (according to [21]).

For the chosen particular conditions of Fig. 5b, a film of 1 μm thickness corresponds to a maximum resistance with respect to expansion and thicker or thinner films provide much smaller resistance. However, thicker films will increase their elasticity by expansion and, therefore, they are much more stable than films with smaller thicknesses the elasticity of which will decrease continuously upon expansion because of the depletion effect.

The nominal concentration c_{ef} is an important parameter influencing the film properties. It depends on the prehistory of a film. Its value does not change during a rapid film expansion but changes gradually with time due to film drainage and evaporation. The nominal concentration is strongly influenced also by the conditions of film formation. In freshly formed ("young") films obtained under particular constant conditions it can be assumed uniform over the film what simplifies their analysis [3, 45].

It is often supposed that nominal concentration c_{ef} in freshly formed films is equal to the concentration c_0 in the solution from which the film is formed [18, 45]. However, it can be concluded from the consideration of the film formation process that it should be higher: during film formation a new film is pulled out of the meniscus (Plateau border), the surfaces on the meniscus are expanded which initiates diffusion to supply additional surfactant from the meniscus bulk to the place of film formation and restores the initial concentration at the interfaces. After the film formation a mechanical equilibrium is established very fast which requires uniform surface tension over the meniscus and film surfaces (only relatively small surface tension gradient results due to drainage). Taking adsorption equilibrium at the surfaces into account one has to conclude that the bulk concentration in the film should be close to that in the meniscus, and the nominal concentration just after film formation can be approximated by [46]

$$c_{ef} = c_0 + \frac{2\Gamma_0}{h_0} \tag{13}$$

where c_0 and Γ_0 refer to the solution remaining in the meniscus and h_0 is the thickness of the resulting film. According to Frankel's law the thickness h_0 depends on the velocity v by which the film is drawn from the solution [2, 47, 48]

$$h_0 = 1.88 \frac{(\mu v)^{2/3}}{(\rho g)^{1/2} \gamma^{1/6}} \tag{14}$$

where μ is the dynamic viscosity, ρ is the solution density and g is the acceleration due to gravity. Though Eq. (14) was derived with the assumption of negligible surface extensibility [2], it describes experimental data sufficiently well. It is clear, that the nominal concentration c_{ef} should depend on the rate of film formation. It will be closer to the initial solution concentration for thicker films formed at higher rates.

Eq.(13) ignores the effect of drainage. However, due to drainage a small difference of surface tensions establishes between the meniscus and the film surfaces. Therefore the bulk concentration in the film should be slightly smaller than in the meniscus, and c_{ef} should be also smaller. For a micellar solution even small surface tension differences between the meniscus and film can result in a much smaller concentration c_{ef} than according to Eq.(13) [46].

When a limited volume of solution is taken for film formation, the solution remaining in the meniscus can be depleted on surfactant due to its redistribution between meniscus and film, and the concentration c_0 can be smaller than the initial concentration. In case the whole solution volume is expanded and forms a film, the nominal concentration will be equal to the initial solution concentration. However, in practice a significant part of solution usually remains in the meniscus and the nominal concentration becomes larger than the initial solution concentration. This is confirmed by estimations made from experimental data of Ref. [18] as discussed in [21].

When the film is formed from mixed surfactant solution its composition will be different from the solution composition. As the lower surface active components are present in the solution in a relatively higher proportion than at the interfaces, they will be faster transported to the place of film formation and the film will be enriched by these components.

With time, due to viscous drainage, the film thickness decreases and becomes non-uniform, therefore the nominal concentration gradually increases and also becomes non-uniform over the film. Only in freshly formed films the nominal concentration can be uniform [3, 45].

D. EFFECT OF THE SURFACE EQUATION OF STATE AND ADSORPTION ISOTHERM PARAMETERS

It is seen from the discussion presented above that a better understanding of the role of rheological properties in film stability requires much more detailed analysis of the effect of different parameters included in the surface equation of state and adsorption isotherm. In Fig.6 the effect of the interaction (Frumkin) constant a on the film elasticity vs. film thickness dependence is shown as calculated from Eqs.(5) and (11) for the 2D compressibility model. It is seen that at high constants a, the maximum film elasticity increases, but a minimum appears left of the maximum. At $a = 2$ the film elasticity is zero in the minimum. This means that the films prepared under considered conditions will become very unstable when their thickness will decrease up to a point close to the elasticity minimum. After that the films are expanded practically without a resistance. A very small external disturbance can result in a fast film thinning and rupture. Thus, at high constants a the films can become unstable far before their thickness decreases up to a critical one (determined by the capillary wave or other mechanisms), and the lifetime of such films should be essentially lower.

The decrease of the nominal concentration c_{ef} results in a decrease of the elasticity maximum and in a shift of the maximum to higher thicknesses (dashed line in Fig.6). The increase of the adsorption constant b leads to an increase of the elasticity maximum in the film elasticity vs. thickness dependencies. However, usually the larger is the constant b, the smaller is the surfactant concentration in the solution.

During the formation of films from such solutions the nominal surfactant concentration in the film c_{ef} will be generally smaller.

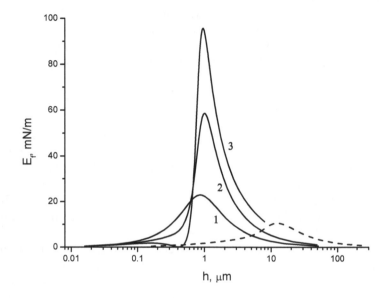

Fig.6. Film elasticity vs. film thickness dependencies for $c_{ef} = 10$ mol/m^3 and interaction constants: $a = 0$ (curve 1), 1.2 (curve 2) and 2 (curve 3) (full lines, 2D compressibility isotherm, $\Omega_0 = 2.5 \cdot 10^5$ m^2/mol, b = 1 m^3/mol, and $\varepsilon = 0.01$ m/mN); dashed line – the same dependence for $c_{ef} = 1$ mol/m^3 and interaction constant $a = 1.2$.

An increase of the adsorption constant b and a simultaneous decrease of c_{ef} results in a shift of the whole elasticity curve to higher thicknesses (Fig.7). Qualitatively this behaviour is confirmed by the experimental data of Bianco and Marmur for sodium dodecyl sulfate (SDS) and cetyltrimethylammonium bromide (CTAB) solutions [18]. Hence, the expansion of films, produced from surfactants with higher activity, should result in a fast lost of elasticity already at such high thicknesses, which exceed the critical thickness by more than one order of magnitude. The effect of rupture at high film thicknesses is often observed in "wet" foams ([5], p.558 and the references therein). This is a consequence of the depletion effect in the solution, which is much more important for low soluble surfactants [3]. Films with a thickness larger than at the elasticity maximum are more stable. However, it should be also taken into account that thicker films have much larger weight, and, therefore, they are much more expanded due to gravity.

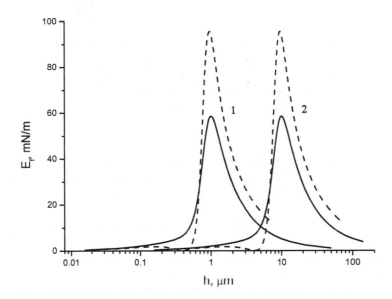

Fig.7. Film elasticity vs. film thickness dependencies for $c_{ef} = 10$ mol/m^3 and $b = 1$ m^3/mol (1) and $c_{ef} = 1$ mol/m^3 and $b = 10$ m^3/mol (2) (2D compressibility isotherm, $\Omega_0 = 2.5 \cdot 10^5$ m^2/mol, $a = 1.2$ (full lines) and 2 (dashed lines), and $\varepsilon = 0.01$ m/mN).

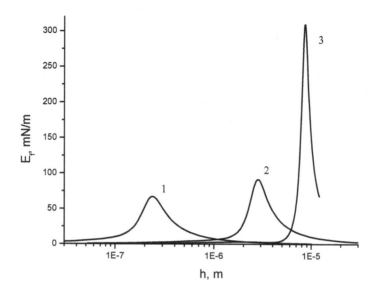

Fig.8. Film elasticity vs. thickness dependencies for normal alcohols: n-hexanol (1), n-octanol (2), n-decanol (3) calculated according to Frumkin isotherm with parameters given in Table 1.

In Fig.8 the film elasticity vs. thickness dependencies for the homologue series of normal alcohols are presented. It is seen that the elasticity maximum of films formed from the less soluble n-decanol solutions is higher. However, only two-fold expansion of such films (corresponding to a decrease in thickness from 10 to 5 μm) should lead to a dramatic decrease of their elasticity. After such expansion these films become very unstable. In contrast, the same expansion of films from n-hexanol and n-octanol does not lead to a significant change of their stability. A similar significant decrease of the elasticity requires an approximately ten-fold expansion for n-octanol and hundred-fold expansion for n-hexanol. This result correlates with the smaller stability of films and foams formed from low soluble surfactants.

Table 1. The properties of n-alcohols used for calculations of dependencies in Fig. 8.

C_nOH	Ω_0, 10^5 m²/mol [a]	a [a]	b, l/mol [a]	C_{ef}, mol/l [b]
C_6OH	1.72	1.04	$1.23 \cdot 10^2$	$4.67 \cdot 10^{-2}$
C_8OH	1.65	1.37	$1.17 \cdot 10^3$	$4.24 \cdot 10^{-3}$
$C_{10}OH$	1.67	1.8	$1.21 \cdot 10^4$	$1.38 \cdot 10^{-3}$

[a] The values of Ω_0, a and b are from [39].
[b] The values of C_{ef} are obtained from Eq.(13) for films with $\Pi_0 = 42$ mN/m and $h_0 = 10$ μm.

E. NON-UNIFORM FILMS

Liquid films are usually non-uniform in thickness and, consequently, in their other properties (elasticity, etc.). For freshly formed vertical films the thickness profile can be directly obtained from the balance of gravity force and surface tension gradient established due to partial extension of a film [3, 45]. This thickness profile changes with time due to liquid drainage and evaporation [2]. In the gravitational field the mechanical quasi-equilibrium in a film is achieved due to variations of the film tension with the height. The appropriate tension distribution along the film is established as the result of local extension of film elements which is accompanied by a local redistribution of surfactant molecules between the film interfaces and the bulk solution.

The mechanical equilibrium establishes very quickly after the formation of a film, much quicker than drainage occurs. Under mechanical equilibrium conditions the weight of a film element in a vertical film is balanced by twice the difference of surface tensions above and below this element [3, 45, 49]:

$$2\Delta\gamma = \rho g h \Delta z \tag{15}$$

where $h = h(z)$ is the local film thickness, Δz is the height of the film element and z is the vertical coordinate. Thus, the surface tension (as well as the film tension) increases with the height. The maximum possible difference of surface tensions is limited by the difference between surface tension of surfactant solution in the meniscus and the surface tension of pure solvent. Hence, the height of a film is also limited [3]. The thicker is the film the smaller is the maximum possible height.

In thin films the diffusion and adsorption equilibrium establishes very fast in any cross-sections of the film. However the surfactant redistribution along the film due to diffusion and liquid drainage is very slow, and, therefore, the difference of surfactant concentrations along the film remains during a long time. Taking into account the adsorption equilibrium and using the Gibbs equation $d\gamma / dc = -RT\,\Gamma / c$ one can find the concentration difference

$$\Delta c = \left(\frac{d\gamma}{dc}\right)^{-1}\Delta\gamma = -\frac{\rho g h c}{2RT\Gamma}\Delta z \tag{16}$$

Substitution of h according to Eq. (11) one obtains

$$\Delta c = -\frac{\rho g}{RT}\frac{c}{c_{ef} - c}\Delta z \tag{17}$$

As mentioned above in freshly formed films obtained under definite constant conditions (i.e. pulled out at constant velocity and without solution depletion in the meniscus) the effective concentration c_{ef} can be assumed constant. For this case the integration of Eq.(17) yields the concentration profile in the film

$$z - z_1 = -\frac{RT}{\rho g}\left(c_{ef}\,ln\frac{c}{c_1} - c + c_1\right) \tag{18}$$

where $c_1 = c(z_1)$ is the concentration in a reference point (this can be a point just above the meniscus). Being combined with Eq.(11) and with the surface tension isotherm $\gamma(c)$ Eq.(18) allows to obtain the film thickness profile $h(z)$ and the surface tension profile $\gamma(z)$ in a freshly formed film. Numerical examples of such profiles are given in [3, 45]. It is interesting to note that by a rapid lifting of freshly formed films (c_{ef} = const) the concentration c, adsorption Γ and surface tension γ should be the same for the film elements with a given thickness h before and after extension, independent of varying z (as they are related to each other via Eq.(11) and surface tension and adsorption isotherm equations). And when the concentration in the reference point $c(z_1)$ is also not changed the whole film thickness profile preserves after extension up to the level of the previous film height (by such extension a new part of film is pulled out from the meniscus).

Ageing of a film results in a variation of the effective concentration c_{ef} over the film, and Eq.(18) is not valid anymore. Generally the concentration profile can be obtained by a numeric integration of Eq.(17). For this aim, however, the dependence $c_{ef}(z)$ should be found first.

Let us now consider the mechanical quasi-equilibrium in a spheroidal film (soap bubble) formed at the capillary tip under gravity. The weight of a film element enclosed between the polar angles θ and $\theta+d\theta$ (cf. Fig.9) is

$$dw = 2\pi \rho g h R_f \, sin\theta \cdot dl \tag{19}$$

where h = $h(\theta)$ is the local film thickness, R_f = $R_f(\theta)$ is the radial coordinate of the film element, and dl is the width of the film element which can be expressed as

$$dl = \left[R_f^2 + \left(dR_f / d\theta \right)^2 \right]^{1/2} \cdot d\theta \tag{20}$$

Here the centre of the spherical coordinate system is chosen in the symmetry axis in the plane of maximum horizontal bubble cross-section (Fig. 9). Due to the very small film thicknesses the difference between the internal and external radial coordinates R_f of the film can be ignored.

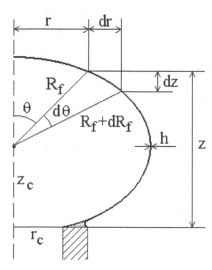

Fig. 9. Thin spheroidal film formed at the tip of a capillary of inner radius r_c.

The tangential component of the weight is equilibrated by the film tension difference

$$d\gamma_f = 2d\gamma = -\frac{dw \cdot \cos\varphi}{2\pi R_f \sin\theta} \tag{21}$$

where φ is the angle between the tangent to the film surface and the vertical direction which is given by

$$\cos\varphi = -\frac{dz}{dl} = \frac{R_f \sin\theta - (dR_f / d\theta)\cos\theta}{\left[R_f^2 + (dR_f / d\theta)^2\right]^{1/2}} \tag{22}$$

Here $z = R_f \cos\theta + z_C$ is the height of the film element above the plane of the capillary tip, z_C is the height of the centre of the coordinate system. From Eqs. (19) - (22) one obtains

$$\frac{d\gamma}{d\theta} = -\frac{1}{2}\rho g h \left[R_f \sin\theta - \left(\frac{dR_f}{d\theta}\right) \cos\theta \right] \tag{23}$$

In Eq. (23) (as well as in Eq.(15)) the contribution of the viscous flow produced by capillary suction is neglected because the corresponding pressure gradient is negligibly small in the nearly flat and parallel parts of the film having a very small thickness. This contribution is, however, significant in the transition zones between the film and the meniscus.

As the interfacial tension $\gamma = \gamma(c)$ and adsorption $\Gamma = \Gamma(c)$ are functions of the local surfactant concentration $c = c(\theta)$, one obtains according to Eqs. (11) and (23)

$$\frac{dc}{d\theta} = -\frac{\rho g \Gamma}{c_{ef} - c} \left(\frac{d\gamma}{dc}\right)^{-1} \left[R_f \sin\theta - \left(\frac{dR_f}{d\theta}\right) \cos\theta \right] \tag{24}$$

Integration of Eq. (24) using the respective surface equation of state $\gamma(\Gamma)$ and the adsorption isotherm $\Gamma(c)$ yields the concentration profile $c(0)$ in a film. Using the Gibbs equation one can integrate Eq.(24) in its general form

$$\frac{RT}{\rho g} \left[c_{ef} \, ln \frac{c(\theta)}{c(\pi/2)} - c(\theta) + c(\pi/2) \right] = -R_f(\theta) \cos\theta \tag{25}$$

where $c(\pi/2)$ is the concentration in the equatorial plane (at $\theta = \pi/2$). Then, with account for the surface equation of state and the adsorption isotherm one obtains the interfacial tension $\gamma(\theta)$ and surfactant adsorption $\Gamma(\theta)$ distributions along the film, and with Eq. (11) we get the film thickness profile $h(\theta)$.

For a complete description of the foam bubble shape the dependency $R_f(\theta)$ is necessary. This dependency can be obtained from the normal force balance at the bubble surface. The surface area of the film element is

$$dA = 2\pi R_f \sin\theta \cdot dl \tag{26}$$

Using Eq.(19) and the Laplace equation the pressure balance at the bubble surface can be expressed as

$$\Delta P = 2\gamma K + \rho g h \sin\varphi \tag{27}$$

where ΔP is the excess pressure in the bubble (independent of θ), and K is the local mean curvature of the film. From differential geometry the curvature K can be expressed as

$$K = \frac{2R_f^2 + 3\left(\dfrac{dR_f}{d\theta}\right)^2 - R_f\dfrac{dR_f}{d\theta}\cot\theta - \dfrac{1}{R_f}\left(\dfrac{dR_f}{d\theta}\right)^3\cot\theta - R_f\dfrac{d^2R_f}{d\theta^2}}{\left[R_f{}^2 + (dR_f/d\theta)^2\right]^{3/2}} \tag{28}$$

Eqs. (22), (27) and (28) yield a non-linear second-order differential equation with respect to the function $R_f(\theta)$

$$\frac{d^2R_f}{d\theta^2} = 2R_f + \frac{3}{R_f}\left(\frac{dR_f}{d\theta}\right)^2 - \cot\theta\frac{dR_f}{d\theta} - \frac{\cot\theta}{R_f{}^2}\left(\frac{dR_f}{d\theta}\right)^3$$

$$-\frac{1}{2\gamma R_f}\left[\Delta P - \rho g h\frac{R_f\cos\theta + (dR_f/d\theta)\sin\theta}{\left[R_f{}^2 + (dR_f/d\theta)^2\right]^{1/2}}\right]\left[R_f{}^2 + \left(\frac{dR_f}{d\theta}\right)^2\right]^{3/2} \tag{29}$$

Solving the set of Eqs. (11), (25) and (29) with a respective surface equation of state $\gamma(\Gamma)$ and adsorption isotherm $\Gamma(c)$ one obtains the exact bubble shape, thickness profile and surfactant distribution in the film. Because of its complexity the equations has to be integrated numerically.

The theory presented above contains only one unknown parameter – the nominal (effective) surfactant concentration c_{ef}. By fitting the theory to an experimental bubble shape we can obtain this parameter. Having this information the change of the bubble shape due to an expansion can be predicted and compared with experiments. The simultaneous measurement of the variation in thickness along the film will provide additional independent and direct information to be compared with the theory. This gives also the possibility to study the local film elasticity (instead of average values).

The bubbles shapes shown in Fig. 10 are calculated according to the theory presented above for the surface equation of state and adsorption isotherm accounting for 2D compressibility in the adsorption layer. It is seen that the bubble formed by a thin film (h = 0.3 μm) is practically not deformed by gravity whereas that with a thick film (h = 8 μm) is considerably deformed.

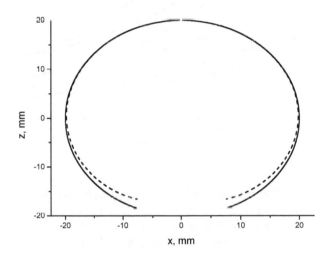

Fig.10. Shape of bubbles obtained according to Eqs. (11), (25) and (29) for an adsorption layer described by a 2D compressibility model of (Eqs. (6), (7) and (9), $\Omega_0 = 2.5 \cdot 10^5$ m²/mol, b = 1 m³/mol, a = 1.2, and $\varepsilon = 0.01$ m/mN); film thickness at the bubble top: $h_0 = 0.3$ μm (full line) and $h_0 -$ 8 μm (dashed line); radius of curvature at the bubble top $R_{c0} = 20$ mm; $c_{ef} = 6$ mol/m³ (according to [21]).

Fig. 11 demonstrates the profiles of film thickness. The thickness of the film decreases from the bottom of the bubble to its top because of the expansion in the gravitational field. As a consequence, the surfactant concentrations in the bulk of the film and at its surfaces decrease and the surface tension increases with height (Fig. 12).

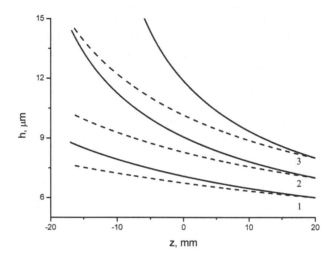

Fig.11. Film thickness profiles obtained for the same models and the same sets of parameters as in Fig.4; film thickness at the bubble top: $h_0 = 6$ μm (1), 7 μm (2), 8 μm (3); radius of curvature at the bubble top $R_{c0} = 20$ mm; $c_{ef} = 6$ mol/m^3 (according to [21]).

The resulting surface tension gradients counteract the weight of the films. This behaviour is similar to that for flat vertical films [3, 45, 49]. From Figs. 11 and 12 one can also see the effect of the surface equation of state and adsorption isotherm on the film characteristics.

Because of the varying film thickness and concentration the film elasticity should be also non-uniform over the film. Fig. 13 gives some examples of film elasticity variation with height. For thick films the elasticity increases whereas for thin films it decreases with height. However, in thin films the elasticity is more uniform because the deformation due to gravity is less.

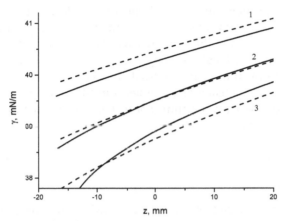

Fig.12. Surface tension profiles obtained for the same models and the same sets of parameters as in Fig.4; film thickness at the bubble top: $h_0 = 6$ μm (1), 7 μm (2), 8 μm (3); radius of curvature at the bubble top $R_{c0} = 20$ mm; $c_{ef} = 6$ mol/m^3 (according to [21]).

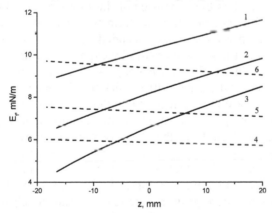

Fig.13. Variation of film elasticity with height obtained for adsorption layers described by a 2D compressibility model with the same set of parameters as in Fig.10; film thickness at the bubble top: $h_0 = 6$ μm (1), 7 μm (2), 8 μm (3), and $h_0 = 0.6$ μm (4), 0.7 μm (5), 0.8 μm (6); radius of curvature at the bubble top $R_{c0} = 20$ mm; $c_{ef} = 6$ mol/m^3 (according to [21]).

Thus, the elasticity of non-uniform films is determined by the local concentration of surfactants and film thickness. Upon stretching the different parts of the film should have different expansion inversely proportional to their local elasticity. Hence, the film elasticity measured for a non-uniform film is an averaged characteristic. To obtain the total area increase by expansion of a non-uniform film one has to integrate the local area increase over the film. For a flat rectangular film of constant width one can write

$$\Delta A = 2H\Delta\gamma \int\limits_{x_1}^{x_2} \frac{dx}{E_f} \tag{30}$$

where $\Delta\gamma$ is the surface tension change caused by stretching, H is the film width, x is the coordinate along the stretching direction, x_1 and x_2 are the initial coordinates, $A = H(x_2 - x_1)$ is the initial film area.

Upon expansion of non-uniform films, such as in Fig.13, the local expansion is most significant in those parts where the elasticity is smaller: in thick films the expansion of the lower parts is larger (curves 1-3), and the film becomes more uniform by expansion, whereas in thin films the upper parts are most expanded (curves 4-6), and the non-uniformity in film thickness increases by expansion, what can lead to its fast rupture (or black film formation) as discussed in [3]. The film ruptures in a place where the critical thickness is reached at first. Those parts in the films are most sensitive to external disturbances. Hence, the local elasticity is, probably, a much more important factor for the stability of non-uniform films than their average elasticity.

The curves 4-6 in Fig.13 describe films the elasticity of which decreases with decreasing thickness, i.e. they correspond to positions left from the elasticity maximum in Figs.5-8. As discussed above, for low soluble surfactants these positions are reached after much smaller initial expansions than for more soluble surfactants (cf. Fig.8). The further expansion of the film should lead to increasing non-uniformity in thickness and, therefore, to a faster film rupture. As for low soluble surfactants such positions correspond to sufficiently high thicknesses, the required initial expansion can be produced due to the effect of gravity just after the film is formed. This behaviour is consistent with a smaller stability of films and foams formed from low soluble surfactants. It is clear, however, that the possibility of film rupture due to such mechanism should depend also on the conditions of film formation and the external disturbances.

F. MECHANICAL DISTURBANCES OF FILMS

The possibility of film rupture under external influence depends on the magnitude and kind of the disturbance. The disturbances can be classified as slow and rapid, mechanical and non-mechanical (e.g. heat-and/or mass-exchange), symmetrical and asymmetrical, long-wave and short-wave, finite and infinitesimal, local and non-local [5,6,40,49]. We will consider here mechanical disturbances. Depending on the particular conditions a mechanical disturbance can be characterized by

– a definite increase in the film area ΔA (for example, film expanded in a frame or soap bubble grown at the capillary tip to a required size),

– a definite increase in the film tension $\Delta \gamma_f$ (by increasing external force, for example, in a vertical film under gravity conditions),

– a definite energy increase ΔF (for example, due to an external impulse of a certain kinetic energy).

In the first case, when an area increase is given, the respective film tension increase can be found through the film elasticity: $\Delta \gamma_f = E_f (\Delta A / A)$ (for small relative area increase, $\Delta A / A \ll 1$, and uniform film). Similarly, when a film tension increase is given, the respective area increase can be easily obtained via $\Delta A = A \Delta \gamma_f / E_f$ (or Eq.(30) – for non-uniform films). In the third case the surface energy increases from $A\gamma f$ to $(A + \Delta A)(\gamma f + \Delta \gamma f)$, i.e. the energy increase is

$$\Delta F \approx A\Delta \gamma_f + \gamma_f \Delta A = \left(E_f + \gamma_f\right)\Delta A \qquad (31)$$

For large relative extensions the respective integrals should be taken.

The films in a foam are continuously disturbed either due to external factors (e.g. vibrations) or due to rupture of neighbouring films and rearrangements of the foam structure (because of redistribution of liquid in channels and films, gas diffusion between the bubbles, etc.) [5]. These processes lead to a local disturbance of film properties (surface tension, pressure, concentrations, etc.). Accordingly, all three considered types of mechanical disturbances are possible in foams.

Film rupture happens when the critical thickness is reached elsewhere in the film at its expansion. This is possible when the magnitude of the disturbance (ΔA, $\Delta \gamma_f$ or ΔF) exceeds a critical value for a particular film. To predict these critical values one has to analyze the local elastic (visco-elastic) properties of the film, which are determined by the local film thickness and concentrations, which in turn depend on the prehistory of a film, as discussed above. However this is a rather complicated problem for foams, which is one of the reasons why it is so difficult to establish correlations between foam stability and elasticity of films.

G. KINETIC ASPECTS BY FILM STRETCHING

Up to now only the regimes with slow film expansions were considered, when the deviations from local diffusion and adsorption equilibrium in the film cross-sections can be ignored. However, liquid films can be subjected also to fast expansions (or compressions), when the equilibrium cannot be established instantaneously. In such cases the resistance of a film to expansion will be larger and can depend on the characteristic time of the disturbance [5, 6].

For the case of fast film expansions a generalised equation can be derived for the elasticity modulus of films [50, 51]

$$E_f = 2E_0\left[1+\frac{dc}{d\Gamma}\sqrt{\frac{D}{i\omega}}\,tanh\left(\frac{h}{2}\sqrt{\frac{i\omega}{D}}\right)\right]^{-1} \tag{32}$$

For relatively slow oscillations ($\omega \to 0$) or very thin films ($h \to 0$) the frequency (time) dependence disappears and the known expression for the film elasticity, Eq.(5), follows directly. In the opposite limiting case (ω and h are large) one obtains the known expression for the dilational surface elasticity of a single interface $E_d = \dfrac{E_f}{2} = E_0 \left[1 + \dfrac{dc}{d\Gamma} \sqrt{\dfrac{D}{i\omega}} \right]^{-1}$. In

the former case the diffusion layer thickness $\sqrt{D/\omega}$ is larger than the film thickness h, and the diffusion layers at the two film interfaces strongly overlap, while in the last case the diffusion layers do not overlap and the interfaces do not influence each other during the expansion. Thus, the expressions for film elasticity, Eq.(5), and dilational surface elasticity are simply two particular cases of the more general Eq.(32).

When the film is expanded very slowly, the surfactant molecules have sufficient time to adsorb at the interfaces. Then the film elasticity is described by Eq.(5) (or more complicated equations still assuming equilibrium surfactant distribution), and in this case it is usually called the Gibbs elasticity [3, 5, 6, 11, 15, 18, 24, 52]. Note, very often, however, the value $E_0 = -d\gamma/d\ln\Gamma$ is also called the Gibbs elasticity [4, 12, 29, 50]. To avoid contradictions it is proposed to call the value E_0 limiting [19, 24, 50] or intrinsic [11] surface elasticity (compare also Chapters 1 and 2 of this book).

In contrast, when the expansion is much faster than the characteristic time for diffusion and adsorption, then the adsorption layers behave like insoluble ones, and the film elasticity approaches its limiting value $E_f = 2E_0 = -2d\gamma/d\ln\Gamma$, which is usually called dynamic or Marangoni elasticity [5, 6, 44, 52].

According to Eq.(32), there are two characteristic times, $\tau_h = h^2/D$ and $\tau_S = (d\Gamma/dc)^2/D$, which determine the $E_f(\omega)$ dependence. Their correlation with the characteristic time of the disturbance $t_d = 2\pi/\omega$ lead to the following particular cases ($\tau_S < \tau_h$):

a) $\tau_S < \tau_h < t_d$, the film elasticity is described by Eq.(5);

b) $\tau_S < t_d < \tau_h$, the film elasticity is twice the elasticity of an individual adsorption layer E_d;

c) $t_d < \tau_S < \tau_h$, the film elasticity tends to its limiting value $E_f = 2E_0$.

In very thin films the contribution of surfactants dissolved in the bulk is negligibly small, therefore, for $\tau_h < \tau_S$ (h < $d\Gamma/dc$) the film elasticity is always close to its limiting value $E_f = 2E_0$ for any time of disturbance.

Fig.14 demonstrates the effect of characteristic disturbance time on film elasticity for n-octanol solution at the same conditions as in Fig.8.

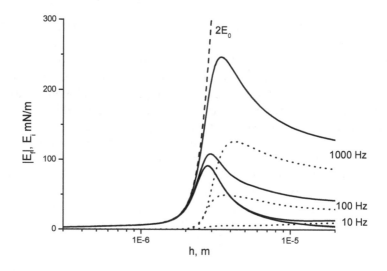

Fig.14. Effect of characteristic disturbance time on film elasticity modulus $|E_f|$ (full lines) and imaginary part E_i (dotted lines) for n-octanol solution at the same conditions as in Fig.8 and frequencies ($f = \omega/2\pi$): 0 (infinitely slow expansion), 10, 100 and 1000 Hz.

In thicker films ($d\Gamma/dc < h < \sqrt{D/\omega}$) at slow expansion/compression rates the elasticity is close to the equilibrium (Gibbs) elasticity (case a). When the disturbance time decreases the film elasticity becomes a complex value and its modulus increases (case b). For very fast disturbances and in thinner films we obtain the limiting (Marangoni) elasticity, which decreases rapidly when film thinning occurs (case c).

In the intermediate case b both the real and imaginary parts contribute to the elasticity modulus (Fig.14). In fact, they represent the effect of surface dilational elasticity and viscosity. Their increase with increasing expansion rate leads to a strong stabilizing effect for films with respect to fast disturbances. With increasing surface dilational elasticity and viscosity a larger external force and larger work are required to expand the film.

For films obtained from low soluble surfactants this stabilizing effect is much smaller. In this case a relatively small initial expansion of the film results in a strong depletion in the bulk solution, and the surface dilational elasticity and especially viscosity strongly decrease with expansion. For example, in case of a n-decanol solution at the same conditions as in Fig.8 the surface dilational viscosity decreases practically to zero by an initial film expansion from 10 to 8 μm. Therefore the films formed from low soluble surfactant solutions are less stabilized with respect to fast disturbances.

It should be taken into account that liquid films are in mechanical equilibrium with their Plateau borders. Under expansion of a film the surfaces of Plateau borders are also expanded. The ratio between the relative expansions of the Plateau border and film is inversely proportional to their elastic moduli [3]

$$\frac{(\Delta \ln A)_f}{(\Delta \ln A)_{Pb}} = \frac{2E_{Pb}}{E_f} \tag{33}$$

When the film elasticity is near its maximum (Fig.5b) and the expansion is slow, so that the surface tension at the Plateau border surfaces deviates only slightly from equilibrium, then the relative expansion in border will be much larger than that in the film. During this expansion a new film is pulled out of the border, the thickness of which is determined by the pull-out velocity (Eq.(14)). However, when the film elasticity is far left from the maximum (where it is small because of the solution depletion) and the expansion is fast, then the expansion will mainly occur in the film what can lead to its fast rupture. Accordingly, the properties of the whole film depend strongly on the processes within the Plateau borders (and within the transition zones between the borders and the film). This produces additional difficulties in studies of film properties.

H. CONCLUSIONS

The stability of films in foams and emulsions is a very complicated problem because it depends on many factors of rather different nature. The present analysis is not aimed to consider the problem of film stability in general but only attempts to underline the important contribution of surface and film rheology. Therefore it does not discuss such problems as film drainage and evaporation, formation of black films, marginal regeneration, etc. which are considered in the relevant literature. The problem of rheology effects on film stability is much less understood. It is clear, however, that there is no simple direct correlation between film elasticity and stability of foams and emulsions. The given analysis is focused on the understanding of effects of surfactant exchange between the bulk of a film and the interfaces and, in this respect, the effects of surfactant properties.

The surfactant conservation by expansion of a film is of primary importance for its mechanical properties. The local variation of the bulk concentration with thickness affects the film elasticity and, therefore, its stability. Upon expansion the concentration in the bulk of a film can decrease by more than one order of magnitude. This depletion effect has strong influence on the film elasticity. However it depends on the surfactant properties. For low soluble surfactants the state of depletion is easily reached what leads to a fast decrease of the elasticity already after a relatively small initial film expansion. This explains, at least partially, the decreasing stability of foams from low soluble homologues, though their surface elasticity is higher. Also the depletion effect results in a lower stability of films from low soluble surfactant solutions with respect to fast disturbances.

Better understanding of the role of rheological properties requires a much more detailed analysis of the effect of different parameters of the surface equation of state and adsorption isotherm. One of the problems in rheological studies stems from the fact that the rheological properties are far more sensitive to state and interactions of adsorbed molecules than the equilibrium surface tension data. Measurements of only equilibrium surface tensions are therefore insufficient to obtain accurate information about the surface and film elasticity and to prove their effect on foam/emulsion stability.

Considering the rheology of films it is also necessary to take into account that the bulk surfactant concentration in a film depends on the way in which this film is produced, and, therefore, cannot be easily obtained from the initial solution concentration. The knowledge of the precise balance of a surfactant within the film and over the whole foam or emulsion is very important. In particular, a surfactant depletion can occur also in the Plateau borders when the film is pulled out of it, especially in dry foams. In this respect it should be noted that the effect of rheological properties have to be analysed simultaneously with other processes in a film, such as drainage or evaporation. In addition, the non-uniformity of real films should be taken into account.

ACKNOWLEDGEMENTS

The work was financially supported by projects of the European Space Agency (FASES MAP AO-99-052) and the DFG (Mi418/16-1). The authors express their thanks to Emmie Lucassen-Reynders and Jaap Lucassen for their helpful discussions.

I. REFERENCES

1. J.W. Gibbs, The scientific papers, V.1, Dover Publ., NY, 1961.
2. K.J. Mysels, K. Shinoda, S. Frankel, Soap films. Studies of their thinning, Pergamon Press, London, 1959.
3. J. Lucassen, Dynamic properties of free liquid films and foams, in: Anionic surfactants. Physical chemistry of surfactant action. Surfactant Science Ser., Vol. 11, E.H. Lucassen-Reynders (Ed.), Marcel Dekker Inc., NY, 1981, 217-265.
4. I.B. Ivanov (Ed.); Thin liquid films: Fundamentals and applications. Surfactant Science Series Vol. 29, Marcel Dekker Inc., NY, 1988.
5. P.M. Kruglyakov and D.R. Exerowa, Foams and Foam Films, in Studies of Interface Science, Vol. 5, D. Möbius and R. Miller (Editors), Elsevier, Amsterdam, 1997.

6. V.V. Krotov, A.I. Rusanov, Physicochemical Hydrodynamics of Capillary Systems, Imperial College Press, London, 1999.
7. B.P. Radoev, D.S. Dimitrov, I.B.Ivanov, Colloid Polymer Sci., 252 (1974) 50.
8. D.S. Dimitrov, Progress in Surface Science, 14 (1983) 295.
9. I.B. Ivanov, D.S. Dimitrov, P. Somasundaran, R.K. Jain, Chem. Eng. Sci., 40 (1985) 137.
10. A. Sharma, E. Ruckenstein, Colloid Polymer Sci., 266 (1988) 60.
11. A.A. Sonin, A. Bonfillon, D. Langevin, J. Colloid Int. Sci. 162 (1994) 323.
12. K.D. Danov, D.S. Valkovska, I.B. Ivanov, J. Colloid Int. Sci., 211 (1999) 291.
13. K.J. Mysels, M.C. Cox, J.D. Skewis, J. Phys. Chem., 65 (1961) 1107.
14. M. van den Tempel, J. Lucassen, E.H. Lucassen-Reynders, J. Phys. Chem., 69 (1965) 1798.
15. A. Prins, C. Arcuri and M. van den Tempel, J. Colloid Interface Sci., 24 (1967) 84.
16. V.V. Krotov, A.I. Rusanov and N.A. Ovrutskaya, Kolloid. Zh., 34 (1972) 528.
17. V.V. Krotov, A.I. Rusanov and N.D. Rjasanova, Kolloid. Zh., 34 (1972) 534.
18. H. Bianco, A. Marmur, J. Colloid Int. Sci. 158 (1993) 295.
19. E.H. Lucassen-Reynders, Surface elasticity and viscosity in compression/dilation, in: Anionic Surfactants, Physical Chemistry of Surfactant Action, Surfactant Science Ser., Vol. 11, E.H. Lucassen-Reynders (Ed.), Marcel Dekker Inc., NY, 1981, p.173-216.
20. Y. Jayalakshmi, L. Ozanne and D. Langevin, J. Colloid Interface Sci., 170 (1995) 358.
21. V.I. Kovalchuk, A.V. Makievski, J. Krägel, P. Pandolfini, G. Loglio, L. Liggieri, F. Ravera and R. Miller, Colloids Surfaces A, 261 (2005) 115.
22. V.I. Kovalchuk, R. Miller, V.B. Fainerman and G. Loglio, Adv. Colloid Interface Sci., 114-115 (2005) 303.
23. V.V. Krotov and A.I. Rusanov, Kolloid. Zh., 34 (1972) 81.

24. E.H. Lucassen-Reynders, A. Cagna, J. Lucassen, Colloid Surfaces A, 186 (2001) 63.
25. E.H. Lucassen-Reynders, J. Lucassen, P.R. Garrett, D. Giles, F. Hollway, Dynamic surface measurements as a tool to obtain equation-of-state data for soluble monolayers, in: Advan. Chem. Ser. Monolayers, E.D. Goddard (Ed.), 144 (1975) 272-285.
26. J. Lucassen and R.S. Hansen, J. Colloid Interface Sci., 23 (1967) 319.
27. C. Stenvot, D. Langevin, Langmuir, 4 (1988) 1179.
28. A. Bonfillon and D. Langevin, Langmuir, 9 (1993) 2172.
29. K-D. Wantke, H. Fruhner, J. Fang and K. Lunkenheimer, J. Colloid Interface Sci., 208 (1998) 34.
30. K-D. Wantke and H. Fruhner, J. Colloid Interface Sci., 237(2001) 185.
31. V.I. Kovalchuk, J. Krägel, A.V. Makievski, G. Loglio, F. Ravera, L. Liggieri and R. Miller, J. Colloid Interface Sci, 252 (2002) 433.
32. V.I. Kovalchuk, G. Loglio, V.B. Fainerman and R. Miller, J. Colloid Interface Sci., 270 (2004) 475.
33. V.I. Kovalchuk, J. Krägel, A.V. Makievski, F. Ravera, L. Liggieri, G. Loglio, V.B. Fainerman and R. Miller, J. Colloid Interface Sci., 280 (2004) 498.
34. V.B. Fainerman, R. Miller, V.I. Kovalchuk, Langmuir, 18 (2002) 7748.
35. V.B. Fainerman, R. Miller, V.I. Kovalchuk, J. Phys. Chem. B, 107 (2003) 6119.
36. V.B. Fainerman, V.I. Kovalchuk, E.V. Aksenenko, M. Michel, M.E. Leser and R. Miller, J. Phys. Chem., 108 (2004) 13700.
37 V.B. Fainerman, V.I. Kovalchuk, M.E. Leser and R. Miller, in "Colloid Stability. The role of Surface Forces. Part 1.", Colloid and Interface Science Series, V.1, Th. Tadros (Ed.), John Wiley, 2007, p. 307-333.
38. C. Stubenrauch, V.B. Fainerman, E.V. Aksenenko and R. Miller, J. Phys. Chem., 109 (2005) 1505.
39. V.B. Fainerman, R. Miller, E.V. Aksenenko, A.V. Makievski, Surfactants – chemistry, interfacial properties and application,

V.B. Fainerman, D. Möbius, R. Miller (Editors), in: Studies in Interface Science, Vol. 13, Elsevier, Amsterdam, 2001, pp. 189– 286.

40. A.I. Rusanov, V.V. Krotov, Progr. Surface Membrane Sci., 13 (1979) 415.

41. A.I. Rusanov, Mendeleev Commun., (2002) 218.

42. A.I. Rusanov, Colloids Surfaces A, 239 (2004) 105.

43. A.I. Rusanov, J. Chem. Phys., 120 (2004) 10736.

44. J.A. Kitchener, Nature, 194 (1962) 676.

45. P.G. de Gennes, Langmuir, 17 (2001) 2416.

46. J.A. Kitchener, Nature, 195 (1962) 1094.

47. K.J. Mysels, M.C. Cox, J. Colloid Sci., 17 (1962) 136.

48. J. Lyklema, P.C. Scholten, K.J. Mysels, J. Phys. Chem., 69 (1965) 116.

49. W.E. Ewers, K.L. Sutherland, Australian J. Sci. Res., 5 (1952) 697.

50. P.Joos, Dynamic surface phenomena, Dordrecht, VSP, 1999.

51. A.V. Makievski, V.I. Kovalchuk, J. Krägel, M. Simoncini, L. Liggieri, M. Ferrari, P. Pandolfini, G. Loglio and R. Miller, Microgravity - Science and Technology Journal, 16-1 (2005) 215.

52. J.A. Kitchener, Foams and free liquid films, in Recent progress in surface science, J.F. Danielli, K.G.A. Pankhurst, and A.C. Riddiford, eds.), V.1, Academic Press, NY, London, 1964, 51-93.

INTERFACIAL RHEOLOGY OF BIOLOGICAL LIQUIDS: APPLICATION IN MEDICAL DIAGNOSTICS AND TREATMENT MONITORING

V.N. Kazakov[1], V.M. Knyazevich[1], O.V. Sinyachenko[1],
V.B. Fainerman[1] and R. Miller[2]

[1]Donetsk Medical University, 16 Ilych Avenue, 83003 Donetsk, Ukraine
[2]Max-Planck-Institut für Kolloid- und Grenzflächenforschung, 14424
Potsdam, Germany

Contents

A. INTRODUCTION

Human biological liquids contain various surface active molecules which adsorb at liquid interfaces leading to variations in surface tension. The work of Polányi in 1911 [1] is probably one of the earliest research on surface tensions of such biological liquids (cerebrospinal liquid). First surface tension measurements of blood serum were performed by Morgan and Woodward [2] in 1913, and Künzel published in 1941 [3] a first systematic study on the surface tensions of blood serum and cerebrospinal liquid. Lately, publications concerning studies of surface tension of blood, cerebrospinal and amniotic liquid, gastric juice, saliva, expired air condensate and other human biologic liquids become more frequent [4-11]. However, these studies are still incomprehensive, and the methods used are often not reliable enough. For the time being, only one branch of medical science, namely the studies of lung surfactant systems, can be regarded to as one where interface tensiometry is widely applied. In this particular area, significant clinical progress was already achieved in the treatment of the pathologies related to disorders in the lung surfactant system [12, 13]. In recent years the experimental methods were essentially improved and an increasing interest has been observed in interfacial studies of biologic liquids dealing with surface tension and also interfacial rheology of various systems, such as blood serum and plasma, cerebrospinal and amniotic liquid, gastric juice, saliva, pulmonary surfactants and other, and the results were summarised in books and reviews [14-17].

The interrelation between the dynamic surface tension of blood serum and other biologic fluids, and the pathologic characteristics of various diseases were studied so far mainly by the maximum bubble pressure method [14,15,18], and it was shown that have a large potential for differential diagnosis and monitoring of the efficiency of therapies. Via correlations between interfacial tensiographic parameters and the contents of surface active compounds it is possible to indicate some surfactants which affect the surface tension characteristics of serum, urine, synovial fluid and liquor. In turn, dynamic surface tensiometry of biological liquids for different diseases is capable of reflecting rapidly and rather accurately the total contents and general composition of surfactants, including pathological proteins and other compounds formed and accumulated during the development of a disease.

It is important to note that for example the dynamic surface tension of blood serum is essentially more sensitive to various pathologies than the corresponding equilibrium surface tension values [15,17]. Yet more sensitive with respect to the influence of various surface active admixtures to protein solutions is the interfacial (dilational or shear) rheology [16, 19-24]. Shear experiments, applied to interfacial layers formed by mixtures of proteins and surfactants, yield qualitative information on the interfacial structure. In contrast, the dilational rheology has a direct interrelation to the mechanisms of adsorption dynamics and the thermodynamic state of the corresponding interfacial layer [25].

Any biological liquid is essentially a mixture of hundreds of inorganic and organic substances, including tens of surfactants (proteins, fatty acids, phospholipids etc.). For such mixtures the dilational rheology cannot be regarded as merely the cumulative from each component and can be regarded as a complement to results of biochemical, spectroscopic and other quantitative and qualitative analyses. Interfaces in the human organism are extremely extended; therefore one can expect that the results of dynamic surface tensiometry and dilational rheology reflect the actual composition of human interfacial layers and chemical and exchange processes which take place at these interfaces. Various diseases affect the composition of interfacial layers and the physico-chemical processes within these layers. Therefore, dynamic surface tensiograms and dilational rheology can be regarded as a comprehensive indicator of some pathologic disturbances.

The present review discusses problems concerning the dynamic surface tension and dilational rheology of some biological liquids, mainly blood serum, urine, breath condensate and cerebrospinal liquid, with special reference to rheumatic, neurological, pulmonary and some other internal diseases. In the present study we also present an analysis of the influence of added non-ionic surfactant on the dilational rheological characteristics of human serum albumin (HSA), serving as a suitable model system [26].

B. EXPERIMENTAL METHODS

Among all known methods, which can be used for the measurements of dynamic surface tension γ in the very short lifetime range, the maximum bubble pressure method (MBPM) is the most appropriate one for medical applications involving biological liquids. The main advantages of MBPM are the small sample volume (1 ml or less), and the wide range of surface lifetime (0.001 s to 100 s) [14,15]. Very reliable data of dynamic surface tension in the short time range obtained for such systems by using the bubble pressure tensiometers BPA-1P (SINTERFACE Technologies, Germany), and MPT2 (Lauda, Germany) tensiometers [14,15,26-28], and these were used in the presented experiments as well, all performed at 25°C.

The surface tension experiments for long time data are made in an optimum way by using the drop profile analysis tensiometry. Here, the drop of studied liquid is rapidly formed (during 1-2 s) at the tip of a Teflon capillary, and the time dependent surface tension measured every 10 s in the time interval of up to 1200-1800 s. The selection of optimum parameters of the dynamic surface tensiograms was discussed in detail in [14,15,18]. The most informative selected surface tension values are $\gamma_{0.01}$, γ_1, γ_{100} and γ_∞, obtained at times t =0.01 s, 1 s, 100 s and extrapolated to t $\rightarrow\infty$ respectively. Also the final slope of a tensiogram $\lambda = (d\gamma / d(1/\sqrt{t}))_{t \rightarrow \infty}$ is used.

At the time of 1200-1800 s after drop formation has elapsed, the drop was subjected to stress deformations, i.e. a rapid increase of the surface area by about 5-8 %, and the visco-elasticity modulus related to the stress experiment E of the surface layer was calculated as:

$$E = \frac{A\Delta\gamma}{\Delta A} \tag{1}$$

where A is the initial surface area of the drop, $\Delta\gamma$ is the initial surface tension jump caused by the stepwise increase of the drop surface area ΔA. The variation in surface tension induced by the stress deformation obeys the exponential equation [14]:

$$\Delta\gamma_t = \Delta\gamma \, exp(-\Delta t / \tau) \tag{2}$$

where Δt is the time expired since the deformation, τ is the corresponding relaxation time.

For relaxation experiments with harmonic oscillations the bubble and drop profile analysis tensiometry (for example PAT 1 from SINTERFACE Technologies, Germany) is suitable. This method can additionally be equipped with a so-called coaxial double capillary [29], which then allows formation of a defined drop at the capillary tip of the solution by supplying the solution through the external channel. Evaporation of the solvent results in a decrease of the drop volume during the experiment, and any liquid loss can be compensated by injection of pure solvent through the central channel of the coaxial capillary. To study the dilational elasticity, after having reached the adsorption equilibrium the solution drop can be subjected to harmonic oscillations at frequencies f between 0.005 Hz and 0.2 Hz, with oscillation amplitudes of 7-8%. Figure 1 illustrates as an example the time dependence of surface tension $\gamma(t)$ and drop surface area $A(t)$ during harmonic oscillations of a drop at a frequency of 0.01 Hz with and amplitude of about 10% for an aqueous 0.5 μmol/l HSA solution in the presence of 1 μmol/l $C_{14}EO_8$ [26].

The results of experiments with harmonic oscillations of small surface area amplitudes ΔA can be analysed using the Fourier transformation:

$$E(i\omega) = A_0 \frac{F[\Delta\gamma]}{F[\Delta A]}, \tag{3}$$

where A_0 is the initial area of the drop surface. The initial experimental data were first filtered to exclude scattering errors. Also a constant drift of the surface tension with time, caused by a yet small deviation of the system from equilibrium when starting harmonic oscillations, was eliminated.

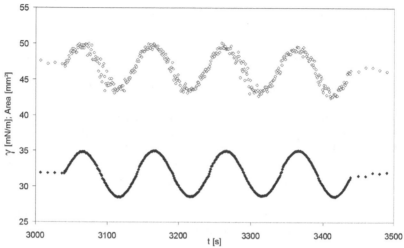

Fig. 1. Time dependence of surface tension ($\Diamond$) and drop surface area ($\blacklozenge$) during the harmonic oscillations: frequency 0.01 Hz, mixed solution of 0.5 µmol/l HSA and 1 µmol/l $C_{14}EO_8$.

C. SURFACE ELASTICITY OF MIXED PROTEIN/SURFACTANT

SURFACE LAYERS

Addition of surfactants can modify adsorbed protein layers at the liquid/liquid and liquid/gas interfaces. Lucassen-Reynders was possibly the first to analyse the dilational rheology of surfactant mixtures [30]. Later Garrett and Joos [31] generalised the theory by Lucassen and van den Tempel [32,33] to surfactant mixtures, which was later on further developed [34-37] and then refined for application to protein containing mixed solutions [38].

The surface dilational modulus E for harmonic area oscillations can be defined by the expression (1) in the differential form:

$$E = \frac{d\gamma}{d \ln A} \tag{4}$$

The measured modulus E consists of an elastic part accounting for the recoverable energy stored in the interface, and a viscous contribution reflecting the loss of energy through relaxation processes.

The complex elasticity modulus of mixed surface layers derived for harmonic oscillations at frequency f and small magnitude ΔA is given by the expression [38] (compare also the details given in Chapter 9 of this volume):

$$E = \frac{1}{B}\left(\frac{\partial \Pi}{\partial \ln \Gamma_1}\right)_{\Gamma_2}\left[\sqrt{\frac{i\omega}{D_1}}a_{11} + \sqrt{\frac{i\omega}{D_2}}a_{12}\frac{\Gamma_2}{\Gamma_1} + \frac{i\omega}{\sqrt{D_1 D_2}}(a_{11}a_{22} - a_{12}a_{21})\right] +$$
$$\frac{1}{B}\left(\frac{\partial \Pi}{\partial \ln \Gamma_2}\right)_{\Gamma_1}\left[\sqrt{\frac{i\omega}{D_1}}a_{21}\frac{\Gamma_1}{\Gamma_2} + \sqrt{\frac{i\omega}{D_2}}a_{22} + \frac{i\omega}{\sqrt{D_1 D_2}}(a_{11}a_{22} - a_{12}a_{21})\right] \tag{5}$$

where $\Pi = \gamma_0 - \gamma$ is the surface pressure, γ_0 is surface tension of solvent, Γ_j, c_{sj} and D_j are the adsorption, subsurface concentration and diffusion coefficient of the component j, respectively, $\omega = 2\pi f$ is the angular frequency of the generated area variations, $a_{ij} = (\partial \Gamma_i / \partial c_j)\big|_{c_{k \neq j}}$ are the partial derivatives to be determined from the adsorption isotherm, and $B = 1 + \sqrt{i\omega/D_1}\,a_{11} + \sqrt{i\omega/D_2}\,a_{22} + (i\omega/\sqrt{D_1 D_2})\cdot(a_{11}a_{22} - a_{12}a_{21})$.

The discussion of the visco-elasticity of surface layers by complex numbers is only a mathematical procedure, giving easier access to the rheological quantities important for our study here. Reminding that the complex dilational modulus

$$E(i\omega) = E_r + i\,E_i \tag{6}$$

we get the related quantities such as the visco-elasticity modulus $|E|$ and the phase angle ϕ between stress $(d\gamma)$ and strain (dA) as

$$|E| = \sqrt{E_r^2 + E_i^2}, \quad \phi = arctg(E_i/E_r), \tag{7}$$

Assuming a pure diffusion relaxation, the rheological dilational characteristics of the surface layer for single surfactant solutions were given by Lucassen and van den Tempel [32,33] as:

$$|E| = E_0 (1 + 2\zeta + 2\zeta^2)^{-1/2}, \quad \phi = arctg[\zeta/(1+\zeta)], \qquad (8)$$

with

$$E_0 = d\Pi/d\ln\Gamma, \quad \zeta = \left(\frac{\omega_D}{\omega}\right)^{1/2} \text{ and } \omega_D = \frac{D}{2}\left(\frac{dc}{d\Gamma}\right)^2$$

The surface visco-elastic parameters, as given by Eq. (8) for single surfactants and Eqs. (5) to (7) for mixtures, contain two and six derivatives, respectively. To determine these derivatives, certain models of the surface layer (equation of state and adsorption isotherm) have to be applied. When we deal with model systems, three types can be considered: (i) adsorption of individual protein; (ii) adsorption of individual surfactant; and (iii) simultaneous adsorption of protein and surfactant from a mixed solution. The numerical procedure for the solution of Eqs. (5)-(8) was given in [38] and also discussed in all details in Chapter 9 of this volume. The theoretical models used in the calculations were:
- for individual surfactant solutions we assume internal compressibility of molecules in the surface layer [39-41];
- for solutions of individual proteins we assume a possible variation of the molecular area depending on the surface pressure [42,43];
- for the simultaneous adsorption from non-ionic surfactant/protein mixed solutions we use a theoretical model described recently in [44].

Figure 2 illustrates the dependencies of the visco-elasticity modulus $|E|$ and phase angle ϕ on the oscillation frequency for a 0.2 µmol/l HSA solution in absence and presence of various added amounts of $C_{14}EO_8$ (as modelling surfactant) at concentrations between 1 µmol/l and 5 µmol/l [26]. Figures 3 and 4 show similar dependencies for HSA concentrations of 0.5 and 1.0 µmol/l, respectively.

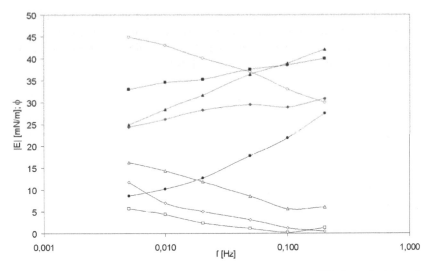

Fig. 2.   Dependencies of the visco-elasticity modulus (filled symbols) and phase angle (open symbols) on the oscillation frequency for the individual HSA solution (□,■) with concentration 0.2 μmol/l, and for the solution with various admixtures of $C_{14}EO_8$: 1, (◇,◆); 2, (△,▲); and 5 μmol/l (○,●).

These results clearly indicate the essential influence of small amounts of surfactant on the dilational rheological characteristics of HSA. While for pure protein layers the characteristic feature is a rapid decrease of the phase angle (in the frequency range studied) down to zero, small admixtures of surfactant (1 to 3% of the protein mass) lead to a significant increase in the phase angle and remarkable changes of the visco-elasticity modulus. For the relatively high admixture of 5 μmol/l surfactant even the shape of the phase angle dependence on oscillation frequency is altered. This behaviour is caused by the competitive adsorption of $C_{14}EO_8$ accompanied by partial displacement of HSA from the surface layer (as predicted by the theory [14]); in this case the rheological characteristics of the mixture become similar to those of the surfactant. Therefore, the data presented in Figs. 2-4 show that the dilational rheology can give an essential information input to studies of blood serum in pathological cases which typically lead to the presence of various surfactants in the blood and due to molecular interactions to variations in the interfacial behaviour of HSA.

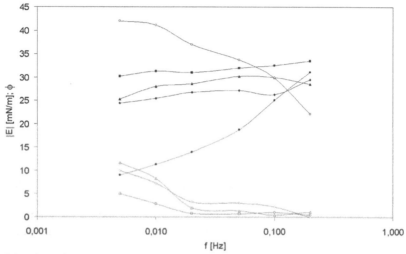

Fig. 3.   The same as in Fig. 2 for the HSA solution with concentration 0.5 µmol/l.

Note that the results for individual HSA solutions presented in Figs. 2-4 agree well with data obtained in [45] for bovine serum albumin (BSA), a substances of similar molecular mass and structure. The data reported in [45] were discussed in [43,46] on the basis of the theoretical model proposed in [42] and applied to the dilational rheology of mixtures of β-lactoglobulin with non-ionic surfactants in [38]. The experimental data in [45] agree very well with the theoretical predictions [42,43].

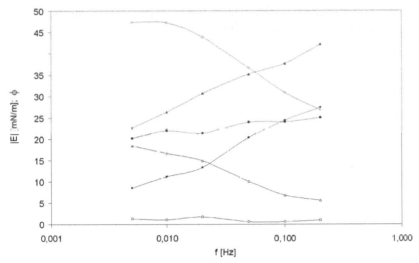

Fig. 4.   The same as in Fig. 2 for the HSA solution with concentration
1.0 μmol/l.

For comparing the experimental data presented in Figs. 2 to 4 with
the theoretical model given by Eqs. (3)-(6), the following model
parameters were used:
- for HSA [43, 46]: $\omega_0 = 2.5 \cdot 10^5$ m²/mol, $\omega_{min} = 3.0 \cdot 10^7$ m²/mol,
    $\omega_{max} = 7.5 \cdot 10^7$ m²/mol, $a = 1$, $b_j = 3 \cdot 10^5$ l/mol   (or
    $\Sigma b_j = 5.43 \cdot 10^7$ l/mol for the molecule as a whole);
- for $C_{14}EO_8$ [47]: $\omega_{0S} = 3.8 \cdot 10^5$ m²/mol, $a = -1.7$, $b = 3.0 \cdot 10^8$ l/mol, and
    $\varepsilon - 0.007$ m/mN.

Here ω_{min} and ω_{max} are the minimum and maximum molar areas of
the protein, ω_0 is the molar area increment, chosen equal to the molar
area of water as the solvent $\omega_{min} \gg \omega_0$, a is a Frumkin-type
intermolecular interaction parameter, b_j is the adsorption equilibrium
constant for the protein in the jth state, ω_{0S} is the molar area of the
surfactant at $\Pi = 0$, b is the adsorption equilibrium constant for the
surfactant, ε is the relative two-dimensional compressibility coefficient

of adsorbed surfactant molecules in the surface layer. Some results of model calculation are shown in Fig. 5.

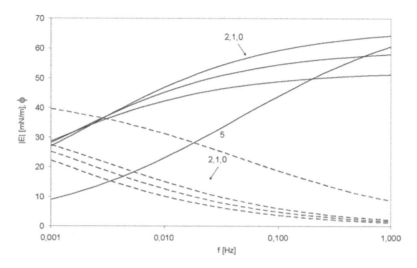

Fig. 5. Dependencies of the visco-elasticity modulus (solid lines) and phase angle (dotted lines) on the oscillations frequency for HSA solutions with concentration 0.2 µmol/l with various $C_{14}EO_8$ additions (curves labelled by concentration values 0, 1, 2 and 5 µmol/l).

One essential feature of the calculations should be noted, as the rheological characteristics were measured using the oscillating drop method (finite amount of surface active molecules in a single drop). The parameters listed above correspond to the actual equilibrium [46], i.e. to the final concentration in the single drop after the adsorption equilibrium has been established. Therefore the initial concentration of surfactant in the solution c has to be recalculated into the equilibrium concentration c_0 using the following surfactant mass balance equation:

$$\Gamma_S = \frac{V_D}{A_D}(c - c_0),$$

(9)

where A_D and V_D are the area and volume of the drop. The surfactant adsorption Γ_S was calculated according to the respective theoretical model, i.e., the Γ_S and c_0 values were determined for a given concentration using Eq. (9) and the model given in [38,44] by an iterative procedure. For example, for an initial surfactant concentration of 1 µmol/l in a single drop the equilibrium concentration was calculated to be 0.12 µmol/l, and for an initial concentration of 5 µmol/l the equilibrium concentration results to 2 µmol/l. The diffusion coefficients used here were $5 \cdot 10^{-10}$ m²/s for the surfactant and $1 \cdot 10^{-10}$ m²/s for the protein [25]. Note, such estimations can be done for systems of known composition, while for biological liquids, under discussion below, no accurate information exist and therefore, often only effective values or estimations for particular parameters can be obtained.

It is seen from the comparison of Figs. 2 and 5 that the theoretical model [38,44] provides a satisfactory description of the experimental dependencies at low initial $C_{14}EO_8$ concentrations, Moreover, even at a $C_{14}EO_8$ concentration of 5 µmol/l, the qualitative agreement between theory and experiment is rather good. This agreement can be even improved by selecting optimum values for the protein and surfactant diffusion coefficients. At HSA concentrations above 0.2 µmol/l a two-dimensional condensation of the protein or a bilayer formation can occur, and the surface pressure isotherm exhibits a characteristic plateau. A further increase of the HSA concentration does not lead to a noticeable surface pressure increase, in spite of a significant increase in the adsorption [42]. The theory developed in [38] disregards these phenomena, and is therefore unable to analyse the respective data in Figs. 3 and 4 quantitatively. The examples of the good agreement of the theory [38] with experiment for adsorption and rheological behavior of protein/surfactant mixtures are presented in [48,49].

D. RHEOLOGICAL CHARACTERISTICS OF BIOLOGIC LIQUIDS
FOR SOME PATHOLOGIES

In the Physicochemical Centre of the Donetsk Medical University, screening experiments were performed using dynamic surface tensiometry (maximum bubble pressure and drop shape methods) and surface rheometry (primarily based on stress experiments, and for the obstructive jaundice cases also the drop surface harmonic oscillations method) with more than 2000 patients suffering from rheumatic, neurological, pulmonary and some other internal diseases, and for more than 200 healthy persons. The dynamic surface tension and rheological parameters of various human liquids, such as blood serum, urine, breathing air condensate, cerebrospinal liquor and others, were measured, giving a total of more than 10000 experiments. The experimental results were analysed statistically, and the mean values (M), mean deviations (m), correlation coefficients (r) and the calculated reliability of the statistical parameters (the probability of a wrong conclusion p).

1. Rheumatoid arthritis

For the rheumatoid arthritis, the rheological characteristics of breath condensate (stress experiment) were found to depend on the patients' gender and age, and the extent of the pathologic process activity. With the increase of the disease duration, a decrease of both the pH and visco-elasticity modulus E of the breath condensate was observed. The breath condensate relaxation time τ, obtained from the stress experiments, exhibits an inverse correlation with both the fibrinogenemia parameters and the level of nitrites/nitrates in the blood. The development of arterial hypertension is accompanied by increasing E values of the breath condensate. The corresponding properties depend on the presence and extent of the respiratory pathology: in absence of pheumopathy, $E = 29.8 \pm 1.7$ mN/m and $\tau = 486 \pm 60$ s; otherwise $E = 21.3 \pm 1.8$ mN/m and $\tau = 667 \pm 145$ s. The catadrome of the rheumatoid pheumopathy is accompanied by a decrease of the visco-elasticity modulus of the breath condensate. A value for E < 15 mN/m appears to be indicative of severe rheumatoid pheumopathy.

For patients with rheumatoid arthritis an increase in the elasticity modulus E and relaxation time τ of the urine were found in 95% and 68%, respectively, and for the nephropathic patients (suffering from glomerulonephritis, interstitial nephritis, amyloidosis) the respective values were 100% and 55%. It was shown that by interfacial tensiometry and rheometry 95% (p < 0.001) of the kidney pathogenic changes were detected, while the traditional examination methods were able only in 23% of the cases. For patients with nephropathy the observed values of E and τ of urine are lower as compared with other patients suffering from rheumatoid arthritis, see Table 1.

Table 1. Urine rheological properties for patients with rheumatoid arthritis (M±m)

Parameter	Group of patients		Reliability of differences (p)
	Nephropathy absent	Nephropathy present	
E, mN/m	26.9±0.8	24.3±0.8	0.028
τ, s	290±12.2	238±9.8	0.002

The modulus E of urine correlates inversely with the duration of a disease directly with the degree of the activity of the pathologic process. The severity of the nephropathy affects the urine relaxation time τ. A correlation exist between the extent of kidney injury and the physico-chemical parameters of urine. In particular, E and τ values inversely correlate with the nephropathy severity, which is of high practical significance.

To summarise, the rheumatoid arthritis leads to a decrease of the expiratory exudation of moisture. In particular, the exudation is most pronounced when the pneumopathy is strongly developed and correlates with the E and pH values of the breath condensate. The rate and extent of moisture exudation affect the rheological characteristics of breath condensate, and correlate with the parameters of hemodynamics in a lesser circulation and external respiratory function. The development of rheumathoid pneumopathy is accompanied by an increase E modulus of the breath condensate. Also, for patients with rheumatoid arthritis, for urine changes in the values of E and τ were observed, depending on the nosological form of the nephropathy, the severity and duration of the disease, and the extent of the pathologic process activity.

2. Juvenile rheumatoid arthritis

For juvenile rheumatoid arthritis, a remarkable increase of the plasma viscosity (PV) as compared with that of healthy children was observed. Also the dilational surface stress experiments indicate a significant decrease of the E and τ values of blood serum, as one can see in Table 2. It is important to note that the equilibrium surface tension of blood serum is almost the same for ill and healthy children (45.7 ± 0.9 and 45.9 ± 1.0 mN/m, respectively). This fact reveals essentially higher sensitivity of the surface rheological methods used here to study the interfacial characteristics of human biologic fluids.

Table 2. Rheological characteristics of blood serum for patients with juvenile rheumatoid arthritis and healthy children for compare (M±m)

Parameters	Examined groups		Reliability of differences (p)
	Ill	Healthy	
PV, mPa·s	2.1±0.08	1.6±0.03	<0.001
E, mN/m	26.5±0.6	33.1±1.5	<0.001
τ, s	104.1±3.4	127.4±4.1	0.001

The univariate dispersion and regression analysis demonstrates a direct correlation between the activity of juvenile rheumatoid arthritis and the parameters of plasma viscosity and relaxation time τ. The visco-elastic properties of blood serum depend on the patients' gender: in the boys group lower E values were observed. The ankylosing of joints (characteristic for the stage IV of the disease) is related to change of the visco-elastic properties of blood serum. The univariate dispersion analysis shows that the existence of rheumatoid nodules and heart pathology lead to changes of the blood serum relaxation time τ. The concentration of serum fibrinogen is positively correlated with the PV and negatively correlated with the E values. In addition, a significant direct correlation between PV and the erythrocytes aggregation index, and the inverse correlation between the modulus E and the deformability index were found.

To summarise, the juvenile rheumatoid arthritis leads to an increase of the erythrocytes' aggregation and deformability indices, and the thrombocytes aggregation index. In this case the rheological characteristics of blood serum depend on the duration of the disease, its activity and stage. The juvenile rheumatoid arthritis results in a decreased visco-elasticity modulus E and relaxation time τ of the blood serum. These parameters depend on the pathologic process activity, the extent and distribution of the arthritis, the extent of the injury of joints, the existence of arthroankylosing and subcutaneous nodules, the injury of heart, liver or other symptoms not directly related to joints, and on the extent of the development of the hyperviscous syndrome.

3. Reactive arthritis

For patients with reactive arthritis, the inverse correlation was found between the dynamic surface tension of blood serum and the erythrocytes' aggregation index. It was reliably shown that the E and τ decrease with the plasma viscosity increase. In general, an increase of plasma viscosity was found for 64% of reactive arthritis patients, while an increase of blood serum equilibrium surface tension was detected for 40% of patients, and a decrease of the parameters E and τ by 73% and

62%, respectively. The E and τ values were shown also to depend on the activity of reactive arthritis.

The modulus E depends on the degree to which the tendovaginitis or autonomic imbalances are expressed. A direct correlation was found between E and the degree of angioneurosis. The extent of ophtalmopathy inversely correlated with E and τ, and also with the blood serum equilibrium surface tension. The development of a pathology of the aortic valve leads to changes in the blood serum equilibrium surface tension γ_∞. From the obtained results it was concluded that for the reactive arthritis any values $\gamma_\infty > 50$ mN/m affect unfavourably the aortic valve.

For patients with reactive arthritis a nephropathy results in increasing dynamic surface tensions and E modules of urine, accompanied by decreased values of equilibrium surface tension γ_∞ and τ. The rheological properties of urine do not depend on the patients' age and total activity of the disease, while at the same time these parameters are determined by the duration of the disease. The development of a nephric pathology results in decreased values of $\gamma_{0.01}$, γ_{100} and γ_∞ and simultaneously increase E, as shown in Table 3.

Table 3. Rheological properties of urine for patients with reactive arthritis and kidney damage but without nephropathy (M±m)

Parameters	Examined groups		Reliability of differences (p)
	Nephropathy absent	Nephropathy present	
$\gamma_{0,01}$, mN/m	72.6±0.30	70.7±0.38	<0.001
γ_{100}, mN/m	62.0±0.58	59.0±0.53	<0.001
γ_∞, mN/m	49.8±0.40	46.3±0.46	<0.001
E, mN/m	16.3±0.43	20.0±0.53	<0.001
τ, s	160±5.5	148±3.9	0.070

The values of $\gamma_{0.01}$ and γ_∞ inversely correlate with the total proteinuria, leukocytouria and fibronectynuria levels. Also, correlations exist for albumin concentration in the urea with γ_{100} and E modulus, of leukocytes concentration with γ_{100}, and fibronectin concentration with τ. The rheological properties of urine do not depend on the extent of the arthritis and its functional severity. It should be noted that, while individual extrarenal clinical symptoms of reactive arthritis do not significantly correlated with the levels of total protein, leukocytes, erythrocytes, fibronectin and nitrites/nitrates in the urine, the urine surface tension depends on the presence of an urogenital pathology.

To summarise, for patients with reactive arthritis a nephropathy results in a significant decrease of the dynamical surface tension and τ and an increase of E of the urine.

4. Gout

For gout it was shown that there is a significant decrease of E and τ of the blood serum: the relaxation time decreases for 85% of the patients, while the elasticity modulus decreases for 79% of the patients. The dynamic surface tension of blood serum depends on the development of metabolic syndrome, the body mass index, the levels of glycaemia and insulinemia, and on the presence of concrements in the kidneys. The value of the E modulus depends on the concentration of glucose in the blood and the concentration of calcificates in the articular tissue, while τ is governed by the insulin resistivity and the depositions of calcium in the articular tissue. The purine exchange parameters were found to be related only to the surface visco-elastic properties of the blood serum. In particular, the dispersion analysis has revealed that E is affected by the activity of xanthine oxidase and 5-nucleotidase. The values of γ_∞ and τ reflect the qualitative and quantitative composition of surfactants related to the synthesis of cytokines and eicosanoids in the blood serum.

The increase of $\gamma_{0.01}$ and γ_{100} of the blood serum for patients with chronic gouty arthritis was detected in 87% and 53% of cases, respectively. Decreased values of $d\gamma / d(1/\sqrt{t})$ for blood serum were found for 73% of screened patients with an intermittent form of the articular syndrome, while for the chronic form this value becomes 97%. The severity degree of the articular syndrome is closely related to the surface tension $\gamma_{0.01}$ of the blood serum: its decrease indicates a most probable unfavourable prognosis of the development of gouty arthritis.

There exist opposite correlations of the relaxation time τ for breath condensate with the rate (positive correlation) and volume (negative correlation) of respiratory moisture discharge. The rheological properties of breath condensate depend on the patients' age and duration of the disease. For gout patients who do not suffer from chronic bronchitis the surface tension of breath condensate is significantly higher than that for healthy persons. In this case the τ values become lower, while E remains virtually the same. The parameter γ_∞ of breath condensate depends on the activity of xanthine oxidase in the blood. This fact strongly indicates the involvement of this purine exchange enzyme in the development of a pulmonary pathology. The equilibrium value of breath condensate γ_∞ directly correlates with the pulmonary clearance of uric acid, while for τ an inverse correlation exists.

For patients with gout, the dynamic surface pressure of urine is reliably higher than that for healthy persons, while the relaxation time τ is lower for ill persons. The rheological properties of urine depend on the patients' gender, but is independent of the age, duration of disease, the type of gouty nephropathy, the type and severity of the arthritis. The levels of total albumin and fibronectin in urine affect the dynamic and equilibrium surface tensions of this biologic liquid. The content of albumins in the urine does not affect significantly the value of E, but the value of τ is closely related with the albuminuria level. With the increase of the nitrites/nitrates concentration, the dynamic surface tension of urine becomes higher, because the nitrites/nitrates are surface inactive.

The surface activity and rheological properties of urine depend on the activity of the purine exchange enzymes: xanthine oxidase, adenosine deaminase and 5-nucleotidase. The uricuria parameters affect the equilibrium surface tension of urine. The γ_1 value of urine correlates with the concentration of uric acid in the blood, γ_{100} correlates with the uric acid and activity of xanthine oxidase, γ_∞ correlates with the contents of uric acid and oxypurinol in blood.

Of major interest are data related to the dynamic surface tension of blood serum and urine for gout patients with unaffected and deteriorated function of the kidneys. With the development of renal insufficiency we observed a significant decrease of the dynamic surface tensions of blood serum and urine, which was especially significant in the short time range. This fact could be ascribed most probably to the increase of the concentration of surfactants in the blood and urine. It should be noted that the slope $d\gamma / d(1/\sqrt{t})$ of the tensiograms (of both blood serum and urine) for patients with an unaffected renal function decreases more significantly than that for patients with insufficient renal function.

To summarise, gout leads to a decrease of the parameters E and τ for the blood serum which are correlated to the metabolic syndrome, to the type of articular and renal pathology, to the state of purine exchange and endothelial function of vessels. Also changes in the physico-chemical characteristics of breath condensate (decrease of γ_∞ and τ) were observed, caused by the patients' age and the duration of the disease, by the changes of the xanthine oxidase activity and by the pulmonary excretion of uric acid. For the patients with gouty nephropathy the increase of the dynamic surface tension and decrease of the relaxation time τ for urine were observed.

5. Systemic lupus erythematosus

The physico-chemical parameters of breath condensate for patients and healthy persons are shown in Table 4. Systemic lupus erythematosus leads to a decrease of pH of the breath condensate and to a significant (by more than a factor of two) decrease of τ, while the E value in the stress experiment remains almost unchanged.

The development of lupous pneumopathy is accompanied by a decrease in pH and E of the breath condensate. While traditional methods are able to detect pulmonary changes in 39% of the cases, the decrease values of pH, γ_∞ and τ for the breath condensate were found for 63%, 80% and 97% of patients, respectively. Therefore, physico-chemical studies of breath condensate provide a more reliable criterion for early diagnostics of pneumopathy, because the deviation of parameters from normal values could be found for the majority of patients. As one could expect, the severity of pulmonary lesion affects the rheological characteristics of breath condensate ($p < 0.001$). The dispersion analysis shows that a pulmonary pathology leads to changes in pH, γ_∞, E and τ of the breath condensate.

Table 4. Acidity and rheological parameters of breath condensate for patients with systemic lupus erythematosus and healthy persons (M±m)

Physicochemical parameters	Examined groups		Reliability of differences (p)
	Ill	Healthy	
pH	7.23±0.022	7.39±0.020	<0.001
E, mN/m	22.4±1.2	24.8±2.1	0.287
τ, s	214.8±10.6	462.6±25.5	<0.001

The systemic lupus erythematosus even in its early stages of pathologic changes in the lungs activates the pulmonary surfactant system with excessive accumulation of surfactants at the alveolar surface. The increased surfactant concentration in the lungs leads to a decrease of the interalveolar pressure. The physico-chemical characteristics ($\gamma\infty$, E and τ) of breath condensate for patients with systemic lupus erythematosus depend on the activity of the disease and the peculiarities of the pathologic processes ($p < 0.001$). The systolic pressure in the pulmonary artery correlates inversely with the pH, $\gamma\infty$ and τ values of the breath condensate.

Pulmonary hypertension affects the rheological properties of breath condensate for patients with systemic lupus erythematosus ($p < 0.001$). Significantly increased concentrations of nitrites/nitrates, urea and uric acid in breath condensate were detected. The development of pneumopathy is accompanied by increasing nitrites/nitrates concentration, which depends also on the severity of pulmonary pathology. The level of nitrites/nitrates in breath condensate is interrelated with the severity of pulmonary pathology. This fact indicates that a corresponding test should be included into the complex procedure performed for the diagnostics of systemic lupus erythematosus and the estimation of the severity of the disease. The increased nitrites/nitrates contents in breath condensate leads to decreased τ value, while the amount of ammonia results in a similar decrease of $\gamma\infty$. A direct correlation exists between the pulmonary clearances of these nitrous products and surface rheological parameters.

The disorders of adaptation processes, excretive and detoxification functions caused by systemic lupus erythematosus can lead to increasing concentrations of toxic substances in the blood. These substances are compounds of medium molecular mass (300-5000 Da), such as peptides, peptide hormones, regulators of physiologic processes (fibronectin, chromostatin), immune active peptides etc. The development of pneumopathy does not lead to concentration changes of medium molecular mass compounds in the respiratory fluid and blood. On the other hand, the concentration of compounds with medium molecular mass depends on the respiratory fluid discharge and determines the rheological properties (E and τ) of breath condensate (p < 0.001).

Dynamic interfacial tensiometry and rheometry of blood and cerebrospinal fluid can be successfully applied to cases of systemic lupus erythematosus and a number of neurological diseases, such as vascular, infectious and traumatic tumours. In our studies, these methods were used in cases of neurolupus. The systemic lupus erythematosus is characterised by increased values of dynamic surface tension and $d\gamma / d(1/\sqrt{t})$ of the blood serum tensiograms. Such changes in the adsorption layer properties were found for 70% and 40% of the patients screened, respectively. Also, significantly decreased of E and τ values were detected. When the peripheral nervous system is affected, also the dynamic surface tension of blood serum in the time range below 1 s is lowered. It was shown that the values γ_{100} > 60 mN/m are indicative of a damage to the central nervous system, and values $\gamma_{100} < 55$ mN/m reflect the presence of peripheral nervous disorders. The parameters of dynamic interfacial tensiometry and rheometry of blood serum for patients with neurolupus correlate with the plasma bulk viscosity. In particular, correlations between plasma viscosity and the dynamic and equilibrium surface tensions of blood serum and the relaxation time τ were observed.

We have analysed also the dynamic surface tension and rheological characteristics of the cerebrospinal liquid for patients with cerebral neurolupus. The development of neurolupus is accompanied by increasing values of E and τ of the blood serum. For systemic lupus erythematosus, strong direct correlations exist between the dynamic surface tension of blood serum and cerebrospinal liquid. At the same time, for persons of the reference group, no dispersion, regression and correlation links exist between the physico-chemical characteristics of these liquids (including E and τ values). We have analysed the correlations between the physico-chemical parameters of blood serum and cerebrospinal liquid and the patients' age, duration and severity of the disease. It was shown that correlations exist between the severity of the disease and the dynamic surface tension of blood serum and cerebrospinal liquid, and the E value of the blood. The anti-phospholipid syndrome leads to significantly increased values of E and τ for blood serum and $\gamma 0.01$ for cerebrospinal liquid.

For the lupous nephritis, an increase of E value by a factor of two, and a significant increase of the surface tension of urine in the range of short and medium surface ages was observed, while the long time surface tension decreased (cf. see Table 5).

The patients' age and blood arterial pressure level do not affect the surface rheological parameters of urine, while the patients' gender affects the values of γ_1 and τ of urine. The nephrotic syndrome is related with the $\gamma_{0.01}$, while the renal function has its influence on the majority of physico-chemical parameters, including the dynamic surface tension, E and τ. The rheological parameters of urine exhibit unambiguous reaction to the deteriorating renal function: relaxation times $\tau < 170$ s indicate a negative prognosis of the lupous nephritis development.

Therefore, dynamic surface tension and rheology data of urine can be used in the curative treatment efficiency control for patients suffering from systemic lupus erythematosus with a nephrotic syndrome and renal insufficiency.

Table 5. Physico-chemical parameters of urine for patients with lupous nephritis and healthy persons (M±m)

Parameters	Examined groups		Reliability of differences (p)
	Ill	Healthy	
$\gamma_{0.01}$, mN/m	71.9±0.18	69.5±0.33	<0.001
γ_1, mN/m	67.9±0.28	66.3±0.56	0.010
γ_∞, mN/m	49.1±0.59	51.6±0.66	0.005
E, mN/m	31.3±1.70	16.0±0.58	<0.001

The morphological form of lupous glomerulonephritis has its own peculiarities relative to the changes of the tensiometric and rheometrical parameters of urine. Our statistical analysis has revealed that values $E > 40$ mN/m and $\tau < 170$ s correspond to the mesangium capillary glomerulonephritis, while $E < 25$ mN/m and $\tau > 240$ s are characteristic to the mesangium proliferative glomerulonephritis. The dispersion analysis has shown that the lesion extent of the renal glomerules, tubules, stroma and vessels affects the dynamic and equilibrium surface tensions, E and τ values of urine. The results show that the extent of the lesion of renal structures is directly reflected by the surface tension parameters of urine in the short and medium surface lifetime, while vascular lesions determine the relation between the γ_∞ values of urine and blood. Values $\gamma_1 > 70$ mN/m for urine indicate a negative prognosis of the development of lupous nephritis.

To summarise, the systemic lupus erythematosus is characterised by a decrease in pH, equilibrium surface tension and τ of breath condensate, depending on the extent of pulmonary pathology which determines the pressure in the pulmonary artery. The lesion of the nervous system is interrelated with changes of the content of electrolytes and lipids in the blood, its dynamic surface tension, E and τ values, and is accompanied by changes in the rheological properties of the liquor. Lupuous nephritis leads to increased values of E and dynamic surface tension of urine. The development of a nephrotic syndrome affects the majority of rheological parameters. The morphological form of glomerulonephritis and the extent of renal lesion are also reflected by changes of the interfacial tensiometric parameters of urine and blood serum.

6. Dermatic lupus erythematosus

The dermatic lupus erythematosus leads to a significant decrease of the E and τ parameters for blood serum as compared to those of healthy persons. The rheological properties of blood depend ($p < 0.001$) on the dermatic lupus erythematosus patients' gender, age and the duration of the disease. All these factors affect the equilibrium surface tension and E values, while the $\gamma_{0.01}$ and γ_1 values depend only on the duration of the disease. Note that the surface relaxation properties of blood depend neither on the patients' gender nor the age, or the duration of the disease. A dependence exists of the rheological characteristics and the number of recurrences and the duration of a subsequent remissions, which reflect the severity of the pathologic process.

The tensiometric characteristics of blood serum reflect the influence of the intensity of discoid focuses, skin atrophy, erythema, bullous elements and trophic changes of appendages of skin (see Figs. 6 and 7 as examples). The severity of each type of skin pathology affects the values of γ_{100}, γ_∞ and E. It is evident from the dispersion analysis that the values of $\gamma_{0.01}$, γ_1 and τ depend on the presence of discoid focuses and telangiectasises. The severity of the discoid focuses is correlated with γ_{100} and γ_∞, while the bullous elements affect the E values and the diffuse alopecia affects the τ values. The results of the studies indicate that values $\gamma_\infty < 40$ mN/m and $E < 25$ mN/m are indicative of a negative prognosis with respect to the course of telangiectasises for patients with dermatic lupus erythematosus.

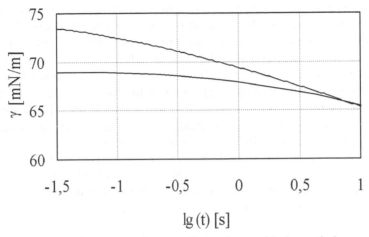

Fig. 6. Blood serum tensiograms for patients with dermatic lupus erythematosus; lower curve - discoid focuses; upper curve - telangiectasises.

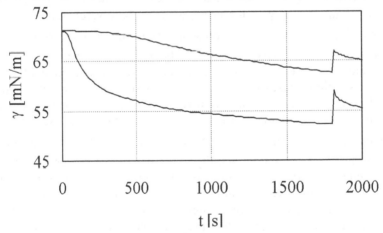

Fig. 7. Blood serum tensiograms for patients with dermatic lupus erythematosus; lower curve - diffuse alopecia; upper curve - bullous elements; after 1800 s the drop of blood serum was subjected to a stress deformation.

The rheological characteristics of blood serum, in particular, the E values, depend on the localisation of the pathologic process. Also of interest are some peculiar features of the changes in the dynamic tensiometric parameters. In particular, skin lesion of the back, thorax, neck and head leads to changes in $\gamma_{0.01}$, γ_1, γ_{100} and γ_∞, respectively. The number of skin lesion regions affects the values of $\gamma_{0.01}$, γ_{100}, γ_∞ and τ. In turn, the number of skin lesion regions and lesion focuses determine the E value. It is shown that the relaxation time value $\tau < 90$ s indicates an unfavourable prognosis for the extent of pathologic processes.

To summarise, the dermatic lupus erythematosus is accompanied by changes in the rheological properties of blood serum. These characteristics depend on the severity of various clinical symptoms of the disease and the extent of the pathologic processes.

7. Scleroderma systematica

Scleroderma systematica leads to increased values of $\gamma_{0.01}$ and γ_1 for blood serum, and of $\gamma_{0.01}$ for urine, while γ_1 is lowered for urine. The values of $d\gamma/d(1/\sqrt{t})$ for blood serum and urine increase or decrease, respectively. These changes in the dynamic interfacial tensiometric parameters are specific to the scleroderma systematica in contrast to other systemic diseases of the connective tissue. Of important practical significance are the correlations between the dynamic surface tension parameters of biological fluids and the activity of scleroderma systematica In particular, values $d\gamma/d(1/\sqrt{t}) > 10$ mN·m^{-1}s$^{1/2}$ for blood serum indicate a continuing activity of the pathologic processes during a pathogenetic therapy, in spite of the fact that biochemical and immunologic tests can exhibit the tendency to normalisation. The dynamic surface tension of blood serum is influenced by the lesion of the heart and digestive system, while the parameters of urine are affected by the articular and renal syndrome. We have shown that the decrease of $\gamma_{0.01}$ below 70 mN/m for blood serum can indicate a cardiac pathology or a pathology of the esophagus or gastrointestinal tract.

One pathogenetic factor is the change of the rheological properties of blood serum related to the immune imbalance, mainly due to the enhanced synthesis of immunoglobulins. A clinical demonstration of this fact is the hyperviscous syndrome. The increase of the plasma viscosity is caused by changes of the molecular composition and configuration of albumins. Correlations exist between the immunoglobulinemia and the interfacial tensiometry and rheometry parameters. Thus, the results of the tensiometric studies of blood serum enable us to draw definite conclusions about disturbances in the albuminous and adipose metabolism.

For patients suffering from scleroderma systematica the pH of breath condensate is decreased, and the γ_∞ and τ values of this fluid become significantly lower. The rheological properties of breath condensate depend on the severity of the pulmonary pathology, which affects the values of γ_∞, E and τ. The physico-chemical state of breath condensate is influenced by the restrictive and obstructive bronchopulmonary disturbances, and also by the hemodynamics conditions within the lesser circulation.

8. Purpura rheumatica

The patients with purpura rheumatica exhibit statistically significantly increased values of γ_1, γ_{100}, decreased plasma viscosity, and decreased values (by almost 30%) of E and τ as compared to healthy persons (cf. Table 6). The regression analysis reveals a significant increase in the relaxation time τ with the duration of the disease. The activity of pathologic processes correlates directly with the value of γ_∞, while there is an inverse correlation with $\gamma_{0.01}$ and γ_1. It was shown that values $\gamma_\infty > 50$ mN/m are indicative for a negative prognosis of the development of the disease.

The visco-elastic and relaxation properties of blood serum are influenced by the articular, muscular and cardiac lesions. The plasma viscosity correlates with the extent of nephropathy, and τ correlates with the extent of skin lesion. The dynamic surface tension of blood serum depends on extent of the affection of skin, liver and central nervous system. The presence and level of cryoglobulinemia affects significantly the thrombocytes aggregation index; in this case it correlates positively. Also, an inverse correlation exists between the cryoglobulin parameters in blood serum and the E value. We have shown that thrombocytes aggregation indices > 45% combined with E < 25 mN/m can suggest the presence of cryoglobulins in the patients' blood.

Table 6. Rheometric and tensiometric parameters of blood serum for patients with purpura rheumatica and healthy persons (M±m)

Parameters	Examined groups		Reliability of differences (p)
	Ill	Healthy	
E, mN/m	24.5±0.3	33.7±1.1	<0.001
τ, s	101±2.1	128±3.5	<0.001
γ_1, mN/m	69.0±0.2	68.3±0.1	0.002
γ_{100}, mN/m	59.9±0.3	58.2±0.25	<0.001

To summarise, for patients with purpura rheumatica the changes in the aggregation state of thrombocytes and erythrocytes occur. The blood plasma viscosity, E and τ values, and dynamic surface tensions are determined by the activity and severity of clinical symptoms of the disease, by the presence of cryoglobulinemia, hyperuricemy and hypercreatininemia, and by the concentration of certain albumins, lipids and lipoproteins in the blood.

9. Chronic rheumatic heart disease

For this pathology, a significant (by almost 50%) decrease of the relaxation time of breath condensate is observed, accompanied by increased E values (Table 7). The decrease in τ is characteristic both for male and female patients.

Table 7. Rheological parameters of breath condensate for patients with chronic rheumatic heart disease and healthy persons (M±m)

Parameters	Examined groups		Reliability of differences (p)
	Ill	Healthy	
E, mN/m	32.8±3.0	24.8±2.2	0.031
τ, s	281±8.2	463±25.5	<0.001

The dispersion analysis has shown that the activity of rheumatic processes affects E. For breath condensate the τ value depends on the rheumatism activity and degree of blood circulation inefficiency. The peripheral vascular resistance correlates with E and τ.

The physico-chemical parameters of breath condensate are affected by the aortal valve lesion, the hypertrophy of the right ventricle and the interventricular septum. There is correlation between the rheological parameters and hemodynamic parameters in the lesser circulation. In particular, the blood velocity in the pulmonary artery inversely correlates with E and τ, and E directly correlates with the final systolic size of the left ventricle. The rheological properties of breath condensate are also related to the systolic function of the left ventricle. In particular, the myocardial contractility correlates with E and τ, and the left ventricular ejection fraction correlates with E. The total physico-chemical state of breath condensate in patients with chronic rheumatic heart disease is related to the blood circulation velocity, pressure gradient in the pulmonary artery, and the size of the right ventricle. The τ value depends on the ratio of the size of pulmonary artery to the size of aorta, and E depends on the parameters of blood circulation rate in pulmonary artery, pressure gradient in the pulmonary artery, pulmonary vascular resistance and contractive ability of the right ventricle.

It was shown that dynamic interfacial tensiometry and rheometry data of breath condensate provide information on the hemodynamics in the lesser circulation: values $E > 45$ mN/m for patients with chronic rheumatic heart disease are indicative of a pulmonary hypertension.

10. Diabetic nephropathy

For diabetes mellitus a characteristic feature is the dynamic surface tension decrease of blood serum and urine, which reflects the accumulation of certain surfactants in these fluids. Certain sexual differences exist in the measured changes caused by diabetes mellitus: while for male patients the values of $d\gamma / d(1/\sqrt{t})$ of blood serum become lower, for female patients the this parameter increases. For diabetes mellitus patients without pronounced symptoms of diabetic nephropathy the values of $d\gamma / d(1/\sqrt{t})$ for blood serum is significantly higher as compared with healthy persons (e.g. in 81% of female cases), while for urine it becomes lower (in 84% of male patients).

The renal lesion caused by diabetes mellitus is accompanied by high values of $d\gamma / d(1/\sqrt{t})$. For patients with diabetic nephropathy the surface tension of urine becomes lower, but in cases with chronic pyelonephritis the equilibrium surface tension and the dynamic surface tension γ_1 increase. The decrease or increase of the γ_1 value for urine could be a reliable differential diagnostic indicator of diabetic nephropathy vs chronic pyelonephritis for diabetes mellitus. The presence of a nephrotic syndrome affects significantly the γ_1 value for urine. Pyelonephritis in a chronic stage also affects the parameters of dynamic interfacial tensiograms of biologic fluids, see Figs. 8 and 9. It could be suggested that the differences in the β_2-microglobulinuria levels determine the parameters of dynamic interfacial tensiograms for male and female patients without diabetic nephropathy.

Significant correlations were found between the albuminuria and $\gamma_{0.01}$ of urine, and between the β_2-microglobulinuria and γ_{100} of urine. In the analysis of the interfacial tensiometric parameters, deviations in the urine parameters are more frequent. In particular, the deviations of $\gamma_{0.01}$ from its normal value were detected in 52% of all cases (in 72% of male cases), and of γ_1 in 76% of all cases (in 91% of male cases). Therefore, the applicability of the dynamic interfacial tensiometry for urine is superior with respect to traditional parametric albuminuria, β_2-microglobulinuria and fibronectinuria, and monitoring of the tensiometric parameters should be recommended for early diagnostics of diabetic nephropathy and for monitoring the curative treatment efficiency.

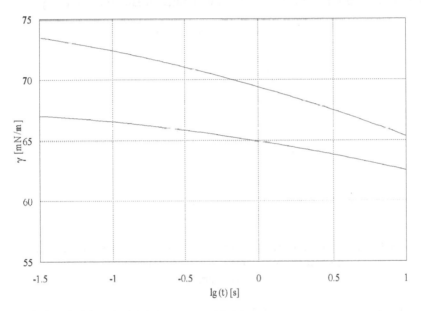

Fig. 8. Blood serum tensiograms for patients suffering from diabetic nephropathy with preserved renal function (upper curve) and with renal insufficiency (lower curve).

For patients with diabetic nephropathy, only the γ_{100} of blood serum and $d\gamma / d(1/\sqrt{t})$ values of urine do not depend on the diabetes mellitus type, i.e. they became significantly lower. Maximum changes in the physico-chemical parameters were observed for patients with renal lesions: the decrease of surface tension of blood serum and urine in the short and medium surface lifetime range. The development of a nephritic syndrome is accompanied by a surface tension decrease for blood serum and urine in the medium surface lifetime range. Therefore, the dynamic surface tension data of biologic fluids for patients with diabetes mellitus provide important information which enable one to make conclusions about the severity of the diabetic nephropathy, to prognosticate the nephrotic syndrome, and to monitor the medical treatment efficiency.

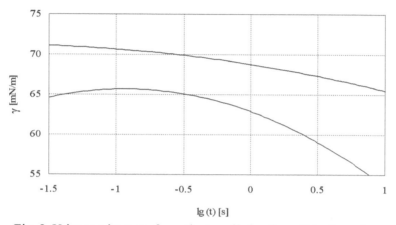

Fig. 9. Urine tensiograms for patients suffering from diabetic nephropathy with preserved renal function (upper curve) and with renal insufficiency (lower curve).

The decrease of the renal function results in changes of the physico-chemical characteristics for blood serum and urine, similar to those caused by the nephrotic syndrome. Therefore, for patients with diabetic nephropathy the γ_1 values of blood serum and urine reflect the severity of the disease: a decrease of this parameter below 67 mN/m and 68 mN/m for blood serum and urine, respectively, indicate a negative prognosis of the course of diabetic nephropathy.

To summarise, the diabetes mellitus is characterised by a surface tension decrease of blood serum and urine in the medium and long surface lifetime range, and these changes are more pronounced for female patients. If no renal lesion occurs, the values of γ_1 and γ_{100} for biologic fluids become lower, while the development of diabetic nephropathy leads to a decrease of γ_{100} and urine γ_1 values for blood serum.

11. Chronic pyelonephritis

For the chronic pyelonephritis decreased equilibrium surface tensions of blood serum were observed for male and female patients. Also, for female patients a decrease in the γ_1 values blood serum and urine is observed. The sexual dimorphism of the disease is illustrated also by the increase of $d\gamma/d(1/\sqrt{t})$ values of urine for male patients while for female patients these values decrease. This parameter correlates directly with the duration of the chronic pyelonephritis. The $\gamma_{0.01}$ value for urine exhibits a certain inverse dependence on the concentrations of albumin, uric acid, oxypurinol and nitrites/nitrates in this biologic fluid. The uricosuria influences also the parameters of dynamic interfacial tensiograms of urine in the medium surface lifetime range.

The deterioration of the renal function is accompanied by a significant increase of $d\gamma/d(1/\sqrt{t})$ of urine. This fact could be used as one of the negative prognostic criteria of the development of this disease. The surface tension of urine is determined by the contents of glycosaminoglycans. For patients with chronic pyelonephritis with preserved renal function the concentration of glycosaminoglycans in the urine becomes increased approximately by a factor of two, while a decrease in the renal function leads to an increase of this factor to a value of three. The increase of the glycosaminoglycans amount, in turn, stimulates the inflammatory process. There is a positive correlation between the level of glycosaminoglycans in urine and the extent of leucocyteuria. The regression analysis shows that the surface tension of urine depends on the glycosaminoglycans level and that this dependence is especially evident for patients with decreased renal function.

It should be noted that pronounced changes in the dynamic interfacial tensiograms of blood serum were detected only for chronic pyelonephritis with renal insufficiency. The changes of the concentration of uric acid, albumins, lipids and electrolytes can significantly contribute to changes in the dynamic surface tension of urine if renal insufficiency exists.

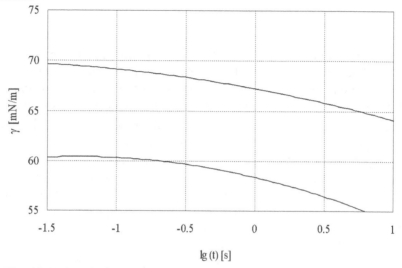

Fig. 10. Urine tensiograms for patients with chronic pyelonephritis. Upper curve, diabetes mellitus; lower curve, diabetes mellitus absent.

The peculiar features for cases of chronic pyelonephritis with respect to the presence or absence of diabetes mellitus are differences in the γ_{100} and $d\gamma/d(1/\sqrt{t})$ values of blood serum, and the surface tensions of urine, as shown in Fig. 10. While for the patients with diabetes mellitus the surface tension increases in the short and medium surface lifetime range, for patients without diabetes mellitus a decrease of this parameter is observed. This fact is of certain diagnostic value, and can be ascribed to the presence of glucosuria.

The diabetic nephropathy combined with chronic pyelonephritis leads to quite unusual changes of the dynamic interfacial tensiograms, typical neither to a chronic pyelonephritis, nor to a isolated diabetic glomerulosclerosis. In future studies, a comparison of surface tension and rheological characteristics should be made for patients with the same renal functions (without the nephrotic syndrome), the same arterial pressure, similar duration of the disease etc. At the moment, it can be maintained that the interfacial tensiometric parameters for cases of diabetic nephropathy are quite informative, and the differential diagnostics based on these parameters is capable for a quite certain prognosis of the development of pathologic processes.

12. Chronic bronchitis and chronic obstructive pulmonary disease

The rheological parameters of breath condensate for healthy persons depend on gender and age. The interfacial tensiometric parameters of breath condensate depend on the presence of lipids, nitrites/nitrates, ammonia, urea and lactic acid. For patients with chronic bronchitis and chronic obstructive pulmonary disease the rheological properties of breath condensate are determined by the character of respiratory moisture excretion. For these two diseases the rate and extent of respiratory moisture excretion affect the parameters E and τ. The patients' gender and age significantly affect the rheological properties of breath condensate. The volume of expired fluid is inversely correlated to the relaxation time τ. A significant influence of patients' age on the rheological state of breath condensate was also found ($p < 0.001$).

The development of chronic obstructive pulmonary disease is accompanied by a significant increase of E. The dispersion analysis data show that the patients' gender and age affect the τ values of breath condensate, while the severity of the disease influences E and τ. Certain physico-chemical parameters of breath condensate correlate with the contents of nitrites/nitrates and lactic acid. The rheological parameters of breath condensate for coalminers suffering from chronic obstructive pulmonary disease are quite similar to those for male patients of other professions. The character of bronchial obstruction affects the values of E and τ, while a bronchial restriction influences only the E value of the respiratory fluid. It should be noted that the forced expiratory volume (which reflects the severity of bronchial obstruction syndrome) correlates strongly with E and τ with $p < 0.001$.

For patients with chronic obstructive pulmonary disease the values of $\gamma_{0.01}$, E and τ of blood serum are significantly higher than the values typical for healthy persons. In these cases the forced expiratory volume correlates significantly with the dynamic and equilibrium surface tension of blood serum.

For patients suffering from chronic bronchitis accompanied by a coronary heart disease, a significant direct correlation of E and τ values of breath condensate with the patients' age, duration of the disease and the extent of the respiratory disorders. For patients with chronic bronchitis, τ of the breath condensate becomes lower, for chronic obstructive pulmonary disease this value becomes higher, and if both pathologies are present, an increase in E is observed. From the analysis of the data it was concluded that for the the breath condensate values $E > 40$ mN/m for chronic bronchitis, and $E > 50$ mN/m for chronic obstructive pulmonary disease accompanied by coronary heart disease, are indicative of a negative prognosis for the development of the disease. The rheological properties of breath condensate for this combined pathology depend on the contents of cholesterol, ammonia, uric and lactic acids, which affect the parameters E and τ of this biologic fluid.

The erythrocyte aggregation index directly correlates with the τ value of blood serum. An essential arterial hypertension for patients with chronic bronchitis leads to an increasing plasma viscosity and decreasing τ values. The dispersion analysis has shown that for this set of patients the E values of blood serum depend on the amount of chlorine and uric acid in blood, the τ values depend on the presence of calcium and phospholipids, and the equilibrium surface tensions depend on the contents of phosphorus, chlorine, calcium, phospholipids, and nitrites/nitrates. It is shown that for a isolated essential arterial hypertension the relaxation time values $\tau < 100$ s indicate a negative prognosis of the disease development, because these values can be related to a deterioration of the nitrogen-excretory renal function.

13. Peptic gastroduodenal ulcer

For patients with peptic gastroduodenal ulcer the rheological parameters of gastric juice are: $E = 17.8 \pm 1.5$ mN/m, $\tau = 240 \pm 13$ s, and a dependence of dynamic and equilibrium surface tensions on patients' gender and age was found. The blood serum parameters for these patients are $E = 42.8 \pm 1.4$ mN/m, $\tau = 166 \pm 9$ s. For peptic gastroduodenal ulcer the E and τ parameters are higher than those for healthy persons. Significant correlations exist between the relaxation times of gastric juice and blood serum. The dispersion analysis has revealed that the rheological characteristics of blood serum for this group depend on the patients' gender, duration of the disease, the existence of the duodenal erosions, gastroesophagal and duodenogastral refluxes, gastritis and duodenitis.

In cases of gastroesophagal reflux a decrease of τ values of gastric juice is observed, while for duodenogastral reflux the dynamic surface tension of this biologic fluid becomes lower. If a duodenitis is present, an increase of both parameters occurs. For patients with duodenogastral reflux a decrease of the E value of gastric juice was observed, while duodenitis leads to a decrease of γ_∞. If the disease is accompanied by cholecystitis and biliary dyskinesia, the values E and τ become lower.

For patients with peptic ulcer the rheological parameters of gastric juice depend significantly on the hyperemia of mucous tunic of esophagus and stomach, on the friability and patchiness of this tunic, and on the presence of duodenal or gastric erosions. The dynamic surface tension of gastric juice inversely correlates with the extent of fibrogastroscopic features. The E and τ values of gastric juice inversely correlate with the extent of morphologic changes of mucous tunic of stomach and duodenum.

14. Chronic periodontitis

For chronic periodontitis a significant increase of the dynamic surface tensions, E and τ values of oral fluid as compared with those for healthy persons was observed. At the same time, the equilibrium surface tension became lower, as shown in Table 8. The increase of the values of $\gamma_{0.01}$, γ_1 and τ of oral fluid was found in 78%, 79% and 59% of patients, respectively, while a decrease in the equilibrium surface tension γ_∞ was detected in 56% of the patients.

Table 8. Tensiometric and rheological parameters of oral fluid for patients with chronic periodontitis and healthy persons (M±m)

Parameters	Examined groups		Reliability of differences (p)
	Ill	Healthy	
$\gamma_{0.01}$, mN/m	55.9±1.2	66.4±0.5	<0.001
γ_1, mN/m	54.2±1.0	64.1±0.5	<0.001
γ_{100}, mN/m	48.9±0.8	53.5±0.8	0.005
γ_∞, mN/m	45.7±0.7	41.4±0.3	<0.001
E, mN/m	22.1±1.9	26.0±0.7	0.019
τ, s	91.6±9.8	148±3.9	<0.001

The multifactor dispersion analysis reveals an effect of the kind of accompanying pathology on the rheological properties of oral fluid. The patients' age and the duration of the disease affect the tensiometric and rheological parameters of oral fluid, while gender related effects are absent. The presence of an accompanying somatic pathology influences the E value. The severity of chronic periodontitis affects the rheological properties of oral fluid. It was shown that values $\gamma_{100} > 60$ mN/m and $\tau > 160$ s, and also $\gamma_\infty < 40$ mN/m indicate a grave course of the disease, which is of large practical importance. Values of $d\gamma / d(1/\sqrt{t}) < 130$ mN·m^{-1}·s$^{1/2}$ of oral fluid are helpful for early diagnostics of the disease.

15. Microcholelithiasis

The microcholelithiasis is the initial stage of the cholelithiasis. The study of the interfacial dilational rheometry of bile has shown that the visco-elasticity modulus E and the relaxation time τ for patients with this pathology are 12.5±2.8 mN/m and 274.3±33.5 s, respectively, while for healthy persons and for the microcholelithiasis patients after medical treatment the E values were zero within the experimental error (±0.2 mN/m). The fact that E is close to zero can be attributed to an extremely rapid exchange between the solution bulk and surface layer, which is usual for adsorption layers of low molecular surfactants, if their concentration in the bulk is high. On the contrary, the E and τ values for microcholelithiasis patients before medical treatment indicate much lower surfactants concentration in the bile. These results agree with the biochemical parameters of the bile of patients suffering from microcholelithiasis: the cholesterol is increased while the amount of bile acids is lowered. Therefore, the visco-elasticity modulus value could be used in the microcholelithiasis diagnostics and for monitoring the medical treatment.

16. Obstructive jaundice

The group screened was composed of patients with choledocholithiasis-based benign obstructive jaundice. The main method used for re-canalisation of the biliary tract was the endoscopic papillosphincterotomy with mechanical lithoextraction. Figure 11 illustrates the dependencies of the visco-elasticity modulus |E| and phase angle φ on the oscillation frequency f for two blood serum samples, taken from the same patient (female, age 69) before and after successful re-canalisation of the biliary tract.

The re-canalisation results in an essential decrease of the phase angle and increase of the visco-elasticity modulus. These changes in the rheological characteristics can be attributed to the fact that the obstructive jaundice leads to the increase of the amount of bilirubin, bile acids and other surfactants in the blood serum, and also to changed properties of the main surface-active component, the blood serum albumin. Figure 12 illustrates the same results but plotted as frequency dependencies of the real (elasticity) and imaginary components of the complex visco-elasticity modulus. The two constituents of the complex visco-elasticity modulus in the frequency range studied depend linearly on the logarithm of the oscillation frequency ($\omega = 2\pi f$):

$$E_r = \alpha_1 + \beta_1 \, lg\,\omega, \quad E_i = \alpha_2 + \beta_2 \, lg\,\omega \qquad (10)$$

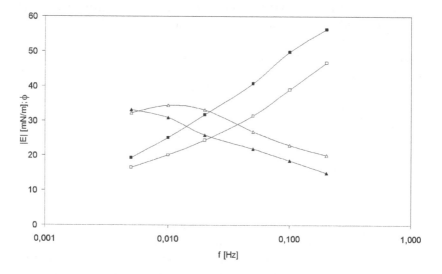

Fig. 11. Dependencies of the visco-elasticity modulus ($\square$,$\blacksquare$) and phase angle ($\triangle$,$\blacktriangle$) on the oscillation frequency for blood serum samples: before the operation ($\square$,$\triangle$) and after the re-canalisation of biliary tract ($\blacksquare$,$\blacktriangle$).

The constant terms α_i in the linear equations (10) correspond to $\omega = 1$ rad/s and hence become approximately equal to the elasticity and viscosity at the maximum studied frequency of 0.2 Hz. This data presentation is convenient to be used for a statistical processing of experimental studies on the surface rheology of blood serum, because all characteristics of the investigated samples can be described by the given four parameters only.

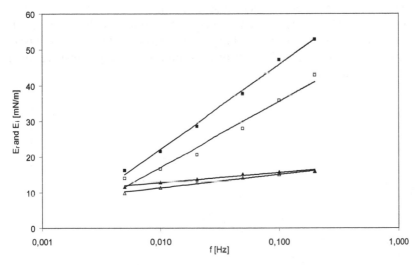

Fig. 12. Dependencies of the real (□,■) and imaginary part of the modulus (△,▲) on the oscillation frequency for the same blood serum samples as in Fig. 11: before the operation (□,△) and after re-canalisation of the biliary tract (■,▲). Lines were calculated from linear regression equations.

Another example of a successful re-canalisation of the biliary tract (female patient, age 22) is illustrated in Fig. 13, where the dependencies of elasticity and reduced viscosity are plotted *vs* angular frequency ω. The regression coefficients in equations (10) are:

before the surgery:

$\alpha_1 = 40.6$ mN/m, $\beta_1 = 17.1$ mN/m, $\alpha_2 = 16.5$ mN/m, $\beta_2 = 3.6$ mN/m;

after re-canalisation of the biliary tract:

$\alpha_1 = 50.0$ mN/m, $\beta_1 = 16.9$ mN/m, $\alpha_2 = 17.5$ mN/m, $\beta_2 = 1.97$ mN/m.

Note that if the re-canalisation of the biliary tract is unsuccessful, the elasticity and viscosity after surgery can become even lower than those observed before the operation.

Thus, the experiments with serum samples taken from patients suffering from choledocholithiasis-based benign obstructive jaundice before and after re-canalisation of the biliary tract demonstrate that the visco-elasticity of the respective adsorption layers is very sensitive of a successful surgery.

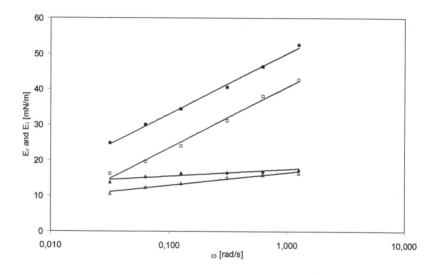

Fig. 13. Dependencies of the real ($\square$,$\blacksquare$) and imaginary part of the modulus ($\triangle$,$\blacktriangle$) on the angular oscillation frequency for a second blood serum samples: before the operation ($\square$,$\triangle$) and after the successful re-canalisation of biliary tract ($\blacksquare$,$\blacktriangle$). Lines were calculated from linear regression equations (10).

E. CONCLUSIONS

With regard to the above analysis of correlations between surface tensiometric and rheological parameters and the contents of surfactants it is now possible to indicate some surfactants which affect the surface tension of different biological liquids. In turn, dynamic surface tensiometry and rheometry of biological liquids for different diseases are capable of providing rapid and rather accurate reflection of the total composition of surfactants, including pathological proteins and other compounds formed and accumulated during the development of the disease. For this reason the study of dynamic surface tension and dilational visco-elasticity of biological liquids are of significant practical interest, due to its capability of providing a differential diagnosis and monitoring of the efficiency of a therapy.

In this chapter described extensive studies of dynamic surface tensions and rheological parameters of various human liquids, such as blood serum, urine, breathing air condensates, cerebrospinal liquor and others. The results for healthy persons are compared with data obtained for people suffering from various diseases. Strong correlations are statistically proofed for nephrological, rheumatological, pulmonary and other diseases. Explanations are given for changes in the dynamic surface tension and rheological parameters caused by differences in the course of the inherent pathological processes. Examples are also given for the evolution of measured data of biological liquids under the influence of medical treatments. It becomes evident from these results that the new type of measurements, namely a combination of dynamic surface tensions and dilational rheology parameters, are very useful for the medical practice.

F. REFERENCES

1. M. Polányi, Biochem. Zeitsch., 34 (1911) 205.
2. J.L.R. Morgan and H.E. Woodward, J. Am. Chem, Soc., 35 (1913) 1249.
3. O. Künzel, Ergeb. Inneren Med. Kinderheil, 60(1941)565

4. E. Hrncir and J. Rosina, Physiol. Res., 46(1997)319
5. H.L. Brydon, R. Hayward, W. Harkness and R. Bayston, British J. Neurosurgery, 9(1995)645
6. M. Efentakis and J.B. Dressman, Euro. J. Drug Metabol. and Pharmacokin., 23(1998)97
7. E.A. Joura, C. Kainz, E.M. Joura, R. Bohm, W. Gruber and G. Gitsch, Zeitschrift für Geburtshilfe und Neonatologie, 199(1995)78
8. J.T. Fell and H.A.H. Mohammad, Intern. J. Pharmac., 125(1995)327
9. E. Adamczyk, T. Arnebrant and P.O. Glantz, Acta Odontol. Scand., 55(1997)384
10. D. Boda, E. Eck and K. Boda, J. Perinat. Med., 25(1997)146
11. E. Manalo, T.A. Merritt, A. Kheiter, J. Amirkhanian and C. Cochrane, Pediatr. Res., 39(1996)947
12. Pulmonary Surfactant: From Molecular Biology to Clinical Practice, Eds. B. Robertson, L.M.G. Van Golde and J.J. Batenburg, Elsevier, Amsterdam, 1992.
13. U. Pison, R. Herold and S. Schürch, Colloid Surfaces A, 114(1996)165
14. V.N. Kazakov, O.V. Sinyachenko, V.B. Fainerman, U. Pison and R. Miller, *Dynamic Surface Tension of Biological Liquids in Medicine*, Elsevier, Amsterdam, 2000.
15. V.N. Kazakov, A.F. Vozianov, O.V. Sinyachenko, D.V. Trukhin, V.I. Kovalchuk and, U. Pison, Adv. Colloid Interface Sci. 86(2000) 1-38
16. R. Wüstneck, J. Perez-Gil, N. Wüstneck, A. Cruz, V.B. Fainerman and U. Pison, Adv. Colloid Interface Sci., 117(2005)33-58.
17. A. Krishnan, A. Wilson, J. Sturgeon, C.A. Siedlecki and E.A. Vogler, Biomaterials, 26 (2005) 3445.
18. V.N. Kazakov, O.V. Sinyachenko, D.V. Trukhin and U. Pison, Colloids Surfaces A, 143(1998)441
19. E. Dickinson, J.A. Hunt and D.G. Dalgleish, Food Hydrocolloids, 4 (1991) 403
20. E. Dickinson, in C. Gallegos (Ed.) Interactions in Protein-Stabilized Emulsions. Progress and Trends in Rheology IV; Proc of the Fourth European Rheology Conference, Sevilla, (1994) 227

21. D.C. Clark, P.J. Wilde, D.R. Wilson and R. Wüstneck, Food Hydrocolloids, 6 (1992) 173.
22. Food Emulsions and Foams: Interfaces, Interactions and Stability, E. Dickinson and J.M. Rodríguez Patino (Eds.), Special Publication No. 227, Royal Society of Chemistry, 1999.
23. Food Colloids 2000 – Fundamentals of Formulation, E. Dickinson and R. Miller (Eds.), Special Publication No. 258, Royal Society of Chemistry 2001.
24. F. Ariola, A. Krishnan and E.A. Vogler, Biomaterials, 27 (2006) 3404.
25. R. Miller and V.B. Fainerman, in "Emulsions: Structure, Stability and Interactions", D.N. Petsev (Ed.), Interface Science and Technology Series, Vol. 4, Elsevier, 2004, p. 61-90.
26. V.N. Kazakov, V.B. Fainerman, P.G. Kondratenko, A.F. Elin, O.V. Sinyachenko and R. Miller, Colloids Surfaces B, 62 (2008) 77–82.
27. V.B. Fainerman, V.N. Kazakov, S.V. Lylyk, A.V. Makievski and R. Miller, Colloids Surfaces A, 250 (2004) 97-102
28. V.B. Fainerman and R. Miller, Adv. Colloid Interface Sci., 108-109 (2004) 287-301.
29. J.K. Ferri, R. Miller and A.V. Makievski, Colloids Surfaces A, 261 (2005) 39-48
30. E.H. Lucassen-Reynders, J. Colloid Interface Sci., 42 (1973) 573.
31. P.R. Garrett, P. Joos, J. Chem. Soc. Faraday Trans. 1, 72 (1976) 2161.
32. J. Lucassen, M. van den Tempel, Chem. Eng. Sci. 27 (1972) 1283.
33. J. Lucassen, M. van den Tempel, J. Colloid Interface Sci. 41 (1972) 491.
34. Q. Jiang, J.E. Valentini, Y.C. Chiew, J. Colloid Interface Sci. 174 (1995) 268.
35. P. Joos, *Dynamic Surface Phenomena,* VSP, Dordrecht, The Netherlands, 1999.
36. B.A. Noskov, G. Loglio, Colloids Surfaces A, 141 (1998) 167.
37. I.B. Ivanov, K.D. Danov, K.P. Ananthapadmanabhan, A. Lips, Adv. Colloid Interface Sci. 114–115 (2005) 61.

38. E.V. Aksenenko, V.I. Kovalchuk, V.B. Fainerman and R. Miller, Adv. Colloid Interface Sci., 122 (2006) 57-66

39. V.B. Fainerman, R. Miller, V.I. Kovalchuk, J. Phys. Chem. B., 107 (2003) 6119.

40. V.B. Fainerman, V.I. Kovalchuk, E.V. Aksenenko, M. Michel, M.E. Leser and R. Miller, J. Phys. Chem. B., 108 (2004) 13700.

41. V.B. Fainerman, R. Miller and V.I. Kovalchuk, Langmuir, 18 (2002) 7748.

42. V.B. Fainerman, E.H. Lucassen-Reynders and R. Miller, Adv. Colloid Interface Sci., 106 (2003) 237.

43. E.H. Lucassen-Reynders, V.B. Fainerman, R. Miller, J. Phys. Chem., 108 (2004) 9173.

44. V.B. Fainerman, S.A. Zholob, M.E. Leser, M. Michel, R. Miller, J. Colloid Interface Sci., 274 (2004) 496.

45. J. Benjamins, Static and Dynamic Properties of Proteins Adsorbed at Liquid Interfaces, Thesis, Wageningen University, 2000.

46. J. Benjamins, J. Lyklema and E.H. Lucassen-Reynders, Langmuir, 22 (2006) 6181.

47. V.B. Fainerman, S.A. Zholob, J.T. Petkov and R. Miller, Colloids Surfaces A, 323 (2008) 56.

48. R. Miller, V.S. Alahverdjieva and V.B. Fainerman, Soft Matter, 4 (2008) 1141.

49. V.S. Alahverdjieva, D.O. Grigoriev, V.B. Fainerman, E.V. Aksenenko, R. Miller and H. Möhwald, J. Phys. Chem., 112 (2008) 2136.

INFLUENCE OF SURFACE RHEOLOGY ON PARTICLE-BUBBLE INTERACTION IN FLOTATION

S.S. Dukhin[1], V.I. Kovalchuk[2], E.V. Aksenenko[3], L. Liggieri[4], G. Loglio[5] and R. Miller[6]

[1] New Jersey Institute of Technology, Newark, USA
[2] Institute of Bio-Colloid Chemistry, 03142 Kiev, Ukraine
[3] Institute of Colloid Chemistry and Chemistry of Water, 03142 Kiev, Ukraine
[4] CNR - Istituto per l'Energetica e le Interfasi, 16149 Genova, Italy
[5] Department of Organic Chemistry, University of Florence, Sesto, Italy
[6] MPI für Kolloid- und Grenzflächenforschung, D-14424 Potsdam, Germany

Contents

A. INFLUENCE OF SURFACE RHEOLOGY ON HYDRODYNAMICS AND DYNAMIC ADSORPTION LAYER OF RISING BUBBLES

1. Qualitative approach

The motion of bubbles and drops in a liquid is influenced by the kinetics of adsorption and desorption of surfactant molecules at the liquid surface. After a certain time the motion rate levels off and a steady state is reached in which the hydrodynamic field around the bubble, and in particular the motion of its surface, controls the steady state process. In contrast to the adsorption or desorption caused by a respective deviation from equilibrium between the adsorption layer and the subphase in a certain moment, the total amount of adsorbed substance does not change after the stationary state of buoyant bubbles has been established. Thus, rather than a time dependence of adsorption, a qualitatively different characteristic parameter determines the specific nature of adsorption kinetics caused by the hydrodynamic field of a buoyant bubble. Steady-state motion of bubbles induces adsorption-desorption exchange with the subsurface, with the amount of substance adsorbed on one part of the bubble surface being equal to the amount desorbed from the other part. Consequently, the surface concentration varies along the surface of a buoyant bubble taking a maximum value at the rear stagnation point and a minimum at the leading pole [1].

The difference in surface concentration between the poles of a bubble is due to its movement. The difference increases with faster movements and disappears in the case of a resting bubble. Therefore, the state of the adsorption layer on a moving bubble surface is qualitatively different from that on a resting one. Such adsorption layers are called dynamic adsorption layers [2]. Thus, a dynamic adsorption layer (DAL) is an analogy of the time dependence (kinetics) of adsorption or desorption initiated by a deviation from equilibrium of a surface layer.

The problem of adsorption kinetics in the case of a buoyant bubble can be transformed into the problem of a dynamic adsorption layer. As a result of such a transformation, the adsorption kinetics is coupled with other processes caused by the angular dependence of adsorption. Surface tension gradients generating the Marangoni effect are connected with a non-uniform adsorption and a feedback arises in this case. The hydrodynamic field around a bubble initiates the adsorption-desorption exchange and leads to a dynamic adsorption layer which retards the motion of the surface and acts on the hydrodynamic field and the adsorption-desorption exchange.

The surface state of a floating bubble depends on its size. Surfaces of reasonably large bubbles are mobile. As a result adsorbed surfactants are pulled down to the rear of the bubble, i.e. even under steady-state conditions the value of adsorption on a mobile bubble surface is different from that on an immobile one, Γ_0 (at the same surfactant bulk concentration).

According to Levich [3] the leading part of the mobile surface of a floating bubble is stretched, the lowest part is compressed. Newly created sections of the surface are being filled with adsorbed substance, while in the compressed part of the surface the substance desorbs. The surface concentration on the leading surface of the buoyant bubble is lower than Γ_0 which induces a continuous flux of surfactant (or adsorbing inorganic ions) from the bulk to the stretched surface. The surface concentration on the rear part of the buoyant bubble is higher than Γ_0 which initiates desorption of surfactant molecules. Thus, $\Gamma(\theta)$ increases in a direction opposite to the bubble motion, i.e. from the leading pole ($\theta = 0$) to the rear cap ($\theta = \pi$), where the angle θ is counted from the leading pole as shown in Fig. 1.

The transport of surfactant to the leading bubble surface and withdrawal of desorbing surfactant into the bulk from the lower half is governed by diffusion and leads to the formation of a so-called diffusion boundary layer adjacent to the surface. Its thickness δ_D is much smaller than the bubble radius a_b,

$$\delta_D = a_b / \sqrt{Pe},\tag{1}$$

where Pe is the Peclet number defined by

$$Pe = a_b \cdot v / D,\tag{2}$$

while Re is the Reynolds number defined by

$$Re = 2a_b \cdot v / \nu, \tag{3}$$

Here v is the bubble velocity of buoyancy; $\nu = \eta/\rho$; ρ and η are density and viscosity of the liquid, respectively, D is the diffusion coefficient of surfactant molecules.

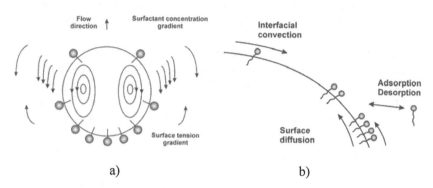

a)   b)

Fig. 1 Schematic of processes at rising bubble surfaces: a - surface concentration gradient created by interfacial convection; b - transport process at the bubble surface; according to [3].

The concentration in the diffusion layer on the leading surface of the bubble is less than in the bulk and it increases from the leading to the rear pole like in a dynamic adsorption layer. At a given value θ, the local equilibrium is preserved between the surface concentration $\Gamma(\theta)$ and the bulk concentration adjacent to the surface $c(a_b, \theta)$,

$$\Gamma(\theta)/c(a_b, \theta) = \Gamma_0/c_0 = K_H, \tag{4}$$

where Γ_0 is the equilibrium concentration on an immobile bubble surface at a surfactant bulk concentration of c_0, K_H is the so-called Henry constant and a measure of the surface activity of the surfactant.

The situation is rather fine balanced since the motion of the surface has an effect on the formation of the dynamic adsorption layer, and vice versa. The surface concentration increases in the direction of the liquid motion while the surface tension decreases, which results in the appearance of forces directed against the flow and retards the surface motion. Thus, the dynamic layer theory should be based on the common solution of the diffusion equation, which takes into account the effect of surface motion on adsorption-desorption processes and of hydrodynamic equations combined with the effect of adsorption layers on the liquid interfacial motion [3].

If the surface of a drop or bubble is immobile for any reason and the coordinate system is defined as moving together with the bubble, the floating velocity is the same as that of a solid sphere. In particular, at small Reynolds numbers, the bubble movement can be described by Stokes' equation,

$$v_{St} = \frac{2a_b^2 \rho}{9v} g,$$ (5)

where g is the acceleration due to gravity, and ρ is liquid density. If the drop surface is mobile, a velocity distribution also arises inside the drop. The velocity distribution over the drop surface can be found by a collective solution of the respective Navier-Stokes equation both inside the drop and in the surrounding liquid. The continuity conditions must be fulfilled both for the velocity and the tensor of viscous stresses when passing the phase boundary. Under these boundary conditions, the Stokes equation (linearised Navier-Stokes equation at very small Reynolds numbers Re « 1) for drops floating up in a liquid was solved independently by Rybczynski [4] and Hadamard [5]. Of course, this solution applies also to the case of a floating bubble when its viscosity is set equal to zero (neglecting the effects of the order of the ratio of the gas and liquid viscosities). According to the Hadamard-Rybczynski approximation the velocity of floating bubbles can be expressed by the equation

$$v = \frac{1}{3} \frac{a_b^2 \rho}{\eta} g,$$ (6)

where the small density of the gas is omitted.

Data on the velocity of rising bubbles in aqueous surfactant solutions (sodium dodecyl sulphate SDS) published by Okazaki [6] are presented in Fig. 2. It illustrates the influence of surface rheology on the process of bubble rising. Small bubbles rise as solid spheres, even in thoroughly cleaned liquids. It is typical that even at low concentrations such as μM the velocity of bubbles is essentially decelerated, when $a_b < 0.03$ cm (Re < 36). Due to the Marangoni effect, the hydrodynamics of small bubbles is almost identical to that of solid spheres. Surface rheology effects the hydrodynamics for larger rising bubbles as well. The difference is that a higher surfactant concentration is necessary for their retardation.

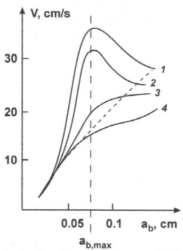

Fig. 2 Velocity of bubbles floating up in aqueous SDS solution at different concentrations: $1 - 0$; $2 - 10^{-6}$; $3 - 1.2 \cdot 10^{-5}$; $4 - 10^{-3}$; dotted curve - solid bubble in distilled water; according to [6]

The DAL study is more realistic for larger bubbles and large Re because trace concentrations of surface active impurities cannot retard the bubble surface movement completely. Therefore, different modes of DAL can exist at large Re: the retardation of the entire bubble surface, or the retardation of only a small part of the bubble surface in the vicinity of rear pole, while almost the entire bubble surface remains mobile. The last situation is called rear stagnant cap mode. The same modes can also arise if Re is not large, but super-clean water is used with the addition of a very low amount of surfactant. At the same time, significant simplifications of

DAL theory are possible if Re $\ll 1$. These simplifications allowed the development of this theory initially only for small Re, and only later it turned out that the hydrodynamic modes, which were quantified for small Re, are relevant to large Re as well.

Three ranges of Reynolds numbers are discussed in this respect: small Reynolds numbers (Re < 1), intermediate Re (1 < Re < 200) and large Re (Re > 200). At large Re bubble the buoyancy is complicated by the bubble shape deformation and bubble path instabilities. Due to bubble surface deformation the hydrodynamic resistance increases and at rather high Reynolds numbers the terminal velocity decreases with increasing bubble dimension (cf. Fig. 2).

2. The rear stagnant region of a buoyant bubble

At low surfactant concentrations the surface of a bubble is not uniformly stagnant. Usually the main part of the bubble surface can be weakly stagnant, and only a narrow truly stagnant region exists in the vicinity of the rear pole. The existence of a stagnant cap was confirmed experimentally in a number of studies extensively referred to by Dukhin *et al.* [7, 8].

It can be assumed that the surface of the bubble is completely free at $\theta < \pi - \psi$ and completely stagnant at $\theta > \pi - \psi$. In addition, it is assumed that in the former region the adsorption is much lower and in the latter much higher than the equilibrium value Γ_0. This allows us to assume that surfactants adsorb at the free surface, and desorb from the surface to the solution bulk with $c_0 \approx 0$ in the stagnant region. The size of the strongly stagnant region is determined by the balance of the total adsorption flux to the bubble surface and the total desorption flux from the bubble surface as shown in Fig. 3.

The non-uniform surface concentration Γ' establishes a non-uniform surface tension γ' which resists tangential shearing at the interface,

$$\tau_{r\theta}\big|_{r=a_b} = -\frac{1}{a}\frac{\partial \gamma}{\partial \theta} = -\frac{1}{a}\frac{\partial \gamma}{\partial \Gamma}\frac{\partial \Gamma}{\partial \theta}, \tag{7}$$

where $\tau_{r\theta}$ is the shear stress.

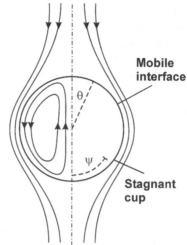

Fig. 3 Stagnant cap model, according to [7, 8].

The balance of Marangoni and viscous stresses, given by Eq. (7) as the basis for the determination of the surfactant distribution along the surface, was used later in the stagnant cap theory. This stress balance is often characterised qualitatively by a dimensionless number, the Marangoni number (cf. [9]),

$$Ma = RT\Gamma_0/\eta v, \tag{8}$$

which characterises the ratio of the surface pressure that surfactant molecules exert against the viscous forces tending to compress the surfactant layer.

As a purely hydrodynamic problem, the velocity field due to a stagnant cap at the rear of a moving drop was solved exactly in terms of an infinite series of Gegenbauer polynomials with constants depending on the cap angle ψ. From this series, an analytical solution $F(\psi)$ for the drag exerted on the drop can be obtained. This yields the terminal velocity once the external force on the drop is resolved,

$$F(\psi) = 4\pi\eta V a_b \left\{ \frac{\eta}{4\pi(\eta + \eta')} \left[2\psi + \sin\psi - \sin 2\psi - \frac{1}{3}\sin 3\psi \right] + \frac{2\eta + 3\eta'}{2\eta + 2\eta'} \right\}. \tag{9}$$

For the limiting case of $\psi = 0$ (no surfactant), equation (9) is transformed into the Hadamard-Rybczynski equation. For $\psi = \pi$ (completely stagnant interface), a solid sphere behaviour results.

If the drop viscosity becomes infinitely large ($\eta' \to \infty$), again the drag of a solid sphere is obtained. The drag force for the special case of a bubble is easily obtained from (9) by setting $\eta' \to 0$. The cap angle is determined by the adsorbed amount of surfactant necessary to cause a surface pressure that balances the compression due to the viscous shear on a cap of angle ψ.

In 1983 Sadhal and Johnson [10] derived a dimensionless equation for the difference in shear stresses exerted on the interface in the cap region by the surrounding fluid phases

$$\left(\tau^{(2)}_{z\theta(s)} - k\tau^{(1)}_{z\theta(s)}\right) = h(\theta, \psi)/\lambda, \tag{10}$$

where the tangential stress is scaled by $\eta V / a_b$, k is the ratio of the droplet to the continuous phase viscosity, and $h(\theta, \psi)$ is a rather complicated function.

The balance of Marangoni and viscous stresses (10), reformulated in terms of Γ, was integrated to obtain the surfactant distribution Γ as a function of ψ. The surfactant distribution can be integrated over the cap region to obtain the total amount M on the surface. This variable M is also computed independently via the surfactant conservation equations by equating the two expressions for ψ. For a given ψ the drag coefficient and terminal velocity can be calculated.

3. Dynamic adsorption layer at slow desorption kinetics

When we assume a slow desorption kinetics, discussed in literature in different terms such as barrier limited or kinetic controlled [11, 12, 13], the surfactant exchange between the surface and the bulk is controlled by the surfactant transport from the sublayer to the surface and in opposite direction. In this case it is not necessary to solve the convective diffusion equation because the transport across the diffusion layer is faster and the concentration change is small across the diffusion layer. This allows to estimate the surfactant concentration $c(a_b, \theta)$ within the sublayer as approximately equal to the bulk concentration c_0. The total adsorption rate at a unit surface is

$$j_n = -P(\Gamma) + Q[c(a_b,\theta)]$$
$$= -k_{des}\Gamma(\theta) + k_{ad}c(a_b,\theta)(1-\Gamma(\theta)/\Gamma_\infty) \quad (11)$$
$$= -k_{des}\Gamma(\theta) + k_{ad}c_o(1-\Gamma(\theta)/\Gamma_\infty)$$

where Γ_∞ is maximum surface concentration.

A theory of dynamic adsorption layer of rising bubble at weak surface retardation and at surfactant transport controlled by sorption kinetics was developed by He *et al.* [9]. The relation for the total amount adsorbed can be obtained by multiplying Eq. (11) by θ and integrating from 0 to π. Under steady-state conditions the net diffusive flux is equal to zero,

$$\int_0^\pi \sin\theta \frac{\partial c}{\partial r}\bigg|_{r=0} d\theta = 0, \quad (12)$$

Multiplying both sides of Eq. (12) by $1/\Gamma_\infty$ and expressing the flux density by Eq. (11), one obtains

$$-\int_0^\pi \frac{\Gamma(\theta)}{\Gamma_\infty} \sin\theta d\theta + \int_0^\pi C\left[1-\frac{\Gamma(\theta)}{\Gamma_\infty}\right] \sin\theta d\theta = 0, \quad (13)$$

with

$$C = \frac{k_{ad}}{k_{des}} \frac{c_o}{\Gamma_\infty} \quad (14)$$

being the so-called non-dimensional bulk concentration,

$$-M + 4\pi C - CM = 0, \quad \text{or} \quad M = M_0 = \frac{4\pi C}{1+C}, \quad (15)$$

and $M = 2\pi \int_0^\pi \frac{\Gamma(\theta)}{\Gamma_\infty} \sin\theta d\theta$ is the dimensionless total adsorbed amount.

With increasing surfactant density, the finite size of the adsorbed molecules gives rise to strong repulsions between surfactant molecules and generates surface pressures that vary non-linearly with the surface concentration. The concentration of surfactant in the stagnant cap depends principally on the amount adsorbed and the degree of compression exerted by the viscous forces. If these forces are large enough, they can compress the surfactant to a density high enough so that the assumption typical for gaseous states does not apply. Since Sadhal and Johnson [10] used the constitution equation for gaseous states to determine the surface pressure they underestimated the cap angle and consequently the drag coefficient. He *et al.* [9] obtained a more realistic value for the cap angle by allowing for non-linear interactions. They used the Szyszkowsky-Langmuir equation of state

$$\gamma_0 - \gamma = -RT\Gamma_\infty \, ln(1 - \Gamma/\Gamma_\infty).$$ (16)

to formulate an expression for a non-linear surface pressure.

From Eq. (16) and the Marangoni stress balance Eq. (10), a differential equation for the surfactant distribution can be obtained

$$\frac{Ma}{(1-\Gamma)}\frac{\partial \Gamma}{\partial \theta} = h(\theta,\psi)/\lambda.$$ (17)

where Ma is the Marangoni number defined as $Ma = RT\Gamma_\infty/(\eta v)$, h is the tangential component of the gradient of liquid velocity distribution [10], λ is the drag coefficient, Γ is the surface concentration normalized by Γ_∞. The surfactant distribution is obtained by integrating Eq. (17) from an angular position in the cap to ψ. From the second integration the total (non-dimensional) amount on the surface, $M(\psi)$ is obtained. Thus,

$$\Gamma(\theta) = 1 - exp\left[\frac{1}{Ma\lambda}\int_\theta^\psi h(\theta',\psi)d\theta'\right],$$ (18)

$$M(\psi)/2\pi = \int_0^\psi \sin\theta\left\{1 - exp\left[\frac{1}{Ma\lambda}\int_\theta^\psi h(\theta',\psi)d\theta'\right]\right\}d\theta.$$ (19)

Eq. (18) is obtained under the condition $\Gamma(\psi) = 0$. Combination of Eqs. (15) and (19) results in an implicit equation for the cap angle ψ:

$$\frac{1}{(1+k)} = \frac{1}{2}\int_0^\psi \sin\theta\left\{1-\exp\left[\frac{1}{Ma\lambda}\int_0^\psi h(\theta',\psi)d\theta'\right]\right\}d\theta. \qquad (20)$$

As the inner integral in equation (20) can be evaluated analytically, solutions for ψ as a function of k and Ma may be obtained by fixing Ma and ψ and numerically evaluating the outer integral in order to solve for $k \equiv 1/C$ (note, that the coefficients k in Eqs. (20) and (10) are different).

4. Rear stagnant cap (RSC) at higher Reynolds numbers

The problem is an analogue of the investigations by Sadhal and Johnson [10] for large Re. For a long period of time it was the main unsolved problem of bubble hydrodynamics at large Re. Fdhila and Duineveld [14] elaborated a numerical algorithm for this problem for $50 < Re < 200$. They used the vorticity-stream function in spherical polar co-ordinates (r, θ) and a finite difference technique for solving the two coupled non-linear difference equations, avoiding linearisation and decoupling assumptions. The ω values for the vorticity were obtained by using a polynomial fitting of the stream function in the vicinity of the interface. Their elaborated numerical algorithm was verified by comparison with known theoretical results. The analytical solutions for the Stokes and potential flow were obtained for very low Re values and $Re = 200$, respectively. Also, comparison with some models for the drag coefficient and the separation angle gave satisfactory results

Elaboration of a numerical algorithm for the transfer of momentum creates the prerequisite for the extension of the stagnant zone model to large Re. Fdhila and Duineveld [14] united their numerical algorithm with the model of He and Maldarelli for surfactant kinetic controlled adsorption (section 1.3). The specificity of the hydrodynamic flow manifests itself through the function $h(\theta,\psi)$ used in Eqs. (17) to (20) for the case of creeping flow in the framework of the theory of Sadhal and Johnson [10]. Based on the numerical algorithm developed in [14], Eq. (20) can be extended to Re in the range $50 < Re < 200$.

$$\Gamma(\theta) = 1 - \exp\left(-\frac{1}{Ma}\int_{\pi-\phi}^{\theta}\omega d\theta\right).$$ (21)

Through (21) this equation depends on the interfacial vorticity and the outer flow. After simplifications are introduced by the assumption of a certain cap-angle, the flow problem depends only on three non-dimensional parameters: the Reynolds number Re, the Marangoni number Ma and the dimensionless bulk concentration C.

The numerical results obtained in [14] allow to establish the main features of the DAL structure and surface retardation for $50 < Re < 200$. The tangential velocity of clean bubbles is positive and reaches a maximum near the equator. When a part of the bubble surface is contaminated by surface active molecules, the velocity at the clean part is nearly unchanged, while it slows down very sharply at the leading edge of the stagnant cap (Fig. 4).

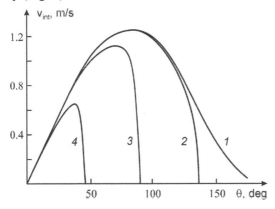

Fig. 4 The interfacial velocity distribution $v_{int}(\theta)$ at $Re = 200$: 1, $\psi = 0°$; 2, $\psi = 45°$; 3, $\psi = 90°$; 4, $\psi = 135°$; according to Fdhila and Duineveld [14]

In Fig. 5 the cap-angle is given as a function of the dimensionless bulk concentration C. Specific surfactant properties, Γ_∞ and k_{des}/k_{ad}, need to be known from experiments in order to verify model results. These constants are obtained from the equilibrium surface tension isotherm and are used in the calculation of the concentration dependence of the bubble rising velocity (Fig. 6). Excellent agreement between the experimental and numerical critical concentrations was achieved for example for Triton X-100.

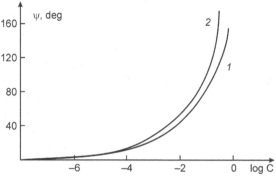

Fig. 5 The cap-angle versus the non-dimensional concentration in the bulk: 1, $a_b = 0.3$ mm; 2, $a_b = 0.4$ mm; according to [14]

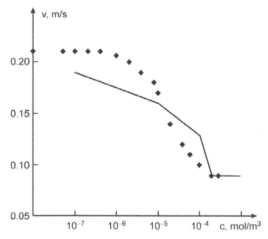

Fig. 6 The rise velocity of a bubble with an equivalent radius $a_b = 0.4$ mm, versus the Triton X-100 surfactant concentration in the bulk; solid line – calculated, symbols – measured; according to [14].

The interest in surfactant effects on rising bubbles increased recently vey much, initiated by numerical simulations of bubble motion in water at high Re [15, 16, 17, 18, 19, 20]. In particular, the simulation in [19] revealed that the critical Re (Re_c) at which a bubble behaves as a solid particle, can be estimated, and the average adsorption of surfactant coincides with the equilibrium adsorption corresponding to a zero net adsorption flux for Re<Re_c, and inversely, the averaged adsorption exceeds the equilibrium adsorption for Re>Re_c.

The most systematic investigation of the stagnant cap regime in the motion of spherical surfactant laden bubbles is described in [20]. Previous theoretical research on the stagnant cap regime has not studied in detail the competing roles of bulk diffusion and kinetic adsorption in determining the size of the stagnant cap angle. The numerical calculations for the cap angle and drag coefficient as a function of the bulk concentration of surfactant are accomplished, including both bulk diffusion and adsorption kinetics. For the case of diffusion-limited transport (infinite rate of adsorption kinetics), it is shown that very small bulk concentrations can immobilize the entire surface, and the critical concentrations, which immobilize the surface as a function of the surfactant parameters is calculated. The effect of kinetics reduces the cap angle (hence reduce the drag coefficient) for a given bulk concentration of surfactant. The work presented in [20] quantitatively correlates simulations with measurements. The experimental results on the drag of a bubble rising in a glycerol-water mixture, as a function of the dissolved concentration of a polyethoxylated non-ionic surfactant, whose bulk diffusion coefficient and a lower bound on the kinetic rate constants have been obtained separately by measuring the dynamic tension. The measured drag coefficients are intermediate between the drag coefficients of a bubble with a tangentially mobile and a completely immobilized surface. Simultaneously, the importance of surface rheology in bubble rising applications attained a larger recognition. Two examples are the significance for flotation [21,22] and for the detection of organic contaminants in water [23].

5. Dynamic adsorption layer effects on bubble buoyancy at high Re

It is possible to determine the conditions of realisation of different states of dynamic adsorption layer formation for a buoyant bubble using the concept of hydrodynamic and diffusion boundary layers, having a thickness δ_G and δ_D respectively, and independent of angle θ. The evaluations are simplified under the assumption of either a strong

$$\left.\frac{\delta v}{\delta z}\right|_{z=a_b} \approx \frac{v}{\delta_G}, \tag{22}$$

or a weak retardation of the surface

$$div_s[\Gamma(\theta)v_\theta(\theta)] \approx \frac{\Gamma(\theta)v_0}{a_b}. \tag{23}$$

where v_0 characterizes the surface velocity $v_\theta(\theta)=v_0 \sin\theta$.

Let us first consider the conditions at which the surfactant adsorption at the bubble surface is strongly retarded,

$$|\Gamma(\theta) - \Gamma_0| \ll \Gamma_0; \quad v_0/v \ll 1. \tag{24}$$

This estimation given by Dukhin $et\ al.$ [7, 8] results in:

$$|\Gamma(\theta)-\Gamma_0|/\Gamma_0 \approx \frac{\eta v}{RT\Gamma_0}\frac{a_b}{\delta_G} \ll 1 \tag{25}$$

which allows an approximation of the bubble surface velocity under the conditions

$$D\frac{|c(a_b)-c_0|}{\delta_D} \approx D\frac{|\Gamma-\Gamma_0|}{\delta_D}\frac{c_0}{\Gamma_0} \approx \frac{\Gamma_0 v_0}{a_b}. \tag{26}$$

From Eqs. (25) and (26) the second necessary condition of the considered state can be obtained,

$$\frac{v_0}{v} \approx \frac{\eta D c_0}{RT\Gamma_0^2}\frac{a_b}{\delta_D}\frac{a_b}{\delta_G} \ll 1, \tag{27}$$

and the following estimation for the retardation coefficient χ_b [7, 8] at Re $\gg$ 1 can be obtained,

$$\chi_b \approx \frac{RT\Gamma_0^2}{Dc_0} \frac{\delta_D}{a_b} \frac{\delta_G}{a_b}. \tag{28}$$

New conditions for the formation of the second state of the DAL formation by non-ionic surfactants is formulated under conditions where the surface concentration slightly deviates from the equilibrium state, Γ_0, and the bubble surface is weakly retarded,

$$|\Gamma(\theta) - \Gamma_0| / \Gamma_0 \ll 1 \text{ at } v_0/v \approx 1. \tag{29}$$

The conditions of a slight deviation of an adsorption layer from equilibrium at $Re \gg 1$ are derived in the same way as at $Re \ll 1$ [7, 8]. The second necessary condition is that the viscous stresses on the surface of a rising bubble should be much smaller than the characteristic value of a strong surface retardation:

$$\eta \left[\frac{\delta v_0}{\delta z} - \frac{v_0}{a_b} \right]_{z=a_b} = \frac{1}{a_b} RT \frac{\delta \Gamma}{\delta \theta} \ll \eta \frac{v}{\delta_G}. \tag{30}$$

After rearrangements we obtain

$$\frac{\eta D c_0}{RT\Gamma_0^2} \frac{a_b}{\delta_D} \frac{a_b}{\delta_G} \gg 1. \tag{31}$$

Note that the left hand sides of the inequalities (27) and (31) are identical which indicates that the regions, where the first and second conditions of DAL formation are valid, do not overlap.

Now the conditions for the formation of a third state of dynamic adsorption layers are considered. In this case the adsorbed surfactants are almost completely dislocated to the lower pole and the main part of the bubble surface (except for a narrow rear zone) is weakly retarded,

$$|\Gamma(\theta) - \Gamma_0| / \Gamma_0 \approx 1 \text{ at } v_0/v \ll 1. \tag{32}$$

The derivation of the condition of strong variation of surface concentration along the bubble surface at $Re \gg 1$ is carried out in the same way as at $Re \ll 1$ [7, 8]. The second necessary condition is Eq. (30) which can be rewritten in the form,

$$\frac{\eta v}{RT\left[\Gamma(\pi)-\Gamma(0)\right]}\frac{a_b}{\delta_G} >> 1.$$

(33)

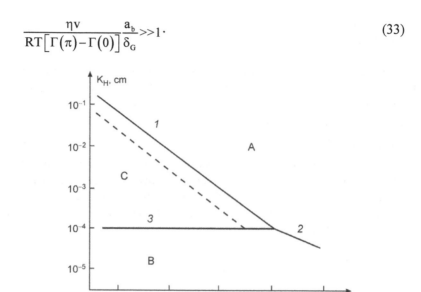

Fig. 7 Conditions of realisation of different states of dynamic adsorption layer formation of nonionic surfactant. Estimates are given for bubbles with a radius of 0.05 cm. The regions of parameters c_0 and Γ_0/c_0 are as follows: A, slight deviation of surface concentration from equilibrium and strong surface retardation; B, slight deviation of surface concentration from equilibrium and weak surface retardation; C, almost complete displacement of adsorbed surfactant to the rear stagnation pole and a weak retardation of the main part of the surface, according to [7, 8].

These conditions are presented in Fig. 7 in a graphic form. The regions of the three states of dynamic adsorption layer are separated by curves 1, 2 and 3 which are given by the following respective three equations:

$$\frac{\eta v}{RT\Gamma_0}\frac{a_b}{\delta_G} = 1,$$

(34)

$$\frac{\eta}{\chi_b} = 1,$$

(35)

$$\frac{\Gamma_0}{c_0 \delta_D} = 1 \cdot \tag{36}$$

Regions A, B and C are separated from each other not by lines but rather by wide bands as there is a change from "»" into the condition "«", and vice versa.

6. Surface rheology and rising bubble retardation

For slow adsorption kinetics and low surfactant concentration the surfactant adsorption practically does not affect the initial stage of bubble rising after detachment. During this stage the rise velocity can establish which corresponds to an initially clean bubble surface. For small Reynolds numbers $(Re < 1)$ this velocity value is given by the Hadamard-Rybczynski relationship. During the next stage adsorbed surfactant molecules modify the bubble surface. The interface becomes step by step partially or even fully stagnant because of the Marangoni effect, and the hydrodynamic field and the rising bubble velocity change during this period of time accordingly. Thus, time dependent adsorption of the surfactant gives rise to a time dependent bubble velocity.

When adsorption is the slowest process in the system the adsorption kinetics is described by the equation

$$\frac{\partial \Gamma}{\partial t} + div_s (v_s \Gamma) = -k_{des}\Gamma + k_{ad}c_0 (1 - \Gamma / \Gamma_\infty) \tag{37}$$

which has to be used instead of Eq. (11). Integration of Eq. (37) along the bubble surface, similarly to Eqs. (13) to (15), yields [24]:

$$\frac{dM}{dt} = \frac{1}{\tau}(M_0 - M) \tag{38}$$

where M_0 is given by Eq. (15) and $\tau = \dfrac{1}{k_{des}(1+C)}$ is the characteristic relaxation time. With the help of the initial condition $M(0) = 0$ Eq. (38) yields

$$M = M_0 \left[1 - exp\left(-\frac{t}{\tau} \right) \right]. \tag{39}$$

Thus, under non-steady state conditions one obtains instead of Eq.(20):

$$\frac{1}{1+k}\left[1-exp\left(-\frac{t}{\tau}\right)\right]=\frac{1}{2}\int_{0}^{\Psi}sin\theta\left\{1-exp\left[\frac{1}{Ma\lambda}\int_{0}^{\Psi}h(\theta',\Psi)d\theta'\right]\right\}d\theta \qquad (40)$$

After an analytical theory given in [24] a further development was achieved by he application of numerical calculations [25], without the constrains of slow adsorption kinetics. Also the bulk diffusion was quantified allowing now for any desorption rate, and also ellipsoidal and dimpled ellipsoidal-cap regimes were discussed.

The solution of this equation gives the cap angle ψ as a function of time [24]. With time the cap angle ψ increases, and, accordingly, the bubble rise velocity decreases. Thus, there is a transition period of the motion of a bubble after detachment due to the non stationary process of the formation of a surface stagnant zone.

The terminal velocity of a single bubble rising in a column of 4 m length of pure water and surfactant solutions was recently studied by Sam et al. [26]. It was shown that the axial rising velocity varies with height after release. Initially the bubble accelerates and then reaches a maximum velocity in less than 0.5 s. After this the bubble begins to decelerate due to adsorption of surface active molecules (impurities in the case of water or surfactants in case of solution).

Systematic studies in this area were performed in [27, 28,29,30, 31]. Bubble motion as a function of distance from a point of detachment and phenomena occurring during the bubble approach in pure water and solutions of various surface active substances were described and discussed and it was shown that the presence of surfactants has a profound influence on the terminal velocity and on the profiles of the local velocity. At low surfactant bulk concentrations, there are three distinct states in the bubble motion: (i) a rapid acceleration, (ii) a maximum velocity followed by a monotonous decrease, and (iii) attainment of a terminal velocity. In contrast, at high concentrations and in pure water (in absence of surfactants) there are only stages (i) and (iii). It is shown that the bubble terminal velocity decreases rapidly at low surfactant concentration, but some characteristic concentrations (adsorption layer coverage) can be found above which the velocity almost ceases to decrease. Immobilisation of the bubble surface caused by surfactant adsorption (surface tension gradient inducement) lowers the bubble velocity by more than two times. The presence of a maximum in the local velocity profiles is an indication that a stationary non-uniform distribution of adsorbed molecules (necessary for immobilisation of the bubble interface) was not established there.

B. INFLUENCE OF SURFACE RHEOLOGY ON MICROFLOTATION

1. General

Flotation of small particles represents an independent scientific problem inasmuch as the transition from coarse to fine grinding may be accompanied by qualitative changes in the mechanism of an elementary flotation act considered as the interaction of a particle with the surface of a bubble.

The traditional treatment of flotation pays mainly attention on the formation of bubble-particle aggregates and on the physico-chemistry of flotation reagents. Such an approach is insufficient to solve quite a number of technological problems in flotation, in particular of flotation technology of small particles (less than 20-40 μm in size). As applied to water purification, the behaviour of such small particles is of interest in the elementary flotation act.

Since flotation of small particles by small bubbles is a qualitatively new process, it was quite natural that Clarke and Wilson introduced the special term microflotation [32]. Unlike the classical flotation where the main elementary act is complicated by the impact of inertia and the accompanying deformation of the bubble surface, microflotation is a completely colloid chemical process and it can be described in terms of modern colloid chemistry such as orthokinetic heterocoagulation [33].

In general, in the elementary flotation act, one may distinguish a stage at which the particle approaches a bubble and a stage of attachment of the particle onto the bubble surface. Passing from large to small particles, the mechanism changes qualitatively, both in the stages of approach and attachment, as it was suggested by Derjaguin and Dukhin [33, 34] almost five decades ago. An analysis of DAL effect on microflotation is complicated as it has an effect both on the transport stage and the stage of attachment.

Sufficiently large particles move linearly under the effect of inertia forces until collision with the bubble surface, which takes place if the target distance $b < a_b + a_p$ (Fig. 8), where a_b is the radius of the bubble and a_p is the radius of the particle. The liquid flow envelops the bubble surface, and the particles are entrained to a greater or a lesser extent by the liquid. The smaller the particles and the lesser the difference of their density relative to that of the medium is, the weaker are the inertia forces acting upon them and the closer are the particle trajectories to the liquid streamlines. Thus, at the same target distance fairly large particles move almost linearly (Fig. 8, line 1), while fairly small particles move essentially along the corresponding liquid flow lines (line 2). The trajectories of particles of intermediate size are located in between the lines 1 and 2. As the size of particles decreases, the trajectories shift from line 1 to line 2 and the probability of collision decreases.

For estimating the flotation efficiency, the dimensionless parameter of collision efficiency is useful,

$$E = b_{cr}^2 / a_b^2 , \tag{41}$$

b_{cr} is the maximum radius of the cylinder of flow around the bubble encompassing all particles deposited on the bubble surface (Fig. 8). All particles moving along the streamline at a target distance $b < b_{cr}$ are deposited on the bubble surface (Fig. 9, as indicated by a dashed line), otherwise they are carried away by the flow.

From Fig. 9 it is evident that calculations are essentially reduced to determine the so-called 'grazing trajectory' (continuous curve) and, correspondingly, the target distance. A similar approach has long been used in aerosol science [35]. The bubble surface moves together with liquid and at Reynolds numbers of about 200 the flow of liquid around the rising bubble is a potential flow if the surface motion is not retarded by surfactants [3].

In order to understand the mechanism of inertia deposition of particles on a rising bubble, the particle inertia path 1 is introduced, which is defined as the distance at which the particle is able to pass in the presence of viscous resistance of the liquid at an initial velocity v_∞,

$$l = \frac{2}{9} \frac{v_\infty a_p^2 \rho_p}{\eta},$$
(42)

where ρ_p is the density of the particle.

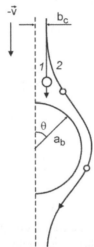

Fig. 8 Influence of the inertia of particles on their trajectory in the vicinity of the floating bubble. Trajectories of large (inertia, line 1) and small (inertia-free, line 2) particles at the same target distance b; according to [33].

Since the bubble surface is impermeable to liquid, the normal component of the liquid velocity at the surface is zero. As the distance from the bubble surface increases, the normal component of the liquid velocity also increases. The thickness of the liquid layer, in which the normal component of the liquid velocity decreases due to the effect of the bubble, is of the order of the bubble radius. The particle crosses this liquid layer due to inertia whereby its deposition depends on the dimensionless Stokes parameter,

$$St = \frac{2}{9} \frac{\rho_p V a_p^2}{a_p \eta} .$$

(43)

For St > 1 inertia deposition is obviously possible, yet calculations have shown that it appears to be possible also at St < 1 as long as St is not too small.

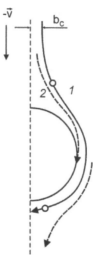

Fig. 9. Continuous lines refer to the concept of grazing trajectories of particles, dashed lines indicate the trajectories of the particle at b<b$_{cr}$ and b>b$_{cr}$, according to [34].

In processes involving an inertialess approach of particles to the bubble surface, their size also plays an important role. At the equatorial plane the closest approach of the streamline to the bubble surface is attained. In Fig. 10 the broken line (curve 1) represents the liquid streamline having a distance from the bubble surface in the equatorial plane identical to the particle radius. Some authors erroneously believe that this liquid streamline is limiting for particles of that radius. The error consists in that the short-range hydrodynamic interaction (SRHI) is disregarded.

Under the influence of the SRHI the particle is displaced from the liquid streamline 1 so that its trajectory (curve 2) in the equatorial plane is shifted apart from the surface with a separation larger than its radius. Therefore, no contact with the surface occurs and, correspondingly, $b(a_p)$ is not the critical target distance.

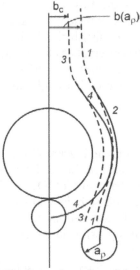

Fig. 10 Influence of the finite dimension of particles in inertia-free flotation on their trajectory in the vicinity of a floating bubble. The liquid flow lines corresponding to target distances $b(a_p)$ and b_c are indicated by dashed lines. The continuous lines are characteristic of the deviation of the trajectory of particles from the liquid flow lines under the influence of short-range hydrodynamic interaction; according to [34].

Due to the SRHI the distance from the particle to the surface in the equatorial plane is greater than the distance from the surface to the liquid streamline with which the trajectory of the particle coincides at large distances from the bubble. It may thus be concluded that $b_{cr} < b(a_p)$. The grazing liquid streamline (curve 3) is characterised by the fact that the particle trajectory (curve 4) which branches off under the influence of the SRHI, lies in the equatorial plane at a distance a_p from the bubble surface.

For a potential flow regime [36] we have

$$E_{0P} = 3a_p/a_b, \tag{44}$$

where the subscript P corresponds to the potential flow regime around the bubble. With Re < 1, the liquid flow becomes viscous and the motion of the bubble surface is usually completely retarded by the adsorption layer of surfactants so that the velocity distribution is described by Stokes' formula. Moreover, according to the observations by [6], at Re < 40 the bubble rises like a solid sphere even in the absence of surfactant.

In the Stokes regime an expression similar to Eq. (44) is given by

$$E_0 = \frac{3}{2}\left(\frac{a_p}{a_b}\right)^2. \tag{45}$$

It was shown by Derjaguin and Dukhin [33] that inertialess flotation is facilitated by the effect of surface forces. Two cases may be encountered here. It is known from the work of Derjaguin and Zorin [37] that interphase films may lose their stability and spontaneously disintegrate as the critical thickness h_{cr} is attained in the course of thinning. Thus, inertialess flotation becomes possible in the presence of flotation reagents provided $h_{cr} \neq 0$.

The molecular interaction forces of a spherical particle with a flat surface at fairly small h is expressed by [38],

$$F = A_H a_p/6h^2, \tag{46}$$

where A_H is the Hamaker constant. Hence the attractive force [39, 40] increases more rapidly with decreasing h than the resistance of the viscous interlayer to thinning [33]. Accordingly, the attachment of small particle is possible without particle-bubble contact (contactless microflotation) within primary or secondary minimum according to the DLVO theory. The formulas derived by using the Stokes and potential distribution of velocities in the liquid enveloping the bubble have the form

$$E_{cS} = E_{0S}f_S, \tag{47}$$

$$E_{cP} = E_{0P}f_P, \tag{48}$$

where f_P and f_S are the functions which reflect the influence of the SRHI on the elementary act of flotation and depend on the dimensionless parameter $H_{cr} = h_{cr}/a_p$.

In the interval $5 < Re < 100$ these formulas can be approximated by expressions derived by Rulyov [41],

$$E_{cP} \approx \frac{a_p^{0.8}}{a_b}, \tag{49}$$

$$E_{cS} \approx \frac{a_p^{1.4}}{a_b^2} \tag{50}$$

Thus, collision efficiency at a given bubble size can strongly change with the degree of mobility of the surface, i.e. the effect of surface rheology on collision efficiency is very strong. It is important to characterise this effect not only for limiting states of DAL, as researchers usually restrict their studies to.

To analyse the effect of surface retardation on the radial velocity of liquid at the bubble surface, we consider the expression for velocity divergence

$$\frac{1}{z\sin\theta}\frac{\partial}{\partial\theta}(v_\theta \sin\theta)+\frac{\partial v_z}{\partial z}=0. \tag{51}$$

Integrating from a_b to z under the assumption that at $z - a_b \ll a_b$ the changes in the tangential velocity are not significant, the radial velocity decreases inversely proportional to the retardation coefficient χ_b,

$$v_z(z,\theta)\Big|_{z=a_b+a_p} \simeq \overline{v}\frac{\eta}{\chi_b}\frac{(z-a_b)}{a_b}\Big|_{z=a_b+a_p} \cos\theta = \overline{v}\frac{\eta}{\chi_b}\frac{a_p}{a_b}\cos\theta. \tag{52}$$

At moderate values of χ_b, the deposition of particles happens under the effect of the radial liquid velocity, while at strong retardation it is due to sedimentation if the density of the particles differs remarkably from the density of the liquid medium.

This important statement can be approximated by the ratio of sedimentation velocity of a particle to the velocity of its inertialess movement together with the liquid according to Eq. (52) leading to the dimensionless parameter

$$\Lambda = \frac{\Delta\rho}{\rho} \frac{a_p}{a_b} \frac{\chi_b}{\eta}.$$

(53)

Large Λ values correspond to the predomination of sedimentation; at small Λ values the influence of a residual mobility on collision can predominate, even at a strongly retarded bubble surface. If the normal velocity component is proportional to $\cos(\theta)$, the angular dependence diminishes in Eq. (53) because the normal component of the sedimentation velocity is also proportional to $\cos(\theta)$.

The ratio $\Delta\rho/\rho$ changes from values in the range 0-0.2 for emulsions to 2-7 for minerals. Taking into account strongly retarded bubble surfaces, we can conclude that sometimes sedimentation can predominate. For emulsions the influence of sedimentation is restricted and the surface retardation can play a significant role.

Since in the non-retarded state the collision efficiency E_0 is described by Sutherland's formula (44), the effect of DAL can be approximated by the expression,

$$E \approx E_0 \frac{\eta}{\chi_b}.$$

(54)

Although Eq. (54) is an approximation of low accuracy, it is very valuable for predicting the very strong effects of bubble radius, the surface activity and concentration of the used surfactant. Depending on these parameters E can vary over two orders of magnitude which is much more essential than any errors even by 100%, which may have arisen when deriving Eq. (54).

At high surface activity due to the residual surface mobility, a zone of low surface concentration can appear near the front stagnation point. In other words the angle ψ characterising the dimension of the stagnant cap can be slightly less than π and

$$\pi - \psi \ll \pi.$$

(55)

For the sake of simplicity we introduce the special term front stagnation free zone (FSFZ). The Marangoni stress balance can be used as the basis for this estimation,

$$\frac{1}{a_b}\frac{\partial\gamma}{\partial\theta} \approx \eta\frac{v}{a_b}\sqrt{Re} \cdot \qquad (56)$$

Integrating this equation between $\theta = 0$ and $\theta = \theta^*$, the left hand side can be estimated as $\Delta\gamma$, while the right hand side gives $\eta v Re^{1/2}a_b$. The existence of an FSFZ means that $\gamma(0) = \gamma_0$, where γ_0 is the surface tension of pure water.

The angle θ^* corresponds to that part of the adsorption layer which is in equilibrium with the bulk solution. It means that $\gamma_0 - \gamma(\theta^*)$ equals the decrease of the equilibrium tension in the contaminated water. The value changes in the range 1-10 mN/m in distilled, tap or river water. The approximate result of the integration on the right hand side of Eq. (56) can be represented by introducing a multiplier a_b

$$\gamma_o - \gamma(\theta^*) \approx \eta v\sqrt{Re} \, . \qquad (57)$$

The right hand side of Eq. (57) depends very strongly on bubble size. It can exceed the left hand side of the equation at high Reynolds number such that the viscous stress compresses the adsorption layer to such a degree that the FSFZ appears. On the other hand, the left hand side can predominate at sufficiently high surface tension changes and not very big bubbles. For example for Re = 100 a FSFZ can be absent. In the intermediate range of Re the appearance of an FSFZ in contaminated water is probably impossible.

These conclusions can change for micellar surfactant solutions. If the surfactant concentration exceeds the critical concentration of micellisation (CMC) the surface tension does not change with surfactant concentration. It means that the Marangoni-Gibbs effect and the surface retardation of a bubble also disappear.

Both CMC and surface activity increase with higher homologues. The residual mobility of a bubble influences the microflotation at high Reynolds numbers and say $K_H < 10^{-3}$ cm. At $K_H > 10^{-3}$ cm a residual mobility in contaminated water can be provided by micellisation. Indeed this high surface activity is possible for sufficiently high homologues, corresponding to rather low values of the CMC, for example 10^{-4} - 10^{-3} M. Thus the mobility of a bubble surface can preserve at $K_H > 10^{-3}$-10^{-2} cm and $c_0 < 10^{-4}$-10^{-3} M. However, this conclusion relates to sufficiently big bubbles and the formation of a FSFZ, only.

Particles can be deposited on the FSFZ and on the stagnant cap. Let us estimate that part of collision efficiency connected with the deposition on the FSFZ. It means that the coordinates $(\pi - \psi, a_b + a_p)$ must be substituted into the equation of the stream function for the potential flow,

$$E_P (\pi - \psi) = E_0 \left[\sin(\pi - \psi) \right]^2 \cong 3 \frac{a_p}{a_b} (\pi - \psi)^2 . \tag{58}$$

Note that the substitution of the coordinates $(\pi/2, a_b + a_p)$ yields Sutherland's equation (44).

2. Influence of dynamic adsorption layer on attachment of small particles to bubble surfaces

More often a situation exists when particles and bubbles carry charges of the same sign and the electrostatic barrier is localised so that the secondary potential well is not sufficiently deep to fix small particles, or is practically absent. At the same time a particle attachment in the near potential well is possible by overcoming the barrier.

The addition of cationic surfactant for changing the sign of the bubble surface charge provides attachment of negatively charged particles. Note, air bubbles are usually slightly negatively charged due to adsorption of OH^- ions. Thus, the adsorption of cationic surfactants results in the change of the sign of the bubble surface charge. In such a case the electrostatic barrier for attachment of negatively charged particles would disappear. However, when the bubble surface is mobile and the surfactant's surface activity is high, adsorption is strongly decreased at the leading bubble surface. Therefore too large Γ/c values are undesirable because the surfactant may be removed from the front pole due to high surface mobility, i.e. the surface rheology should be accounted for. On the contrary, at too small Γ/c values a change of the sign of the bubble surface charge is not provided.

3. Dynamic adsorption layer effect on small particles detachment

The surface rheology affects the surface velocity of a rising bubble, and, therefore, also the normal component of the velocity of liquid in a close vicinity of the surface. Accordingly, both the particle attachment and the possibility of its subsequent detachment should be affected by the surface rheology as well.

Here we consider a situation, typical for contactless flotation. Similarly to DLVO theory we assume that the particle is attached in a force minimum due to the balance of attraction and repulsion forces, and a very thin film of liquid remains between the particle and bubble surface. Contactless microflotation often occurs, when the electrostatic attraction force counteracts the molecular repulsion force. This may occur when the signs of charge for the bubble and particles are different, what may be achieved by using cationic surfactants. This may also occur at the same negative sign of charges, when the difference in their absolute values is larger than 3 and when the electrolyte concentration is sufficiently low [42]. Recently, this prediction was experimentally confirmed by Pushkarova and Horn [43]. Molecular repulsion forces are also often observed in flotation.

The mechanism of the influence of the surface mobility of a rising bubble on particle detachment was considered in detail by Mishchuk et al. [42], however, in absence of any surfactant. The normal velocity of liquid is zero at the bubble surface, but it is not zero at small distances from the bubble surface. Thus, a particle located close to the surface becomes involved into the liquid flow and moves either towards or outwards the bubble surface, depending on the velocity direction. In case of a potential velocity distribution (for a bubble rising at large Re) the liquid moves towards the mobile bubble surface above the bubble equator, while it moves outwards below the equator. Accordingly, the normal component of the liquid flow carries particles to the upper part of a rising bubble surface and promotes their attachment. On the contrary, the normal component of the liquid flow promotes the detachment of particles attached below the bubble equator and carries them away from the surface. The detachment occurs, when the hydrodynamic detachment force F_{det} exceeds the maximum attractive forces carrying the particle back to the energy minimum, which may be calculated from DLVO theory [42].

The magnitude of the hydrodynamic force applied to an attached particle [44, 45] is determined by the normal velocity of liquid at the distance, which is equal to the distance between the particle centre and the bubble surface. This distance is about the particle radius a_p, which is usually very small in comparison with the bubble radius a_b: $a_p/a_b \ll 1$. For a potential flow and Stokes flow regimes, the liquid velocity v_z at a distance a_p from the bubble surface is approximately $v(a_p/a_b)$ and $v_{st}(a_p/a_b)^2$, respectively [7]. For large Re and a potential flow regime the bubble velocity exceeds that for smaller bubbles rising in a Stokes regime by an order of magnitude. Consequently, the velocity $v_z(a_p)$, estimated by the products above, in the case of large bubbles (large Re) is about two orders of magnitude larger than in the Stokes case. Accordingly, the hydrodynamic detachment force is also by two orders of magnitude larger. One can conclude that the phenomenon of particle detachment from a rising bubble-particle aggregate manifests itself mainly in the case of potential flow of large rising bubbles [42]. An equation for the critical particle radius as a function of bubble velocity and bubble radius, of the particle and the bubble surface potentials, and of the electrolyte concentration was derived there. The particles with dimensions larger than a critical value detach from the bubble.

This conclusion needs a refinement, because the usual presence of surfactants and their influence on the surface rheology is not accounted for in this paper as it is rather difficult, while a qualitative analysis is possible and was given by Dukhin et al. ([7], section 10.7.).

The presence of surfactants can lead to a surface retardation at Re>40. For low surface active molecules the whole surface is retarded almost uniformly (section 1.5) which may prevent particles from detachment. There is a possibility to quantify the balance between electrostatic attraction and hydrodynamic detachment forces. The Eq. (52) yields an estimation for the normal component of the flow velocity, which can be used to estimate the hydrodynamic detachment force.

At high surface activity an increase of surfactant concentration occurs only near the rear stagnant cap, while the rest of the surface is not strongly retarded. One can assume that the detachment of small particles from the stagnant cap is unlikely. The bubble surface at the rear stagnant cap is strongly retarded, and the normal component of liquid velocity is much lower here than at the upper not retarded part. This can prevent detachment of sufficiently small particles from the stagnant cap, while it is possible above this region. Two cases have to be discriminated: the rear stagnant cap is small (its boundary is located below the equator) or it is large (in the opposite case). In the first case the surface is mobile between the rear stagnant cap and the equator. The integration of Eq. (51) within this zone

$$v_z = \frac{1}{a_b \sin\theta} \cdot \frac{\partial}{\partial\theta}(v_\theta \sin\theta)(z - a_b) \tag{59}$$

shows that the normal velocity is rather large here, because v_θ varies from its maximum value near the equator down to almost zero within the rear stagnant cap. We can conclude that detachment of particles may occur within this zone.

Eq. (59) can be used also for the estimation of the normal velocity above the equator. Above the equator the expression $\frac{\partial}{\partial\theta}(v_\theta \sin\theta)$ changes its sign, i.e. v_z changes its direction. Near the front pole the normal velocity component is directed towards the bubble surface. Approaching the boundary of the rear stagnant cap the normal component changes its sign. This means that particle detachment may occur in the vicinity of the stagnant cap boundary even when this boundary is located above the equator. Whether it occurs or not depends on the location of the stagnant cap boundary. It occurs, when this boundary is not far from the bubble equator, and it does not occur, when this boundary is not far from the bubble front pole. It means that the detachment may be eliminated, when the main part of the bubble surface is retarded, while near the front pole the FSFZ is formed (see [7], and Section 2.1).

The possibility of detachment arises for sufficiently large particles, because the hydrodynamic detachment force is proportional to the particle dimensions. For smaller particles the electrostatic attraction forces may exceed the hydrodynamic detachment force, as it was investigated in [42] without accounting for surface rheology.

The condition of low electrolyte concentration (10^{-4}-10^{-3} mol/l), necessary for the onset of electrostatic attraction for negative charges of both bubble and particle and for the onset of microflotation, is often not satisfied in industrial conditions, while microflotation may occur even under this conditions. The reason is the so called hydrophobic forces. In particular, the surfactant adsorption layer may promote these forces. It means that the surfactant distribution along the bubble surface may be important also when the particle attachment is controlled by hydrophobic forces.

Recent investigations [46, 47, 48, 49] revealed that microbubbles attached to the surface of hydrophobic particles promote long range interaction, which was earlier prescribed to hydrophobic interaction. Now it is called pseudo-hydrophobic interaction.

Van der Waals interaction between the bubble and microbubbles is attractive, while the bubble-particle interaction is repulsive. Consequently, the particle coverage with microbubbles decreases the height of the electrostatic barrier (provided the electrostatic force is repulsive) and increases the depth of the energy well. Thus, a coverage of the particles by microbubbles enhances their attachment and suppresses their detachment. Nevertheless, a possibility of detachment preserves. This means that the influence of DAL and surface rheology exists also when the microflotation is enhanced due to microbubbles. However, the mechanism of this influence attains new features. Although the van der Waals forces between the bubble and attached microbubbles become attractive, the total surface force changes less than its van der Waals component, because simultaneously the electrostatic force becomes repulsive, while it could be attractive at different absolute values of surface potentials of the same sign, as in the case of particle/bubble interaction. The electrostatic interaction between the bubble and microbubbles is repulsive, because it is interaction between identical surfaces what is always repulsive.

Hence, the influence of the surfactant adsorption layer on the bubble/microbubble interaction arises through the influence on the electrostatic interaction. It depends on the increased surface charge within the rear stagnant cap in case of ionic surfactants. When a particle enters the diffusion layer near the cap, the surfactant diffuses from the cap to the microbubble surface, which attains rapidly almost the same values of surface concentration and surface charge. Hence, the surface rheology, which affects the DAL structure, affects also microflotation in the case, when particles are covered by microbubbles.

4. Surface rheology and microflotation in contaminated water. Residual mobility of bubble surface and collision efficiency

Since surface tension of tap water and even distilled water is lower than expected for really clean water, it is commonly accepted that under conditions of industrial flotation the bubble surface is completely retarded due to water contamination. Justification is based on the fact that the bubble rising velocity is close to that calculated for solid spheres. The latter conclusion is not unambiguous. A small residual mobility of the bubble surface can remain without any effects on the rising velocity but strongly increases collision efficiency.

Residual mobility of a bubble surface can substantially intensify microflotation of emulsions since sedimentation of emulsion droplets on a bubble surface is very slow. On the contrary, in microflotation of suspensions it can be insignificant since the rate of sedimentation of dispersed particles of high density on bubble surfaces is one or two orders of magnitude higher than of emulsions. Aggregated suspensions represent an intermediate case due to the aggregates' density.

Let us consider two additional factors which can strongly increase bubble surface mobility and intensify the second stage of microflotation in contaminated water but are usually not taken into account:

i) Experiments by Loglio et al. [50] and Sam et al. [26] have shown that the stationary rising velocity of large bubbles in sea and tap water is reached after a path of the order of several meters. This means that on the initial section of about one meter, the surface of sufficiently large bubbles cannot be completely retarded. Thus, we can conclude that a high initial collision efficiency corresponding to an originally clean surface can be preserved on a sufficiently long path of rising as long as the process of accumulation of impurities on bubble surface is not too fast. At not too high water contamination, the microflotation process can largely proceed on the 'initial' path of rising where the initial collision efficiency can exceed the final one by up to two orders of magnitude.

ii) In microflotation the processes of deposition of disperse particles and adsorption of molecular contaminations on the bubble surface proceeds in parallel. This means that in microflotation the level of impurities immobilising the bubble surfaces decreases and that the residual mobility of the bubble surface and also the role of DAL in microflotation increases.

C. INFLUENCE OF SURFACE RHEOLOGY IN FLOTATION

1. General

In the classical flotation process, large bubbles of about 1 mm in diameter are used, in contrast to microflotation for which smaller bubbles (with radius of about 100 μm) are recommended [7]. Accordingly, bubble surface immobilisation predominates in microflotation. On the contrary, the formation of a FSFZ can often occur for large bubbles in flotation.

The influence of surface rheology leads to the existence of two different modes of particle-bubble interaction in the flotation process with large bubbles. These modes correspond either to the immobilisation of entire bubble surface or to the FSFZ formation. The difference between the modes is caused by the enhancement of inertial forces due to the surface mobility. As it was stressed in Section 2, for a potential flow the radial component of liquid velocity is larger by a factor of approximately a_b/a_p as compared with Stokes' flow. This difference exists also for large Re numbers. At the same time, this radial component determines the inertial deposition of particles on the surface of rising bubble. As the increase of this flow component is not as strong as it is for the case of potential flow, one could not expect any qualitative differences in the regularities of particle-bubble interactions if large bubbles are considered instead of small ones, when the bubble surface is immobilised. The systematic outline of the theory for particle-bubble collision is presented in the recent monograph by Nguyen and Schulze [51] for the case of completely immobilised bubble surface. In contrast, the enhancement of inertial interaction due to surface motion results in new phenomena: particle rebound and centrifugal forces.

2. Qualitative change in the flotational collision stage at FSFZ

Depending on the extent of inertia forces effects, the hydrodynamic stage of the elementary act of flotation proceeds very differently. Four quantitatively different conditions of particle deposition on bubbles can be differentiated depending on the Stokes number [52]:
1. $St \ll 0.1$: Inertia forces have practically no effect on the motion of particles and can be considered as inertia-free.
2. $St < 0.1$: Inertia forces can impede particle deposition on a bubble [7].
3. $0.1 < St < 1$: An inelastic inertia impact of particles on a bubble surface is characteristic and, as will be demonstrated below, a major part of kinetic energy of the particles get lost both during the approach to the bubble and at the impact itself, when a liquid interlayer is formed between the surfaces of particle and bubble.
4. $St > 3$: The trajectory of a particle deviates very slightly from a straight line and the energy change of the particle when it approaches the bubble and collides is so small that the impact can be considered quasi-elastic.

The characteristic feature of the latter two cases is the availability of inertia impact of a particle on a bubble surface. As result, a thin liquid layer is formed between them and the dynamics of its thinning and the corresponding energy dissipation determine to a large extent the likelihood of particle attachment to the bubble or its jump back from the surface (Fig. 11). Particles are repelled from the bubble surface if the film has no time, during collision, to flow out to a critical value at which it spontaneously ruptures and a three-phase wetting perimeter is formed. It was found experimentally that the attachment of a particle on a bubble can happen not only after the first, but also at a repeated rebound - once the particle has lost a considerable part of its kinetic energy [53, 54].

There is a theory made for a more accurate calculation of the minimum thickness of a liquid film h_{min} to be reached during a collision [7]. The estimated value of h_{min} in the range between 0.2 μm and 1 μm exceeds the critical thickness h_{cr} for a spontaneous rupture of a thin water film. This means that attachment by collision is impossible for particles of adequately smooth surfaces.

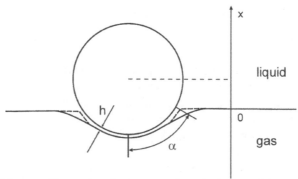

Fig. 11 Diagram illustrating the impact of a spherical particle on a bubble surface with the formation of a liquid interlayer; according to [53].

The theory of repetitive collision shows that the minimum thickness of the liquid interlayer during a second collision exceeds h_{cr} many times. Thus, the attachment by a second collision is also impossible with particles at smooth surfaces. The derived equation of the particle trajectory between the first and the second collision is restricted to Stokes numbers $St < 1$ and only one repetitive collision is possible under this condition. An additional restriction is given by the difference between St and St_{cr} which must not be too small. The notion of St_{cr} arises from the deposition theory, which considers a particle as a material point, i.e. without account for its dimension and interception role. Due to this simplification, the deposition is impossible at $St < St_{cr}$.

The attachment by a first or repetitive collision is not excluded in the theory for particles having a special surface roughness or shapes favourable for the rupture of comparatively thick water films.

The attachment of particles with smooth or slightly rough surfaces is possible only by sliding. Particle rebounds cannot be neglected as they govern the initial conditions of the sliding process which starts after the first or second recoil. The length of the second recoil l_r determines the initial distance to the surface for a sliding process. The joint action of particle rebound and centrifugal forces make a second collision impossible when the first collision takes place too far from the front pole, i.e. at an angle $\theta > \theta_{0cr}$ (Fig. 12). A more complete description with account for time necessary for the extension of the three phase contact is given in Fig.13.

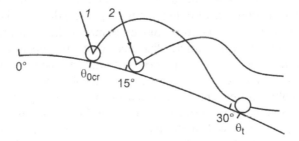

Fig. 12 Mechanism of deposition prevention in the zone $\theta_{0cr} < \theta < \theta_t$ due to the joint action of inertia reflection of a particle from the bubble surface and centrifugal forces: 1, grazing trajectory at a single reflection; 2, impossibility of deposition at $\theta > \theta_{0cr}$.

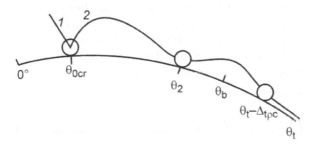

Fig. 13 Flotation mechanism by attachment by sliding after the second collision. θ_{0cr} is the critical angle of first collision, θ_2 is the angle at the end of the second rebound and at the beginning of sliding, $\theta_t - \Delta_{tpc}$ is the angle at which the film ruptures and the t.p.c. extension begins, θ_t is the maximum angle of sliding restricted by centrifugal force influence.

The equation for θ_{0cr} and the collision efficiency is obtained by taking into account the joint negative influence of particle rebound and centrifugal forces on particle-bubble collision. The experiments of Hewitt *et al.* [55] show the dependence of flotation rate on the diameter of highly hydrophobic particles and large contact angles. This and the weak sensitivity to hydrophobicity and to the magnitude of contact angle means that the extreme in the dependence of flotation rate on particle diameter is caused by the collision stage. At subcritical Stokes numbers the measured capture efficiency increases with particle diameter while for supercritical St the opposite tendency is observed in agreement with the theory. The onset of particle rebound and the enhancement of centrifugal forces at St_{cr} cause the decrease of collision frequency with increasing particle size [7, 55].

One can conclude that surface rheology and DAL influence practically all sub-processes involved in the attachment processes by collision and sliding. Surface retardation influences the hydrodynamic field around the bubble and the centrifugal forces and length of the particle recoil. Surface retardation by the DAL affects the bubble velocity and consequently the bubble-particle inertial hydrodynamic interaction. It also affects the drainage and thereby the minimum thickness of liquid interlayers established during first or second collision and sliding. As the result the particle trajectory between the first and second collision as well as the collision efficiency are very sensitive to the DAL structure. There is a big difference in the influence of the DAL on the attachment process of particles to bubble surfaces between the two conditions:

$$\Gamma_0/c_0 \gg \delta_D \text{ and } \Gamma_0/c_0 \ll \delta_D. \tag{60}$$

3. Influence of dynamic adsorption layer on detachment

A decrease of detachment forces can be arranged through the control of the DAL. It relates to big bubbles because the lower part of their surface cannot be always retarded beyond a narrow RSC. A large normal component of the liquid velocity at the mobile part of the lower surface can cause strong detachment forces. This difficulty can be avoided by increasing surfactant concentration which extends the rear stagnant cap.

The contact angle within the rear stagnant cap can strongly differ from its equilibrium value because surface concentration exceeds the equilibrium value. In particular, the difference between surface concentration within the RSC and the equilibrium value causes a difference in the Stern potential too. Here, the absolute value of the Stern potential can exceed its equilibrium value because the surface concentration of OH⁻ is higher than at equilibrium at pH values close to neutral. This holds true as long as the presence of ionic surfactants does not complicate the situation.

This phenomenon becomes even more important after the results published by Li and Somasundaran [56, 57] who stated that the absolute value of the negative charge of the water-air interface is sufficiently high at 10^{-2}-10^{-1} M NaCl and at sufficiently high pH value. From a comparison with the DAL theory we can draw the conclusion that surface concentration of OH⁻ within the RSC exceeds equilibrium. Consequently, the Stern potential can also exceed its equilibrium value. Due to the high sensitivity of the contact angle to the Stern potential, this difference can be important. Thus, it is possible that OH⁻ ions can serve as ionic collectors if all conditions are satisfied (sufficiently high pH, not too high electrolyte concentration and definite electrical surface properties of the particles). A quantitative evaluation of this phenomenon appears to be impossible because of lack of a DAL theory for large Reynolds numbers.

At adsorption of cationic surfactants (under the same conditions as just discussed above) the absolute value of the potential Ψ_1 of a water-air interface can be smaller than that of the particle Ψ_2, i.e. $|\Psi_1| < |\Psi_2|$. Within the RSC a significant decrease in Ψ_1 is possible and simultaneously $|\Psi_1 - \Psi_2|$ increases. Taking into account the fourth degree in Eq. (10D.6) in [7], even a small increase in $|\Psi_1 - \Psi_2|$ can cause a big increase of the contact angle within the RSC. Thus, the mechanism of the supporting influence of cationic surfactants on microflotation and flotation can be different. In microflotation the electrostatic barrier can be decreased, in flotation the contact angle can be increased. Naturally, both effects manifest themselves simultaneously. Li and Somasundaran [56, 57] observed a bubble recharge due to adsorption of multivalent inorganic cations. Thus, their application is recommended in order to increase the contact angle and to stabilise bubble-particle aggregates. Naturally, selective adsorption of multivalent ions at the water-air interface is important. But even in the absence of adsorption selectivity under equilibrium conditions a deviation from equilibrium can happen due to the increase in surface concentration within the RSC. This is important for the estimation of the contact angle increase caused by cation adsorption.

Recently the strong influence of surfactant on bubble-particle adhesion was determined by direct measurements [58, 59]. This confirms the importance of surface rheology, which affects the formation of RSC and the increased surface concentration in it, which in turn affects the strength of bubble-particle aggregates, similar to processes occuring in wet foams during flotation, when a larger rising bubble collides with a smaller bubble. The outcome of this collision is affected by the elasticity of the formed foam film [60] and the rate of its drainage [61, 62].

D. CONCLUSIONS

Steady-state motion of bubbles induces adsorption-desorption exchange with the subsurface, with the amount of substance adsorbed on one part of the bubble surface being equal to the amount desorbed from the other part. Accordingly, the dynamic adsorption layer forms with a surface concentration varying between the poles of a bubble due to its movement. On the other hand the dynamic adsorption layer retards the motion of the surface and influences the hydrodynamic field around a bubble. Due to coupling between the adsorption-desorption processes and hydrodynamics a variety of different dynamic regimes is possible for a bubble rising in surfactant solution which depend on the particular conditions (specific surfactant properties, bubble size, etc.).

In non-ionic surfactants solutions three different characteristic states of dynamic adsorption layer are possible depending on the surfactant activity and concentration (Fig.7): (A) slight deviation of surface concentration from equilibrium and strong surface retardation; (B) slight deviation of surface concentration from equilibrium and weak surface retardation; (C) almost complete displacement of adsorbed surfactant to the rear stagnation pole (RSC formation) and a weak retardation of the main part of the surface.

The dynamic adsorption layer formation reveals itself in time dependent velocity profiles of bubbles after detachment. At low surfactant bulk concentrations a sequence of three distinct states in the bubble motion can be observed: (i) a rapid acceleration, (ii) a maximum velocity followed by a monotonous decrease because of formation of the surface tension gradient, and (iii) attainment of a terminal velocity. In contrast, at high concentrations and in pure water (in absence of surfactants) there are only the stages (i) and (iii).

The performed analysis shows that the effect of the surface rheology, what is closely related to the state of DAL, can be significant for all flotation regimes: either microflotation or flotation of large bubbles. This effect is rather complicated as it concerns both the particle transport stage and the stage of attachment (equally detachment). In particular, the collision efficiency at a given bubble size can strongly change with the degree of mobility of the surface. Therefore it is important to characterise this effect not only for limiting states of DAL.

The collision efficiency depends on the radial velocity of liquid in a close vicinity of the bubble surface, which according to Eq. (52) decreases inversely proportional to the retardation coefficient χ_b. At moderate values of χ_b, the deposition of particles happens under the effect of the radial liquid velocity, while at strong retardation it is due to sedimentation if the density of the particles differs remarkably from the density of the liquid medium. The influence of a residual mobility on collision can predominate, even at a strongly retarded bubble surface. At $K_H > 10^{-3}$ cm a residual mobility in contaminated water can be provided by micellation. A small residual mobility of the bubble surface can remain without any effects on the rising velocity but can strongly increase collision efficiency.

A large normal component of the liquid velocity at the mobile part of the lower surface can cause strong detachment forces. This difficulty can be avoided by increasing surfactant concentration which extends the rear stagnant cap. In the RSC the bubble surface is strongly retarded and the normal component of liquid velocity is much lower than that at the upper not retarded part. This can prevent detachment of sufficiently small particles from the stagnant cap.

The bubble surface immobilisation predominates in microflotation. On the contrary, the formation of a FSFZ can often occur for large bubbles in flotation. The radial component of liquid velocity determines the inertial deposition of particles on the surface of rising bubble. In contrast, the enhancement of inertial interaction due to surface motion results in new phenomena: particle rebound and centrifugal forces.

One can conclude that surface rheology and DAL influence practically all sub-processes involved in the attachment processes by collision and sliding. Surface retardation influences the hydrodynamic field around the bubble and the centrifugal forces and length of the particle recoil. Surface retardation by the DAL affects the bubble velocity and consequently the bubble-particle inertial hydrodynamic interaction. It also affects the drainage and thereby the minimum thickness of liquid interlayers established during first or second collision and sliding.

The non-uniform surfactant distribution along the mobile surface of rising bubbles (section A) is the prerequisite of the surface rheology influence on microflotation and flotation (sections B and C). Recently large progress was achieved in both theoretical and experimental investigations of the dynamic adsorption layer of rising bubble. This is an indirect confirmation of the predicted role of the surface rheology in microflotation and flotation. A direct confirmation requires to overcome larger experimental difficulties and will yet take longer time.

ACKNOWLEDGEMENTS

The work was financially supported by projects of the European Space Agency (FASES MAP AO-99-052), and the DFG SPP 1273.

E. REFERENCES

1. Frumkin, R.B. and Levich, V.G., Zh. Phys. Chim., 21(1947)1183
2. Dukhin, S.S., Thesis, Moscow, Institute of Physical Chemistry, 1965
3. Levich, V.G., *Physicochemical Hydrodynamics*, Prentice-Hall, Englewood Cliffs, 1962

4. Rybczynski, Bull. de Cracovie A, (1911)40
5. Hadamard, J.S., Comp. Rend., 152(1911)1735
6. Okazaki, S., Bull. Chem. Soc. Japan, 37(1964)144
7. Dukhin, S.S., Kretzschmar G., and Miller, R., *Dynamics of Adsorption at Liquid Interfaces*, in Studies in Interface Sci., D. Möbius and R. Miller (Eds.), Elsevier, Amsterdam, 1995.
8. Dukhin, S.S., Miller, R., and Loglio, G., Physicochemical Hydrodynamics of Rising Bubble, *Drops and Bubbles in Interfacial Research*, Studies in Interface Science, Vol. 6, D. Möbius and R. Miller (Eds), Elsevier, Amsterdam, 1998.
9. He, Z., Maldarelli, C. and Dagan, Z., J. Colloid Interface Sci., 146(1991)442
10. Sadhal, S.S. and Johnson, R.E., J. Fluid Mech., 126(1983)126, 237
11. Fainerman, V.B., Kolloidn. Zh., 36 (1977) 113.
12. Kretzschmar, G. and Miller, R., Adv. Colloid Interface Sci., 36 (1991) 65
13. Ravera, F., Ferrari, M., Miller, R. and Liggieri. L., J. Phys. Chem. B, 105 (2001) 195.
14. Fdhila, R.B. and Duineveld, P.C., Phys. Fluids, 8(1995)310
15. Wang, L., Yoon, R.-H., Int. J. Miner Process, 85(2008)101
16. Wang, L., Yoon, R.-H., Colloids and Surfaces, 282-283(2006)84
17. Stevenson, P., Colloids and Surfaces, 305(2007)1
18. Mc Laughlin, J.B., J. Colloidal Interface Sci., 184(1996)614
19. Cuenot B., Magnaudet J., Spennato B., J. Fluid Mechanics, 339(1997)25
20. Takagi, S., Matsumoto, Y., American Society of Mechanical Engineers, Fluids Engineering Division (Publication) FED, 251(2000)667
21. Liao, Y., McLaughlin, J.B., J. Colloid Interface Sci., 224 (2000)297
22. Takemura, F., Physics of Fluid, 17(2005)048104-1
23. Wang,.Y., Papageorgiou, D.T., Maldarelli, C., J. Fluid Mechanics, 390(1999)251
24. Zholkovskij, E.K., Kovalchuk, V.I., Dukhin S.S., and Miller, R., J. Colloid Interface Sci., 226(2000) 51
25. Zhang, Y., Finch, J.A., J. Fluid Mechanics, 429(2001)63

26. Sam, A., Gomez, C.O., and Finch, J.A., Int. J. Mineral Processing, 47(1996)177

27. Krzan M. and Malysa K., Psychic Probl. Miner. Process. 36(2002)65

28. Krzan M., Lunkenheimer K., and Malysa K., Langmuir, 19(2003)6586

29. Krasowska M., Krzan M. and Malysa K., Psychic Probl. Miner. Process. 37(2003)37

30. Krzan M., Lunkenheimer K., and Malysa K., Colloids Surfaces A, 250(2004)431

31. Malysa, K., Krasowska, M., and Krzan, M., Adv. Coll. Interface Sci., 114-115(2005)205.

32. Clarke, A.N. and Wilson, J, *Foam Flotation*, Marcel Decker, 1983

33. Derjaguin, B.V. and Dukhin, S.S., Trans. Inst. Mining Met., 70(1960)221

34. Derjaguin, B.V. and Dukhin, S.S., Izv. Akad. Nauk SSSR, Otdel. Metal 1. Topi., 1(1959)82

35. Langmuir, J. and Blodgett, K., *Mathematical Investigation of Water Droplet Trajectories*, Gen, Elec. Comp. Rep., July, 1945

36. Sutherland, K.L., J. Phys. Chem., 58(1948)394

37. Derjaguin, B.V. and Zorin, Z.M., Zh. Fiz. Khim., 29(1955) 1010, 1755

38. Mahanti, J. and Ninham, B.V., *Dispersion Forces*, Academic Press, 1976

39. Derjaguin, B.V., Chem. Scr., 9(1976)97

40. Derjaguin, B.V., Dukhin, S.S., Rulyov, N.N. and Semenov, V.P., Kolloidn. Zh., 38(1976)258

41. Rulyov, N.N., Kolloidn. Zh., 40(1978)898

42. Mishchuk, N.A., Koopal, L.K. and Dukhin, S.S., J. Colloid Interface Sci., 237 (2001) 208

43. Pushkarova, R.A. and Horn, R.G., Colloids Surfaces A, 261 (2005) 147

44. Derjaguin, B.V., Dukhin, S.S. and Rulyov, N.N., in *Surface and Colloid Science*, E. Matijevich (Ed.), Vol. 13, Wiley Interscience, New York, 1984, p.71

45. Dai, Z., Dukhin, S.S., Fornasiero, D. and Ralston, J., J. Colloid Interface Sci., 197 (1998) 275

46. Dai, Z., Fornasiero, D. and Ralston, J., J. Chem. Soc., Faraday Trans., 94 (1998) 1983

47. Attard, P., Langmuir, 16 (2000)4455

48. Mishchuk, N.A., Ralston, J. and Fornasiero, D., J. Phys. Chem. A, 106(2001)689

49. Mishchuk, N.A., Colloids Surfaces A, 267 (2005) 139

50. Loglio, G., Degli-Innocenti, N., Tesei, U. and Cini, R., Il Nuovo Cimento, 12(1989)289

51. Nguyen, A.V. and Schulze, H.S., *Colloidal Science of Flotation*, Elsevier, 2005

52. Dukhin, S.S., Rulyov, N.N. and Dimitrov, D.S., *Coagulation and Dynamics of Thin Films*, Naukova Dumka, Kiev, 1986

53. Stechemesser, H.J., Freiberger Forschungshefte, A790 Aufbereitungstechnik, (1989)

54. Bergelt, H., Stechemesser, H. and Weber, K., Intern. J. Miner. Process., 34(1992)321

55. Hewitt, D., Fornasiaro, D. and Ralston, J., Mining Eng. Minerals Engineering, 7(1994)657

56. Li, Ch. and Somasundaran, P., J. Colloid Interface Sci., 146(1991)215

57. Li, Ch. and Somasundaran, P., J. Colloid Interface Sci., 148(1992)587

58. Neethling, S., Cilliers, J., Chemical. Engineer, 765(2005)26

59. Zawala, J., Swiech, K., Malysa, K., Colloids and Surface, 303(2007)293

60. Tasoglu, S., Demisci, U., Maradoglu, M., Physics of Fluids, 20(2008) 040805

61. Spyridopoulos, M.T., Simons, S.J.R., Chemical Engineering Research and Design, 82(2004)490

62. Preuss, M., Butt, H.J., Langmuir, 14,12(1998)3164

INTERFACIAL RHEOLOGY IN FOOD SCIENCE AND TECHNOLOGY

Philipp Erni[1,2], Erich J. Windhab[1] and Peter Fischer[1]

[1] Laboratory of Process Engineering, Institute of Food Science and Nutrition, ETH Zürich, Switzerland.
[2] Hatsopoulos Microfluids Laboratory, Department of Mechanical Engineering, Massachusetts Institute of Technology, Cambridge MA. USA.

Contents

A. INTRODUCTION – FOODS AS SOFT MULTIPHASE MATERIALS

The foods we consume on a daily basis belong to a wide range of materials: some are simple liquids (a clear soft drink may be nothing more than an aqueous solution of sugar and some water-soluble flavor), some are solid (sugar or salt crystals) - but the vast majority of food materials belong to the category of 'soft condensed matter' [1], sometimes also called 'complex fluids'; indeed, most foods are heterogeneous materials, and they are often some of the most complex examples of soft matter systems [1, 2]. From the agricultural production of the raw materials, to the physiology of flavor perception and nutrient uptake, food materials often involve physicochemical phenomena occuring on a wide variety of length scales and time scales. The length scales range from those associated with the molecular conformation of food proteins adsorbed at liquid interfaces, to the convective mixing length scales of the order of meters in industrial scale fermentations. Relevant time scales may be in the submillisecond regime during the generation of foam bubbles in an industrial whipping device, up to months or even years associated with the the long-term shelf life of a food emulsion (Fig. 1). Many food raw materials are synthesized and assembled by plants and animals and depend on external factors such as soil quality, climate and farming standards - it is therefore not surprising that common analytical and synthetic approaches to colloid and polymer science are often doomed to fail for food materials. However, understanding the physicochemical properties of food materials is essential for numerous aspects of food science and technology, such as the standardized characterization of raw materials and innovative products, or for optimized conservation and industrial processing. Consider the following examples: (i) for the sensory scientist, knowledge about the interaction between food and human perception of flavor and texture depends on understanding the transport properties of the food material; (ii) for a process engineer, design of industrial processing equipment is only possible if the microstructure and rheology of the materials involved can be modeled or quantified experimentally; (iii) nutritional functionality of food is determined by a 'materials bottleneck' - ingestion of food is a complex series of physical transport processes and biochemical reactions, many of which are controlled by physicochemical material properties. Naturally, this is relevant for materials that are in a multiphase liquid state with a high area-to-volume ratio, such as foams, emulsions, and composite materials. Notice that this not only includes the final products a consumer gets to see, but also any intermediate product or ingredient, anywhere along the agro-food chain involving raw materials, processing and consumption. Some random examples of high-interface materials in food science are beer foam [3], carbonated beverages [4], ice cream [5], fermentation media in industrial bioreactors, whipped cream, or espresso foam. In most of these fluids, stabilizing biopolymers (for example, proteins and peptides) are adsorbed to the liquid interface; in others, adsorbed particles and/or lipid droplets may play an equally important role [6]. Interestingly, some of those components have been among the most popular model systems in the more fundamental studies on interfacial rheology; in particular, the popularity of purified proteins as model systems has been due their outstanding capacity to impose strong viscoelastic properties onto interfaces, which has in many cases has been empirically known and applied for thousands of years.

Interfacial rheology plays a role in several aspects of food science: (i) Investigation of the interfacial rheology surfactants, biopolymers and particles at gas/liquid and liquid/liquid interfaces: a great number of food emulsifiers have been investigated extensively in the last two decades [7-20]; the literature is somewhat fragmented into shear and dilational studies, but also includes work on combined efforts in both deformation modes [21-24]. (ii) Elucidating the relation between interfacial rheology, interface structure, and macroscopic material properties of food materials remains a major challenge: not a single application of food science involves isolated, ideal interfaces; different levels of complexity must ultimately be understood and modeled in order to ascribe physical and engineering significance to interfacial rheological parameters, from interfaces, to thin films (i. e. two interfaces separated by a thin fluid layer), to actual foams and emulsions.

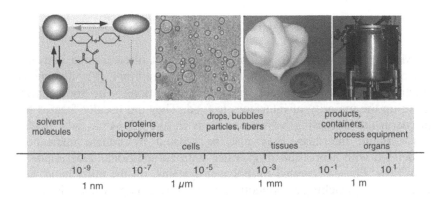

Fig. 1. Length scales relevant for food materials. Top row images from left to right: cartoon of protein adsorption at oil/solution interfaces; micrograph of an oil-in-water emulsion; a stable, highly concentrated foam; container used for industrial-scale mixing and fermentations.

In this contribution, we discuss the role of interfacial rheology in the context of food science and technology. We focus on the role of biopolymer adsorption layers in foams and emulsions, and we assess which aspects of interfacial rheology will be useful to explain the action of food emulsifiers. Our goal here is to point the reader towards those fundamentals that in our view are essential to apply concepts of interfacial viscoelasticity not only to food materials, but also to consumer products, pharmaceuticals and biomaterials in general. The fundamentals of interfacial rheometry itself are covered elsewhere in this monograph; in contrast, we will discuss methods and techniques that are necessary to understand the interplay between interfacial rheology and the macroscopic properties in complex fluids. These methods include single drop fluid dynamics, coalescence studies, optical rheometry of emulsions, interfacial rheo-optics, modified atomic force microscopy techniques, as well as microfluidic methods. Finally, we provide examples for successful studies linking interfacial rheology with macroscopic properties such as foam and emulsion formation and stability.

B. INTERFACES IN FOOD MATERIALS

Mobile interfaces in food materials are most often formed between one of the three following pairs of fluids: (i) gas/liquid; (ii) oil/aqueous solution; (iii) phase-separated biopolymer solutions. The former two make up the majority of interfaces in food. They are usually formed by mechanically disrupting bubbles or drops in hydrodynamic flow fields; gas / liquid interfaces may also be created by bubble nucleation associated with a pressure drop, often with an additional superimposed bulk flow. Phase-separated biopolymer systems, on the other hand, are obtained by quenching an aqueous mixture of two biopolymers dissolved in water, resulting in two distinct phases with drop or bicontinuous morphologies [25]. The interfacial tensions and density differences in those 'water-in-water' emulsions are orders of magnitude lower than in the traditional oil/water emulsions, resulting in more prominent roles for the interfacial bending rigidity and for mass transfer across the phase boundary. Due to the already low interfacial energy in those systems, the relevance of adsorbed surfactants or polymer is less pronounced than oil/water or gas/water systems. On the other hand, gas/liquid and oil/water interfaces must be stabilized by adsorbed molecules and particles to provide sufficient kinetic stability needed for foods emulsions and foams; the most common species used to stabilize liquid interfaces are:

Surfactants, in particular monoglycerides, phospholipids, and sugar esters [16, 22, 26]. Notice that the surfactants used in the food industry are usually mixtures of surfactants; for example, a 'soluble non-ionic emulsifier' may contain trace amounts of charged and/or poorly soluble contaminants (e.g., fatty acids), affecting the phase behavior of the surfactant both at the interface and in the bulk [27].

Proteins and partially hydrophobic biopolymers. Examples include the milk proteins β-lactoglobulin [12, 28] and the caseins [13], or ovalbumins from hen egg white [29-32]. Acacia gum, or 'gum arabic', is a complex natural hybrid biopolymer containing both peptide and polysaccharide components [21]; it is frequently used in the beverage and flavor industries [33].

Particles: liquid interfaces may also be stabilized by colloidal particles adsorbed at the interface [34]; in this case, the three-phase contact angle between the particle material and the two fluids strongly influences the fluid morphology and the macroscopic interface curvature; additionally, colloidal interactions must be accounted for carefully. A number of interfacial rheological and flow studies have been performed on particle-covered interfaces [28, 35]. In the context of food science, it is interesting to expand the concept of what constitutes a 'particle': solid-like entities with dimensions between tens of nanometers up to micrometers may be formed by surfactants if an appropriate location in the phase diagram is chosen; similarly, globular proteins may to some extent be considered as 'colloidal particles', depending on the degree of interfacial denaturation and unfolding. For example, β-lactoglobulin spread at the air/solution interface has been shown to behave as a colloidal soft glass rather similar to polystyrene particle monolayers [28].

Notice that the term 'emulsifier' used in the food industry may be defined in different ways and may often refer to any of the three categories of interfacially active species mentioned above. For example, a food-grade emulsifier of the polyglycerol fatty ester family might be categorized as a surfactant based on its molecular structure, including hydrophobic and hydrophilic portions. However, these surfactants can self-assemble into supramolecular aggregates exhibiting complex phase behavior [2,27]; therefore, a 'food emulsifier' might be present both as a self-assembled, surface-active particle and dissolved or adsorbed surfactant molecules. Additionally, while commercial food emulsifiers may be subject to rigorous inspection to comply with industrial quality specifications and worldwide food legislation, they are mostly complex mixtures, often containing fatty esters with a broad distribution of hydrocarbon chains [27].

Some typical examples of interfacial rheological data found in food materials are shown in Fig. 2a-b.

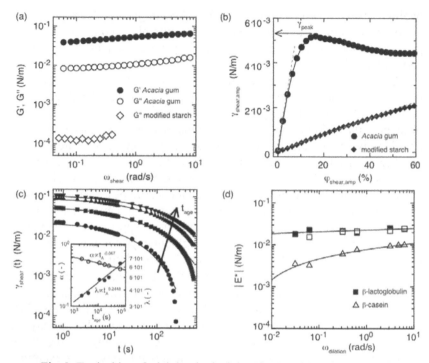

Fig. 2. Typical interfacial rheological data for food biopolymers at liquid interfaces. (a) Interfacial shear rheology: frequency-dependent interfacial shear moduli G' and G" of *Acacia* gum and hydrophobically modified starch at the oil/water interface [21]. (b) Stress-strain curves measured from the amplitudes of oscillatory shear stress $\gamma_{shear,amp}$ and shear strain $\varphi_{shear,amp}$ for *Acacia* gum and modified starch; notice that the plant gum layer exhibits non-linear stress-strain behavior with a critical peak stress ('yield stress' (c) Looped step strain experiments (interfacial shear deformation) of β-lactoglobulin; four individual responses of the shear stress relaxation are shown, each performed on the same interface at different interface ages t_{age}. A stretched exponenential model [22] is fitted to each relaxation curve; the inset shows the progression of two model parameters, the relaxation time λ and the stretch exponent α with increasing t_{age}). (d) Magnitude of the dilational modulus E* of a globular protein (β-lactoglobulin) and a flexible protein (β-casein) (phosphate buffer, pH 6.7; all oil/water interfaces with carvene oil).

A comparison is shown for a typical globular protein (β-lactoglobulin), a flexible random coil protein (β-casein), a hybrid biopolymer, *Acacia* gum (or 'gum arabic'), and octenyl-succinated waxy maize starch. *Acacia* gum is a hybrid polyelectrolyte containing both protein and polysaccharide subunits. Here we show the interfacial shear moduli of its adsorption layers at the oil/water interface and compare it with adsorbed layers of the hydrophobically modified starch, which for economic and political reasons is often used as a substitute for *Acacia* gum in technological applications [21]. In dilational experiments, the viscoelastic response of the starch derivative is just slightly weaker than for *Acacia* gum, whereas we observe pronounced differences in shear flow: the interfaces covered with the plant gum flow like a rigid, solid-like material, with large storage moduli and a linear viscoelastic regime limited to small shear deformations, above which we observe apparent yielding behavior. In contrast, the films formed by hydrophobically modified starch are predominantly viscous and the shear moduli are only weakly dependent on the deformation. Concerning their most important technological use as emulsion stabilizers, the dynamic interfacial responses imply not only distinct interfacial dynamics, but also different stabilizing mechanisms for these two biopolymers.

C. INTERFACIAL RHEOMETRY OF FOOD AND BIOMATERIALS

Surfactant and biopolymer layers in food and biological materials are amenable to interfacial rheometry just like synthetic surfactants, polymers, and particles, as described in detail elsewhere in this monograph, i.e. they do not generally require specialized methods. However, unlike those latter systems, the vast majority of food grade emulsifiers and biopolymers are often complex mixtures with wide molecular weight distributions and varied phase behavior. A tendency has emerged to (i) compare 'dirty', industry grade surfactants and polymers with the corresponding purified entities, and (ii) to study the behavior of mixtures of polymers and surfactants, including both interfacial structure and rheology. For example, the interfacial rheological response of whey protein concentrate has been interpreted in terms of the individual whey proteins, e.g. β-lactoglobulins, lactalbumin etc. [12, 13, 17]).

The large number of interfacial rheological studies on proteins is not only due to their biological and industrial importance, but also because proteins generally give strong interfacial rheological responses compared to other surfactants, and they were some of the first molecules studied by interfacial rheology [36-38]. To some extent, proteins are 'model' surfactants for interfacial rheometry: in shear deformation, interfacial rheological properties can only be detected if there is some degree of interaction between the adsorbed species. For most small molecular weight surfactants adsorbed to the interface from a solution, shear viscosity or elasticity is simply not measurable. For example, sodium dodecyl sulfate has been shown to give measurable interfacial shear responses only in the presence of significant amounts of dodecanol, whereas the interfacial shear viscosity of the neat surfactant is difficult to detect [39]. Seen from a different angle, this insight also means that sometimes impurities (dodecanol) are more important for the surface rheological response than the surfactant itself; this further emphasizes the need to study surfactant mixtures. (ii) In dilational rheology, the slow time scales associated with protein diffusion, adsorption and desorption give access to measurements for which small molecular surfactants are too fast; however, this usually comes at the expense of more complicated surface equations of state for proteins and polyampholytes.

Many fundamentals of what is known today about the rheology of adsorbed proteins have been reported in a series of papers by Graham and Phillips [7-11]. They noted the frequency-dependence of the dilational modulus of β-casein, attributed to loop/train relaxation, whereas the globular proteins were rigid and frequency-independent. The shear responses, measured in creep tests, were strongly viscoelastic for bovine serum albumin (BSA) and lysozyme, whereas the values for β-casein were very small and close to the detection limit. Most importantly, these authors stated that protein films can be seen as a thin interfacial gel layer. Developments after this study have been summarized by Dickinson [12, 13], van Vliet et al. [14], Damodaran [15], and Möbius and Miller [17]. Interfacial shear rheology can provide information about protein-protein interactions [40]; however, steady shear experiments are now also known to cause fracture in rigid protein layers [41]. Therefore, suitable experiments for proteins should be performed either at very small deformations and stresses [32], or the deformation-dependence itself must be taken into account, for example using stress-strain curves or strain amplitude sweeps [21]. The non-linear, 'yielding' properties have received increased attention, and it is now established that (globular) protein layers can form 'gels' or 'soft glasses', depending on the structural arrangement and correlation length of the adsorbed film. Particle network models [42-44] have been proposed to simulate the deformation response of these systems. Both the shear and dilational properties can be strongly time-dependent (see Fig. 2c). Ageing has been observed in both deformation modes [23, 45] on time scales much longer than those commonly needed to reach surface pressure equilibrium (if there is any).

Since the first measurements of dilational rheological parameters with oscillating pendant drops have been reported [46], this technique has become very popular and has been applied to a number of proteins, including lysozyme [24], bovine serum albumin (BSA) [46, 47], β-lactoglobulin [28, 29, 48], β-casein [24, 49-52] α-gliadin (a wheat protein) and pea globulins [53] (see example in Fig. 2d). A comprehensive study on wheat proteins, including surface rheology and thermodynamics, ellipsometry. surface force measurements and FRAP mobility measurements (FRAP: fluorescence recovery after photobleaching) was performed by Örnebro et al. [54]. Renault et al. [30] studied native and S-ovalbumin and found that the interfacial rheology is correlated with the development of an intermolecular β-sheet network at the air-water interface. In another study with ovalbumin, the net electric charge of the protein was established as an important factor for the surface rheology [31], a result that is also relevant to other globular proteins [29].

Further focus areas in interfacial rheology for food-related systems are: (i) surface interactions of small molecular weight surfactants with proteins or other polyelectrolytes [55-60]; (ii) 'fluidization' of protein layers by 'competitive' adsorption with small molecular weight surfactants [61-63]; and (iii) chemical or enzymatic interfacial cross linking of proteins [64-67]. Recently, advances have been made in the connection of interfacial rheological properties of adsorbed proteins and intrinsic stability of the corresponding proteins [68]. Apparently, the hardness (internal cohesion) of protein molecules is strongly related to the macroscopic rheology of the films. Cascão-Perreira et al. [47] investigated conformers of BSA with different structural stability, i.e. structural isomers of the same protein with different intrinsic 'strength' of the single molecule. They found that the rigid N-state conformer of adsorbed BSA gives more elastic rheological responses than the other conformers.

D. RELATING INTERFACIAL STRUCTURE TO INTERFACIAL RHEOLOGY

1 Interfacial rheo-optics

2D rheo-optical methods are useful to simultaneously measure of surface rheological properties and structural changes induced by the applied deformation or stress fields. The most frequently used rheo-optical techniques for liquid interfaces are Brewster-angle microscopy (BAM) [69, 70], ellipsometry, surface linear dichroism (SLD), surface circular dichroism (SCD), surface quasi-elastic light scattering (SQELS), total internal reflection microscopy/spectroscopy (TIRFM/S) and tracer microrheology performed at liquid interfaces (where the displacement of interfacial probes is observed by optical methods, either light or fluorescence microscopy). A review and further references can be found in van der Linden et al. [71]. BAM has been enormously popular in various areas of surface chemistry, including the fields of food and biological materials. For example, Rodriguez-Patino et al. [18-20] used BAM to study the relation between structure and dynamic properties for water-soluble lipids and milk proteins (β-casein, caseinate and whey protein isolate (WPI)) adsorbed at fluid/fluid interfaces, as well as mixtures of monoglycerides and WPI). Murray et al. [72] measured the rheological response to rapid area expansions of WPI, sodium caseinate and β-lactoglobulin films at the air/water interface along with BAM imaging. The Rehage group studied the structure and rheological properties of ultra-thin membranes used for the synthesis of microcapsules [73]. Garofalakis and Murray [26] characterized the morphology and dilational rheology of sugar ester monolayers.

A family of techniques related to 2D rheo-optics are surface particle tracking methods. Optical microscopy of particles or single molecules can be used to visualize interfacial flows. Here, information can be obtained either from macroscopic flows, or (at least in principle) from Brownian motion of the particles via a modified Stokes-Einstein relation. However, interpretation of particle displacement data in terms of interfacial rheological parameters is not trivial [74-76]. Flow geometries suitable for particle tracking experiments include for example 2D-channels [77], 2D Couette [78], viscous traction devices (cylinder with rotating floor) [79], oscillating magnetized needles [80], or simple compression/expansion in the Langmuir film balance. Koehler et al. [81, 82] used confocal imaging to monitor liquid drainage in foam films; these authors have recorded velocity profiles within individual foam channels (Plateau borders) and include surface viscosities into their modeling of the results. Olson and Fuller [83] used 2D versions of contraction and expansion flow cells based on a Langmuir trough. Maldarelli's group [84] has used a custom-built flow cell in combination with in-situ tensiometry and fluorescence microscopy to study the phase behavior of weakly soluble surfactants at the air/water interface. In particular, they were able to follow the dynamic assembly induced by the flowing subphase by tracking the morphology and phase diagrams starting from an initially clean interface. Azadani et al. [85] studied the flow-enhanced interfacial crystallization of the protein streptavidin at an air/water interface covered with a lipid monolayer, visualized by optical microscopy in an open-cylinder apparatus with a rotating floor [79].

In Fig. 3, we show two examples of surface shear flows visualized by Brewster-angle microscopy. Here, the optical technique is used *in situ* for real-time imaging and direct visualization of the structure, phase behavior and orientation effects due to flow processes in Langmuir films. A laser beam (He-Ne Laser, 632.8 nm) is guided through a Glan-

Thompson polarizer under the Brewster angle condition for the air/water interface, arctan(n $_{water}$/n$_{air}$) = 53.1°.

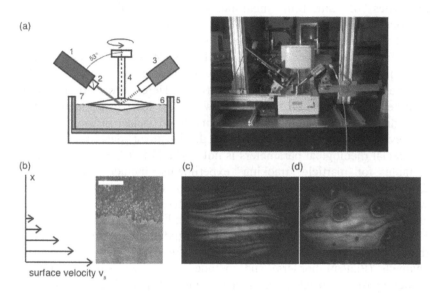

Fig. 3. Morphology of liquid interfaces during shear flow visualized by Brewster-angle microscopy (BAM). (a) Schematic of a rotating disk interfacial rheometer combined with BAM. 1: laser, 2: polarizer, 3: CCD camera with microscope lens, 4: rotating disk of surface rheometer, 5: rheometer cup and jacket, 6, 7: liquid interface. (b) Visualization of localized shear in an ike interfacial layer of the surfactant sorbitan tristearate at the air-water interface [22]; the schematic velocity profile based on video images of the surface flow indicates yielded and unyielded surface regions. (c) Surface of a supersaturated solution of myristic acid (C$_{14}$ fatty acid, pH 2, 4mg/l, 25 °C) at the air/water interface in shear flow; flow direction from left to right, with the rotating disk located at the bottom of the image. (d) Same surface, 60 seconds after cessation of the flow.

For the empty surface, polarized light is annihilated and no reflection is observed. In the presence of surface-active molecules (or impurities) the Brewster angle condition is no longer met, and partial reflection occurs. A CCD camera equipped with a long working distance microscope lens captures the reflected laser beam, providing video images to study deformations, wall slip and structural transformations, such as yielding transitions, in interfacial films. A Brewster-angle micrograph obtained during a large deformation shear experiment performed with spread layers of the surfactant sorbitan tristearate is shown in Fig. 3b. Above a critical deformation, the surface layer is disrupted, similar to a yield stress fluid in three dimension. The micrograph shown here shows the coexistence of yielded and unyielded regions; localization of the surface shear flow is illustrated with the schematic surface velocity profile on the left. BAM imaging is commonly used to monitor phase domains and phase transitions in surfactant monolayers. Fig. 3c demonstrated how these morphologies can be strongly altered by flows in the surface and/or the subphase. For the surface structures shown here, observed on the surface of a supersaturated solution of myristic acid (C_{14} fatty acid), the domains of different intensities are strongly elongated under flow conditions and relax back to circular shapes upon cessation of the flow (Fig. 3d), similar to the surface tension driven relaxation of drops and bubbles in three dimensions.

Besides microscopy methods, infrared reflection-absorption spectroscopy (IRRAS) and neutron or X-ray reflection measurements have been used in the past to obtain complementary information about the interfacial layer structures. For example, IRRAS has been applied to monolayers containing fatty acids, lipids, and proteins [86-89]. Meinders et al. [90] showed that in combination with spectral simulation it is also possible to determine the protein concentration and conformation in thicker adsorbed interfacial layers. For example, Martin et al. [91] were able to show that for β-casein and glycinin no conformational changes occur during adsorption to interfaces while for β-lactoglobulin and ovalbumin, some change in the secondary protein structure are observed. In a similar approach Fang and Dalgleish [92] followed the change in conformation for β-lactoglobulin upon adsorption at an oil/water interface by FTIR for over 72 h and observed ongoing changes in secondary structure over the whole duration of the experiment.

Neutron reflectivity is sensitive to the neutron refractive index profile normal to the interface and yields structural features including the thickness of adsorbed layers or multilayers. In order to use neutron reflection to study liquid-liquid systems the neutron beam must pass through one of the liquid phases to reach the interface. Due to the strong attenuation of the neutron beam by any liquid containing hydrogen, oil layers in combination with null reflecting water are the most promising way to study the aggregation of proteins and mixed systems at the liquid-liquid interface [96-99]. Neutron reflection has been used to study protein mixtures and a number of other systems such as soaps and adsorbed copolymers [63, 93-102] while recent X-ray scattering studies of the liquid-liquid interface have demonstrated the ability to probe molecular ordering and phase transitions [103, 104].

2. Nano- and micromechanics at liquid interfaces

Gunning et al. [105] developed a method to attach oil drops to the cantilever of an atomic force microscope (AFM) and to immobilize droplets on a glass substrate. This allows to monitor droplet-droplet interactions in aqueous solution as a function of the interdroplet separation. These droplet deformability measurements using AFM appear to be sensitive to interfacial rheology.

Anguelouch et al. [106] performed experiments with rotating ferromagnetic nanowires confined to the air/liquid interface and suggested their use in interfacial shear rheology. These authors discuss rotational drag coefficients for interracial films of various thicknesses from 30 nm to 150 nm. As is true for other microprobe-based techniques, where circumference-to-area ratios are large, a full analysis of the interfacial fluid dynamics including interface compressibility and accounting for local concentration-dependent contact angles remains a formidable challenge; however, if drag coefficients at gas/liquid or liquid/liquid interfaces can be calculated properly, interfacial microrheometry may well be a promising 2D equivalent of the corresponding bulk (3D) technique [107]. In particular, the non-homogeneity of composite surface layers often found in food materials is an important aspect, which can be quantified with multiple particle methods [108, 109].

Monteux et al. [110] measured the mechanical response of particle-laden fluid interfaces by recording the internal pressures of drops coated with particle layers as a function of the drop volume and found fluid, jammed and buckled (out-of-plane) states; in consequence, their results suggest that in addition to the common interfacial rheological parameters described elsewhere in this monograph (i. e. the interfacial viscosities and moduli in both shear and dilation), it is often necessary to consider additional mechanical characteristics if the interfacial layer is solid-like. We will discuss this further below for the case of emulsion drops covered with a protein layers and for liquid capsules covered with a solid membrane.

A full picture of the interfacial phenomena involved in the flow and stability of foams and emulsions should also include thin film forces. An example for a combined study involving both thin film forces and surface rheology can be found in Stubenrauch and Miller [111].

E. FROM INTERFACIAL RHEOLOGY TO MACROSCOPIC FLUID PROPERTIES

1. Linking interfacial rheology with bulk rheology

While considerable progress has been made in the interfacial rheological characterization of adsorption layers at gas/solution and oil/solution interfaces, their actual role for the structure and macroscopic dynamics of bulk materials is still elusive. Examples for work providing such correlations include coalescence studies at planar oil/water interface [112] in which a positive correlation of coalescence stability with the interfacial shear viscosity was established, a discussion of the influence on the dilational modulus on Ostwald ripening [113] and experiments on the deformation and breakup behavior of emulsion drops covered with a protein emulsifier [114-117]. Van Hemelrijck et al. [118] performed measurements on compatibilized immiscible polymer blends and attributed a slow relaxation process found in small amplitude oscillatory shear to in-plane interfacial relaxation; they interpreted their results in terms of a modified Palierne emulsion model provided by Jacobs et al. [119]. Some models for the rheology of (dilute) emulsion involving interfacial rheology have been proposed [120, 121]; an early model for emulsion rheology accounting for interfacial rheology was by proposed by Oldroyd already in 1955 [122]. Stone and coworkers [123] investigated aspects of interfacial rheology in foam drainage; interface/bulk correlations are also discussed are also given in the monograph by Edwards et al. [124] and in two articles by Lucassen-Reynders [125] and Lucassen-Reynders and Kuijpers [126]. In the remainder of this chapter, we provide a number of examples for studies linking interfacial rheology with macroscopic properties of drops, foams and emulsions.

2. Clean drops and bubbles in hydrodynamic flow fields

The behavior of emulsion drops, bubbles, or capsules in rheometric flows [127] and the critical conditions for their breakup [128] provide a wealth of information about the interplay between interfacial stresses and bulk hydrodynamic stresses [129] relevant for the analysis of emulsification and foaming processes. Taylor was the first to rigorously analyze the problem of a neutrally buoyant drop in simple base flows [130], providing solutions for the internal and external velocity fields at low capillary numbers and infinite viscosity ratios, and giving first order expressions for the drop deformation and orientation angle. A drop with undeformed radius R and viscosity η is considered. The macroscopic behavior of the drop is governed by external viscous forces (deforming) and internal viscous and interfacial forces (shape-conserving). The drop shape is characterized by a deformation measure $D = (L-B)/(L+B)$, involving the lengths of the major and minor axes L and B of a spheroid drop projected into the velocity-shear rate plane, and the orientation angle α between the major drop axis and the perpendicular to the flow direction (see Fig. 4).

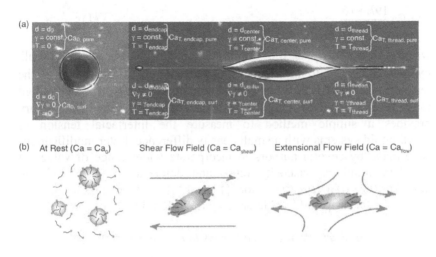

Fig. 4.(a) A single emulsion drop deforming in the four-roll mill, a flow cell suitable for combined shear and elongational flows [127-130]. Left: in the absence of flow; right: drop centered above the stagnation point of a plane hyperbolic flow. Inset text indicates localized characteristic parameters at the center of the drop, at the end caps, and in the thread connecting the two: the local capillary numbers Ca, the local Marangoni stresses (surface tension gradients) $\nabla\gamma$, bulk shear stresses T, and diameters d. (b) Cartoon of surfactant-covered drops at rest (left), deforming in simple shear flow (middle), and in combined shear and elongational flow (right).

Boundary conditions for Taylor's analysis are (i) steady flow, (ii) incompressible Newtonian liquids, (iii) negligible inertia and buoyancy, (iv) the velocity and tangential stresses are continuous at the interface, and (v) the interfacial stress boundary condition is comprised only of the interfacial tension balancing the normal stresses. Taylor's results for the drop shape in the case of small Capillary numbers are

$$D = Ca \frac{19\lambda + 16}{16\lambda + 16} \qquad (1)$$

where $Ca = T R/\gamma$ is the Capillary number, i.e. the ratio of bulk hydrodynamic (T) to interfacial stresses (γ/R), R is the radius of the undeformed drop. If all other parameters are known from independent measurements performed using different techniques, the theory provides a simple method to measure the interfacial tension of liquid/liquid systems with a small density difference. Later modifications were made by several authors to incorporate wider range of values of Capillary numbers, viscosity ratios and deformations. [127-133], to describe the transient behavior [134, 135] and to predict critical conditions for drop and bubble breakup [128, 136-138].

3. Surfactant-covered drops in rheometric flows

Surfactants, compatibilizers, and proteins influence the dynamics of drops and bubbles by reducing the interfacial tension upon adsorption, and in many cases by forming an adsorption layer with distinct local fluid mechanical and rheological properties. The interfacial stress boundary condition necessary to describe drop deformation is now no longer governed by the static interfacial tension alone, as for clean drops, but depends on a variety of additional parameters. Surface active components may diffuse, adsorb, and desorb from the bulk liquids to the interface and vice versa, inducing a local net change of the surfactant concentration when drops are deformed and fragmented. Many studies in a first approximation only consider insoluble surfactants, i.e. constant total surfactant coverage and no net adsorption and desorption processes. Under flow then two competing processes, surfactant dilution and surfactant convection within the interface, are considered for drop deformation. The former is due to the increase of interfacial area as the drop deforms, which tends to decrease surfactant concentration and results in increased interfacial tension and hence smaller drop deformation. Surfactant convection along the interface increases its concentration at the tips of the drop, lowering the interfacial tension there and locally facilitating the deformation and eventual breakup (Fig. 4a).

All those events are controlled by several factors, including the magnitude of Marangoni stresses (i.e., stresses due to local gradients in surface tension), curvature effects, the initial interfacial tension and coverage, the ratio of the fluid viscosities, and the capillary number [129, 139-142]. Ignoring interfacial rheological properties, Marangoni stresses may be sufficient to describe the surface tension driven redistribution on the deforming interface. Such additional stresses enter into the tangential component of the interfacial stress balance, meaning that the pressure jump across the liquid interface is no longer balanced by the static interfacial tension alone.

Notice that even if interfacial rheology is not considered, Marangoni stresses may be included in an effective dilational interfacial viscosity [124, 144]. This latter quantity has the advantage of being an experimentally measurable parameter that can be used to directly compare experimental data with theory or simulation [124, 145]. Besides a number of analytical approaches [123, 140] valid for small deviations from the spherical drop shape at rest, numerical methods are vital to describe the motion of surfactant-covered drops [139, 141-143, 146-153], always in combination with a suitable interface description containing an interfacial equation of state $\gamma = f(\Gamma)$ and, in the case of soluble surfactants, an adsorption isotherm [139, 143, 147, 153]. All these models differ with respect to the interfacial equation of state used, ranging from linear relations to the Frumkin equation of state [146]. Furthermore, the results strongly depend on how fast or slow the surfactant transport processes are with regard to the flow timescale, expressed in the Péclet number [140, 141, 143, 147, 154], with the extreme cases being those of insoluble surfactants [142] and of fully soluble, small molecular weight surfactants dominated by diffusion-controlled transport [143, 155].

4. Drops with interfacial rheology

Flumerfelt [144] has extended Cox' earlier small-deformation theory [131], formulated for clean drops, to a more general expression for the interfacial stress boundary condition, accounting for dynamic interfacial effects by replacing the thermodynamic interfacial tension by the Boussinesq–Scriven interfacial stress law [156], including both shear and dilational stresses at the interface. Here, spatial variations in concentrations are assumed to be small, therefore gradients in interfacial tension (Marangoni stresses) are accounted for as an apparent dilational interfacial viscosity. The interfacial viscosities in the Flumerfelt model lead to additional interfacial viscosity ratios, and the expression for the drop deformation is then given by $D = f(Ca, \lambda, \lambda_\eta, \lambda_\zeta)$, where the additional ratios λ_η, and λ_ζ are dimensionless groups relating the interfacial shear viscosity and the effective dilational viscosity (postulated as a composite quantity containing both 'intrinsic' dilational and surface-tension gradient components) to the viscosity of the matrix liquid. An experimental study of the model for various soluble surfactants was performed by Phillips et al. [145]. Their results show that interfacial shear viscosity and apparent dilational viscosity can be obtained from drop deformation experiments, and that these quantities play a significant role in the behavior of drops in shear fields. Just as the thermodynamic interfacial tension can be obtained for surfactant-free systems from Taylor's theory, the Flumerfelt theory can be used to calculate the interfacial shear viscosity and an apparent dilational viscosity from drop deformation data if one of the properties is independently measured. In a later publication, Pozrikidis [157] applied the numerical Boundary Integral Method (BIM) to the same problem.

The breakup behavior of protein-covered emulsion drops has been shown to be influenced by the rheological properties of the adsorption layer [114-115]. Using an optical flow cell, we have demonstrated the effect of an adsorbed protein layer on the deformation induced by a hydrodynamic flow field onto oil drops [116]. Despite a considerable decrease in the static interfacial tension, an emulsion drop covered with a globular protein is less deformed under identical hydrodynamic stresses due to the presence of the macromolecular adsorption layer.

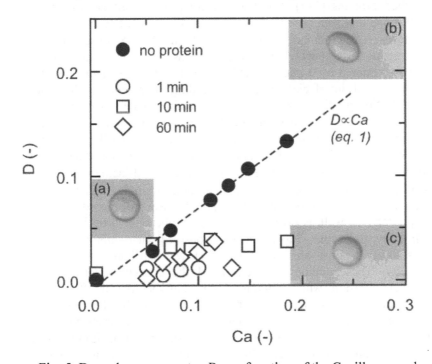

Fig. 5. Drop shape parameter D as a function of the Capillary number in simple shear flow for a single emulsion drop before and after adsorption of β-lactoglobulin to the interface after different incubation times of the drop in a protein solution. Ca is based on the transient values of the interfacial tension from pendant drop tensiometery measured at the respective interface ages (γ_{1min} = 19 mN/m, γ_{10min} = 16 mN/m, γ_{60min} =11.3 mN/m). The insets show sample images of the identical drop (R = 0.74 mm) with and without adsorbed protein layer (a - no flow; b - uncovered drop deforming at a shear rate of 6.8 s^{-1}; c – same drop at the same shear rate, but after incubation in the protein solution for 10 min.).

We have also investigated the anisotropy in dilute emulsions under flow using rheo-small angle light scattering (SALS) in the presence of adsorbed proteins [117]. Emulsions were prepared either with excess small molecular weight surfactant (sodium dodecyl sulfate, SDS) or with a surface-active globular protein (β-lactoglobulin). SDS is used far above the aqueous CMC, hence the interfacial stress condition of the drops can be approximated by a pseudo equilibrium interfacial tension, i.e. shear and dilational interfacial stresses, including those due to interfacial concentration gradients, are balanced on a time scale much faster than our observation time. In contrast, the deformation of emulsion drops stabilized with β-lactoglobulin is controlled by the solid-like behavior of the globular protein layer. Williams et al. [114] studied the effect of two structurally different surface-active proteins, β-lactoglobulin and β-casein, on the breakup behavior of mm-sized single emulsion drops. They used a concentric cylinder optical shear to record the drop morphology as well as critical Capillary numbers for drop breakup. In their study, β-casein and low concentrations of β-lactoglobulin increased the stability against breakup; this effect was attributed to interfacial elasticity. These authors also discussed the concept of an 'effective interfacial tension' which contains interfacial rheological effects an is higher than the static interfacial tension. On the other hand, in the case of high concentrations of β-lactoglobulin the found the drops to be less stable than expected, and discussed the role of interfacial rigidity, which makes the drop more susceptible to tangential stresses, as a possible reason. This latter case with a rigid protein layer results in breakup behavior being practically independent of the viscosity ratio, with the layer properties completely dominating the behavior. In a similar study, these results were later confirmed on micrometer-sized emulsion drops in a concentric cylinder shear cell [115], in combination with force measurements performed on protein adsorption layers.

5. Composite interfaces

Deformable capsules surrounded by a membrane have been the subject of numerous fluid mechanical and rheological studies [110, 158-

166] mostly motivated by the relevance of the mechanical response of cells [167-169] and lipid vesicles in flow fields [170, 171]. A fundamental difference to drop deformation theories is the absence of a static interfacial tension: the membrane is seen as a thin layer of a (visco-)elastic solid, expressed in terms of constitutive laws borrowed from solid material mechanics, for example the Hooke, Skalak, and Mooney–Rivlin laws [159-167]. Solid-stabilized ('Pickering') emulsions [172] have been known since thousands of years - mustard, milk-based emulsions or fluids used in oil recovery are only a few examples. In the context of interfacial rheology, Lucassen discussed the importance of non-homogeneous 'composite' interfaces [173]. Particle-stabilized drops and can now be produced in a controlled manner in microfluidic devices [174, 175]. For the behavior of such 'armored' drops or bubbles the relevant physicochemical parameters are far more diverse than the 'simple' static interfacial tension of a clean emulsion drop [176]; it appears that the 'classical' interfacial rheological quantities in both shear and dilational interface deformations need to be complemented first with non-linear interfacial viscoelastic material functions, but also with additional mechanical quantities such as the Young's modulus [160, 177] and the bending stiffness [164] for a full characterization of composite or particle-stabilized interfaces [110, 179].

For example, Chang and Olbricht [179] showed that capsule squeeze experiments can be used to obtain values for the membrane modulus in liquid capsules with a synthetic polymer membrane; it is likely that this method is also suitable for the case of solid-like interfacial layers (e.g. globular proteins, particle or composite layers).

6. Emulsions with interfacial rheology

Rheological data of emulsions are often analyzed in terms of the Palierne model [180, 181] for the complex shear modulus (linear viscoelastic regime) G_E^* of an emulsion, G_E^* being a function of the complex moduli of the continuous and drop phases, the interfacial tension, and the drop size distribution. Modifications of the model use average drop diameters rather than the entire distribution [182]. The Palierne model provides the complex bulk modulus $G_E^*(\omega)$ of a dilute emulsion

$$G_E^*(\omega) = G_M^*(\omega) \frac{1 + 3\sum_i \phi_i A_i / B_i}{1 - 2\sum_i \phi_i A_i / B_i} \qquad (2)$$

if the following are known: (i) the drop and continuous phase rheology, expressed in the frequency-dependent linear viscoelastic moduli $G_C^*(\omega)$ and $G_D^*(\omega)$; (ii) the morphology of the emulsion, expressed in the drop size distribution (here: volume fractions ϕ_i of drops with radius R_i); (iii) the static interfacial tension, γ. All these parameters are summarized in the coefficients $A_i, B_i = f(\omega, R_i, \gamma, G_C^*(\omega), G_D^*(\omega))$. If a single mode Maxwell fluid is assumed for the continuous phase, the Palierne model yields additional 'relaxation shoulders' in the storage modulus $G'(\omega)$ of the emulsion, centered above the shape relaxation time scale of individual drop sizes. Therefore, polydisperse emulsions or foams yield a distribution of relaxation times associated with the shape maxima of individual drop or bubble fractions. In another approach, introduced by Gramespacher and Meissner [183], the relaxation time spectrum of the emulsion is utilized to obtain information about (shape) relaxation processes of the drops. This extension of the Choi-Schowalter theory [184] shows that the relaxation time spectrum of the emulsion is a combination of the spectra of the individual phases, and an additional peak in the spectrum can then be attributed to the shape relaxation (due to the constant interfacial tension) of the drops. Their expression for the emulsion modulus contains a characteristic drop shape relaxation time scale

$$\tau_D = \left(\frac{\eta_c R}{\gamma}\right) \frac{(19\lambda + 16)(2\lambda + 3)}{40(\lambda + 1)} \qquad (3)$$

which is adopted from the Oldroyd emulsion model [122] and often used in the analysis of drop relaxation experiments [185]; if emulsions are considered [181], Eq. 3 additionally contains a term with the drop phase volume fraction (notice that in the literature on polymer blends the term 'interfacial viscoelasticity' is sometimes used to describe the observed additional 'elastic' effect due to the presence of deformed drops. In that case, the term does not refer to anything related to interfacial rheology, but it simply describes the 'elasticity' provided to the system by the tendency of a deformed drop to spring back to its spherical shape). Generally, by monitoring the time evolution of a characteristic drop shape parameter, an experimental time constant can be obtained and compared to τD. For example, emulsion drops can be deformed in well-defined flow fields within microfluidic devices; this allows to routinely measure interfacial tensions at oil/water interfaces based on the shape relaxation of emulsion drops flowing through a microchannel [186]. However, careful consideration must be given to the appropriate choice of flow geometries, shear stresses, and the accessible time scales. To emphasize this, assume that the factor containing the disperse/continuous phase viscosity ratio λ on the right-hand side of Eq. 3 has a value close to unity (which is often the case). The relaxation time scale of a droplet from the deformed state back to the spherical rest state then scales with the ratio $\eta c R/\gamma$, i.e. the ratio of the continuous phase bulk viscosity ηc to the capillary pressure γ/R. Assume an emulsion droplet with a radius of 10 μm in a continuous phase, both with viscosities close to that water, and an interfacial tension of the order of 10^{-2} Nm^{-1} - the corresponding time constant is of the order of 1 μs, which of course will be difficult to resolve with common instrumentation. Similar considerations can be made using Eq. 1 to obtain an order of magnitude for the hydrodynamic stresses necessary to obtain significant drop deformations. Therefore, both the drop shape relaxation time and the extent of deformation are most strongly influenced by the drop radius, the bulk phase viscosity, the flow rate, and the interfacial tension. Notice that elongational flows have a stronger ability to impose large deformations onto drops or bubbles [138] due to their irrotational character; such flows can easily be generated in microfluidic channels, for example using contraction/expansion geometries [187]. Alternatively, drops can also be forced into highly elongated shapes by confinement in a narrow gap or microchannel.

The model expressions given above for emulsions with a constant interfacial tension have been extended to include interfacial viscoelastic moduli by Friedrich's group [119]; in their model, the functions A_i and B_i additionally contain expressions involving an interfacial shear modulus G and an interfacial dilatational modulus E. Corresponding experiments with compatibilized blends of immiscible polymers were performed by Hemelrijck et al. [118]; they observed secondary relaxation shoulders in the storage modulus of the emulsions, as was predicted by the Jacobs modification of Palierne's emulsion model. Since these measurements were performed on polymer systems with very large bulk viscosities, it has not yet been possible to perform independent measurements of the same interface to obtain the interfacial viscoelastic moduli of the interfacial layer; the authors attribute the in-plane relaxation to an unspecified in-plane relaxation mechanism, which is different from the drop shape relaxation due to the static interfacial tension. Friedrich and Antonov [188] have investigated the influence of different amounts of symmetric block copolymers on the shape and interfacial relaxation times of a polymer blend. They combined predictions of the extended Palierne model with assumptions on the changes of interfacial tension with the amount of compatibilizer. Based on the modified Palierne model with interfacial viscoelasticity [119], these authors chose the Gibbs elasticity as the relevant interfacial rheological parameter and derived a scaling relation for the interfacial relaxation time.

On the other hand, experiments with emulsion drops stabilized by globular proteins (lysozyme, β-lactoglobulin) with predominantly solid-like interfacial viscoelastic properties (interfacial shear moduli $G' > G''$ at all frequencies) [116, 117] have shown a different role of the protein layers: rather than adding additional relaxation time scales to an emulsion, drop deformation is simply suppressed or strongly reduced in the presence of the protein, and the rheology of the emulsion approaches that of a suspension of rigid spheres, coated with a polyampholyte, in a viscous or viscoelastic medium. It appears that for small drop or bubble radii, high continuous phase viscosities and high values of the interfacial moduli this would the rule rather than the exception; in those cases, the bulk properties are described well by soft glassy rheology or paste models [189-191]. On the other hand, effective viscosities in highly concentrated systems can exceed the viscosities of the component fluids by orders of magnitude; this, along with the increased role of the osmotic pressure in compressed systems can lead to strongly deformed interfaces even in the case of a low viscosity continuous phase [192, 193].

G. CONCLUDING REMARKS

To conclude this overview, we will point out some of the challenges in interfacial rheology and its relation to the study of (bulk) transport processes of complex fluids. Since the interplay of momentum and mass transfer to/from and within liquid interfaces with the hydrodynamics of the bulk liquids is still poorly understood, there is a demand for both experimental and theoretical insight into the role of surfactants at liquid interfaces. Adsorption layers with strong viscoelastic properties are highly relevant in the food and consumer industries (e.g. beverage emulsions, milk-based foams, encapsulated active ingredients). Additionally, the relationship between interfacial rheology macroscopic fluid properties of foams and emulsions continues to offer a number of challenging open problems to both physical chemists and chemical engineers.

For example, Fig. 6 shows a comparison of foam lifetime with the apparent surface shear viscosity of a protein solution. Whereas the more stable foams are accompanied by higher values of the surface shear viscosity, it would be misleading to postulate a direct relation between the two quantities – indeed, many surface-active molecules readily stabilize foams or emulsions despite relatively low surface (shear) viscosities. An obvious first remedy to the problem is to account for both shear and dilational rheological parameters, rather than just focusing on one deformation mode alone. A second step would include consideration of the large deformation or far-from-equilibrium behavior in either deformation mode, since the strains occurring in foaming are always large with respect to the length scale of the interface. Additionally, liquid drainage in foam films or bubble coalescence is influenced by thin film forces, measured for example using the thin film balance (TFB) or modified surface force apparatus (SFA) methods. Since each of these aspects can be challenging and requires dedicated instrumentation, combined studies involving all of the above methods are still the exception rather than the rule [111], but they will ultimately be necessary to provide a full and quantitative picture of the role of interfacial rheology for foams and emulsions.

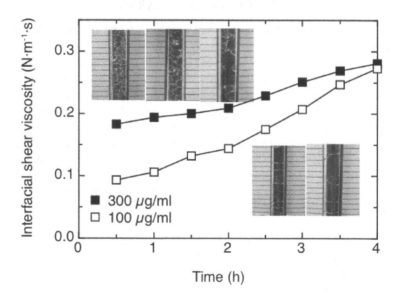

Fig. 6. Comparison of the time-dependent apparent interfacial shear viscosity with foam lifetime for solutions of ovalbumin at two different solution concentrations. Protein solutions were kept for 2h in a large surface-area container before foaming on a frit mounted to a transparent tube. Line spacing is 10 mm. Photographs are taken 5 and 20 seconds (bottom row) and 20, 80 and 100 seconds (top row).

Finally, it is also worth noting that the considerable number of studies of the interfacial rheology of protein layers is not only due to their relevance in biology and technology (in particular in food and pharmaceutical materials) – a more profane reason is the simple fact that the viscoelasticity of many adsorbed protein layers is *measurable* at all! Imagine we prepare a solution of a common small-molecular weight soluble surfactant, say SDS, or a sorbitan ester, at a concentration relevant for practical applications, say $c \sim 1\%$ w/w. We perform a comprehensive series of surface rheological experiments with an interfacial shear rheometer – and the result will most probably be disappointing! The measured interfacial viscoelasticity in shear deformation, if there is any, may simply be undetectable. Still hopeful, we then proceed with measurements on the dilational rheology of the same system, for example with the oscillation pendant drop or Langmuir trough methods – and again, we end up frustrated, because as much as we attempt to compress and expand our surfactant adsorption layer, the surface tension simply does not respond on the time scale of our measurement because any perturbation in interfacial concentration is rapidly equilibrated by diffusion relaxation into or from the bulk liquid. If surface rheological effects are measurable at all with such surfactants (low molecular weight, easily soluble in the bulk phase, etc.) they are often due to the presence of trace amounts of impurities (for example: SDS may contain small amounts of dodecanol from the production process). Besides having a look at the vast amount of literature available on the surface rheology of a great number of surfactants and polymers, it is usually helpful to first obtain surface equations of state of a surfactant, i.e. to measure the static interfacial tension as a function of the interfacial concentration (or area) using a Langmuir trough or a pendant drop tensiometer. This often gives a good indication on the regimes in which surface rheological measurements will make sense. Fortunately, in the specific fields such as food science and engineering, pharmaceutical technology, biomedical materials, or oil recovery, those classes of surfactants, polymers and particles amenable to interfacial rheology have been investigated intensively in the last three decades. Therefore, in these cases the focus in future research efforts is likely to involve increased attention to an improved physical understanding of microstructural, processing and stability aspects and their relation with interfacial rheology.

ACKNOWLEDGEMENTS

The authors acknowledge the Swiss National Science Foundation (SNF) and ETH Zurich for funding some of the work described in this contribution. The technical staff and students at the Process Engineering Laboratory in the Institute of Food Science and Nutrition at ETH Zurich are acknowledged for contributions to some of the results presented here: Patrick Heyer, Thomas Beeler, Christoph Eschbach, Muriel Graber, Rok Gunde and Bruno Pfister.

REFERENCES

1. A.Donald, Nat.Mater., 3 (2004) 579.
2. R.Mezzenga, P.Schurtenberger, A.Burbidge and M.Michel, Nat.Mater., 4 (2005) 729.
3. S.N.E.van Nierop, D.E.Evans, B.C.Axcell, I.C.Cantrell and M.Rautenbach, J.Agric.Food Chem., 52 (2004) 3120.
4. G.Liger-Belair, R.Marchal, B.Robillard, T.Dambrouck, A.Maujean, M.Vignes-Adler and P.Jeandet, Langmuir 16 (2000) 1889.
5. M.D.Eisner, H.Wildmoser and E.J.Windhab, Colloid Surface A, 263 (2005) 390.
6. N.E.Hotrum, M.A.C.Stuart, T.van Vliet and G.A.van Aken, Colloid Surface A, 240 (2004) 83.
7. D.E.Graham and M.C.Phillips, J.Colloid Interface Sci., 70 (1979) 403.
8. D.E.Graham and M.C.Phillips, J.Colloid Interface Sci., 70 (1979) 415.
9. D.E.Graham and M.C.Phillips, J.Colloid Interface Sci., 70 (1979) 427.
10. D.E.Graham and M.C.Phillips, J.Colloid Interface Sci., 76 (1980) 227.
11. D.E.Graham and M.C.Phillips, J.Colloid Interface Sci., 76 (1980) 240.
12. E.Dickinson, J.Chem.Soc.Faraday Trans., 94 (1998) 1657.
13. E.Dickinson, Colloid Surface B, 15 (1999) 161.
14. T.van Vliet, A.H.Martin and M.A.Bos, Curr.Opin.Colloid Interface Sci., 7 (2002) 462.

15. S.Damodaran, Curr.Opin.Colloid Interface Sci., 9 (2004) 328.
16. B.S.Murray, Curr.Opin.Colloid Interface Sci., 7 (2002) 426.
17. D.Möbius and R.Miller, Proteins at Liquid Interfaces.Elsevier, Amsterdam, 1998.
18. J.M.Rodríguez Patino, M.R.Rodríguez Niño and C.Carrera Sánchez, Langmuir, 18 (2002) 8455.
19. J.M.Rodríguez Patino, M.R.Rodríguez Niño, C.Carrera Sánchez, C.Cejudo Fernández and J.M.Navarro García, Chem.Eng.Comm., 190 (2003) 15.
20. J.M.Rodríguez Patino and C.Carrera Sánchez, Langmuir, 20 (2004) 4530.
21. P.Erni, E.J.Windhab, R.Gunde, M.Graber, B.Pfister, A.Parker and P.Fischer, Biomacromolecules, 8 (2007) 3458.
22. P.Erni, P.Fischer and E.J.Windhab, Langmuir, 21 (2005) 10555.
23. E.M.Freer, K.S.Yim, G.G.Fuller and C.J.Radke, J.Phys.Chem.B, 108 (2004) 3835.
24. E.M.Freer, K.S.Yim, G.G.Fuller and C.J.Radke, Langmuir, 20 (2004) 10159.
25. E.Scholten, J.Sprakel, L.M.C.Sagis and E.van der Linden, Biomacromolecules, 7 (2006) 339.
26. G.Garofalakis and B.S.Murray.Colloid Surface B, 12 (1999) 231.
27. N.Dürr-Auster, J.Kohlbrecher, T.Zürcher, R.Gunde, P.Fischer and E.J.Windhab, Langmuir, 23 (2007) 12827.
28. P.Cicuta, E.J.Stancik and G.G.Fuller, Phys.Rev.Lett., 90 (2003) 236101.
29. T.Lefevre and M.Subirade, J.Colloid Interface Sci., 263 (2003) 59.
30. A.Renault, S.Pezennec, F.Gauthier, V.Vié and B.Desbat, Langmuir, 18 (2002) 6887.
31. S.Pezennec, F.Gauthier, C.Alonso, F.Graner, T.Croguennec, G.Brulé and A.Renault, Food Hydrocoll., 14 (2000) 463.
32. P.Erni, P.Fischer, E.J.Windhab, V.Kusnezov, H.Stettin and J.Läuger, Rev.Sci.Instrum., 74 (2003) 4916.
33. S.Friberg, K.Larsson and J.Sjöblom (Eds.), Food Emulsions, Fourth Edition.Marcel Dekker, New York, 2004.

34. B.P.Binks, Curr.Opin.Colloid Interface Sci., 7 (2002) 21.
35. S.Reynaerts, P.Moldenaers and J.Vermant, Phys.Chem.Chem.Phys., 9 (2007) 6463.
36. J.Boyd, J.R.Mitchell, L.Irons, P.R.Musselwhite and P.Sherman, J.Colloid Interface Sci., 45 (1973) 478.
37. N.W.Tschoegl, Austral.J.Biol.Sci., 14 (1961) 288.
38. N.W.Tschoegl, Trans.Soc.Rheol., 6 (1962) 384.
39. D.Li and J.C.Slattery, J.Colloid Interface Sci., 125 (1988) 190.
40. E.Dickinson, S.E.Rolfe and D.G.Dalgeish, Int.J.Biol.Macromolecules, 12 (1990) 189.
41. N.E.Hotrum, M.A.C.Stuart, T.van Vliet and G.A.van Aken, Langmuir, 19 (2003) 10210.
42. C.M.Wijmans, E.Dickinson, Langmuir, 14 (1998) 7278.
43. C.M.Wijmans, E.Dickinson, J.Chem.Soc.Faraday Trans., 94 (1998) 129.
44. C.M.Wijmans, E.Dickinson, Phys.Chem.Chem.Phys., 1 (1999) 2141.
45. E.Dickinson, B.S.Murray and G.Stainsby, J.Colloid Interface Sci., 106 (1985) 259.
46. J.Benjamins, A.Cagna and E.H.Lucassen-Reynders, Colloid Surface A, 114 (1996) 245.
47. L.G.Cascão-Pereira, O.Theodoly, H.W.Blanch and C.J.Radke, Langmuir, 19 (2003) 2349.
48. N.Wüstneck, B.Moser and G.Muschiolik.Colloid Surface B, 15 (1999) 263.
49. R.Miller, V.B.Fainerman, E.V.Aksenenko, M.E.Leser and M.Michel, Langmuir, 20 (2004) 771.
50. J.Maldonado-Valderrama, V.B.Fainerman, M.J.Gálvez-Ruiz, A.Martin-Rodriguez, M.A.Cabrerizo-Vílchez and R.Miller, J.Phys.Chem.B, 109 (2005) 17608.
51. J.Maldonado-Valderrama, V.B.Fainerman, E.V.Aksenenko, M.J.Gálvez-Ruiz, M.A.Cabrerizo-Vílchez and R.Miller, Colloid Surface A, 261 (2005) 85.
52. P.Cicuta and I.Hopkinson, J.Chem.Phys., 114 (2001) 8659.
53. V.Ducel, J.Richard, Y.Popineau and F.Boury, Biomacromolecules, 5 (2004) 2088.

54. J.Örnebro, T.Nylander and A.C.Eliasson.J.Cereal.Sci., 31 (2000) 195.

55. C.Monteux, C.E.Williams, J.Meunier, O.Anthony and V.Bergeron, Langmuir, 20 (2004) 57.

56. C.Monteux, G.G.Fuller and V.Bergeron, J.Phys.Chem.B, 108 (2004) 16473.

57. P.A.Gunning, A.R.Mackie, A.P.Gunning, N.C.Woodward, P.J.Wilde and V.J.Morris, Biomacromolecules, 5 (2004) 984.

58. B.S.Murray, A.Ventura and C.Lallemant, Colloid Surface A, 143 (1998) 211.

59. J.Zhao, D.Vollhardt, G.Brezesinski, S.Siegel, J.Wu, J.B.Li and R.Miller, Colloid Surface A, 171 (2000) 175.

60. P.Wilde, A.Mackie, F.Husband, P.Gunning and V.Morris, Adv.Colloid Interface Sci., 108-109 (2004) 417.

61. E.Dickinson, D.S.Horne and R.M.Richardson, Food Hydrocoll., 7 (1993) 497.

62. A.R.Mackie, A.P.Gunning, J.Wilde and V.J.Morris, Langmuir, 16 (2000) 2242.

63. R.J.Green, T.J.Su, J.R.Lu, J.Webster and J.Penfold, Phys.Chem.Chem.Phys., 2 (2000) 5222.

64. E.Dickinson and Y.Matsumura, Int.J.Biol.Macromolecules, 13 (1991) 26.

65. A.I.Romoscanu and R.Mezzenga, Langmuir, 21 (2005) 9689.

66. M.Faergemand and B.S.Murray, J.Agric.Food.Chem., 46 (1998) 884.

67. Y.Chanyongvorakul, Y.Matsumura, A.Sawa, N.Nio and T.Mori, Food Hydrocoll., 11 (1997) 449.

68. A.H.Martin, M.A.C.Stuart, M.A.Bos and T.van Vliet, Langmuir, 21(2005) 4083.

69. S.Hénon and J.Meunier, Rev.Sci.Instrum., 62 (1991) 936.

70. D.Hönig and D.Möbius, J.Phys.Chem., 95 (1991) 4590.

71. E.van der Linden, L.M.C.Sagis and P.Venema, Curr.Opin.Colloid Interface Sci., 8 (2003) 349.

72. B.S.Murray, B.Cattin, E.Schüler and Z.O.Sonmez, Langmuir, 18 (2002) 9476.

73. H.Rehage, B.Achenbach and F.-G.Klaerner, Langmuir, 18 (2002) 7115.
74. M.Sickert and F.Rondelez, Phys.Rev.Lett., 90 (2003) 126104.
75. T.M.Fischer, Phys.Rev.Lett., 92 (2004) 139603.
76. M.Sickert and F.Rondelez, Phys.Rev.Lett., 92 (2004) 139604.
77. D.K.Schwartz and C.M.Knobler, Phys.Rev.Lett., 73 (1994) 2841.
78. R.S.Ghaskadvi and M.Dennin, Rev.Sci.Instrum., 69 (1998) 3568.
79. M.J.Vogel, R.Miraghaie, J.M.Lopez and A.H.Hirsa, Langmuir, 20 (2004) 5651.
80. C.F.Brooks, G.G.Fuller, C.W.Frank and C.R.Robertson, Langmuir, 15 (1999) 2450.
81. S.A.Koehler, S.Hilgenfeldt and H.A.Stone, J.Colloid Interface Sci., 276 (2004) 420.
82. S.A.Koehler, S.Hilgenfeldt, E.R.Weeks and H.A.Stone, J.Colloid Interface Sci., 276 (2004) 439.
83. D.J.Olson and G.G.Fuller, J.Non-Newtonian Fluid Mech., 89 (2000) 187.
84. M.L.Pollard, R.Pan, C.Steiner and C.Maldarelli, Langmuir, 14 (1998) 7222.
85. A.A.Azadani, J.M.Lopez and A.H.Hirsa, Langmuir, 23 (2007) 5227.
86. R.Zangi, M.L.de Vocht, G.T.Robillard and A.E.Mark, Biophys.J., 83 (2002) 112.
87. M.L.de Vocht, K.Scholtmeijer, E.W.van der Vegte and O.M.H.de Vries, Biophys.J., 74 (1998) 2059.
88. C.R.Flach, F.G.Prendergast and R.Mendelsohn, Biophys.J., 70 (1996) 539.
89. H.Lavoie, B.Desbat, D.Vaknin and C.Salesse, Biochemistry, 41 (2000) 13424.
90. M.B.J.Meinders, G.G.M.van den Bosch and H.H.J.de Jongh, Eur.Biophys.J., 30 (2000) 256.
91. A.H.Martin, M.B.J.Meinders, M.A.Bos, M.A.Cohen Stuart and T.van Vliet, Langmuir, 19 (2003) 2922.
92. Y.Fang and D.G.Dalgleish, J.Colloid Interface Sci., 196 (1997) 292.
93. J.S.Phipps, R.M.Richardson, T.Cosgrove and A.Eaglesham, Langmuir, 9 (1993) 3530.

94. T.Cosgrove, J.S.Phipps and R.M.Richardson, Colloid Surface, 62 (1992) 199.
95. A.Zarbakhsh, A.Querol, J.Bowers and J.R.P.Webster, Faraday Discuss., 129 (2005) 155.
96. D.J.Cooke, C.C.Dong, J.R.Lu, R.K.Thomas, E.A.Simister and J.Penfold, J.Phys.Chem.B, 102 (1998) 4912.
97. I.P.Purcell, J.R.Lu jr., R.K.Thomas, A.M.Howe and J.Penfold, Langmuir, 14 (1998) 1637.
98. D.S.Horne, P.J.Atkinson, E.Dickinson, V.J.Pinfield and R.M.Richardson, Int.Dairy J., 8 (1998) 73.
99. T.Gutberlet, B.Kloesgen, R.Krastev and R.Steitz, Adv.Eng.Mat., 6 (2004) 832.
100. A.P.Maierhofer, M.Brettreich, S.Burghardt, O.Vostrowsky, A.Hirsch, S.Langridge and T.M.Bayerl, Langmuir, 16 (2000) 8884.
101. B.J.Clifton, T.Cosgrove, R.M.Richardson, A.Zarbakhsh and J.R.P.Webster, Physica B, 248 (1998) 289.
102. A.A.Van Well and R.Brinkhoff, Colloids Surface, A 175 (2000) 17.
103. M.Li, A.M.Tikhonov, D.J.Chaiko and M.L.Schlossman, Phys.Rev.Lett., 86 (2001) 5934.
104. M.L.Schlossman, Curr.Opin.Colloid Interf.Sci., 7 (2002) 235.
105. A.P.Gunning, A.R.Mackie, P.J.Wilde and V.J.Morris, Langmuir, 20 (2004) 116.
106. A.Anguelouch, R.L.Leheny and D.H.Reich, Appl.Phys.Lett., 89 (2006) 111914.
107. F.C.MacKintosh and C.F.Schmidt, Curr.Opin.Colloid Interface Sci., 4 (1999) 300.
108. T.Savin and P.S.Doyle, Phys.Rev.E., 76 (2007) 021501.
109. V.Prasad, S.A.Koehler and E.R.Weeks, Phys.Rev.Lett., 97 (2006) 176001.
110. C.Monteux, J.Kirkwood, H.Xu, E.Jung and G.G.Fuller, Phys.Chem.Chem.Phys., 9 (2007) 6344.
111. C.Stubenrauch and R.Miller, J.Phys.Chem.B, 108 (2004) 6412.

112. E.Dickinson, B.S.Murray and G.Stainsby, J.Chem.Soc.Faraday Trans., 84 (1988) 871.

113. S.Mun and D.J.McClements, Langmuir, 22 (2006) 1551.

114. A.Williams, J.J.M.Janssen and A.Prins, Colloid Surface A 125 (1997) 189.

115. D.B.Jones and A.P.J.Middelberg, AIChE J., 49 (2003) 1533.

116. P.Erni, P.Fischer and E.J.Windhab, Appl.Phys.Lett., 87 (2005) 244104.

117. P. Erni, P. Fischer, V. Herle, M. Haag and E.J. Windhab, ChemPhysChem, 9 (2008)1833.

118. E.Van Hemelrijck, P.van Puyvelde, S.Velankar, C.W.Macosko and P.Moldenaers, J.Rheol., 48 (2004) 143.

119. U.Jacobs, M.Fahrlander, J.Winterhalter and C.Friedrich, J.Rheol, 43 (1999) 1495.

120. A.Nadim, Chem.Eng.Comm., 150 (1996) 391.

121. K.D.Danov, J.Colloid Interface Sci., 235 (2001) 144.

122. J.G.Oldroyd, Prod.Royal Soc.London Ser.A, 232 (1955) 567.

123. H.A.Stone, S.A.Koehler, S.Hilgenfeldt, and M.Durand, J.Phys.-Cond.Mat., 15 (2003) S283.

124. D.A.Edwards, H.Brenner, and D.T.Wasan, Interfacial Transport Processes and Rheology.Butterworth-Heinemann, Stonheam, Massachusetts, 1991.

125. E.H.Lucassen-Reynders, Food Struct., 12 (1993) 1.

126. E.H.Lucassen-Reynders and K.A.Kuijpers, Colloid Surface, 65 (1992) 175.

127. HA.Stone, Annu.Rev.Fluid Mech., 26 (1994) 65.

128. H.P.Grace, Chem.Eng.Comm., 14 (1982) 225.

129. P.Fischer and P.Erni, Curr.Opin.Colloid Interface Sci., 12 (2007) 196.

130. G.I.Taylor, Proc.Royal Soc.London - Ser.A, 146 (1934) 501.

131. R.G.Cox, J.Fluid Mech., 37 (1969) 601.

132 C.E.Chaffey and H.Brenner, J.Colloid Interface Sci., 24 (1967) 258.

133 E.J.Hinch and A.Acrivos, J.Fluid Mech., 98 (1980) 305.

134. P.L.Maffettone and M.Minale, J.Non-Newton.Fluid Mech.78 (1998) 227.
135. P.L.Maffettone and M.Minale, J.Non-Newton.Fluid Mech., 84 (1999) 105.
136. S.Torza, R.G.Cox and S.G.Mason, J.Colloid Interface Sci., 38 (1972) 395.
137. V.Cristini, S.Guido, A.Alfani, J.Blawzdziewicz and M.Loewenberg, J.Rheol., 47 (2003) 1093.
138. K.Feigl, S.F.M.Kaufmann, P.Fischer and E.J.Windhab, Chem.Eng.Sci., 58 (2003) 2351.
139. K.Feigl, D.Megias-Alguacil, P.Fischer and E.J.Windhab, Chem.Eng.Sci., 62 (2007) 3242.
140. H.Stone and L.Leal, J.Fluid Mech., 220 (1990) 161.
141. X.Li and C.Pozrikidis, J.Fluid Mech., 341 (1997) 165.
142. Y.W.Kruijt-Stegeman, F.N.van de Vosse and H.E.H.Meijer, Phys Fluids, 16 (2004) 2785.
143. W.J.Milliken, L.G.Leal, J.Colloid Interface Sci., 166 (1994) 275.
144. R.W.Flumerfelt, J.Colloid Interface Sci., 76 (1980) 330.
145. W.J.Phillips, R.W.Graves and R.W.Flumerfelt, J.Colloid Interface Sci., 76 (1980) 350.
146. Y.P.Pawar and K.J.Stebe, Phys.Fluids, 8 (1996) 1738.
147. W.J.Milliken, H.A.Stone and L.G.Leal, Phys.Fluids, A 5 (1993) 69.
148. C.D.Eggleton, Y.P.Pawar and K.J.Stebe, J.Fluid Mech., 385 (1999) 79.
149. C.D.Eggleton, T.-M.Tsai and K.J.Stebe, Phys.Rev.Lett., 87 (2001) 048302.
150. P.Vlahovska, M.Loewenberg and J.Blawzdziewicz, Phys.Fluids, 17 (2005) 103103.
151. I.Bazhlekov, P.D.Anderson and H.E.H.Meijer, J.Colloid Interface Sci., 298 (2006) 369.
152. Y.Renardy, M.Renardy and V.Cristini, Eur.J.Mech.B − Fluids, 21 (2002) 49.

153. J.J.M.Janssen, A.Boon and W.G.M.Agterof, AIChE J., 43 (1997) 1436.

154. D.C.Tretheway and L.G.Leal, AIChE J., 45 (1999) 929.

155. C.D.Eggleton and K.J.Stebe, J.Colloid Interface Sci., 208 (1998) 68.

156. L.E.Scriven, Chem .Eng.Sci., 12 (1960) 98.

157. C.Pozrikidis, J.Non-Newton.Fluid Mech., 51 (1994) 161.

158. U.Seifert, Adv.Phys., 46 (1997) 13.

159. D.Barthès-Biesel and H.Sgaier, J.Fluid Mech., 160 (1985) 119.

160. D.Barthès-Biesel, A.Diaz and E.Dhenin, J.Fluid Mech., 460 (2002) 211.

161. A.Diaz, D.Barthès-Biesel and N.Pelekasis, Phys.Fluids, 13 (2001) 3835.

162. G.Breyiannis and C.Pozrikidis, Theor.Comput.Fluid.Dyn., 13 (2000) 327.

163. S.Ramanujan and C.Pozrikidis, J Fluid.Mech., 361 (1998) 117.

164. C.Pozrikidis, J Fluid.Mech., 440 (2001) 269.

165. A.Walter, H.Rehage and H.Leonhard, Colloid Surface A, 183 (2001) 123.

166. K.S.Chang and W.L.Olbricht, J.Fluid.Mech., 250 (1993) 609.

167. M.Abkarian, M.Faivre and A.Viallat, Phys.Rev.Lett., 98 (2007) 188302.

168. J.M.Skotheim and T.W.Secomb, Phys.Rev.Lett., 98 (2007) 078301.

169. D.E.Discher, N.Mohandas and E.A.Evans, Science, 266 (1994) 1032.

170. K.H.de Haas, D.van den Ende, C.Blom, E.G.Altena, G.J.Beukema and J.Mellema, Rev.Sci.Instrum., 69 (3) (1998) 1394.

171. V.Kantsler and V.Steinberg, Phys.Rev.Lett., 95 (2005) 258101.

172. B.P.Binks, Curr.Opin.Colloid Interface Sci., 7 (2002) 21.

173. J.Lucassen, Colloid Surface, 65 (1992) 139.

174. A.Subramaniam, M.Abkarian, L.Mahadevan and H.Stone, Langmuir, 22 (2006) 10204.

175. A.B.Subramaniam, M.Abkarian and H.A.Stone, Nat.Mater., 4 (2005) 553.

176. S.Arditty, V.Schmitt, F.Lequeux and F.Leal-Calderon, Eur.Phys.J.B, 44 (2005) 381.

177. D.Vella, P.Aussillous and L.Mahadevan, Europhys.Lett., 62 (2004) 212.

178. J.K.Ferri, W.F.Dong, R.Miller and H.Möhwald, Macromolecules, 39 (2006) 1532.

179. K.S.Chang and W.L.Olbricht, J.Fluid Mech.250 (1993) 587.

180. J.F.Palierne, Rheol.Acta, 29 (1990) 204.

181. T.Jansseune, J.Mewis, P.Moldenaers, M.Minale and P.L.Maffettone, J.Non-Newton Fluid Mech., 93 (2000) 153.

182. D.Graebling, A.Benkira, Y.Gallot and R.Muller, Eur.Polym.J., 30 (1994) 301.

183. H.Gramespacher and J.Meissner, J.Rheol., 36 (1992) 1127.

184. S.J.Choi and W.R.Schowalter, Phys.Fluids, 18 (1975) 420.

185. R.G.Larson, The structure and rheology of complex fluids.Oxford University Press, Oxford, UK, 1999.

186. S.D.Hudson, J.T.Carbal, W.J.Goodrum, Jr., K.L.Beers and E.J.Amis, Appl.Phys.Lett., 87 (2005) 081905.

187. T.M.Squires and S.R.Quake, Rev.Mod.Phys., 77 (2005) 977.

188. C.Friedrich and Y.Antonov, Macromolecules, 40 (2007) 1283.

189. J.Bibette, F.Leal-Calderon, V.Schmitt and P.Poulin, Springer Trans.Mod.Phys., 181 (2003) 1.

190. P.Sollich, F.Lequeux, P.Hebraud, and M.E.Cates, Phys.Rev.Lett., 78 (1997) 2020.

191. P.Coussot, Soft Matter, 3 (2007) 528.

192. S.Cohen-Addad, M.Krzan, R.Hohler, and B.Herzhaft, Phys.Rev.Lett., 93 (2004) 028302.

193. H.M.Princen, J.Colloid Interface Sci., 91 (1983) 160.

POBING MECHANICAL PROPERTIES OF LANGMUIR MONOLAYERS WITH OPTICAL TWEEZERS

Z. Khattari[1] and Th. M. Fischer[2]

[1] Physics Department, Hashemite University, Zarqa, Jordan
[2] Institut für Experimentalphysik V, Universität Bayreuth,
Universitätsstr. 30, 95440 Bayreuth, Germany

Contents

A. INTRODUCTION

Langmuir-monolayers at the air/water-interface have been investigated since the first film balance was developed by Langmuir in 1917 [1], and this device remains the central part of any setup used to study monomolecular films at the air/water-interface. Various methods have been developed to investigate their electrical [2], thermodynamic [3], structural [4] and viscoelastic [5] properties. For the last 20 years it has been possible to visualize monolayers either by Brewster angle (BAM) [6,7] or by fluorescence microscopy [8,9].

Various microstructures can be observed in the phase coexistence regions of the quasi two dimensional Langmuir monolayers. This is in contrast to 3d-systems, where the coexisting phases are macroscopically separated, in order to minimize the interfacial energy of the system. This peculiarity of the 2D-system is due to the competition between long-range dipolar interactions of the molecules and the line tension of the phase boundaries. Many of the structures observed on Langmuir-monolayers can be understood using this model.

The flow profile at the interface is determined by the rheological properties of the 2D-monolayer and the 3d-subphase. The ratio of the surface shear viscosity of the monolayer and the viscosity of the subphase defines a characteristic length. Above and below this characteristic length different flow profiles and relaxation dynamics of non-equilibrated two dimensional liquid systems are expected. On the larger length scale the subphase flow dominates, while on the smaller length scale the flow is mainly determined by the rheological properties of the monolayer. Different devices have been developed to investigate the viscosity of Langmuir monolayers. The torsion pendulum knife edge shear rheometer [10] works on a millimeter scale and is sensitive to relatively large surface shear viscosities. Smaller viscosities and structures may be investigated by the shear distortion of liquid condensed (LC) domains in liquid expanded (LE) surroundings using a rotating trough [11]. Heckl, Miller and Möhwald [12] used an inhomogeneous electrical field around a needle to create electrophoretic motion and distort circular LC-domains. The canal flow experiment [13] allows the observation of the circular, two dimensional flow profile driven by surface pressure gradients. None of these techniques can be used to manipulate monolayers on the micrometer length scale while leaving the surroundings undisturbed. The device presented in this work, a combination of optical tweezers with fluorescence and Brewster angle microscopy, enables this kind of handling.

In this study, the monolayer is mechanically manipulated using particles with a higher refractive index than their surroundings in a focussed laser beam. The forces exerted on these particles are in the range of several pN. In addition, local heating by the focused laser causes temperature gradients which disturb monolayer patterns. Various applications of both techniques are presented.

B. EXPERIMENTAL

A schematic drawing of the combination of tweezers with fluorescence microscopy is shown in Figure 1. Two simultaneous operations are performed by a 100x water immersion objective, numerical aperture 1.0, built into the bottom of a temperature controlled film balance. First it projects a fluorescence image of the monolayer onto a SIT camera (Hamamatsu C 3077-01). Fluorescence is excited in the labeled monolayer by a p-polarized argon ion laser (465 nm, 150 mW). The excitation light is blocked by a filter in front of the detector. Second, two IR-laser beams (λ=1064 nm, P=10 mW-4 W), coupled into the optical path using a dichroic mirror (transparent for fluorescence light), are focussed by the objective onto the monolayer. The diameter of the beams is fit to the objective entrance in order to maximize the lateral optical forces. Before the beams are reflected at the dichroic mirror both are focused separately using Galilee optics at the air water interface, where one of the beams can be moved laterally by two rotating mirrors (not shown in Figure 1).

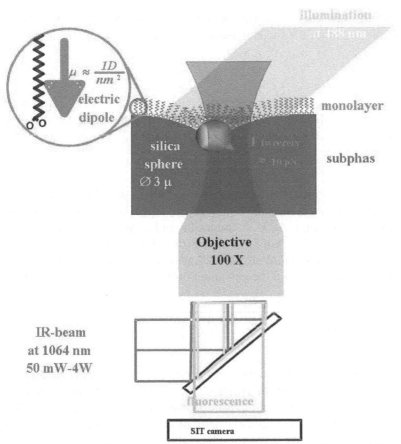

Fig. 1 Sketch of a fluorescence microscopy combined with optical

1. Thermal manipulation

The applied laser power P (10 mW - 4 W) causes an observable local heating. This can melt the low-temperature phase in a locally defined region ($10\mu m \leq r \leq 100\mu m$). In the beam waist of the laser, a fraction of 10^{-6} of the laser power P is converted into heat. Assuming this heat is dissipated into the subphase by heat conduction only, the rise in temperature within the focus is estimated to be

$$\Delta T = \alpha \frac{3P}{4\pi\kappa} \tag{1}$$

Here, $\kappa \approx 0.6$ W/m.K is the heat conductivity of water. Equation 1 predicts that applying 1W laser power creates the local temperature difference of $\Delta T \approx 5$K.

1.1. Cavitation bubbles

Fig. 2 shows a laser induced cavitation bubble of radius $R \approx 8$ μm in a methyloctadecanoate monolayer at an external Temperature of $T_\infty = 37°$C, the area per molecule $A \approx 35$ Å^2/molecule and a laser power of $P = 1.2$W. In previous work [14], we have shown that the cavitation bubble arises due to thermal Marangoni stress leading to a convection roll that opens the bubble. We estimated the cavitation bubble radius as

$$R \approx -\frac{d\sigma_W}{dT} \frac{\alpha r_F P}{\kappa(\pi_\infty - \pi_b)\left(\dfrac{D(\pi_\infty - \pi_b)}{\beta E} + 1\right)} \tag{2}$$

Here $d\sigma_W/dT$ is the slope of the surface tension of water with temperature, r_F is the focal width of the laser focus, π_b is the binodal surface pressure of the liquid expanded/ gas coexistence, D and E are the diffusion constant and Gibb's elasticity of the gaseous phase and β is the temperature diffusivity of water. Note that this equation predicts the cavitation bubble radius to increase as the external pressure is lowered toward the binodal pressure. The closer one is to liquid expanded gas coexistence the larger the cavitation bubble will be. The cavitation bubble in Fig. 2 is far from the coexistence region, and the radius is small. No flow other than always present air convection driven flow is observed in the liquid expanded phase at these conditions. Laser induced cavitation bubbles arise due to thermal Marangoni stress producing a convection roll in the subphase that opens the bubble.

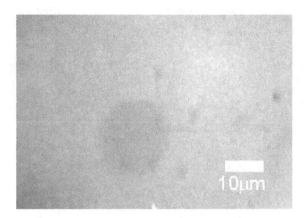

Fig. 2. Laser generated gas bubble at an IR laser power of $P=1.2W$ in liquid expanded phase of methyloctadecanoate labeled with 0.8% of fluorescence dye (BODYPY), $T=37^{\circ}C$ $A=35$ $\mathring{A}^2/molecule$.

1.2. Thermo capillary flow past the cavitation bubble

The situation changes when further increasing the laser power or if we lower the external surface pressure closer to the binodal pressure (expand the monolayer) [14]. A stationary convective flow in an arbitrary direction breaks the radial symmetry of the flow pattern. The convective flow causes the laser induced cavitation bubble to be displaced from the laser focus further downstream by a distance d. We could measure the flow profile of this convective flow by simply increasing the laser power. Increasing the IR laser power results in the formation of smaller satellite bubbles around the cavitation bubble, these satellite bubbles move along the monolayer. We then track the motion of the smaller gas bubbles that move around the laser induced cavitation bubble.

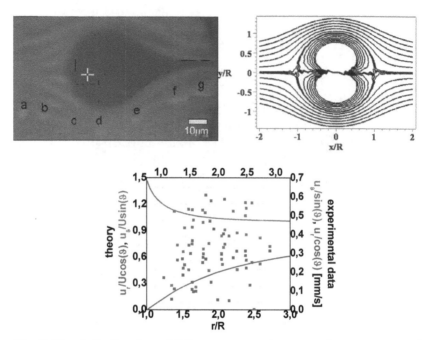

Fig. 3 (*Top left*): Overlay of 7 fluorescence microscope images separated by 80 ms each of a DMPC/DChol mixture illuminated with an IR laser. The black stream line shaped laser induced cavitation bubble is marked with a cross hair that indicates the point where the laser is focused. One dihydrocholesterol rich domain has been marked in green (a-g) in each image so one can see the flow of this domain while time progresses. Video clip 1 shows the motion of the DMPC/DChol liquid expanded mixture around the laser induced cavitation bubble; (*Top right*) Theoretically predicted streamlines for a thermo capillary flow driven by a thermal gradient via the temperature dependence of the liquid gas line tension; (*Bottom*) Normalized theoretical and experimental radial and azimuthal velocity profile plotted as a function of the normalized radial position. The theory predicts that the data should fall on this master curve for any azimuthal position. We extract a value of U=0.47 mm/s for the thermo capillary velocity.

This way of tracking the monolayer flow, however, is accompanied by large distortions of the laser induced cavitation bubble due to the high Peclet number $Pe=\eta R^2 \gamma/\lambda$ (η the viscosity of the subphase , γ the shear rate and λ the line tension between the gaseous and the liquid expanded phase).

Less deformation of the cavitation bubble is observed when using liquid domains in a coexistence region between two liquid phases in a mixed monolayer for the tracking of the motion. This is done in a monolayer mixture of 58% DMPC, 40% dihydrocholesterol, and 2% NBD-HDA at $T_\infty = 23°C$, A= 200 $Å^2$/molecule. Fig. 3 shows an overlay of 7 images separated by a time of 80 ms of this mixture exposed to an IR laser of P=2W.

The symmetry breaking flow around the laser induced cavitation bubble is from the left toward the right and dihydrocholesterol rich domains are swept along the streamlines of the flow. The laser induced cavitation bubble is circular with the exception of the rear of the bubble that is deformed toward a streamlined shape.

Breaking of the radial symmetry leads to a temperature gradient along the liquid expanded / gas phase boundary of the laser induced cavitation bubble. If the displacement d of the bubble from the laser center is small and assuming that the temperature profile around the laser is dominated by heat conduction we estimate the temperature gradient along the bubble boundary as:

$$\frac{dT}{d\vartheta} = -\frac{\alpha P r_F d \sin\vartheta}{2\pi\kappa R^2} \tag{3}$$

We believe that this temperature gradient is the major source for the flow of the monolayer surrounding the laser induced cavitation bubble. Since the line tension $\lambda(T)$ of the liquid expanded/ gas phase boundary depends on temperature, the temperature gradient will cause a line tension gradient

$$\frac{d\lambda}{d\vartheta} = \frac{d\lambda}{dT} \cdot \frac{dT}{d\vartheta} \tag{4}$$

that gives rise to thermo capillary flow. Similar flow is observed for three dimensional bubbles that are immersed in a liquid with a linear temperature gradient [15-20]. In order to estimate the monolayer flow caused by the thermal gradient we consider a 2D circular gas bubble embedded in liquid surroundings subject to a thermal gradient. The 2D-Marangoni flow that opens the laser induced cavitation bubble effects only the flow inside the laser induced cavitation bubble not the flow in the liquid expanded phase that we measure. Therefore in our theoretical derivation we neglect the 2D-Marangoni flow that superposes the thermo capillary flow caused by the line tension gradient and just accept the presence of the bubble. Both the gaseous and the liquid expanded phase are 2D-fluids of negligible surface shear viscosity [21]. Viscous dissipation therefore is governed by the flow of the 3d subphase under the monolayer. A positive temperature dependent line tension λ between the liquid and gaseous phase is associated with the phase boundary. Since the temperature varies as a function of the location at the surface so does the line tension. It is the velocity profile generated by the gradient in line tension that we wish to calculate. The flow of the subphase is described by Stokes equations and the continuity equation:

$$-\nabla p + \eta \nabla^2 \mathbf{u} = 0 \quad and \quad \nabla \cdot \mathbf{u} = 0 \tag{5}$$

where p is the subphase pressure and $\mathbf{u}$ denotes the subphase velocity. The fluid motion of the liquid expanded and gaseous phase obeys the 2D-Stokes equation:

$$-\nabla_s \pi_m + \eta \left. \frac{\partial \mathbf{u}}{\partial z} \right)_s - \frac{\lambda(\vartheta)}{R} \delta(r - R) \mathbf{e}_r = 0 \quad for \quad z = 0 \tag{6}$$

Here π_m is the surface pressure of the monolayer, ∇_s the surface gradient, the second term is the viscous drag force from the subphase and the last term ensures that the dynamic normal surface stress satisfies the Laplace condition. $\lambda(\vartheta)$ is the line tension of the liquid expanded/ gaseous interface and $\delta(r-R)$ is Dirac's delta function. The surface velocity profile solving this equation is:

$$u_r(z=0) = U \cos\vartheta \begin{cases} 1 - \dfrac{1}{2}[K(r/R) - D(r/R) + E(r/R)] & \text{for } r < R \\[2mm] 1 - \dfrac{R}{2r}[K(R/r) - D(R/r) + E(R/r)] & \text{for } r > R \end{cases}$$ (7)

$$u_\vartheta(z=0) = U \sin\vartheta \begin{cases} 1 - \dfrac{1}{2}[K(r/R) - D(r/R) - 2E(r/R)] & \text{for } r < R \\[2mm] 1 - \dfrac{R}{2r}[-2K(R/r) + 2D(R/r) + E(R/r)] & \text{for } r > R \end{cases}$$

where $K(k)$ and $E(k)$ are the complete elliptic integrals of the first and second kind [22], and $D(k)=[K(k)-E(k)]/k^2$. Equation (13) therefore predicts the equatorial velocity to be 3/2U a factor of 3/2 larger than the asymptotic velocity. The normalized radial and azimuthal velocity profiles eq(7) are plotted as a function of the normalized radius in Fig.3 (bottom right). We see that the data extracted from the images in figure 3 top left follows the trend of the theoretical prediction, although with considerable scatter of data [23]. The comparison of experiment and theory yields a thermo capillary velocity of U=0.47 mm/s. The stream function ψ at the surface reads:

$$\psi(z=0) = U \sin\vartheta \left[r - \frac{r^<}{2} \left\{ K(r^</r^>) + D(r^</r^>) - E(r^</r^>) \right\} \right]$$ (8)

$$\text{where } r \le \min(r,R), \quad r \ge \max(r,R),$$

The streamlines of the surface flow are given by $\psi(z=0)=const$ and they are displayed in Fig.3 top right. They correspond quite well to the streamlines seen in figure 3. The relation between the thermo-capillary velocity U and the line tension gradient is:

$$U = \frac{4}{3\pi\eta R \sin\vartheta} \frac{d\lambda}{d\vartheta}$$ (9)

Combining the result in equation (9) with (3) and (4) yields.

$$\frac{d\lambda}{dT} = -\frac{3\pi^2 \eta R^3 U \kappa}{2\alpha Pr_F d} \tag{16}$$

Using the experimental values and estimating $d \approx 1\mu m$ equation (16) predicts a line tension gradient of $d\lambda/dT = 2.4$ pN/mK.

Sufficient heating produces a thermo capillary flow. Gradients in temperature and a temperature dependent line tension propel the liquid surroundings around the laser induced cavitation bubble that is fixed in position. We could compare the experimental flow profile with theoretical predictions with satisfactory agreement

1.3. Laser induced collapse

As we have shown heating of a monolayer at low surface pressure can induce cavitation bubbles. In this section we show that heating of a monolayer at elevated surface pressure may also induce collapse of the monolayer [24]. The phase diagram in figure 4 displays the temperature dependence of the collapse pressure of methyloctadecanoate. The collapse pressure decreases with increasing temperature with a slope $d\pi_c/dT \approx -0.7 \ 10^{-3}$ N/mK that is significantly smaller than the change of the surface tension of the bare air/water interface with temperature $d\sigma_w/dT = -0.17 \ 10^{-3}$ N/mK. A monolayer experiencing a temperature gradient is in mechanical equilibrium, if the gradient in surface tension vanishes $0 = \nabla\sigma = \nabla\sigma_w - \nabla\pi = (d\sigma_w/dT)\nabla T - \nabla\pi$ which occurs if

$$\nabla\pi = \frac{d\sigma_W}{dT}\nabla T \tag{17}$$

where, π and T denote the surface pressure and temperature and σ_w is the surface tension of the bare air/water interface.

Mechanical equilibrium lines are therefore isotension lines that appear as lines with slope $d\sigma_w/dT$ in the surface pressure versus temperature diagrams. If one starts in the liquid condensed monolayer phase and locally heats the monolayer then the system locally moves from its original point in the phase diagram along an isotension toward higher temperatures. We assume that the surface pressure $\pi(\rho_s, T)$ is a state function of the surface density ρ_s and the temperature T. The mechanical equilibration is achieved by a transient surface flow that generates a surface density gradient of surfactants $\nabla\rho_s$ in accordance to equation (17)

$$\nabla\rho_s = -\frac{\dfrac{\partial(\pi - \sigma_W)}{\partial T}\bigg)_{\rho_s}}{\dfrac{\partial\pi}{\partial\rho_s}\bigg)_T}\nabla T \tag{18}$$

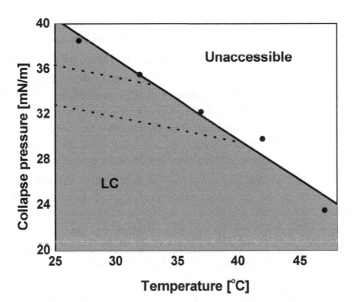

Fig. 4. Collapse pressure versus temperature for methyloctadecanoate, obtained from isotherms at a compression speed of 50cm^2/min. The Solid line represents linear fit for the data points. The dotted lines are lines of constant surface tension.

Generically this transient flow is from the hotter to the colder region ($\nabla \rho_s / \nabla T < 0$). Upon further heating one eventually reaches the intersection point of the isotension with the collapse line. For temperatures higher than this intersection temperature the mechanical equilibrium is no longer possible since there are no equilibrium states beyond the collapse. Rather than further following the isotension, heating will partially convert the monolayer toward a three dimensional collapsed phase. The system follows the collapse line (black line in Figure 2) with a surface tension gradient,

$$\nabla \sigma = -\frac{d(\pi_c - \sigma_W)}{dT} \nabla T, \qquad (19)$$

with due to Gibb's phase rule the collapse pressure $\pi_c(T)$ unlike the surface pressure $\pi(\rho_s, T)$ being a function only of the temperature. If $d\sigma_W / dT > d\pi_c / dT$ the non vanishing surface tension gradient generates a surface flow that points along the temperature gradient (from cold to warm) in the opposite direction than the transient flow that occurs while staying away from the collapse line. Simply speaking, the direction of the surface flow of the monolayer is determined by its collapse line, that is, when the monolayer is at the collapse line and $d\sigma_W / dT > d\pi_c / dT$ the surface flows from colder to hotter regions, on the other hand, when the monolayer is far away from the collapse line the flow is from hotter to colder regions. Figure 5 shows consecutive fluorescence microscopy images of a methyloctadecanoate monolayer at T= 37 °C, $\pi \leq \pi_c = 33$ mN/m locally heated by the NdYag laser ($\lambda = 1064$nm, P= 3.6W). The laser is focused on the central region of the image (cross hair), where part of the light is adsorbed and heats the surrounding water and the monolayer.

Initially the entire monolayer is in the liquid condensed monolayer phase. Upon sufficient heating above a threshold laser power $P > P_c \approx 2$W a radial flow of the surface toward the center sets in. The flow can be measured quantitatively by following the characteristic texture of the monolayer as a function of time.

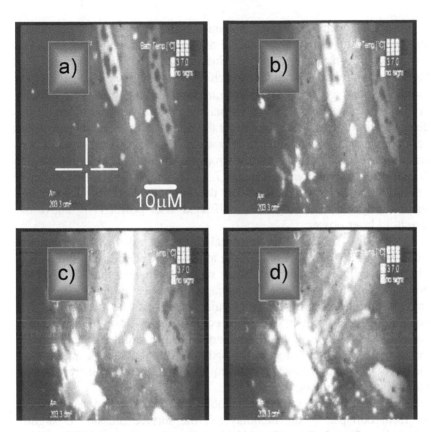

Fig. 5 Fluorescence microscopy images of a methyloctadecanoate monolayer T=37°C π ≤ π_c=33 mN/m showing the collapse of molecules at the focus point of laser (cross hair) at various times after increasing the IR laser to P=3.6W, a) t=0 s; b) t= 1 s; c) t= 2 s; d) t= 7 s. A bright fluorescence arising from the developing 3d aggregate can be seen in the hot spot as time progresses. The velocity profile of the monolayer can be measured by following characteristic points in the monolayer texture as time progresses. Video clip 1 shows the entire time sequence of the local collapse including the recovery of the monolayer after turning off the IR-laser (link to be inserted).

Surfactant material aggregates into 3-dimensional a much denser structures in the hot spot, visualized as highly fluorescent material. The collapsed 3d material grows in radius being fed by the periphery that moves with a radial inward flow reaching from the edge of the aggregate toward regions as far as 500-1000 μm from the hot spot. The velocity profile in the periphery is approximately inversely proportional to the distance from the laser implying that the density of the monolayer in the periphery does not change. It is only at the edge of the aggregate where an abrupt jump in surface density from the monolayer collapse density $1/A_c$ toward the aggregate density $1/A_{aggr}$ occurs. The initial collapse velocity of the inward flow of the monolayer does not change significantly when heating with a different laser power $P>P_c$ above the threshold laser power.

Within few seconds the peripheral flow slows down and eventually stops when the aggregate has grown to a disk of equilibrium radius R_{eq}. The former liquid condensed phase collapses to less than 1% of the original area implying that the average thickness of the 3d aggregate amounts to more than 100 molecular layers of the surfactant. The growths of the aggregate toward its equilibrium radius at different laser powers are shown in figure 5. At short times the radius of the aggregate increases linearly with time until it saturates and reaches the equilibrium radius R_{eq}. Figure 6 shows that the equilibrium radius R_{eq} increases with the applied laser power.

Local heating of a Langmuir monolayer with a focused laser induces the local collapse of the monolayer if the surface pressure is slightly below the collapse pressure and if the collapse pressure of the monolayer decreases faster with temperature than the surface tension of the pure air/water interface. Local collapse is associated with the growth of a 3 dimensional collapse aggregate that is fed by an incompressible inward flow of the two dimensional monolayer surroundings. It should be possible to locally induce collapse in lung surfactant monolayers which would enable time resolved local studies of the monolayer collapse in the lung.

2. Mechanical manipulation

The focus of the laser does not only heat the monolayer but it can be also used as optical tweezers [25]. The force of the optical tweezers results from the sum of all momentum transfers of the photons passing the water/bead and bead/air interfaces and depends on the position of the bead in the tweezers. For an exact calculation of the force the electromagnetic field in the vicinity of the focus in the presence of the bead must be known. With the addition of the air/water-interface near the laser focus, this becomes a difficult problem. The maximum force exerted on such beads in the direction of the beam, in bulk water can be estimated: assuming that approximately 3 percent of the light pressure actually acts on the bead, one obtains a vertical tweezers force of [26]:

$$F_{tweezers} \approx 0.03(n_W P / c),\tag{20}$$

where n_w denotes the refractive index of water and c the speed of light. In the case of beads floating at the air/water-interface, the lateral optical forces obtained are one order of magnitude higher than those predicted by eq 20. For silica beads [$\varnothing$=(3–10) μm], the maximum force exerted by the tweezers is of the order of 1 nN at a laser power of 1 W.

The optical force on the beads depends strongly on the location of the air/water-interface in the laser focus, so it is important to control the immersion depth of the beads. For beads with radii in the order of micrometers, capillary forces ($F_c \approx 100$ nN) dominate and gravitational forces are negligible: the immersion depth of the bead is controlled by its contact angle with the air/water-interface.

Since the tweezers are to be used as an accurate mechanical manipulator their force must be calibrated. As the contact angle at the particle depends on the system, this has to be performed directly on the monolayer under study. In our experiments, the calibration is performed in the LE/LC phase coexistence region of the monolayer, where the surface pressure (contact angle) is roughly independent of the area per molecule. The silica spheres are adsorbed at the 1d-phase boundaries between coexisting monolayer phases due to dipolar interactions of the beads with the monolayer [27]. With a single silica sphere trapped within the focus of the laser beam, the monolayer phase boundary can be manipulated.

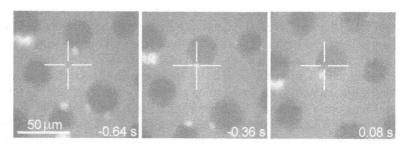

Figure 6 Capture and detachment of a liquid condensed domain of methyl-octadecanoate from the optical tweezers (P=150 mW). (a) The domains are flowing towards the bottom of the image with a velocity of v=(90±20) μm/s. (b) One domain [a=(15±1) μm] is trapped within the tweezers. The other domains continue to flow. A slight increase in the velocity [$v_{rupture}$=(100±20) μm/s, t=0 s] detaches the bead with the domain from the tweezers (c). The calculated optical force is $F_{tweezers}$=(65±15) pN.

In Figure 6 (t=-0.64 s) a silica sphere is adsorbed at a methyl-octadecanoate LC-domain (T=35 °C). The tweezers trap the bead with its domain (t=-0.36 s) while the neighboring domains continue to flow. The force of the tweezers (P=150 mW) may be determined by measuring the velocity of the surrounding LE-phase that is needed to detach the LC-domain with the bead from the tweezers (t=0.08 s). At the detachment velocity (v=100±20 μm/s, t=0 s) the hydrodynamic drag force equals the force of the tweezers. With the viscosity of the subphase (η=0.725 mPa·s [289] and the radius [a=(15±1) μm] of the LC-domain, the hydrodynamic drag force on a solid disk can be calculated [29]:

$$\vec{F} = -8\eta a \vec{v},$$ (21)

which yields a tweezers force of $F_{tweezers}= (9\pm2)$ pN).

3. Measurement of line tension

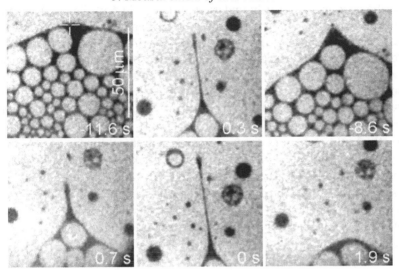

Fig. 7 Fluorescence microscope image of a methyl octadecanoate monolayer in the three phase coexistence region LC (not present in the field of view), LE (bright), G (dark) (T= 27.5 °C, A = 35 Å²/molecule). A silica bead (non fluorescent) of radius a = 2 µm (t = -11.6 s), is trapped within the tweezers (white cross). As the monolayer is flowing (v ≈ 5 µm/s) the LE/G boundary gets deformed (t = -8.6 s) until an elongated stripe, with a droplet shaped head encapsulating the bead is created (t = 0 s). The tweezers are switched off and a shortening of the stripe is observed with time (t = 0.3 s, 0.7 s) until the stripe completely disappears (t = 1.9 s). The shortening enables the measurement of the line tension between the phases.

Fig. 7 shows the deformation and relaxation of a G-region (dark) in LE-phase (bright) carried out at an area per molecule of $A_{mol} = 35$ Å² /

molecule, where the LC-phase covers a minor area fraction of the surface and is not present in the image. One of the silica beads is trapped by the optical tweezers. The monolayer is a subject of a small drift ($v \approx 5$ µm/s) and the bead hits a G/LE-boundary (Fig. 1, t = -11.6 s). The optical force exceeds twice the line tension and the deformation of the boundary by the tweezers increases with time (Fig. 1, t = -11.6 s, -8.6 s and 0 s). An elongated stripe of gaseous phase completed by a droplet shaped head encapsulating the bead in the tweezers is obtained. Assuming the elongation proceeds at constant area of the gas phase an aspect ratio of the stripe shape of 10^3 can be reached. This is possible in two dimensions since there exists no Rayleigh Taylor instability [30], which would break the stripe into a chain of separated droplets in three dimensions. At time t = 0 the tweezers are switched off and the relaxation of the stripe starts. The relaxation is described by the balance of static (line tension F_λ) and viscous forces (drag force of the encapsulated bead F_η). The force of the line tension of both sides $F_\lambda = 2\lambda$ of the stripe is the driving force for the relaxation, while the drag force oppose the shortening of the stripe:

$$F_\eta + F_\lambda = 0 \tag{22}$$

The hydrodynamic drag force on the bead, moved in an air/water interface, is given by $F_\eta = 8a\eta_{sub}\dfrac{dL}{dt}$, where a denotes the radius of the bead, η_{sub} the viscosity of the subphase Resolving for the line tension λ of the LE/gas boundary one finds [31] a line tension of

$$\lambda = 4\eta_{sub}av = 0.5 \text{ pN.} \tag{23}$$

The tweezers also allow to measure the line tension between a condensed and a liquid expanded phase [32]. Figure 8 shows a domain, held in flowing LE-surroundings (flow direction opposite to the tweezers force F_{ext}). The flow deforms the circular shape and after several seconds a stationary shape with a wedge is formed. Here, the viscous force $\vec{F}_\eta = 8R\eta_{sub}\vec{v}$, acting on the domain interface, has to be compensated by F_{ext} at the working point to keep the domain in position.

R denotes the radius of the droplet, η_{sub} the viscosity of the subphase, $\vec{v}$ the velocity of the surroundings. Using R=(21±1) μm, η_{sub}=0.725 mNs/m² [33], v=(58±10) μm/s the drag force on the domain in Fig. 2 is found to be F_η=(7±1.5). For a stationary droplet shape a force balance between F_{ext} and the line forces at the wedge is required:

$$F_{ext} = 2\lambda \cos\frac{\theta}{2} \qquad (24)$$

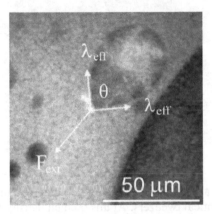

Fig. 8 A LC droplet captured by optical tweezers and deformed by flow. A silica bead adsorbed at the LE/LC boundary is held by the tweezers fixing the position of the domain. The surroundings flow opposite to Fext with velocity v5(58±10) μm/s and deform the originally circular domain (R 5(21±61) μm). From the wedge angle θ=90° an effective line tension λ=(5±1) pN is determined using. .

From the experiment (Fig. 2) a wedge angle θ=90° is measured. Using Eq. (23) and Eq. (24) one finds λ=(5±1) pN.

In Figure 9 the mechanical manipulation of a of bis[8-(1,2-dipalmitoyl-sn-glycero-3-phosphoryl)-3,6-dioxaoctyl]disulfide ("thiolipid") Langmuir-monolayer [34] is shown. In the LC/ LE phase coexistence region, the condensed domains are star shaped (Figure 9a).

A captured silica bead hits the edge of one arm of the star and causes it to rotate (Figures 9a-f). In principal the construction of a two dimensional gearbox, consisting of 10^9 molecules (1fl), is possible.

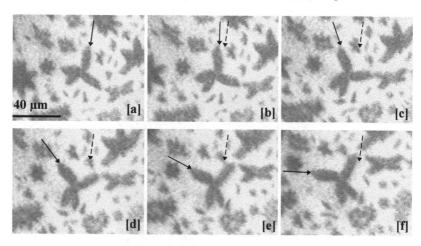

Fig. 9 Rotation of a star shaped condensed domain in a thiolipid Langmuir-monolayer using optical tweezers. At this domain density the individual stars do not mesh. The tweezers are located at the black arrow and the dashed arrow indicates the original location.

In thiolipid films at low densities of LC-domains, direct contact between individual domains rarely occurs. The individual domains are only affected by dipole and hydrodynamic interactions. The flow profile of the domains surroundings around a trapped silica bead is laminar (Figure 10a), as determined by following the trajectories of several domains as they flowed past the trapped bead. A perturbation of the flow of domains is visible in a region reaching approximately 20 μm from the trapped bead.

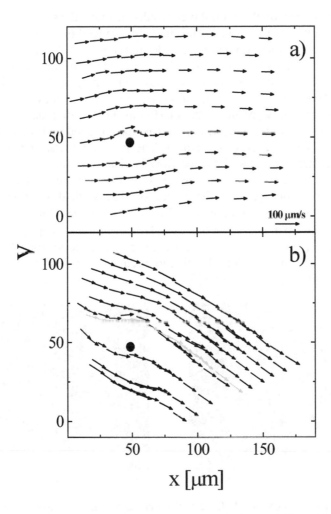

Fig. 10 The transition from laminar to a chaotic flow within a thiolipid monolayer: a) below and b) above the critical meshing density (the trajectory of one of the two domains which switch position as they are interlocked is marked by grey arrows to distinguish it from the second crossing trajectory).

Compression of the monolayer results in more densely packed LC-stars as shown in Figure 11a. The stars are arranged on a regular hexatic structure. Individual domains still do not touch, but the shear flow around the trapped domain now leads to chains of interlocked stars (Figure 11b; one of the chains is marked in white) in the vicinity of a trapped bead. The original hexatic arrangement is destroyed and the corresponding flow profile is chaotic (Figure 11b). Individual trajectories may cross each other when a pair of stars are meshed and rotate around each other (see the gray trajectory). The meshing leads to long range steric interactions and a disturbed flow is visible up to 50 μm from the location of the tweezers. The flow is no longer symmetric and changes its direction after passing the tweezers. The critical fraction of the condensed phase at a velocity $v_\infty = (30 \pm 5)$ μm/s, above which the chaotic flow is observable, is $\phi_c(v) = (30 \pm 5)$ %. In 3d, dense colloidal suspensions, shear flow is known to create chaotic trajectories of individual colloidal particles leading to a reduction in density of colloidal particles within the shear zone [35].

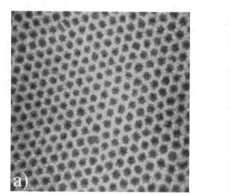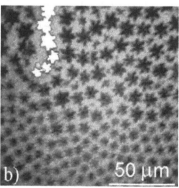

Fig. 11 Textures of the LC/LE coexistence region in a thiolipid monolayer: a) a regular hexatic arrangement of unmeshed domains with the tweezers switched off b) the formation of interlocked chains (one chain is marked white) as the monolayer flows around the trapped bead.

SUMMARY

Local thermal and mechanical manipulation techniques capable of deforming coexisting monolayer phases have been presented. These techniques enable the determination of the nature of the interactions between the different phases, the measurement of surface shear viscosities [23], line tension [21], and dipole density differences [36]. Typical optical forces exerted on trapped particles range from (1-100) pN. The technique allows the analysis of mechanical properties of two dimensional solids, liquid crystals and liquids and might be used for the buildup of two dimensional engines consisting of the order of 10^9 molecules.

ACKNOWLEDGMENTS.

This material is based upon work supported by the National Science Foundation under CHE-0649427.

REFERENCES

1. Langmuir, I., *J. Am. Chem. Soc.* 1917, *39,* 1848.

2. McConnell, H. M. *Annu. Rev. Phys. Chem.* 1991, *42,* 171.

3. Albrecht, O.; Dissertation, University of Ulm, Ulm, 1979.

4. Knobler, C. M.; Desai, R. C. *Annu. Rev. Phys. Chem.* 1992, 43, 207.

5. Edwards, D. A.; Brenner H.; and Wasan D. T. *Interfacial transport processes and rheology* Butterworth-Heinemann: Boston, 1991.

6. Hénon, S.; Meunier, J. *Rev. Sci. Instrum.* 1991, *62,* 936.

7. Hönig, D.; Möbius D. *J. Phys. Chem.* 1991, *95,* 4590.

8. Lösche, M.; Sackmann, E.; Möhwald, H. *Ber Bunsenges Phys. Chem.* 1983, *87,* 848.

9. McConnell, H. M.; Tann, L. K.; Weiss, R. M. *Proc. Natl. Acad. Sci.* (USA) 1984, *81,* 3249.

10. Goodrich, F. C.; Allen, L. H.; and Poskanzer A. *J. Coll. Int. Sci.* 1975, *52,* 201.

11. Benvengnu, D. J.; and McConnell, H. *J. Phys. Chem.* 1992, *96,* 6820.

12. Heckl, W. M.; Miller, A.; and Möhwald, H.; *Thin solid films* 1988, *159,* 125.

13. Schwartz, D. K.; Knobler, C. M.; and Bruinsma, R. *Phys. Rev. Lett* 1994, *73,* 2841.

14. R.M. Muruganathan, Z. Khattari, and Th. M. Fischer; J. Phys. Chem. B 109, 21772-21778 (2005)

15. Young, N. O.; Goldstein, J. S.; and Block, M. J., *J. Fluid Mech.* 1959, 6, 350.

16. Subramanian, R. S., *AIChE J.* 1981 27 (4), 646.

17. Bratukhin, Y. K., Izv. *Akad. Nauk. SSSR. Mekh. Zhidk. Gaza* 5, 156 (1975); NASA *Technical Translation NASA TT* 17093 (June, 1976).

18. Chen, S. H.; *J. Colloid Interface Sci.* 2000, 230, 147.

19. Treuner, M.; Galindo, V.; Gerbeth, G. Langbein, D., Rath, H.,J., *J. Colloid Interface Sci.* 1996, 179, 114.

20. Harper, J. F.; *Adv. Appl. Mech.*, C. S. Yih, Academic press, 1972, *12,* 59

21. Steffen, P.; Heinig, P.; Wurlitzer, S.; Khattari, Z.; and Fischer, Th. M.; *J. Chem. Phys.* 2001, *115,* 994.

22. Gradshteyn, L. S.; and. Ryshik, I. M, *Table of Integrals, Series, and Products,* 6th. Ed. Academic Press, 8.112.

23. The scatter arises due to the high order of the velocity (mm/s) that forces us to measure the velocity by comparing two consecutive frames that are captured by the SIT-camera.

24. R.M. Muruganathan, and Th. M. Fischer; *J. Phys. Chem. B.* 2006, *110,* 22160

25. S. Wurlitzer, C. Lautz, M. Liley, C. Duschl and Th. M. Fischer; *J. Phys. Chem. B* 2001, *105,* 182

26. Block S. M. *Optical tweezers: a new tool for Biophysics*, in Noninvasive Techniques in Cell Biology Wiley-Liss, 1990; 375-402

27. Nassoy, P.; Birch, W. R.; Andelman, D.; and Rondelez, F. *Phys. Rev. Lett.* 1996, *76,* 455.

28. Landolt Börnstein IV/1, 6th. Edition, Springer, Berlin: 1955; 600, 613.

29. Hughes, B. D.; Pailthorpe, B. A.; and White, L. R. *J. Fluid Mech.* 1981, *110,* 349.

30. Rayleigh, *Proc. Lond. Math. Soc.* x, 4 (1887); *Proc. Roy. Soc.* xxix, 71 (1879).

31. S. Wurlitzer, P. Steffen and Th. M. Fischer; *J. Chem. Phys.* 2000, *112,* 591

32. S. Wurlitzer, P. Steffen M. Wurlitzer, Z. Khattari and Th. M. Fischer; *J. Chem. Phys.* 2000, *113,* 3822

33. Landolt Börnstein IV/1, 6th. Edition, Springer, (Berlin) 1955, page 600 and 613.

34. Lang, H.; Duschl, C.; and Vogel, H. *Langmuir* 1994, *10,* 197.

35. Leighton, D., and Acrivos, A. *J. Fluid Mech.* 1987, *181,* 415.

36. Heinig, P.; Wurlitzer, S.; Steffen, P.; Kremer, F.; and Fischer, Th. M. *Langmuir* 2000, *16* 10254

SUBJECT INDEX

9 780367 446055